Functionals of Finite Riemann Surfaces

Functionals of Finite Riemann Surfaces

Menahem Schiffer
&
Donald C. Spencer

Dover Publications, Inc.
Mineola, New York

Bibliographical Note

This Dover edition, first published in 2014, is an unabridged republication of the work originally published as Volume 16 of the *Princeton Mathematical Series,* by Princeton University Press, Princeton, New Jersey, 1954.

Library of Congress Cataloging-in-Publication Data

Schiffer, Menahem.
Functionals of finite Riemann surfaces / Menahem Schiffer and Donald C. Spencer — Dover edition.
p. cm.
Originally published: Princeton, New Jersey : Princeton University Press, 1954.
ISBN-13: 978-0-486-78046-7
ISBN-10: 0-486-78046-5
1. Riemann surfaces. I. Spencer, D. C. (Donald Clayton), 1912–2001, joint author. II. Title.

QA333.S36 2014
515'.93—dc23

2013044992

Manufactured in the United States by Courier Corporation
78046501 2014
www.doverpublications.com

Preface

This monograph is an outgrowth of lectures given by the authors at Princeton University during the academic year 1949—1950, and it is concerned with finite Riemann surfaces — that is to say with Riemann surfaces of finite genus which have a finite number of non-degenerate boundary components.

The main purpose of the monograph is the investigation of finite Riemann surfaces from the point of view of functional analysis, that is, the study of the various Abelian differentials of the surface in their dependence on the surface itself. Riemann surfaces with boundary are closed by the doubling process and their theory is thus reduced to that of closed surfaces. Attention is centered on the differentials of the third kind in terms of which the other differentials may be expressed.

The relations between the functionals of two Riemann surfaces one of which is imbedded in the other are studied, and series developments are given for the functionals of the smaller surface in terms of those of the larger. Conditions are found in order that a local holomorphic imbedding can be extended to an imbedding in the large of one surface in the other. It may be remarked that the notion of imbedding is a natural generalization of the concept of schlicht functions in a plane domain since these functions imbed the plane domain into the sphere.

If a surface imbedded in another converges to the larger surface, asymptotic formulas are obtained which lead directly to the variational theory of Riemann surfaces. A systematic development of the variational calculus is then given in which topological or conformal type may or may not be preserved.

The variational calculus is applied to the study of relations between the various functionals of a given Riemann surface and to extremum problems in the imbedding of one surface into another. By specialization, applications to classical conformal mapping are obtained.

In a final chapter some aspects of the generalization of the theory to Kähler manifolds of higher dimension are discussed.

The first three chapters contain a development of the classical theory along historical lines, and these chapters may be omitted by the specialist. It was felt to be desirable to include these chapters as a means of providing a historical perspective of the field. A more modern treatment is included in Chapter 9 as the special case of a Kähler manifold of complex dimension 1 (a Riemann surface may always be made into a Kähler manifold by the construction of a Kähler metric).

The monograph is self-contained except for a few places where references to the literature are given.

M. SCHIFFER and D. C. SPENCER,
Hebrew University, Jerusalem, and
Princeton University

December, 1951

Acknowledgments

The authors wish to thank the Mathematical Sciences Division, Office of Naval Research, and the Office of Ordnance Research, United States Army Ordnance, since this monograph was written while the authors were engaged in research projects sponsored by these agencies. They also thank Princeton University for funds supplied during the Summer of 1950, and Dr. G. F. D. Duff for his able assistance during this period. Finally, the authors wish to acknowledge their great indebtedness to Dr. Helen Nickerson, who read the manuscript in detail, made many critical suggestions, and in numerous instances recast the proofs to make them more intelligible, and who has made most of the corrections in the galley and page proof. The authors thank the Princeton University Press and its Director, Mr. Herbert S. Bailey Jr., for their patient cooperation during the final stages of preparation of this monograph.

June, 1953

Contents

1. Geometrical and Physical Considerations

1.1. Conformal Flatness. Beltrami's Equation

In the preface to his book [7], F. Klein states that he is not sure that he would ever have reached a well-defined conception of Riemann's theory of functions had not Prym commented to him in the year 1874 that "Riemann surfaces originally are not necessarily many-sheeted surfaces over the plane, but that, on the contrary, complex analytic functions can be studied on arbitrarily given curved surfaces in exactly the same way as on the surfaces over the plane." Klein then goes on to say that he believes the starting point of Riemann's investigations was the physical conception of a steady flow of a fluid, say an electric fluid, over a surface.

In this chapter we bring together various geometrical and physical concepts relating to surfaces which have motivated the development of the theory of Riemann surfaces and of functions of a complex variable. In recent years there has developed a general theory of harmonic tensors on Riemannian manifolds of dimension n. It is illuminating to observe how various features of this theory combine in the case of dimensionality 2 to produce an especially elegant and simple theory. In particular, although a neighborhood of each point of a two-dimensional Riemannian manifold $\mathfrak{V}^2$ cannot generally be mapped isometrically onto a domain of the plane, it can be mapped conformally onto such a domain. In other words, a $\mathfrak{V}^2$ may be said to be locally conformally flat (Euclidean); this property is not generally true for a $\mathfrak{V}^n$, $n > 2$. Furthermore, it turns out that the equations defining harmonic functions and harmonic vectors on a $\mathfrak{V}^2$ are invariant to conformal transformations. These two properties, conformal flatness and the fact that harmonic functions and vectors depend only on the conformal structure, enable us, in the study of harmonic functions and vectors, to disregard the metric of the $\mathfrak{V}^2$ entirely, retaining only the conformal structure. Thus we are led to the axiomatic definition of a Riemann surface as a surface every point of which has a neighborhood which can be mapped conformally onto a region of the plane.

In this chapter we shall discuss briefly the differential geometry of a $\mathfrak{V}^2$, and we define harmonic functions on a general $\mathfrak{V}^2$. Motivation for this definition arises naturally if we consider the flow of an ideal incompressible fluid over the surface.

Consider a two-dimensional Riemannian manifold $\mathfrak{V}^2$ for which the element of arc length ds is given by the formula

$$ds^2 = Ed\xi^2 + 2Fd\xi d\eta + Gd\eta^2, \tag{1.1.1}$$

where ξ and η are curvilinear coordinates and E, F, and G are functions of ξ and η.

Given an arbitrary point of the manifold, we wish to introduce new coordinates $x(\xi, \eta)$, $y(\xi, \eta)$ in some neighborhood of this point so that this neighborhood is mapped conformally onto a region of the Euclidean plane.

We consider the equations

$$\begin{cases} G\left(\dfrac{\partial x}{\partial \xi}\right)^2 - 2F\dfrac{\partial x}{\partial \xi}\dfrac{\partial x}{\partial \eta} + E\left(\dfrac{\partial x}{\partial \eta}\right)^2 = G\left(\dfrac{\partial y}{\partial \xi}\right)^2 - 2F\dfrac{\partial y}{\partial \xi}\dfrac{\partial y}{\partial \eta} + E\left(\dfrac{\partial y}{\partial \eta}\right)^2, \\ G\dfrac{\partial x}{\partial \xi}\dfrac{\partial y}{\partial \xi} - F\left(\dfrac{\partial x}{\partial \xi}\dfrac{\partial y}{\partial \eta} + \dfrac{\partial x}{\partial \eta}\dfrac{\partial y}{\partial \xi}\right) + E\dfrac{\partial x}{\partial \eta}\dfrac{\partial y}{\partial \eta} = 0. \end{cases} \tag{1.1.2}$$

These equations are equivalent to

$$\begin{cases} \dfrac{\partial y}{\partial \xi} = \dfrac{F\dfrac{\partial x}{\partial \xi} - E\dfrac{\partial x}{\partial \eta}}{\sqrt{EG - F^2}}, \\[2ex] \dfrac{\partial y}{\partial \eta} = \dfrac{G\dfrac{\partial x}{\partial \xi} - F\dfrac{\partial x}{\partial \eta}}{\sqrt{EG - F^2}}, \end{cases} \tag{1.1.3}$$

where $\sqrt{EG - F^2} = H$ may have either choice of sign, or to

$$(1.1.3)' \qquad \begin{cases} \dfrac{\partial x}{\partial \xi} = \dfrac{E\dfrac{\partial y}{\partial \eta} - F\dfrac{\partial y}{\partial \xi}}{\sqrt{EG - F^2}}, \\[2ex] \dfrac{\partial x}{\partial \eta} = \dfrac{F\dfrac{\partial y}{\partial \eta} - G\dfrac{\partial y}{\partial \xi}}{\sqrt{EG - F^2}}. \end{cases}$$

Thus x satisfies the partial differential equation

$$(1.1.4) \qquad \frac{\partial}{\partial \xi}\left(\frac{F\dfrac{\partial x}{\partial \eta} - G\dfrac{\partial x}{\partial \xi}}{\sqrt{EG - F^2}}\right) + \frac{\partial}{\partial \eta}\left(\frac{F\dfrac{\partial x}{\partial \xi} - E\dfrac{\partial x}{\partial \eta}}{\sqrt{EG - F^2}}\right) = 0,$$

and y satisfies the same equation. The equation (1.1.4) was first introduced by Beltrami, and bears his name (see [2]).

Let x and y satisfy (1.1.3), and therefore (1.1.2). Then

$$(1.1.5) \qquad \begin{aligned} \frac{\partial(x, y)}{\partial(\xi, \eta)} &= \frac{\partial x}{\partial \xi}\frac{\partial y}{\partial \eta} - \frac{\partial x}{\partial \eta}\frac{\partial y}{\partial \xi} \\ &= \frac{1}{H}\left\{G\left(\frac{\partial x}{\partial \xi}\right)^2 - 2F\frac{\partial x}{\partial \xi}\frac{\partial x}{\partial \eta} + E\left(\frac{\partial x}{\partial \eta}\right)^2\right\} \neq 0 \end{aligned}$$

unless $\partial x/\partial \xi = \partial x/\partial \eta = \partial y/\partial \xi = \partial y/\partial \eta = 0$. For the existence of twice continuously differentiable solutions satisfying (1.1.4) such that at least one of these four quantities is different from zero at each point of a neighborhood of the given point it is sufficient that E, F, and G be twice continuously differentiable or, more generally, continuously differentiable with the first partial derivatives satisfying a Hölder condition, in some neighborhood of the given point. Weaker assumptions (E, F, and G continuous and satisfying a Hölder condition) assure the existence of a continuously differentiable solution of (1.1.3) with non-vanishing Jacobian in a neighborhood of the given point (see [9a, b]). The Hölder condition cannot be dropped in either statement (see [4]). In fact, it is possible to take E, F, and G continuously differentiable in a neighborhood of the given point but such that the only twice continuously differentiable solution of (1.1.4) is $x =$ constant. Similarly, it is possible to choose E, F, and

G continuous but such that the only continuously differentiable solution of (1.1.3) is $x =$ constant, $y =$ constant. In the exceptional cases there can be no non-constant harmonic or analytic functions, respectively, in the neighborhood. If E, F, and G have derivatives of all orders or are real analytic, the same will be true of the solutions x and y of (1.1.4). We remark that the Jacobian has the same sign as H.

If we introduce x and y as local coordinates in this neighborhood we have

$$(1.1.6)\qquad \begin{aligned} d\xi &= \frac{\partial(\xi, \eta)}{\partial(x, y)}\left(\frac{\partial y}{\partial \eta}dx - \frac{\partial x}{\partial \eta}dy\right), \\ d\eta &= \frac{\partial(\xi, \eta)}{\partial(x, y)}\left(-\frac{\partial y}{\partial \xi}dx + \frac{\partial x}{\partial \xi}dy\right), \end{aligned}$$

and by (1.1.2)

$$(1.1.7)\qquad ds^2 = E\,d\xi^2 + 2F\,d\xi\,d\eta + G\,d\eta^2 = \sigma(x, y)(dx^2 + dy^2)$$

where

$$\sigma(x, y) = H\frac{\partial(\xi, \eta)}{\partial(x, y)} > 0$$

using (1.1.5).

Thus the function $z = x + iy$ maps a neighborhood of the given point of the $\mathfrak{V}^2$ one-one and conformally onto a neighborhood in the complex plane with the Euclidean metric $|dz|^2 = dx^2 + dy^2$. The mapping cannot be chosen so as to be isometric, in general, since $\sigma(x, y)$ is not constant.

A continuously differentiable function of the type $z = x + iy$ will be called a local uniformizer of the surface $\mathfrak{V}^2$ in the neighborhood of the given point. If $z' = x' + iy'$ is another uniformizer in the same neighborhood, then x' and y' satisfy equations (1.1.2), which are necessary for a relationship of the form (1.1.7), and by (1.1.3), (1.1.3)′ and (1.1.6) we have

$$(1.1.8)\qquad \frac{\partial x'}{\partial x} = \pm\frac{\partial y'}{\partial y}, \qquad \frac{\partial x'}{\partial y} = \mp\frac{\partial y'}{\partial x},$$

according as z' and z correspond to the same or opposite choices of the sign of H, so z' is an analytic function of z or of its complex conjugate $\bar{z}$ in this neighborhood. Conversely, any analytic function z' of z with $dz'/dz \neq 0$ can be introduced as a uniformizer. In particular,

$$\frac{\partial^2 x'}{\partial x^2} + \frac{\partial^2 x'}{\partial y^2} = 0. \tag{1.1.9}$$

The Cauchy-Riemann differential equations (1.1.8) are the special case of (1.1.3), and (1.1.9) is the special case of (1.1.4), in which $E = G$, $F = 0$. A complex-valued function on the surface will be called an analytic function in a neighborhood of a point of the $\mathfrak{V}^2$ if it is analytic in the ordinary sense as a function of the local uniformizer.

While it is always possible to map a $\mathfrak{V}^2$ locally on the plane with preservation of angles, this is in general not true in the large since, for topological reasons, it is not always possible to map a $\mathfrak{V}^2$ "globally". For example, it is not possible to map a torus onto a subdomain of the plane. However, it follows from the general Uniformization Principle given in Chapter 2 that, if a global map is possible at all, then the $\mathfrak{V}^2$ can be mapped conformally in the large.

Beltrami's equation is the analogue for the surface $\mathfrak{V}^2$ of the Laplace equation (1.1.9). In order to understand this analogy even better, suppose that we consider an incompressible fluid contained between two planes parallel to the x, y plane. The exact nature of this fluid is irrelevant but for many reasons it is convenient to identify it with an electric fluid. Suppose that there is a potential $u = u(x, y)$ (in the case of an electric fluid the electrostatic potential) which gives rise to the streaming, and let $v = v(x, y)$ be the stream-function. Then

$$\begin{gathered} \frac{\partial u}{\partial x} = \frac{\partial v}{\partial y}, \quad \frac{\partial u}{\partial y} = -\frac{\partial v}{\partial x}; \\ \frac{\partial^2 u}{\partial x^2} + \frac{\partial^2 u}{\partial y^2} = 0, \quad \frac{\partial^2 v}{\partial x^2} + \frac{\partial^2 v}{\partial y^2} = 0. \end{gathered} \tag{1.1.10}$$

The curves $u = constant$ are the equipotential curves, and the curves $v = constant$ are the stream-lines. It is clear that u and v are only determined up to additive constants. Moreover, the equations (1.1.10) remain unchanged if we replace u by v and v by $-u$. Corresponding to this we obtain a second system of streaming which Klein calls the conjugate streaming.

From the physical point of view, the Laplace equation

$$\frac{\partial^2 u}{\partial x^2} + \frac{\partial^2 u}{\partial y^2} = 0 \tag{1.1.11}$$

for the potential u expresses the property that as much fluid flows into an element of area per unit time as flows out. The Beltrami equation (1.1.4) has the same physical interpretation when the fluid is considered as streaming over a curved surface $\mathfrak{V}^2$ in space. For let ξ, η be curvilinear coordinates on the surface, and let the arc length be given by (1.1.1). Let u be a function of position on the surface, and let the direction of fluid motion on the surface at every point be perpendicular to the curve $u = constant$ passing through that point. Let the velocity be $\dfrac{\partial u}{\partial n}$, where ∂n is the element of arc drawn on the surface, normal to the curve.

At a point of the surface $\mathfrak{V}^2$ the direction parameters of a ξ-curve (that is, a curve along which η is constant) are

$$\lambda^1 = \frac{1}{\sqrt{E}}, \quad \lambda^2 = 0,$$

and the orthogonal direction is defined by

$$\mu^1 = -\frac{F}{\sqrt{E}}\frac{1}{\sqrt{EG - F^2}}, \quad \mu^2 = \frac{\sqrt{E}}{\sqrt{EG - F^2}}.$$

That is,

$$E\lambda^1\mu^1 + F(\lambda^1\mu^2 + \lambda^2\mu^1) + G\lambda^2\mu^2 = \frac{1}{\sqrt{EG - F^2}}(-F + F) = 0.$$

The velocity of the flow at right angles to the ξ-curve therefore is

$$\frac{\partial u}{\partial \xi}\mu^1 + \frac{\partial u}{\partial \eta}\mu^2 = \frac{1}{\sqrt{E}}\frac{1}{\sqrt{EG - F^2}}\left(-F\frac{\partial u}{\partial \xi} + E\frac{\partial u}{\partial \eta}\right). \tag{1.1.12}$$

By symmetry the velocity of flow across the η-curve is given by

$$\frac{1}{\sqrt{G}}\frac{1}{\sqrt{EG - F^2}}\left(-F\frac{\partial u}{\partial \eta} + G\frac{\partial u}{\partial \xi}\right). \tag{1.1.12$'$}$$

Now consider a small element of the surface $\mathfrak{V}^2$ bounded by the coordinate curves corresponding to the parameter values ξ, $\xi + d\xi$

and η, $\eta + d\eta$. The flow across the coordinate curve extending from the point (ξ, η) to the point $(\xi + d\xi, \eta)$ is

$$\frac{1}{\sqrt{E}}\frac{1}{\sqrt{EG - F^2}}\left(-F\frac{\partial u}{\partial \xi} + E\frac{\partial u}{\partial \eta}\right)\sqrt{E}d\xi = \frac{1}{\sqrt{EG - F^2}}\left(-F\frac{\partial u}{\partial \xi} + E\frac{\partial u}{\partial \eta}\right)d\xi, \tag{1.1.13}$$

while the flow across the segment of the coordinate curve from $(\xi, \eta + d\eta)$ to $(\xi + d\xi, \eta + d\eta)$ is given by

$$\frac{1}{\sqrt{EG - F^2}}\left(-F\frac{\partial u}{\partial \xi} + E\frac{\partial u}{\partial \eta}\right)d\xi + \frac{\partial}{\partial \eta}\left\{\frac{1}{\sqrt{EG - F^2}}\left(-F\frac{\partial u}{\partial \xi} + E\frac{\partial u}{\partial \eta}\right)d\xi\right\}d\eta. \tag{1.1.14}$$

The difference in flow across these two boundary lines of the surface element is therefore equal to

$$\frac{\partial}{\partial \eta}\left\{\frac{1}{\sqrt{EG - F^2}}\left(-F\frac{\partial u}{\partial \xi} + E\frac{\partial u}{\partial \eta}\right)\right\}d\xi d\eta. \tag{1.1.15}$$

Adding the difference in flow across the other two boundary lines and setting the result equal to zero we obtain Beltrami's equation:

$$\frac{\partial}{\partial \xi}\left(\frac{F\frac{\partial u}{\partial \eta} - G\frac{\partial u}{\partial \xi}}{\sqrt{EG - F^2}}\right) + \frac{\partial}{\partial \eta}\left(\frac{F\frac{\partial u}{\partial \xi} - E\frac{\partial u}{\partial \eta}}{\sqrt{EG - F^2}}\right) = 0. \tag{1.1.16}$$

From the form of (1.1.16) it follows that for every u which satisfies (1.1.16) another function v can be found having a reciprocal relation to u. For by (1.1.16) the following equations hold simultaneously:

$$\begin{cases} \dfrac{\partial v}{\partial \xi} = \dfrac{F\frac{\partial u}{\partial \xi} - E\frac{\partial u}{\partial \eta}}{\sqrt{EG - F^2}}, \\[2ex] \dfrac{\partial v}{\partial \eta} = \dfrac{G\frac{\partial u}{\partial \xi} - F\frac{\partial u}{\partial \eta}}{\sqrt{EG - F^2}}. \end{cases} \tag{1.1.17}$$

These equations define v up to an additive constant. From the geometrical meaning of the equations (1.1.17) we see that the systems of curves $u = constant$ and $v = constant$ are in general orthogonal; in fact

$$du^2 + dv^2 = \lambda ds^2 = \lambda(Ed\xi^2 + 2Fd\xi d\eta + Gd\eta^2), \tag{1.1.18}$$

so $u + iv$ is a complex analytic function of position.

If the surface $\mathfrak{B}^2$ is mapped conformally upon another surface with arc length ds_1^2, then

$$ds^2 = \mu ds_1^2.$$

Hence by (1.1.18)

$$du^2 + dv^2 = \lambda\mu ds_1^2 = \lambda_1 ds_1^2,$$

so $u + iv$ is transformed into a complex analytic function in $\mathfrak{B}_1^2$. This property is essentially a consequence of the fact that the equations (1.1.16) and (1.1.17) are homogeneous of degree zero in E, F and G. The stream-lines and equipotential curves on the one surface are mapped into stream-lines and equipotential curves on the other, but the velocity of the flow at corresponding points on the two surfaces is in general quite different. In particular, a harmonic function remains harmonic under a conformal transformation.

This theorem has the following converse: If two complex analytic functions on two surfaces are given and if the surfaces can be mapped onto one another in such a way that at corresponding points of the surfaces the two complex analytic functions have the same values, then the surfaces are mapped conformally onto one another. In fact, this criterion will later be used as the definition of conformal equivalence.

Another approach to Beltrami's differential equation is obtained by considering extremum problems on the $\mathfrak{B}^2$. Let $\varphi(\xi, \eta)$ be a real, twice continuously differentiable function on the surface. We seek the direction of steepest descent of φ at a given point of the surface. For this purpose suppose that we proceed from the point along an arbitrary curve and compute the derivative of φ with respect to the arc length s along this curve. We obtain

$$\frac{d\varphi}{ds} = \varphi_\xi \frac{d\xi}{ds} + \varphi_\eta \frac{d\eta}{ds}. \tag{1.1.19}$$

The fact that s is an arc length is expressed by the condition

$$(1.1.20) \qquad \Phi = E\left(\frac{d\xi}{ds}\right)^2 + 2F\left(\frac{d\xi}{ds}\frac{d\eta}{ds}\right) + G\left(\frac{d\eta}{ds}\right)^2 = 1.$$

We determine the extrema of $\left(\frac{d\varphi}{ds}\right)^2$ under the side condition (1.1.20) by means of the Lagrange multiplier rule. In this way we find

$$(1.1.21) \quad \max\left(\frac{d\varphi}{ds}\right)^2 = (EG-F^2)^{-1}\left[E\left(\frac{\partial\varphi}{\partial\eta}\right)^2 - 2F\left(\frac{\partial\varphi}{\partial\xi}\frac{\partial\varphi}{\partial\eta}\right) + G\left(\frac{\partial\varphi}{\partial\xi}\right)^2\right].$$

In analogy with the terminology of plane geometry, we might call the expression on the right of (1.1.21) $[\text{grad}\,\varphi]^2$. By virtue of its definition, $[\text{grad}\,\varphi]^2$ is independent of the choice of the coordinates ξ and η of $\mathfrak{B}^2$. It is called the first differentiator of Beltrami and can be expressed as follows:

$$(1.1.22) \qquad [\text{grad}\,\varphi]^2 = \frac{-1}{EG-F^2}\begin{vmatrix} E & F & \varphi_\xi \\ F & G & \varphi_\eta \\ \varphi_\xi & \varphi_\eta & 0 \end{vmatrix}.$$

Let φ and ψ be two twice continuously differentiable functions on $\mathfrak{B}^2$ and let μ be a constant. We have

$$[\text{grad}\,(\varphi + \mu\psi)]^2 = (\text{grad}\,\varphi)^2 + 2\mu\,\text{grad}\,\varphi \,.\, \text{grad}\,\psi + \mu^2\,(\text{grad}\,\psi)^2$$

where

$$(1.1.22)' \qquad \text{grad}\,\varphi \cdot \text{grad}\,\psi = \frac{-1}{EG-F^2}\begin{vmatrix} E & F & \varphi_\xi \\ F & G & \varphi_\eta \\ \psi_\xi & \psi_\eta & 0 \end{vmatrix}.$$

Clearly this expression is also invariant to changes of the (ξ, η)-coordinate system.

The operator

$$(1.1.23) \quad \Delta\varphi = \frac{1}{\sqrt{EG-F^2}}\left\{\frac{\partial}{\partial\xi}\left(\frac{G\varphi_\xi - F\varphi_\eta}{\sqrt{EG-F^2}}\right) + \frac{\partial}{\partial\eta}\left(\frac{E\varphi_\eta - F\varphi_\xi}{\sqrt{EG-F^2}}\right)\right\}$$

is called the second differentiator of Beltrami [2, 5]. Its close relation to the first differentiator is exhibited by the identity

(1.1.24) $$\operatorname{grad}\varphi\cdot\operatorname{grad}\psi+\psi\Delta\varphi$$

$$=\frac{1}{\sqrt{EG-F^2}}\left[\frac{\partial}{\partial\xi}\left(\psi\frac{G\varphi_\xi-F\varphi_\eta}{\sqrt{EG-F^2}}\right)+\frac{\partial}{\partial\eta}\left(\psi\frac{E\varphi_\eta-F\varphi_\xi}{\sqrt{EG-F^2}}\right)\right].$$

This identity is a generalization to curvilinear coordinates of the well-known vector identity

$$\operatorname{grad}\varphi\cdot\operatorname{grad}\psi+\psi\Delta\varphi=\operatorname{div}(\psi\operatorname{grad}\varphi).$$

A more detailed study of vector analysis on Riemannian manifolds of higher dimensions will be given in Chapter 9. The reader can regard formula (1.1.24) as an illustration of the general theory.

We wish to apply (1.1.24) as follows: Let ψ vanish on a smooth 1-cycle $\partial\mathfrak{J}$ which bounds a sub-domain $\mathfrak{J}$ of the $\mathfrak{B}^2$. In view of (1.1.24) we have

$$\int_{\mathfrak{J}}\operatorname{grad}\varphi\cdot\operatorname{grad}\psi\sqrt{EG-F^2}\,d\xi d\eta=-\int_{\mathfrak{J}}\psi\Delta\varphi\sqrt{EG-F^2}\,d\xi d\eta,$$

since the divergence integral is equal to an integral over the boundary $\partial\mathfrak{J}$, and this integral vanishes.

The function φ is said to be harmonic in $\mathfrak{J}$ if $\Delta\varphi=0$ in $\mathfrak{J}$. Thus we have proved the following result, namely: If φ is harmonic in $\mathfrak{J}$, and ψ vanishes on $\partial\mathfrak{J}$, then

(1.1.25) $$\int_{\mathfrak{J}}\operatorname{grad}\varphi\cdot\operatorname{grad}\psi\sqrt{EG-F^2}\,d\xi d\eta=0.$$

Let now χ be another twice continuously differentiable function in the closure of $\mathfrak{J}$, which has the same boundary values as the harmonic function φ. The difference $\chi-\varphi$ is a function ψ which vanishes on the boundary. Hence, by (1.1.25)

$$\int_{\mathfrak{J}}(\operatorname{grad}\chi)^2\sqrt{EG-F^2}\,d\xi d\eta=\int_{\mathfrak{J}}(\operatorname{grad}\varphi)^2\sqrt{EG-F^2}\,d\xi d\eta$$
$$+\int_{\mathfrak{J}}(\operatorname{grad}\psi)^2\sqrt{EG-F^2}\,d\xi d\eta.$$

Thus we have verified Dirichlet's Principle in a formal way for the region $\mathfrak{J}$, namely, that any given harmonic function has a smaller integral of the squared gradient than any other smooth function with the same boundary values.

Dirichlet's Principle has served as a starting point for proving the existence of harmonic functions with prescribed boundary values. The corresponding existence proof in Section 2.8 represents a modification of the original Dirichlet's Principle which is due to Kelvin, Riemann, and Hilbert.

1.2. Exterior Differential Forms

So far we have been concerned mainly with the potential function of the fluid flow. We now examine briefly the vector field defined by the fluid velocity. First, however, let us make some remarks of general character.

Let f^1, f^2 be a pair of continuously differentiable functions of the rectangular coordinates x^1, x^2 in a domain σ^2 of the plane. We may consider these functions as defining a change of variables in σ^2. If the Jacobian

$$\frac{\partial(f^1, f^2)}{\partial(x^1, x^2)} = \begin{vmatrix} \dfrac{\partial f^1}{\partial x^1} & \dfrac{\partial f^1}{\partial x^2} \\[2ex] \dfrac{\partial f^2}{\partial x^1} & \dfrac{\partial f^2}{\partial x^2} \end{vmatrix}$$

of f^1, f^2 (in this order) with respect to x^1, x^2 does not vanish over the domain of integration we then have the formula

$$\int df^1 df^2 = \int_{\sigma^2} \frac{\partial(f^1, f^2)}{\partial(x^1, x^2)} dx^1 dx^2, \tag{1.2.1}$$

the first integral being evaluated over the image of σ^2, regarding f^1, f^2 as independent variables. Even if f^1, f^2 do not map σ^2 in a one-one manner, it is a matter of convenient notation to have the formula

$$df^1 df^2 = \frac{\partial(f^1, f^2)}{\partial(x^1, x^2)} dx^1 dx^2. \tag{1.2.2}$$

Since

$$\frac{\partial(f^1, f^2)}{\partial(x^1, x^2)} = - \frac{\partial(f^2, f^1)}{\partial(x^1, x^2)},$$

we should have

$$df^1 df^2 = -df^2 df^1. \tag{1.2.2$'$}$$

We now proceed formally. A differential of degree 1 is an expression of the form

$$\alpha = a_1 dx^1 + a_2 dx^2.$$

If

$$\beta = b_1 dx^1 + b_2 dx^2$$

is another such differential, we define addition of two differentials by the natural formula

$$\alpha + \beta = (a_1 + b_1)dx^1 + (a_2 + b_2)dx^2.$$

If f is a function (or a differential of degree zero), then we define

$$f\alpha = fa_1 dx^1 + fa_2 dx^2.$$

The operator d is defined as usual by the formula

$$df = \frac{\partial f}{\partial x^1} dx^1 + \frac{\partial f}{\partial x^2} dx^2$$

when f has continuous first partial derivatives. A differential of degree 2 is an expression of the form

$$a dx^1 dx^2.$$

The sum of two such differentials, and multiplication by a function are defined in the obvious manner. We now define a multiplication for differentials, called exterior multiplication. The first rules (definitions) are given by the formulas

$$dx^1 dx^2 = -dx^2 dx^1; \; dx^1 dx^1 = dx^2 dx^2 = 0. \tag{1.2.3}$$

Furthermore, the multiplication shall be distributive, associative, and shall reduce to the ordinary scalar multiplication when defined for functions. Scalars (that is, functions or differentials of degree zero) shall commute with all differentials. Thus we have the formulas

$$\alpha(\beta\gamma) = (\alpha\beta)\gamma; \; \alpha(\beta + \gamma) = \alpha\beta + \alpha\gamma; \; (f\alpha)\beta = \alpha(f\beta) = f\alpha\beta$$

where α, β, γ are differentials of any degree, and where f is a function. In other words, we may multiply in the natural way, being careful to use (1.2.3) for simplification purposes. Finally, if

$$\alpha = a_1 dx^1 + a_2 dx^2$$

is a differential of degree 1, we define

$$(1.2.4) \qquad d\alpha = (da_1)dx^1 + (da_2)dx^2.$$

Thus $d\alpha$ is a differential of degree 2. Explicitly carrying out this calculation, we have

$$(1.2.4)' \qquad \begin{aligned} d\alpha &= \left(\frac{\partial a_1}{\partial x^1} dx^1 + \frac{\partial a_1}{\partial x^2} dx^2\right) dx^1 + \left(\frac{\partial a_2}{\partial x^1} dx^1 + \frac{\partial a_2}{\partial x^2} dx^2\right) dx^2 \\ &= \left(\frac{\partial a_2}{\partial x^1} - \frac{\partial a_1}{\partial x^2}\right) dx^1 dx^2, \end{aligned}$$

by (1.2.3) and the rules of calculation.

If f^1 and f^2 are two (continuously differentiable) functions, we have

$$\begin{aligned} df^1 df^2 &= \left(\frac{\partial f^1}{\partial x^1} dx^1 + \frac{\partial f^1}{\partial x^2} dx^2\right)\left(\frac{\partial f^2}{\partial x^1} dx^1 + \frac{\partial f^2}{\partial x^2} dx^2\right) \\ &= \left(\frac{\partial f^1}{\partial x^1}\frac{\partial f^2}{\partial x^2} - \frac{\partial f^1}{\partial x^2}\frac{\partial f^2}{\partial x^1}\right) dx^1 dx^2 \\ &= \frac{\partial(f^1, f^2)}{\partial(x^1, x^2)} dx^1 dx^2 = -df^2 df^1, \end{aligned}$$

which is formula (1.2.2).

The calculus of differential forms which satisfy the above laws was introduced by Grassmann and E. Cartan in a more general form and has been extensively used in the case of manifolds of higher dimension by Bochner, E. and H. Cartan, Hodge, Kodaira, de Rham and others. In the case of 2-dimensional manifolds, it is almost trivial, but it still provides a formal simplification which is useful.

We now return to the velocity field of the fluid flow, but we assume that the flow is over a plane with rectangular coordinates x^1, x^2. Let a_1, a_2 be the components of the fluid velocity in the directions x^1, x^2 respectively. Using the notation (1.2.4)′ we can write the classical Stokes formula in the form

$$(1.2.5) \qquad \int_{\sigma^2} d\alpha = \int_{\partial\sigma^2} \alpha,$$

where $\alpha = a_1 dx^1 + a_2 dx^2$ is a differential form of degree 1 and where $\partial\sigma^2$ denotes the positively oriented boundary of σ^2. The integral on the right in (1.2.5) is the circulation of the fluid around the boundary of σ^2. If this circulation is to be zero for every σ^2, we must have

$$d\alpha = 0, \tag{1.2.6}$$

that is,

$$\frac{\partial a_1}{\partial x^2} - \frac{\partial a_2}{\partial x^1} = 0. \tag{1.2.7}$$

The equation (1.2.7) expresses the property that the vector field a_i is irrotational, and it implies the existence (locally) of a velocity potential $u(x^1, x^2)$ such that

$$du = a_1 dx^1 + a_2 dx^2; \; a_1 = \frac{\partial u}{\partial x^1}, \; a_2 = \frac{\partial u}{\partial x^2}. \tag{1.2.8}$$

If the fluid is incompressible, we have seen in Section 1.1 that the potential function u satisfies (1.1.11); that is, by (1.2.8),

$$\frac{\partial a_1}{\partial x^1} + \frac{\partial a_2}{\partial x^2} = 0. \tag{1.2.9}$$

The equation (1.2.9) is the so-called equation of continuity, and it expresses the property that the divergence of the vector field a_i vanishes or that the vector field is solenoidal.

A vector field a_i which is both irrotational and solenoidal is said to be harmonic. In the case of the flat plane the harmonic character of a vector field (a_1, a_2) is expressed by equations (1.2.7) and (1.2.9). The question arises as to the nature of these equations on a surface which is not flat, and it turns out that the equations which characterize a harmonic vector field on a $\mathfrak{B}^2$ remain unchanged under a conformal transformation. Since a $\mathfrak{B}^2$ is locally conformal to a flat surface, we see then that we may always express the equations characterizing a harmonic vector field in terms of the Euclidean coordinates of the flat plane. In other words, we may use the equations (1.2.7) and (1.2.9) to define a harmonic vector field on a general $\mathfrak{B}^2$.

The fact that the harmonic character of differential forms of degree 1 on a $\mathfrak{V}^2$ depends only on the conformal structure is a special case of a more general result which has often been emphasized by H. Weyl, namely that on Riemannian manifolds of dimension $n = 2p$ the harmonic p-vectors (or harmonic p-forms) depend only on the conformal structure. The proof of this result is based on tensor calculus, and will therefore be omitted.

1.3. Differential Forms on Riemann Surfaces

At the end of Section 1.2 it was pointed out that the conformal structure alone enters into the definition of harmonic functions and harmonic differentials on a $\mathfrak{V}^2$. Since a neighborhood of each point p of a $\mathfrak{V}^2$ can be mapped conformally onto a domain of the plane, the metric may be disregarded entirely in so far as harmonic functions and 1-vectors are concerned. However, this is not true for 2-vectors; and for this reason we shall not consider them further. If x, y are the Euclidean coordinates of the plane which correspond to points of $\mathfrak{V}^2$ in the neighborhood of p, the function $z = x + iy$ is a complex function of position in the neighborhood of p in the sense of Section 1.1, and is called a (local) uniformizer at the point p. We are thus led to consider surfaces which have uniformizers at each of their points, a pair of uniformizers valid over a common neighborhood being related by a conformal mapping. Such a surface is called a Riemann surface, a concept which will be defined more precisely in Section 2.1.

A differential of degree 1 on the Riemann surface is a linear expression of the form

$$\alpha = adx + bdy$$

where a and b are functions (not necessarily analytic) of the local uniformizer $z = x + iy$. There is one such expression for each uniformizer, and the coefficients are supposed to depend on the uniformizer in such a way that α is invariant. We suppose that a and b are continuous together with their first partial derivatives.

A differential of degree 2 on the Riemann surface is an expression

$$cdxdy$$

with a corresponding invariance property.

Finally a function is a differential of degree 0 whose value at a point does not depend on the choice of uniformizer.

An analytic function on a Riemann surface is a complex scalar which, expressed in terms of a local uniformizer, is a power series. In particular, a uniformizer is locally an analytic function. Let $z = x + iy$ be a uniformizer at the point p, and introduce the Wirtinger operators

$$\frac{\partial}{\partial z} = \frac{1}{2}\left(\frac{\partial}{\partial x} - i\frac{\partial}{\partial y}\right), \quad \frac{\partial}{\partial \bar{z}} = \frac{1}{2}\left(\frac{\partial}{\partial x} + i\frac{\partial}{\partial y}\right). \tag{1.3.1}$$

A harmonic function U on the Riemann surface defines an analytic differential

$$dw = \frac{\partial U}{\partial z} dz = \frac{1}{2}\left(\frac{\partial U}{\partial x} - i\frac{\partial U}{\partial y}\right) dz. \tag{1.3.2}$$

Starting on the other hand from an analytic differential $dw = (u + iv)dz$, its real and imaginary parts give the harmonic differentials (in the sense of Section 1.2)

$$\alpha = udx - vdy, \; \beta = vdx + udy. \tag{1.3.3}$$

In fact, the Cauchy-Riemann equations connecting u and v express the solenoidal and irrotational character of the vector fields.

The complex unit i may be regarded as a symbol which distinguishes the positions of the components of the combination $u + iv$. Multiplication of $u + iv$ by i replaces u by $-v$ and v by u. In other words, i sends a harmonic function into its conjugate. We could avoid the use of complex numbers entirely by introducing the following $*$ operator acting on differentials. Given

$$\alpha = a_1 dx^1 + a_2 dx^2,$$

we define

$$*\alpha = -a_2 dx^1 + a_1 dx^2. \tag{1.3.4}$$

We observe that, in (1.3.3), $\beta = *\alpha$; in other words, the real and imaginary parts of an analytic differential are a pair of conjugate harmonic differentials. We have, for general α and β,

$$
(1.3.5) \qquad \begin{cases} **\alpha = -\alpha, \\ \alpha \cdot *\beta = \beta \cdot *\alpha = -*\alpha \,.\, \beta, \\ \alpha \cdot *\alpha = (a_1^2 + a_2^2)dx^1dx^2, \end{cases}
$$

(compare [1]). A differential α with $d\alpha = 0$ is said to be closed. By (1.2.7) and (1.2.9) we see that α is harmonic if α and $*\alpha$ are closed.

In the case of a scalar φ, we set

$$
(1.3.6) \qquad *\varphi = \varphi dxdy
$$

while, for a differential of degree 2, say $cdxdy$,

$$
(1.3.7) \qquad *cdxdy = c.
$$

In particular,

$$
(1.3.8) \qquad **\varphi = \varphi
$$

whenever φ is a scalar or a differential of degree 2.

1.4. Elementary Topology of Surfaces

As the title of this book indicates, we shall be concerned primarily with finite surfaces, that is to say with surfaces which are finite complexes in the sense of topology. A finite surface is one which can be triangulated into finitely many simplexes such that two simplexes are either disjoint, or one is a side of another, or they have a common side which is their intersection. Finite surfaces are completely classified. In fact, a classical theorem states that any finite surface is obtained from the sphere by cutting out holes and attaching handles and cross-caps. In other words, any such surface may be mapped topologically onto the sphere with a finite number of holes, handles and cross-caps. To attach a handle, we may cut out two holes with boundaries C_1 and C_2 and then identify C_1 and C_2 in such a way that, if C_1 is traversed in the positive sense, the boundary C_2 is traversed in the negative sense. To attach a cross-cap we have only to cut out a "circular" hole and then identify diametral points on its boundary. Let m be the number of boundary curves (holes), h the number of handles, and c the number of cross-caps. A surface is orientable if the number c of cross-caps is zero; otherwise it is non-orientable. If it is orientable, the genus is equal to the number h of handles. If it is non-orientable, each handle may be converted into two cross-caps and the genus is defined to be equal to the total number of cross-caps when all handles have been converted, therefore equal to $2h + c$. The surfaces therefore have three topological invariants: orientability or non-orientability, genus, and the number m of boundary curves.

A 2-simplex is oriented by ordering its vertices, say $(P_0P_1P_2)$, a 1-simplex is oriented by ordering its end-points, say (P_0P_1), and a 0-simplex is oriented by attaching a $+$ or $-$ sign. If

$$\sigma^2 = +\,(P_0P_1P_2)$$

is an oriented 2-simplex or "triangle", then

$$-\,\sigma^2 = -\,(P_0P_1P_2) = +\,(P_1P_0P_2).$$

Similarly, if

$$\sigma^1 = +\,(P_0P_1),$$

then

$$-\,\sigma^1 = -\,(P_0P_1) = +\,(P_1P_0).$$

We form chains from the oriented simplexes of the surface. A k-chain ($k = 0, 1, 2$) of a finite surface consists of the finitely many k-simplexes belonging to the surface, each with a definite orientation and with a definite multiplicity. A k-chain C^k is written

$$C^k = u_1\sigma_1^k + u_2\sigma_2^k + \cdots + u_\alpha\sigma_\alpha^k \tag{1.4.1}$$

where the u_k are integers and $\alpha = \alpha^k$ is the number of k-simplexes belonging to the surface. If, for example, $u_1 = 0$, then the simplex σ_1^k does not occur. If $u_1 = 2$, σ_1^k occurs with multiplicity two while, if $u_1 = -\,2$, the oppositely oriented simplex occurs twice. The boundary of σ^k consists of its $(k - 1)$-dimensional sides, each having the induced orientation. We denote the boundary of σ^k by $\partial\sigma^k$. For example,

$$\partial\sigma^2 = \partial(P_0P_1P_2) = (P_1P_2) - (P_0P_2) + (P_0P_1),$$

$$\partial\sigma^1 = \partial(P_1P_2) = (P_2) - (P_1).$$

The boundary ∂C^k of C^k is defined to be

$$\partial C^k = \Sigma u_\lambda \partial\sigma_\lambda^k. \tag{1.4.2}$$

If the boundary of a chain vanishes, the chain is called a cycle. It is easily verified that $\partial^2\sigma = \partial(\partial\sigma) = 0$, and therefore that $\partial(\partial C) = 0$. Thus, any boundary is a cycle. The addition of k-chains is commutative, associative and distributive. The chains

$$C^k = u_1\sigma_1^k + \cdots + u_\alpha\sigma_\alpha^k, \; D^k = v_1\sigma_1^k + \cdots + v_\alpha\sigma_\alpha^k,$$

have the sum

$$C^k + D^k = (u_1 + v_1)\sigma_1^k + \cdots + (u_\alpha + v_\alpha)\sigma_\alpha^k.$$

If a k-cycle C^k is the boundary of a $(k+1)$-chain, it is called a bounding k-cycle or it is said to be homologous to zero; in symbols

$$C^k \sim 0.$$

The residue classes of the group of k-cycles with respect to the subgroup of bounding k-cycles are the homology classes, and they form the elements of the k-dimensional homology group. In the case of a Riemann surface it is only the 1-dimensional homology group which is significant. We remark that on a connected complex (of arbitrary dimension) the 1-dimensional homology group is the Abelianized 1-dimensional homotopy group.

By formula (1.2.5) applied to a 2-chain C^2:

$$\int_{C^2} d\alpha = \int_{\partial C^2} \alpha. \tag{1.4.3}$$

Here $\alpha = a_1dx^1 + a_2dx^2$ is a 1-form. Let

$$P(\alpha, C^1) = \int_{C^1} \alpha \tag{1.4.4}$$

be the period of α around the cycle C^1. From (1.4.3) we see that if α is closed and if $C^1 \sim 0$ (in which case $C^1 = \partial C^2$), then $P(\alpha, C^1) = 0$. Thus the periods of closed differential forms depend only on the homology class of the cycle C^1. In particular, the periods of harmonic differentials depend only on the homology class.

For further details concerning the elementary topology of surfaces we refer the reader to the literature (in particular [8]). We remark that in many applications it is preferable to consider chains whose coefficients are real numbers which are not necessarily integers, for example in applications involving integration over chains on a manifold. Here we require only the simplest results of the classical topology of surfaces.

1.5. INTEGRATION FORMULAS

In this section we bring together the formulas of integration which we shall have occasion to use in the sequel.

The scalar product of two differentials

$$\alpha = a_1 dx + b_1 dy, \ \beta = a_2 dx + b_2 dy$$

over a domain $\mathfrak{F}$ of a Riemann surface is defined to be

$$(1.5.1) \qquad (\alpha, \beta) = \int_{\mathfrak{F}} \alpha \cdot * \beta = \int_{\mathfrak{F}} (a_1 a_2 + b_1 b_2) dxdy = (\beta, \alpha).$$

We observe that

$$(1.5.2) \quad (*\alpha, *\beta) = \int_{\mathfrak{F}} *\alpha \cdot **\beta = -\int_{\mathfrak{F}} *\alpha \cdot \beta = \int_{\mathfrak{F}} \beta \cdot *\alpha = (\alpha, \beta)$$

by (1.3.5). The scalar product of two scalars φ and ψ is defined by the formula

$$(1.5.3) \qquad (\varphi, \psi) = \int_{\mathfrak{F}} \varphi \cdot *\psi = \int_{\mathfrak{F}} \varphi\psi dxdy.$$

Let ψ be a function which is twice continuously differentiable and set

$$(1.5.4) \qquad \Delta\psi = \frac{\partial^2\psi}{\partial x^2} + \frac{\partial^2\psi}{\partial y^2} = 4 \frac{\partial^2\psi}{\partial\bar{z}\partial z}$$

where $z = x + iy$ is a uniformizer. We have

$$(1.5.5) \qquad d\bar{z}dz = (dx - idy)(dx + idy) = 2idxdy.$$

We shall often denote $dx\,dy$ by dA (element of area). Then $dy\,dx = -dA$. It is apparent that

$$(1.5.6) \qquad \Delta\psi dxdy = -2i \frac{\partial^2\psi}{\partial\bar{z}\partial z} d\bar{z}dz$$

is a differential of degree 2 and that

$$(\varphi,\ \Delta\psi) = \int_{\mathfrak{J}} \varphi \Delta \psi dx dy.$$

We observe that

$$\Delta = *d*d.$$

Let φ be a function (scalar), α a differential, both assumed to be continuously differentiable. Then we have the formula

$$(1.5.7) \qquad (d\varphi,\ *\alpha) - (\varphi,\ *d\alpha) = -\int_{\partial\mathfrak{J}} \varphi \cdot \alpha.$$

This follows at once from (1.2.5) by choosing $\sigma^2 = \mathfrak{J}$ and replacing α by $\varphi \cdot \alpha$. Taking $\alpha = *d\psi$, ψ a scalar, we obtain the standard unsymmetrical Green's formula, namely

$$(1.5.8) \qquad (d\varphi,\ d\psi) + (\varphi,\ \Delta\psi) = \int_{\partial\mathfrak{J}} \varphi \cdot *d\psi$$

where, in the more usual notation,

$$(1.5.9) \quad \int_{\partial\mathfrak{J}} \varphi \cdot *d\psi = -\int_{\partial\mathfrak{J}} \varphi \left(\frac{\partial\psi}{\partial y} dx - \frac{\partial\psi}{\partial x} dy\right) = -\int_{\partial\mathfrak{J}} \varphi \frac{\partial\psi}{\partial n} ds.$$

Here ds is an element of arc length on the boundary and $\partial/\partial n$ denotes differentiation with respect to the normal which points to the left of the vector $(dx,\ dy)$.

The symmetrical Green's formula is

$$(1.5.10) \qquad (\varphi,\ \Delta\psi) - (\psi,\ \Delta\varphi) = \int_{\partial\mathfrak{J}} (\varphi \cdot *d\psi - \psi \cdot *d\varphi).$$

Finally, if we take $\alpha = d\psi$, then (1.5.7) becomes, since $d(d\psi) = 0$,

$$(1.5.11) \qquad (d\varphi,\ *d\psi) = -\int_{\partial\mathfrak{J}} \varphi \cdot d\psi.$$

All the above integration formulas are extremely trivial, and may be proved directly. However, the notation we have used has certain advantages.

We shall be concerned mainly with analytic differentials. An analytic differential df has the form

$$df = du + idv \tag{1.5.12}$$

where

$$dv = *du. \tag{1.5.13}$$

Thus

$$df = adx + bdy \tag{1.5.14}$$

where a, b are complex and

$$a = \frac{\partial u}{\partial x} + i\frac{\partial v}{\partial x}, \quad b = \frac{\partial u}{\partial y} + i\frac{\partial v}{\partial y}. \tag{1.5.15}$$

We have

$$d\bar{f} = du - idv = \bar{a}dx + \bar{b}dy \tag{1.5.16}$$

where $\bar{a}$, $\bar{b}$ are the complex conjugates of a, b. By (1.3.4),

$$*d\bar{f} = dv + idu = id\bar{f}, \tag{1.5.17}$$

and it would therefore be natural to define

$$\begin{aligned}(df, dg) &= \int_{\mathfrak{F}} df \cdot *d\bar{g} = i\int_{\mathfrak{F}} df \cdot d\bar{g} \\ &= i\int_{\mathfrak{F}} f'\bar{g}' dz d\bar{z} = 2\int_{\mathfrak{F}} f' \cdot \bar{g}' dxdy\end{aligned}$$

as the scalar product of two analytic differentials. Here $df = f'dz$. However, for formal reasons we take instead

$$(df, dg) = \frac{1}{2}\int_{\mathfrak{F}} df \cdot *d\bar{g} = \int_{\mathfrak{F}} f' \cdot \bar{g}' dxdy. \tag{1.5.18}$$

We observe that

$$(df, dg) = ((dg, df))^{-} \tag{1.5.19}$$

where $(\alpha)^{-}$ denotes the complex conjugate of α. Also

$$(*df, *dg) = (df, dg). \tag{1.5.20}$$

Writing

$$N(df) = (df,\ df), \tag{1.5.21}$$

we have

$$N(df) = \frac{i}{2}\int_{\mathfrak{J}} df d\bar{f} = \int_{\mathfrak{J}} |\,f'\,|^2 dxdy. \tag{1.5.22}$$

In formulas (1.5.18) and (1.5.22), we understand that $dxdy$ is the area element. Thus

$$N(df) \geqq 0. \tag{1.5.23}$$

We note the useful identities

$$\int_{\mathfrak{J}} df d\bar{g} = \int_{\partial\mathfrak{J}} f d\bar{g}, \quad \int_{\mathfrak{J}} df d\bar{g} = -\int_{\mathfrak{J}} d\bar{g} df = -\int_{\partial\mathfrak{J}} \bar{g} df, \tag{1.5.24}$$

and

$$\int_{\mathfrak{J}} df d\bar{g} = 2\int_{\partial\mathfrak{J}} u d\bar{g} = 2i\int_{\partial\mathfrak{J}} v d\bar{g}, \quad df = du + idv. \tag{1.5.25}$$

REFERENCES

1. L. AHLFORS, "Open Riemann surfaces and extremal problems on compact subregions," *Commentarii Math. Helvetici*, 24 (1950), 100—134.
2. E. BELTRAMI, *Ricerche di analisi applicata alla geometria*, Opere I, 107—198, Hoepli, Milan, 1902.
3. R. COURANT, (a) „Über konforme Abbildung von Bereichen, welche nicht durch alle Rückkehrschnitte zerstückelt werden, auf schlichte Normalbereiche," *Math. Zeit.*, 3 (1919), 114—122. (b) "The conformal mapping of Riemann surfaces not of genus zero," *Rev. Univ. Nac. de Tucuman*, Ser. A, II (1941), 141—149. (c) *Dirichlet's Principle, conformal mapping, and minimal surfaces*, Interscience, New York, 1950.
4. P. HARTMAN and A. WINTNER, "On the existence of Riemannian manifolds which cannot carry non-constant analytic or harmonic functions in the small," *Amer. Jour. of Math.*, 75 (1953), 260—276.
5. A. HURWITZ and R. COURANT, *Funktionentheorie*, Springer, Berlin, 1929.
6. O. D. KELLOGG, *Foundations of potential theory*, Springer, Berlin, 1929. (Reprint, Murray Publishing Co.).
7. F. KLEIN, *On Riemann's theory of algebraic functions and their integrals*, Macmillan and Bowes, Cambridge, 1893 (translation by F. Hardcastle of Klein's book *Über Riemanns Theorie der algebraischen Functionen und ihrer Integrale*, Teubner, Leipzig, 1882).

8. S. LEFSCHETZ, *Introduction to topology*, Princeton Univ. Press, 1949.

9. L. LICHTENSTEIN, (a) „Beweis des Satzes, dass jedes hinreichend kleine, im wesentlichen stetig gekrümmte, singularitätenfreie Flächenstück auf einer Teil einer Ebene zusammenhängend und in den kleinsten Teilen ähnlich abgebildet werden kann," *Abh. der Preuss. Akad. der Wiss., Phys.-Math. Cl.* 1911, *Anhang* (1912). (b) „Zur Theorie der konformen Abbildung. Konforme Abbildung nichtanalytischer, singularitätenfreier Flächenstücke auf ebene Gebiete," *Bull. intern. de l'Acad. des Sci. de Cracovie*, Ser. A, 1916, 192—217, (1917).

10. H. WEYL, *Die Idee der Riemannschen Fläche*, Teubner, Berlin, 1923. (Reprint, Chelsea, New York, 1947).

2. Existence Theorems for Finite Riemann Surfaces

2.1. Definition of a Riemann Surface

In this chapter we establish the existence of analytic functions and differentials on Riemann surfaces. The proof of the existence is based on the concept of a Hilbert space of differentials, in the metric of which orthogonal projection is possible. The method of orthogonal projection, which is closely related to the classical Dirichlet's Principle, leads to the existence of harmonic and analytic differentials. It is sufficient to apply these methods to the particularly simple case of a closed orientable Riemann surface. Any compact Riemann surface $\mathfrak{M}$ which has a boundary or is non-orientable may be covered by a symmetric closed orientable Riemann surface called its "double". We shall construct the functionals of $\mathfrak{M}$ (such as the Green's function) in terms of the functionals of its double.

In Section 1.3 we arrived at the concept of a Riemann surface. Now we give a more precise definition. A Riemann surface $\mathfrak{M}$ is a connected complex, to each point p of which there is a neighborhood which is mapped one-one and bicontinuously onto an open domain of the complex z-plane. The variable z belonging to the point p is called a local uniformizing parameter or, more shortly, a uniformizer. If the point q lies in this neighborhood of p and if z' is a uniformizer at q, then z and z' are related by a direct or indirect conformal mapping. That is, z' is an analytic function of z or $\bar{z}$ (the complex conjugate of z) which has a non-vanishing derivative at p.

If the Riemann surface $\mathfrak{M}$ is a finite complex, we say that it is a finite Riemann surface. In this case, corresponding to each boundary point p of $\mathfrak{M}$, we suppose that there is a boundary uniformizer which maps a subdomain of $\mathfrak{M}$ near p onto a domain of the upper half-plane $\operatorname{Im} z > 0$ in such a way that the boundary of $\mathfrak{M}$ goes into a segment of the real z-axis. Every finite Riemann surface is compact and is bounded by m closed curves, $m \geqq 0$. It is proved in topology that a finite Riemann surface $\mathfrak{M}$ is topologically equivalent to the sphere with m holes cut out and with h handles and c cross-caps attached. If $c > 0$ the surface is non-orientable, otherwise it is orientable. If $\mathfrak{M}$ is non-orientable the genus is $2h + c$ while if $\mathfrak{M}$ is orientable the genus is simply equal to h, the number of handles. When $\mathfrak{M}$ is orientable we shall retain only one or the other of the two classes of uniformizers which are related to one another by direct conformal mappings and the other class will be rejected. We remark that a finite Riemann surface cannot have a single isolated point (puncture) as part of its boundary. The unit disc is a finite Riemann surface, but the unit disc punctured at the center is not.

From the point of view of the systematic theory of Riemann surfaces, it would be desirable to define a Riemann surface in a somewhat more general way and to show that this definition is equivalent to the one given above (see [15]). However, since we shall be concerned primarily with functionals of finite Riemann surfaces, we do not enter into these finer topological considerations here.

If $z = x + iy$ is a single-valued analytic function on the Riemann surface and if the value of z at the point p of the surface is a, we say that p lies over the point a of the sphere. In this way we are led to a realization of the Riemann surface as a multi-sheeted surface, spread over the z-sphere, which is conformally equivalent to the original surface. In general, we shall say that two Riemann surfaces are conformally equivalent if there exists a topological mapping of one surface onto the other such that any analytic function or differential on the one is carried into an analytic function or differential on the other. We must distinguish the essential properties common to all realizations from the non-essential properties associated with particular ones. For example, the genus is an essential property while the kind and position of the branch points of a multi-sheeted surface are non-essential properties.

A common realization of a finite Riemann surface of genus zero is a multiply-connected domain of the complex z-plane having m analytic boundary curves. Near a boundary pount z_0 of such a domain we may use a boundary uniformizer s which coincides on the boundary curve with the arc length parameter. A half-neighborhood of the domain at z_0 is mapped by s onto a half-neighborhood bounded by a segment of the real s-axis.

Any finite Riemann surface may be represented as a disc of the complex plane having $2n$ arcs on the boundary of the disc identified in pairs (see end of Chapter 8). It is a classical theorem in the elementary topology of surfaces that any finite surface is equivalent to the disc (topological polygon) with pairs of arcs on the boundary of the disc identified. The identification in topology is accomplished by a topological map of one arc upon the other, a pair of corresponding points in the mapping being regarded as equivalent. In the case of Riemann surfaces we must show that the indentification can be established by one-one conformal maps.

For example, consider the unit disc $|z| < 1$ of the complex z-plane and let the upper half of the circumference $|z| = 1$ be identified with the lower half by the direct conformal mapping which sends the point z into the point $1/z$. The resulting surface is topologically equivalent to the sphere, and it is made into a Riemann surface by introducing appropriate uniformizers. At an interior point of the disc the variable z itself is a uniformizer. At a point p_0 of the surface which corresponds to a pair of identified points z_0, $1/z_0$ on $|z| = 1$, $z_0 \neq 1, -1$, we define a uniformizer in the following way. Let $\mathfrak{N}_1$ be the half-neighborhood at z_0 which is the intersection of the circle $|z - z_0| < \varrho$ with the disc $|z| < 1$, and let $\mathfrak{N}_2$ be a similar half-neighborhood at $1/z_0$. We set

$$t = \begin{cases} z - z_0, & z \in \mathfrak{N}_1 \\ \dfrac{1}{z} - z_0, & z \in \mathfrak{N}_2. \end{cases} \tag{2.1.1}$$

Then the half-neighborhoods $\mathfrak{N}_1$ and $\mathfrak{N}_2$ fit together in the plane of t to form a full neighborhood of $t = 0$, so t is a uniformizer at the point p_0 of the surface. At the point of the surface corresponding to $z = 1$, we define a uniformizer by first mapping the circumference $|z| = 1$ into a straight line by the transformation

$$w = \frac{2}{z+1} - 1 = \frac{1-z}{1+z}$$

in which $z = 1$ goes into $w = 0$ and in which the half-neighborhood

$\mathfrak{N} = [\,|z| < 1] \cap [\,|z-1| < \varrho]$ goes into a half-neighborhood lying to the right of $w = 0$. Set

$$t = w^2 = \left(\frac{1-z}{1+z}\right)^2. \tag{2.1.2}$$

In the plane of t the neighborhood $\mathfrak{N}$ appears as a neighborhood of $t = 0$ cut along the negative real axis, opposite points on the two edges of the cut corresponding to identified points z, $1/z$ near $z = 1$. The identification of points is therefore achieved by erasing the cut to give a full neighborhood of $t = 0$. Thus t is a uniformizer at the point of the surface corresponding to $z = 1$. A similar uniformizer may be defined at $z = -1$. With these definitions of uniformizers, the surface is made into a Riemann surface.

From the topological point of view, the projective plane is represented as the unit disc with diametrically opposite points of its boundary identified. Now let this identification be accomplished by the indirect conformal mapping which sends the point z of $|z| = 1$ into the point $-1/\bar{z}$. Let $\mathfrak{N}_1$ be the half-neighborhood $[\,|z| < 1] \cap [\,|z - z_0| < \varrho]$ at z_0, $\mathfrak{N}_2$ a similar half-neighborhood at $-1/\bar{z}_0$, and define

$$\tau = \begin{cases} z - z_0, & z \in \mathfrak{N}_1 \\ -\dfrac{1}{\bar{z}} - z_0, & z \in \mathfrak{N}_2. \end{cases} \tag{2.1.3}$$

Then τ maps the union of $\mathfrak{N}_1$ and $\mathfrak{N}_2$ onto a complete neighborhood of the origin and is therefore a uniformizer at the point p_0 of the surface corresponding to the pair of identified points z_0, $-1/\bar{z}_0$. With these uniformizers the surface becomes a closed non-orientable Riemann surface which is topologically equivalent to the projective plane.

The simplest example of a non-orientable finite Riemann surface with a boundary is provided by the Möbius strip. The Möbius strip may be obtained by cutting a hole out of the projective plane. Consider also the circular ring $1 < |z| < R$ of the z-plane. The Möbius strip is obtained from this by identifying the points z and $-R/\bar{z}$. This clearly represents a non-orientable surface, because in the identification a small circle oriented clockwise at z goes into a small circle oriented counterclockwise at $-R/\bar{z}$. By cutting the ring along the real axis in the z-plane, and joining the two halves together along corresponding boundaries, it may be verified that the familiar Möbius strip is obtained. Thus the ring $1 < |z| < R$ with points z and $-R/\bar{z}$ identified is a canonical form for the Möbius strip.

2.2. The Double of a Finite Riemann Surface

A finite Riemann surface has a "double", which we presently define. The double of a multiply-connected domain of the plane was first introduced by Schottky [11], and it was later used by Picard (see [9]) and by Klein [5], the latter of whom extended the concept to general Riemann surfaces.

The double $\mathfrak{F}$ of a finite Riemann surface $\mathfrak{M}$ may be defined as follows. Two points of $\mathfrak{F}$ are associated with, or lie over, each interior point of $\mathfrak{M}$ and one point of $\mathfrak{F}$ is associated with each boundary point of $\mathfrak{M}$. Two disjoint neighborhoods of $\mathfrak{F}$ lie over each neighborhood of an interior point of $\mathfrak{M}$. If z is a uniformizer at an interior point of $\mathfrak{M}$, then z is a uniformizer in one of the two associated $\mathfrak{F}$-neighborhoods and $\bar{z}$ (the complex conjugate of z) is a uniformizer in the other. If z is a boundary uniformizer at a boundary point of $\mathfrak{M}$, then a uniformizer at the corresponding point of $\mathfrak{F}$ is given by the variable which is equal to z in one sheet of $\mathfrak{F}$ and equal to $\bar{z}$ in the other. Only direct conformal transformations of the uniformizer are permitted on $\mathfrak{F}$. If $\mathfrak{M}$ is orientable and has a boundary or if $\mathfrak{M}$ is non-orientable, $\mathfrak{F}$ is a closed orientable Riemann surface; if $\mathfrak{M}$ is a closed orientable manifold, $\mathfrak{F}$ consists of two closed orientable surfaces.

To obtain a more complete description of $\mathfrak{F}$, let us assume first that $\mathfrak{M}$ is orientable and that it has a boundary. In this case we suppose (by discarding one class of uniformizers) that two uniformizers which belong to the same neighborhood of $\mathfrak{M}$ are related by a direct conformal mapping. Now let $\tilde{\mathfrak{M}}$ be obtained from $\mathfrak{M}$ by introducing the class of uniformizers which was discarded for $\mathfrak{M}$. The double $\mathfrak{F}$ is obtained by identifying corresponding boundary points of $\mathfrak{M}$ and $\tilde{\mathfrak{M}}$. If p and $\tilde{p}$ are identified boundary points of $\mathfrak{M}$ and $\tilde{\mathfrak{M}}$, let z be a boundary uniformizer of $\mathfrak{M}$ at p. Then $\bar{z}$ is a boundary uniformizer of $\tilde{\mathfrak{M}}$ at $\tilde{p}$, and we take as uniformizer on $\mathfrak{F}$ at the point $p = \tilde{p}$ the variable τ defined by

$$\tau = \begin{cases} z \text{ in } \mathfrak{M} \\ \bar{z} \text{ in } \tilde{\mathfrak{M}}. \end{cases}$$

Second, assume that $\mathfrak{M}$ is a non-orientable finite Riemann surface. To each point of $\mathfrak{M}$ there are two classes of uniformizers, the uniformizers within each class being related to one another by direct conformal transformations. Let $\mathfrak{N}$ be the relatively unbranched two-sheeted orientable covering of $\mathfrak{M}$. Two points of $\mathfrak{N}$ lie over each point of $\mathfrak{M}$ and one class of uniformizers is associated with one of these points, the other class with the other. Then $\mathfrak{N}$ is a connected neighborhood space. For, the curve on $\mathfrak{N}$ lying over a 1-cycle on $\mathfrak{M}$ along which the orientation is reversed leads from one sheet of $\mathfrak{N}$ to the other. In fact, $\mathfrak{N}$ is an orientable finite Riemann surface. If $\mathfrak{M}$ is closed, the double $\mathfrak{F}$ of $\mathfrak{M}$ is defined to be $\mathfrak{N}$. If $\mathfrak{M}$ has m boundary curves, $m > 0$, then $\mathfrak{N}$ has $2m$ boundary curves and two boundary curves of $\mathfrak{N}$ lie over each boundary of $\mathfrak{M}$. If a boundary curve of $\mathfrak{M}$ is oriented and if the two overlying boundary curves of $\mathfrak{N}$ are given corresponding orientations (defined by the continuous mapping of $\mathfrak{N}$ onto $\mathfrak{M}$), then $\mathfrak{N}$ lies to the left of one of the curves and to the right of the other. We now identify the two boundary curves of $\mathfrak{N}$ lying over each boundary curve of $\mathfrak{M}$ (a pair of equivalent points lying over the same boundary point of $\mathfrak{M}$), and we obtain in this way a closed orientable Riemann surface $\mathfrak{F}$ which is defined to be the double of $\mathfrak{M}$.

The two points p and $\tilde{p}$ of $\mathfrak{F}$ which lie over the same point of $\mathfrak{M}$ are called conjugate points of $\mathfrak{F}$. If p corresponds to a boundary point of $\mathfrak{M}$, then $p = \tilde{p}$. Thus at a point p of $\mathfrak{F}$ which corresponds to a boundary point of $\mathfrak{M}$ we may use as uniformizer either the variable z or its complex conjugate $\bar{z}$. The correspondence between conjugate points of $\mathfrak{F}$ defines a one-one indirect conformal mapping of $\mathfrak{F}$ onto itself. If this mapping is S, then $S^2 = I$, the identity mapping. Klein [5] calls a closed orientable Riemann surface symmetrical if, as in the case of $\mathfrak{F}$, there is a one-one involutory indirect conformal mapping of the surface into itself. If we identify conjugate points of a symmetrical surface, we obtain a finite Riemann surface $\mathfrak{M}$ whose double is the given symmetrical surface.

To compute the genus G of the double $\mathfrak{F}$, we assume first that $\mathfrak{M}$

is not closed and orientable and we make a simplicial decomposition of $\mathfrak{M}$. This division of $\mathfrak{M}$ may be carried up (in the usual way) to $\mathfrak{F}$, and we then apply the Euler characteristic formula to $\mathfrak{F}$. Each simplex which is interior to $\mathfrak{M}$ gives rise to two simplexes of $\mathfrak{F}$ while each simplex lying in the boundary of $\mathfrak{M}$ gives rise to one simplex of $\mathfrak{F}$. Since the number of 0-simplexes on any boundary component of $\mathfrak{M}$ is equal to the number of 1-simplexes on that boundary, the boundaries of $\mathfrak{M}$ contribute nothing to the Euler characteristic N of $\mathfrak{M}$ and we see that the characteristic of $\mathfrak{F}$ is equal to $2N$. The genus G of $\mathfrak{F}$ is therefore equal to $N + 1$. But if $\mathfrak{M}$ is equivalent to the sphere with h handles, c cross-caps and m holes we have

$$N = 2h + c + m - 2,$$

so

$$G = 2h + c + m - 1.$$

If $\mathfrak{M}$ is closed and orientable, we define $G = 2h$.

To obtain a formula for G which will be valid in all cases, we let R^0 be the number of components of the double (0-dimensional Betti number). Then R^0 is equal to 1 unless $\mathfrak{M}$ is closed and orientable in which case $R^0 = 2$. The formula

$$G = 2h + c + m + R^0 - 2 \tag{2.2.1}$$

is valid for every finite Riemann surface $\mathfrak{M}$.

The double of a simply-connected domain with boundary is the sphere, while the double of a multiply-connected domain of genus zero with m boundaries is the sphere with $m - 1$ handles. The double of the Möbius strip discussed in Section 2.1 is the torus.

We now define what is meant by an analytic differential dZ^ν of dimension ν on a closed orientable Riemann surface $\mathfrak{F}$. It is a rule which associates to each point of $\mathfrak{F}$ a meromorphic function $g(\tau)$, τ a local uniformizer, such that

$$dZ^\nu = g(\tau)d\tau^\nu, \qquad d\tau^\nu = (d\tau)^\nu, \tag{2.2.2}$$

is invariant to direct conformal transformations of the uniformizer. In other words, if t is another uniformizer at the same point of $\mathfrak{F}$, then

$$\frac{dZ^\nu}{d\tau^\nu} = \frac{dZ^\nu}{dt^\nu}\left(\frac{dt}{d\tau}\right)^\nu, \qquad g(\tau) = g_1(t)\left(\frac{dt}{d\tau}\right)^\nu, \tag{2.2.2$'$}$$

where $g_1(t)$ is the meromorphic function assigned to the uniformizer t. If ν is zero, g is invariant under direct conformal changes of the uniformizer and is called a function of $\mathfrak{F}$.

If $\mathfrak{F}$ is symmetrical and is the double of a finite Riemann surface $\mathfrak{M}$, then to each differential dZ^ν of $\mathfrak{F}$ we can associate a conjugate differential $d\tilde{Z}^\nu$. If p, $\tilde{p}$ are a pair of conjugate points of $\mathfrak{F}$, we define

$$d\tilde{Z}^\nu(p) = (dZ^\nu(\tilde{p}))^-. \tag{2.2.3}$$

In other words, if τ and $\tilde{\tau}$ are uniformizers at p, $\tilde{p}$, then

$$\frac{d\tilde{Z}^\nu(p)}{d\tau^\nu} = \left(\frac{dZ^\nu(\tilde{p})}{d\tilde{\tau}^\nu}\right)^- \left(\frac{d(\tilde{\tau})^-}{d\tau}\right)^\nu. \tag{2.2.3$'$}$$

If $\tilde{\tau} = \bar{\tau}$, the factor $(d(\tilde{\tau})^-/(d\tau))^\nu$ is equal to unity. If $d\tilde{Z}^\nu(p) \equiv dZ^\nu(p)$, we say that dZ^ν is a differential of $\mathfrak{M}$ (of dimension ν). Thus the differentials of $\mathfrak{M}$ are characterized by the property that they take conjugate values at conjugate places of $\mathfrak{F}$ and are real on the boundary of $\mathfrak{M}$. The operation of forming the conjugate of a differential will be called the '$\sim$ operation'. The operation is of order 2 — that is, the square of the operator is the identity.

If $\mathfrak{M}$ is non-orientable, the parameter transformations are both direct and indirect. In fact, at each interior point there are two classes of uniformizers which correspond to conjugate points of $\mathfrak{F}$. If dZ^ν is a differential of $\mathfrak{M}$, then at one point it takes the value dZ^ν and also the value $(dZ^\nu)^-$. We shall therefore agree that in the interior of $\mathfrak{M}$ the differential dZ^ν depends on the uniformizer in such a way that it remains invariant under direct conformal transformations of the uniformizer and goes into the complex conjugate under an indirect conformal transformation of the uniformizer. On a non-orientable surface invariance means "invariance up to the complex conjugate".

As we have already remarked, a differential of dimension 0 is called a function. Differentials of dimension 1, 2, and -1 will be called linear, quadratic and reciprocal differentials respectively. In general, differentials are defined only for integer values of the dimension ν although it is possible to extend the definition to fractional values of ν. An ambiguity then arises in (2.2.2)$'$ owing to the presence of the factor $(dt/d\tau)^\nu$.

There are also differentials of integer dimension which are multi-valued on the manifold. For example, a linear Prym differential is multiplied by a factor of modulus unity when we pass around a closed loop of the manifold which is not homologous to zero (see [11]). However, we shall be concerned for the most part with differentials which are single-valued.

A harmonic function u on a closed orientable surface $\mathfrak{F}$ gives rise to a linear analytic differential of $\mathfrak{F}$, namely

$$\frac{dZ}{dz} = (du)^p_z = \frac{1}{2}\left(\frac{\partial u}{\partial x} - i\frac{\partial u}{\partial y}\right). \tag{2.2.4}$$

If u is single-valued on $\mathfrak{F}$ and has only polar singularities (including logarithmic poles), we say that u is a harmonic function of $\mathfrak{F}$. If $\mathfrak{F}$ is symmetrical and is the double of a surface $\mathfrak{M}$, we say that u is a harmonic function of $\mathfrak{M}$ if it is single-valued and if either dZ or idZ is a linear differential of $\mathfrak{M}$. Suppose that $\mathfrak{M}$ has a boundary and that z in formula (2.2.4) is a boundary uniformizer. If idZ is a differential of $\mathfrak{M}$ we see that $\partial u/\partial x = 0$ on the boundary, while if dZ is a differential of $\mathfrak{M}$ then $\partial u/\partial y = 0$ on the boundary. According as idZ or dZ is a differential of $\mathfrak{M}$, we say that u is a harmonic function of $\mathfrak{M}$ of the first or of the second kind.

If $\mathfrak{M}$ is not a closed orientable surface, the harmonic function of $\mathfrak{M}$ of the first kind which has a source at a point q of $\mathfrak{F}$ and a sink at the conjugate point $\tilde{q}$ will be called the Green's function $G(p, q) = G(p, q, \tilde{q})$ of $\mathfrak{M}$. Let ζ be a uniformizer at q, $\zeta(q) = 0$. Then $\overline{\zeta}$ is a uniformizer at $\tilde{q}$ and we suppose that G is normalized such that

$$\begin{aligned} G &= \log\frac{1}{|\zeta|} + \text{regular terms} \quad (\text{near } q); \\ G &= \log|\zeta| + \text{regular terms} \quad (\text{near } \tilde{q}). \end{aligned} \tag{2.2.5}$$

The harmonic function of the second kind which has a dipole singularity at a point of $\mathfrak{M}$ is of importance in conformal mapping.

2.3. Hilbert Space

A set **H** of elements $f, g, \ldots$ is called a Hilbert space if it satisfies the following postulates:

(1) **H** is a linear space. That is:

(a) There exists a commutative and associative operation, called addition and denoted by $+$, such that $f + g$ belongs to **H** if f and g are both elements of **H**.

(b) There is a field of numbers λ, called multipliers, such that for each element $f \in \mathbf{H}$ and each λ the element λf is defined and belongs to **H**. This scalar multiplication satisfies the usual associative and distributive laws. It is required that $1 \cdot f = f$. If this field is the field of real numbers, **H** is called a real Hilbert space; if it is the field of complex numbers, **H** is called a complex Hilbert space. No fields other than the field of real numbers and the field of complex numbers are considered here.

(c) There exists a null element which, in general, is denoted by zero despite the fact that it is not a number but an element of **H**. It satisfies the requirements

$$f + 0 = f, \qquad \lambda \cdot 0 = 0, \qquad 0 \cdot f = 0.$$

On the left side of the third formula the symbol 0 is a number; elsewhere in the formulas it is to be interpreted as zero element.

(2) **H** possesses a metric which measures the "length" of each element. That is, for each pair of elements f, g of **H** there exists a complex number (f, g) with the following properties:

(a) $$(\lambda f, g) = \lambda (f, g)$$

(b) $$(f_1 + f_2, g) = (f_1, g) + (f_2, g)$$

(c) $$(g, f) = ((f, g))^-$$

(d) $$(f, f) \geqq 0$$

(e) $$(f, f) = 0 \text{ if and only if } f = 0.$$

In the case of a real Hilbert space we assume that (f, g) is real so that (c) becomes $(g, f) = (f, g)$. It is usual to call

$$\| f \| = \sqrt{(f, f)}$$

the norm or "length" of f. However, we shall find it slightly more convenient to use as norm the quantity

$$N(f) = (f, f) = \| f \|^2. \tag{2.3.1}$$

(3) **H** has infinite dimension. To each positive integer n there

exist n elements $f_1, f_2, \cdots, f_n$ which are linearly independent; that is

$$\lambda_1 f_1 + \lambda_2 f_2 + \cdots + \lambda_n f_n = 0$$

only if $\lambda_1 = \lambda_2 = \cdots = \lambda_n = 0$.

(4) **H** is separable. That is, there exists a countable set of elements of **H**, say $f_1, f_2, \cdots, f_n, \cdots$, which is everywhere dense in **H**. Given any element $g \in \mathbf{H}$ and any positive number ε, there exists an element f_ν of the set such that $N(g - f_\nu) < \varepsilon$.

(5) **H** is complete. That is, if a sequence $f_1, f_2, \cdots, f_n, \cdots$ of elements of **H** satisfies the Cauchy criterion that $N(f_\mu - f_\nu)$ tends to zero as μ, ν become infinite, there exists an element $f \in \mathbf{H}$ such that $N(f - f_\mu)$ tends to zero.

The great analogy of the structure of **H** with that of a linear vector space of finite dimension is obvious.

Assume that the space is complex. From (2) we have

$$N(\lambda f + \mu g) = |\lambda|^2 N(f) + 2\,\mathrm{Re}\,\{\lambda\bar{\mu}(f, g)\} + |\mu|^2 N(g) \geqq 0 \tag{2.3.2}$$

for every pair of complex numbers λ, μ. We conclude that

$$|(f, g)| \leqq \sqrt{N(f)}\,\sqrt{N(g)}. \tag{2.3.3}$$

This is the familiar Schwarz inequality. In particular, taking $\lambda = 1$, $\mu = -1$ in (2.3.2), we obtain

$$\begin{aligned} N(f - g) &= N(f) - 2\,\mathrm{Re}\,\{(f, g)\} + N(g) \leqq N(f) + 2\,|(f, g)| + N(g) \\ &\leqq N(f) + 2\sqrt{N(f)}\sqrt{N(g)} + N(g) = \{\sqrt{N(f)} + \sqrt{N(g)}\}^2 \end{aligned}$$

by (2.3.3). Thus we have the triangle inequality

$$\sqrt{N(f - g)} \leqq \sqrt{N(f)} + \sqrt{N(g)}. \tag{2.3.4}$$

The inequality (2.3.3) shows that we may define the angle θ between two elements f, g by the familiar formula

$$\cos\theta = \frac{|(f, g)|}{\sqrt{N(f)}\,\sqrt{N(g)}}. \tag{2.3.5}$$

In particular, we call two elements f, g of **H** orthogonal if $\cos\theta = 0$, that is if $(f, g) = 0$. By (2e), the only element which is orthogonal to itself is the null element.

An example of a real Hilbert space is provided by the real differentials

$$\alpha = adx + bdy$$

which are assumed to be square integrable on a Riemann surface $\mathfrak{M}$. We recall (see Section 1.5) that the scalar product of two differentials α, β is defined to be

$$(\alpha, \beta) = \int_{\mathfrak{M}} \alpha \cdot * \beta. \tag{2.3.6}$$

The completeness of this space follows from the Riesz-Fischer theorem.

We shall find it convenient to use differentials and derivatives simultaneously. A differential df may be written in the form

$$df = f'dz$$

and f' is the derivative connected with the differential df. While we defined scalar multiplication only for differentials by

$$(df, dg),$$

it will sometimes be convenient to denote the same expression also by (f', g').

An example of a complex Hilbert space is provided by the square integrable complex analytic differentials $f'(p)$ on the Riemann surface $\mathfrak{M}$, that is by the analytic differentials $f'(p)$ defined on $\mathfrak{M}$ which possess a finite norm

$$N(f') = \int_{\mathfrak{M}} | f' |^2 dxdy = \frac{1}{2i} \int_{\mathfrak{M}} | f' |^2 d\bar{z}dz. \tag{2.3.7}$$

Let $f'_\mu(p)$ be a sequence of differentials in this Hilbert space which converge to an element $f'(p)$ of the space. Let p_0 be a fixed interior point of $\mathfrak{M}$, and let $z = z(p)$ be a local uniformizer in the neighborhood of p_0 such that $z(p_0) = 0$. We then have the Taylor developments

$$f'_\mu(p) - f'(p) = \sum_{\nu=0}^{\infty} b_{\mu\nu} z^\nu, \tag{2.3.8}$$

each of which converges in a circle $|z| < r$, r being a number which is independent of μ. We have

$$(2.3.9)\quad N(f'_\mu - f') \geqq \pi \sum_{\nu=0}^{\infty} \frac{1}{\nu+1} \, | b_{\mu\nu} |^2 r^{2(\nu+1)} \geqq \pi \frac{| b_{\mu\nu} |^2}{\nu+1} r^{2(\nu+1)}$$

for $\nu = 0, 1, 2, \cdots$. The convergence of f'_μ to f' in the metric of the Hilbert space implies, therefore, the convergence of the coefficients in the local Taylor development of f'_μ to the corresponding coefficients of f'. Thus Hilbert space convergence implies uniform convergence of the differentials in any compact subdomain of $\mathfrak{M}$ lying in its interior. For brevity, we call the latter type of convergence "Vitali convergence".

On the other hand, Vitali convergence does not imply convergence in the metric of the complex Hilbert space even if we assume that

$$(2.3.10)\qquad N(f'_\mu) \leqq M^2$$

where M is independent of μ. However, assuming (2.3.10), we can show that for any fixed differential g' of the space we have

$$(2.3.11)\qquad \lim_{\mu \to \infty} (f'_\mu - f', g') = 0.$$

In fact, let $\mathfrak{M}'$ be a compact subregion lying in the interior of $\mathfrak{M}$. Then f'_μ converges uniformly to a limit f' on $\mathfrak{M}'$ and we have

$$N_{\mathfrak{M}'}(f') = \lim_{\mu \to \infty} N_{\mathfrak{M}'}(f'_\mu) \leqq M^2.$$

Since $\mathfrak{M}'$ is an arbitrary compact subregion, we conclude that

$$(2.3.12)\qquad N_{\mathfrak{M}}(f') \leqq M^2.$$

Now, given $\varepsilon > 0$, let $\mathfrak{M}'$ be chosen such that

$$(2.3.13)\qquad N_{\mathfrak{M}-\mathfrak{M}'}(g') = \int_{\mathfrak{M}-\mathfrak{M}'} | g' |^2 dxdy < \varepsilon^2.$$

Holding $\mathfrak{M}'$ fixed, we choose a number $n_0 = n_0(\varepsilon)$ so large that

$$(2.3.14)\qquad N_{\mathfrak{M}'}(f'_\mu - f') < \varepsilon^2$$

for $\mu \geqq n_0$. By the Schwarz inequality and (2.3.13), (2.3.14), we have

$$(2.3.15)\qquad \begin{aligned} | (f'_\mu - f', g') | &= | (f'_\mu - f', g')_{\mathfrak{M}'} + (f'_\mu - f', g')_{\mathfrak{M}-\mathfrak{M}'} | \\ &\leqq \{ \sqrt{N(g')} + 2M \} \, \varepsilon. \end{aligned}$$

Since ε is arbitrary, the statement (2.3.11) follows.

We are thus led to the concept of weak convergence in the general theory of Hilbert spaces wherein a sequence of elements f_μ is said to converge weakly to an element f if, for each fixed element g of the space, we have

$$\lim_{\mu\to\infty} (f_\mu - f, g) = 0.$$

We recognize that Vitali convergence is a special instance of weak convergence.

In the Hilbert space of analytic differentials $f'(p)$ on an orientable Riemann surface $\mathfrak{M}$, an important but extremely simple inequality deserves to be singled out. Namely, if $f'(p)$ is a given differential which at p_0 has the development

$$f'(p) = \sum_{\nu=0}^{\infty} a_\nu z^\nu, \quad |z| < r,$$

where z is a uniformizer in a neighborhood of p_0, $z(p_0) = 0$, then

$$N(f') \geqq \pi \sum_{\nu=0}^{\infty} \frac{1}{\nu+1} |a_\nu|^2 r^{2(\nu+1)} \geqq \pi |a_0|^2 r^2.$$

Since $a_0 = f'(p_0)$ (differentiation being expressed in terms of the chosen uniformizer z), the inequality may be written in the form

$$|f'(p_0)|^2 \leqq \frac{1}{\pi r^2} N(f'). \tag{2.3.16}$$

This inequality provides a bound for the local value of a differential in terms of its norm. The question arises as to the best possible number $k(p_0)$ in the inequality

$$|f'(p_0)|^2 \leqq k(p_0) N(f'). \tag{2.3.16$'$}$$

Since $f'(p_0)$ is a differential, we see that $k(p_0)\,|dz|^2$ is invariant; in other words, $k(p_0)$ transforms like $|f'(p_0)|^2$. The number $k(p_0)$ will be determined in Chapter 4, and it is connected with an important functional of the Riemann surface.

Let us return to the general Hilbert space $\mathbf{H}$. In it there always exists a complete orthonormal set of elements $f_1, f_2, \cdots$, that is a set such that

$$(f_\mu, f_\nu) = \delta_{\mu\nu} \tag{2.3.17}$$

and such that every element $f \in \mathbf{H}$ can be represented in the form

$$f = \sum_{\nu=1}^{\infty} a_\nu f_\nu, \qquad a_\nu = (f, f_\nu). \tag{2.3.18}$$

In fact, by the separability axiom (4) there exists a countable set of elements φ_ν which is everywhere dense in $\mathbf{H}$. This set can be orthonormalized by the Gram-Schmidt procedure which is a step-by-step process. After $\varphi_1, \varphi_2, \cdots, \varphi_n$ have been replaced by n orthonormal elements $f_1, f_2, \cdots, f_n$, the element f_{n+1} is defined by the formula

$$f_{n+1} = \frac{\varphi_{n+1} - \sum_{\nu=1}^{n} (\varphi_{n+1}, f_\nu) f_\nu}{[N(\varphi_{n+1} - \sum_{\nu=1}^{n} (\varphi_{n+1}, f_\nu) f_\nu)]^{1/2}}. \tag{2.3.19}$$

It is clear that f_{n+1} is orthogonal to all the preceding elements f_ν, $\nu = 1, 2, \cdots, n$, and that $N(f_{n+1}) = 1$. The final set (f_ν) obtained in this way is linearly equivalent to the initial dense set (φ_ν).

Given any element $f \in \mathbf{H}$, let us try to approximate it by linear combinations of the first n elements $f_1, f_2, \cdots, f_n$. In fact, in the combination

$$\sum_{\nu=1}^{n} a_\nu f_\nu,$$

let us determine the a_ν in such a way that the approximation error

$$N(f - \sum_{\nu=1}^{n} a_\nu f_\nu) = N(f) - 2 \operatorname{Re}\{ \sum_{\nu=1}^{n} \bar{a}_\nu (f, f_\nu)\} + \sum_{\nu=1}^{n} | a_\nu |^2 \tag{2.3.20}$$

is a minimum. Since

$$N(f - \sum_{\nu=1}^{n} a_\nu f_\nu) = N(f) - \sum_{\nu=1}^{n} |(f, f_\nu)|^2 + \sum_{\nu=1}^{n} | a_\nu - (f, f_\nu)|^2, \tag{2.3.20$'$}$$

we have

$$N(f - \sum_{\nu=1}^{n} a_\nu f_\nu) \geqq N(f) - \sum_{\nu=1}^{n} |(f, f_\nu)|^2 \tag{2.3.21}$$

with equality if and only if

$$a_\nu = (f, f_\nu). \tag{2.3.22}$$

We observe that the minimizing coefficients a_ν are independent of n. Moreover, by (2.3.21),

$$\sum_{\nu=1}^{\infty} |(f, f_\nu)|^2 \leqq N(f) \tag{2.3.23}$$

(Bessel's inequality). Thus the series

$$\sum_{\nu=1}^{\infty} (f, f_\nu) f_\nu$$

represents an element of **H**. Since its partial sums approximate f better than any other partial sums, and since the f_ν are linearly equivalent to the φ_ν which are everywhere dense in **H**, we conclude that

$$\lim_{n\to\infty} N(f - \sum_{\nu=1}^{n} (f, f_\nu) f_\nu) = 0.$$

That is

$$f = \sum_{\nu=1}^{\infty} (f, f_\nu) f_\nu \tag{2.3.24}$$

and formula (2.3.18) is established.

2.4. Orthogonal Projection

Let **H** be a Hilbert space, **F** a complete linear subspace of **H**. Then **F** is either a unitary space of finite dimension or it is a Hilbert space. We have the fundamental decomposition formula

$$\mathbf{H} = \mathbf{F} + \mathbf{G} \tag{2.4.1}$$

where **G** denotes the orthogonal complement of **F** with respect to **H**.

Let h be an arbitrary element of **H**. Formula (2.4.1) states that

$$h = f + g \tag{2.4.1$'$}$$

where $f \in \mathbf{F}$, $g \in \mathbf{G}$ and $(f, g) = 0$. To prove (2.4.1)′, we seek an element $f \in \mathbf{F}$ which lies nearest to h in the sense of the metric, that is, which minimizes the norm $N(h - f)$.

If d is the greatest lower bound of $N(h - f)$, $f \in \mathbf{F}$, we have the following inequality of B. Levi:

$$\sqrt{N(f_1 - f_2)} \leqq \sqrt{N(h - f_1) - d} + \sqrt{N(h - f_2) - d}. \tag{2.4.2}$$

Here f_1, f_2 are arbitrary elements of **F**. To prove this inequality, let λ and μ be a pair of numbers (real or complex depending on the character of the Hilbert space) such that $\lambda + \mu = 1$. If f_1, f_2 belong to $\mathbf{F}$, so does $\lambda f_1 + \mu f_2$ and we therefore have

$$N(h - (\lambda f_1 + \mu f_2)) \geqq d.$$

That is,

$$N(\lambda(h - f_1) + \mu(h - f_2)) = |\lambda|^2 N(h - f_1) + 2\,\mathrm{Re}\{\lambda\bar{\mu}(h - f_1,\, h - f_2)\} + |\mu|^2 N(h - f_2) \geqq d,$$

or

$$(2.4.3)\quad |\lambda|^2[N(h - f_1) - d] + 2\,\mathrm{Re}\{\lambda\bar{\mu}[(h - f_1,\, h - f_2) - d]\} + |\mu|^2 [N(h - f_2) - d] \geqq 0.$$

The inequality (2.4.3) remains true without the condition that $\lambda + \mu = 1$, and is therefore valid for arbitrary numbers λ, μ. As in (2.3.2), we conclude that

$$(2.4.4)\quad |(h - f_1, h - f_2) - d|^2 \leqq [N(h - f_1) - d] \cdot [N(h - f_2) - d].$$

Hence

$$\begin{aligned} 0 \leqq N(f_1 - f_2) &= N((h - f_1) - (h - f_2)) \\ &= N(h - f_1) - d - 2[(h - f_1, h - f_2) - d] + N(h - f_2) - d \\ &\leqq N(h - f_1) - d + 2\sqrt{[N(h - f_1) - d] \cdot [N(h - f_2) - d]} \\ &\quad + N(h - f_2) - d \\ &= \{\sqrt{N(h - f_1) - d} + \sqrt{N(h - f_2) - d}\}^2, \end{aligned}$$

and this is (2.4.2).

Let f_μ be a sequence of elements of $\mathbf{F}$ such that

$$\lim_{\mu \to \infty} N(h - f_\mu) = d.$$

By (2.4.2) we have

$$\lim_{\mu,\, \nu \to \infty} N(f_\mu - f_\nu) = 0.$$

Thus, since $\mathbf{F}$ is complete, there exists an element $f \in \mathbf{F}$ such that $f = \lim f_\mu$. It is clear that

$$(2.4.5)\quad N(h - f) = d.$$

Finally, let f_0 be an arbitrary element of $\mathbf{F}$. Then $f + \varepsilon f_0$ is also an element of $\mathbf{F}$ and we have

$$(2.4.6)\quad N(h - f - \varepsilon f_0) = N(h - f) - 2\,\mathrm{Re}\{\bar{\varepsilon}(h - f,\, f_0)\} + |\varepsilon|^2 N(f_0) \geqq d.$$

We conclude from (2.4.5) and the arbitrariness of ε that

$$(h - f, f_0) = 0, \qquad f_0 \in \mathbf{F}. \tag{2.4.7}$$

Thus $h - f = g$ is an element of **G**, the orthogonal complement of **F**, and formula (2.4.1)′ is proved.

In the case of a linear vector space of finite dimension, the above process corresponds to orthogonal projection of a given vector h onto a linear subspace **F**. Hence the name "orthogonal projection" has also been applied to the process in a Hilbert space.

2.5. THE FUNDAMENTAL LEMMA

On a Riemann surface differentiation is expressed in terms of a local uniformizer $z = x + iy$. If the derivatives of order k of a function φ on the surface are continuous when expressed in terms of one uniformizer, they are continuous when expressed in terms of any other. A function φ on the surface is said to be of class C^k if it is continuous together with its derivatives up to the order k. In particular, φ is of class C^0 if it is continuous. A differential $\alpha = adx + bdy$ is of class C^k if its coefficients are of class C^k. The class of a differential is clearly independent of the particular uniformizer used in determining it.

Let $\mathfrak{J}$ be a subdomain of the Riemann surface $\mathfrak{M}$. The carrier of a continuous function on $\mathfrak{J}$ is the smallest closed set of points of $\mathfrak{J}$ outside of which the function vanishes. If η is a function on $\mathfrak{J}$ which vanishes outside a compact subdomain of $\mathfrak{J}$, we say (after L. Schwartz and de Rham) that η has a compact carrier.

Let us understand by equality of functions on $\mathfrak{J}$ that the functions differ at most in a set of measure zero. We remark that a set of measure zero in the plane of one uniformizer will appear as a set of measure zero in the plane of any other uniformizer. Measure is understood to be Lebesgue measure.

In the existence proof for harmonic differentials on a Riemann surface based on the method of orthogonal projection, the following lemma is of fundamental importance:

LEMMA 2.5.1. *A scalar φ of class L^2 satisfying the equation*

$$(\varphi, \Delta\eta)_{\mathfrak{J}} = \int_{\mathfrak{J}} \varphi \Delta\eta\, dxdy = 0 \tag{2.5.1}$$

for every scalar η of class C^∞ with a compact carrier is equal to a harmonic function on $\mathfrak{F}$.

If (2.5.1) is satisfied for a wider class of functions η, say η of class C^r, $r \geqq 2$, the conclusion follows a fortiori.

The proof which we give of this lemma follows closely along the lines of the proofs given by Weyl [14b], Kodaira [6] and de Rham [10].

We observe first that the lemma is trivial if $\varphi \in C^2$. For since η has a compact carrier, we have by Green's formula (1.5.10)

$$(\Delta\varphi, \eta) = (\varphi, \Delta\eta).$$

Thus

$$(\Delta\varphi, \eta) = 0$$

for every η of class C^∞ with a compact carrier. Since $\Delta\varphi$ is continuous, we conclude that $\Delta\varphi = 0$, that is φ is harmonic.

Now assume only that $\varphi \in L^2$. Without loss of generality we may suppose that $\mathfrak{F}$ lies in the domain of a local uniformizer $z = x^1 + ix^2$. For simplicity let $x = (x^1, x^2)$, $y = (y^1, y^2)$, and write $dx = dx^1dx^2$. Let ε be a small positive number. We denote the distance between the points x and y by $r(x, y)$ and define

$$\varrho(x, y) = \begin{cases} 0 \text{ if } r(x, y) > \varepsilon \\ 1 \text{ if } r(x, y) < \dfrac{\varepsilon}{2}. \end{cases}$$

For $\varepsilon/2 \leqq r(x, y) \leqq \varepsilon$, we suppose that $0 \leqq \varrho(x, y) \leqq 1$ and that ϱ is chosen in such a way that it is of class C^∞ and symmetric. Let

$$\omega(x, y) = \frac{1}{2\pi}\varrho(x, y) \log \frac{1}{r(x, y)}, \tag{2.5.2}$$

and write

$$\gamma(x, y) = \begin{cases} \Delta_x\omega(x, y), & x \neq y \\ 0 & , x = y \end{cases} \tag{2.5.3}$$

where Δ_x is the Laplace operator with respect to x. Then $\gamma(x, y)$ is of class C^∞, and it is identically zero for $r(x, y) < \varepsilon/2$.

Let $\mathfrak{F}_\varepsilon$ be the subset of $\mathfrak{F}$ whose points are at distance more than ε from the boundary of $\mathfrak{F}$, and let μ be a function of class C^∞ with carrier in $\mathfrak{F}_\varepsilon$. Then

$$\psi(x) = \int_{\mathfrak{F}} \omega(x, y)\mu(y)dy \tag{2.5.4}$$

has a carrier in $\mathfrak{J}$ since $\omega(x, y) = 0$ if $r(x, y) > \varepsilon$. Moreover, ψ is of class C^∞ and we have

$$\Delta_x \psi(x) = -\mu(x) + \nu(x) \tag{2.5.5}$$

where

$$\nu(x) = \int_{\mathfrak{J}} \gamma(x, y)\mu(y)dy. \tag{2.5.6}$$

The formula (2.5.5) is a consequence of Green's formula applied to a small sphere about the point x. We observe that $\nu \in C^\infty$.

Choosing $\eta = \psi$, $\eta \in C^\infty$, we have by hypothesis

$$(\varphi, \Delta\eta) = -\int_{\mathfrak{J}} \varphi \cdot \mu dx + \int_{\mathfrak{J}} \varphi \cdot \nu dx = 0.$$

That is,

$$\begin{aligned} \int_{\mathfrak{J}} \mu\varphi dx &= \int_{\mathfrak{J}} \varphi(x)\left\{\int_{\mathfrak{J}} \gamma(x, y)\mu(y)dy\right\}dx \\ &= \int_{\mathfrak{J}} \mu(y)\left\{\int_{\mathfrak{J}} \gamma(x, y)\varphi(x)dx\right\}dy. \end{aligned} \tag{2.5.7}$$

Since formula (2.5.7) is valid for every choice of the function μ, $\mu \in C^\infty$, which has a carrier in $\mathfrak{J}_\varepsilon$, we readily conclude that

$$\varphi(y) = \int_{\mathfrak{J}} \gamma(x, y)\varphi(x)dx \tag{2.5.8}$$

in $\mathfrak{J}_\varepsilon$. Formula (2.5.8) shows that φ is equal to a function of class C^∞ and hence φ is equal to a harmonic function in $\mathfrak{J}_\varepsilon$. Since ε is arbitrary, we arrive at the lemma.

2.6. The Existence of Harmonic Differentials with Prescribed Periods

We suppose that the Riemann surface $\mathfrak{M}$ is closed and orientable, and we denote by **D** the space of all real differentials

$$\alpha = a_1 dx + a_2 dy \tag{2.6.1}$$

on $\mathfrak{M}$ such that

$$N(\alpha) = (\alpha, \alpha) = \int_{\mathfrak{M}} (a_1^2 + a_2^2)dxdy < \infty. \tag{2.6.2}$$

As remarked in Section 2.3, **D** is a real Hilbert space.

We shall say that a differential α is closed if α is of class C^1 and if $d\alpha = 0$. We shall say that α is exact if there is a single-valued function ψ on $\mathfrak{M}$ of class C^2 such that $\alpha = d\psi$. An exact differential is closed but not conversely.

Let **F** be the linear subspace of **D** composed of differentials α which satisfy

$$(\alpha, *\beta) = 0, \ \beta \text{ exact.} \tag{2.6.3}$$

The condition (2.6.3) is fulfilled by all closed differentials α. For suppose that α is closed. Then by Stokes' formula the period

$$P(\alpha, K) = \int_K \alpha$$

depends only on the homology class of the 1-cycle K. Moreover, there exists locally a function φ of class C^2 such that $\alpha = d\varphi$. The function φ can be continued over the whole of $\mathfrak{M}$, but it will not generally be single-valued on $\mathfrak{M}$. Let h be the genus of $\mathfrak{M}$, and let $K_1, K_2, \cdots, K_{2h}$ be a homology basis of 1-cycles for the surface composed of h dual pairs $K_{2\mu-1}, K_{2\mu}, \mu = 1, 2, \cdots, h$, such that

$$K = K_1 + K_2 - K_1 - K_2 + \cdots + K_{2h-1} + K_{2h} - K_{2h-1} - K_{2h}$$

bounds a subdomain $\mathfrak{J}$ of $\mathfrak{M}$. The subdomain $\mathfrak{J}$ is obtained by cutting in the usual way along the cycles $K_1, K_2, \cdots, K_{2h}$. Writing

$$P_\mu = \int_{K_\mu} \alpha = \int_{K_\mu} d\varphi, \tag{2.6.4}$$

we see that the value of φ at any point of the surface $\mathfrak{M}$ is determined up to a linear combination of periods of the form

$$m_1P_1 + m_2P_2 + \cdots + m_{2h}P_{2h}$$

where the m_μ are integers. By (1.5.11),

$$(\alpha, *\beta) = (d\varphi, *d\psi) = -\int_K \varphi d\psi = -\sum_{\mu=1}^{h} [P_{2\mu}\int_{K_{2\mu-1}} d\psi - P_{2\mu-1}\int_{K_{2\mu}} d\psi]. \tag{2.6.5}$$

Here the right side is zero since all periods of ψ vanish. Thus the condition (2.6.3) is satisfied by closed differentials α. Conversely, if

$\alpha \varepsilon \mathbf{F}$, α of class C^1, then we see at once from (2.6.3) that $d\alpha = 0$.

The subclass of $\mathbf{F}$ composed of differentials α which satisfy

$$(\alpha, *\beta) = 0 \tag{2.6.6}$$

for any closed differential β will be denoted by $\mathbf{E}$. Since $\beta = d\psi$, where ψ is an integral with periods which may be arbitrarily prescribed, we conclude from (2.6.5) that if α is closed, then $\alpha = d\varphi$ where φ is single-valued. Thus (2.6.6) is fulfilled by all exact differentials. Conversely, if $\alpha \varepsilon \mathbf{E}$, α of class C^1, then α is closed, hence exact.

We shall establish, by the method of orthogonal projection, the fundamental decomposition formula (compare [14b])

$$\mathbf{F} = \mathbf{E} + \mathbf{H} \tag{2.6.7}$$

where $\mathbf{H}$ denotes the space of harmonic differentials on $\mathfrak{M}$. It is clear that $\mathbf{E}$ and $\mathbf{H}$ are orthogonal spaces since a harmonic differential χ is characterized by the conditions that χ and $*\chi$ are closed. Hence, if $\alpha \in \mathbf{E}$, we have by (2.6.6)

$$(\alpha, \chi) = -(\alpha, *\chi_1) = 0$$

where $\chi_1 = *\chi$.

If, in particular, we select from $\mathbf{F}$ any closed differential, formula (2.6.7) tells us that this closed differential is equal to a differential of $\mathbf{E}$ plus a harmonic differential. The differential of $\mathbf{E}$ must be of class C^1 since it is equal to the difference of two differentials, one of which is C^1, the other harmonic. Since this differential of $\mathbf{E}$ is closed, it is exact, and we conclude that the harmonic differential has the same periods as the given closed one. In particular, there exist harmonic differentials having prescribed periods.

Let γ be any differential of $\mathbf{F}$. By the method of orthogonal projection (Section 2.4), we see that

$$\gamma = \alpha + \chi \tag{2.6.8}$$

where $\alpha \in \mathbf{E}$ and χ is orthogonal to all elements of $\mathbf{E}$:

$$(\chi, \alpha_1) = 0, \quad \alpha_1 \in \mathbf{E}. \tag{2.6.9}$$

Now let η be a single-valued function of class C^3 on $\mathfrak{M}$ whose carrier lies in the domain of a local uniformizer $z = x + iy$. Let $z' = x' + iy'$ be any other uniformizer valid in a neighborhood contained in this domain, and define

$$(2.6.10)\quad \varphi = \frac{\partial \eta}{\partial x} = \frac{\partial \eta}{\partial x'}\frac{dx'}{\partial x} + \frac{\partial \eta}{\partial y'}\frac{dy'}{\partial x}, \quad \psi = \frac{\partial \eta}{\partial y} = \frac{\partial \eta}{\partial x'}\frac{\partial x'}{\partial y} + \frac{\partial \eta}{\partial y'}\frac{\partial y'}{\partial y}.$$

Since φ and ψ do not depend on the choice of the uniformizer, they are functions on $\mathfrak{M}$ of class C^2 which give rise to the exact differentials $d\varphi$ and $d\psi$. Taking $\alpha_1 = d\varphi$, we have by (2.6.9)

$$(2.6.11)\qquad 0 = (\chi,\ d\varphi) = \int\limits_{\mathfrak{M}} \left(a_1 \frac{\partial \varphi}{\partial x} + a_2 \frac{\partial \varphi}{\partial y}\right) dxdy$$

where

$$(2.6.12)\qquad \chi = a_1 dx + a_2 dy.$$

Taking $\beta = d\psi$, we have by (2.6.3)

$$(2.6.13)\qquad 0 = (\chi,\ *d\psi) = \int\limits_{\mathfrak{M}} \left(-a_1 \frac{\partial \psi}{\partial y} + a_2 \frac{\partial \psi}{\partial x}\right) dxdy.$$

We suppose that the integrands in these integrals are expressed in terms of the particular uniformizer used above to define φ and ψ. Then

$$\frac{\partial \varphi}{\partial y} = \frac{\partial^2 \eta}{\partial x \partial y} = \frac{\partial \psi}{\partial x}.$$

Hence, subtracting (2.6.13) from (2.6.11), we obtain

$$(2.6.14)\qquad \int\limits_{\mathfrak{M}} a_1 \left(\frac{\partial \varphi}{\partial x} + \frac{\partial \psi}{\partial y}\right) dxdy = \int\limits_{\mathfrak{M}} a_1 \Delta \eta dxdy = 0.$$

Applying Lemma 2.5.1, we conclude that a_1 is harmonic. Similarly a_2 is harmonic. In particular, $\chi \in C^1$.

For every exact differential $d\varphi$, we have by (2.6.9) and (2.6.8)

$$(2.6.15)\qquad (\chi,\ d\varphi) = 0,\ (\chi,\ *d\varphi) = 0.$$

Applying (1.5.2) and (1.5.7), we see that these conditions become

$$(2.6.16)\qquad 0 = (*\chi,\ d\varphi) = (\varphi,\ *d\chi),$$

$$(2.6.17)\qquad 0 = (\chi,\ d\varphi) = -(\varphi,\ *d*\chi).$$

Since φ is an arbitrary function of class C^2, we conclude that

$$(2.6.18)\qquad d\chi = 0,\ d*\chi = 0.$$

In other words, χ is a harmonic differential and the decomposition formula (2.6.7) is therefore proved.

We remark that a harmonic differential on a closed surface cannot be exact unless it is zero. For harmonic differentials are orthogonal to the exact ones. We are thus led to the conclusion that a harmonic differential without periods is zero.

2.7. Existence of Single-valued Harmonic Functions with Singularities

In order to obtain harmonic functions and differentials with singularities, it is necessary to modify the preceding argument by choosing, instead of γ, a differential which does not satisfy (2.6.3).

Let p_0 be an arbitrarily chosen point of the closed orientable surface $\mathfrak{M}$, and let $z = x + iy = re^{i\varphi}$ be a particular uniformizer at p_0 which is valid for $|z| \leqq b$, $b > 0$. Let $0 < a < b$, and define (after Weyl [14a])

$$\Phi = \begin{cases} \dfrac{\cos\varphi}{r} + \dfrac{r\cos\varphi}{a^2}, & 0 \leqq r \leqq a, \\ 0, & \text{elsewhere on } \mathfrak{M}. \end{cases} \tag{2.7.1}$$

Now let

$$\theta = \begin{cases} h(r, \varphi), & 0 \leqq r < \dfrac{a}{2}, \\ \Phi, & \text{elsewhere on } \mathfrak{M}, \end{cases} \tag{2.7.2}$$

where $h(r, \varphi)$ is chosen in such a way that θ is of class C^3 in $r < a$. We suppose that Φ and θ behave invariantly with respect to changes of the uniformizer, so both are functions on the surface $\mathfrak{M}$. The differential $d\theta$ is of class C^2 for $0 \leqq r < a$ and elsewhere outside this circle, but it has a discontinuity across the circle $r = a$. Because of this discontinuity, $d\theta$ does not satisfy (2.6.3).

By orthogonal projection (Section 2.4), we find that

$$d\theta = \alpha + \chi, \tag{2.7.3}$$

where $\alpha \in \mathbf{E}$ and

$$(\chi, \alpha_1) = 0, \qquad \alpha_1 \in \mathbf{E}. \tag{2.7.4}$$

We again introduce the functions φ and ψ defined by (2.6.10), and we reason in the same way. However, since $\gamma = d\theta$ does not belong to **F**, we do not have (2.6.13) but have instead only

$$(\alpha,\ *d\psi) = \int_{\mathfrak{M}} \left(-b_1\frac{\partial\psi}{\partial y} + b_2\frac{\partial\psi}{\partial x}\right) dxdy = 0 \tag{2.7.5}$$

where

$$\alpha = b_1 dx + b_2 dy. \tag{2.7.6}$$

Therefore, in place of (2.6.14), we now obtain (since $(\chi,\ d\varphi) = 0$)

$$(d\theta,\ d\varphi) = \int_{\mathfrak{M}} b_1 \Delta\eta\, dxdy. \tag{2.7.7}$$

Let $\mathfrak{J}$ be the carrier of η, and let $\mathfrak{J}_0$ be the uniformizer circle $r \leqq a$. It is sufficient to consider the following cases:

(1) $\mathfrak{J} \cap \mathfrak{J}_0 = 0$. In this case, since $d\theta \equiv 0$ outside $\mathfrak{J}_0$, we have

$$\int_{\mathfrak{M}} b_1 \Delta\eta\, dxdy = 0. \tag{2.7.8}$$

(2) $\mathfrak{J} \cap \mathfrak{J}_0 \neq 0$ but $\mathfrak{J}$ has no point in common with $r \leqq a/2$. Applying Green's formula (1.5.8) to $\mathfrak{J}_0$, we find that

$$(d\theta,\ d\varphi)_{\mathfrak{J}_0} + (\varphi,\ \Delta\theta)_{\mathfrak{J}_0} = \int_{\partial\mathfrak{J}_0} \varphi \cdot *d\theta = -\int_{\partial\mathfrak{J}_0} \varphi \frac{\partial\theta}{\partial n} ds.$$

Since $\partial\theta/\partial n = 0$ on $\partial\mathfrak{J}_0$ (that is, on $r = a$), the boundary integral vanishes and, since $\Delta\theta = 0$ in $\mathfrak{J} \cap \mathfrak{J}_0$, we obtain (2.7.8).

(3) $\mathfrak{J} \subset \mathfrak{J}_0$. Integrating by parts, we find that

$$\begin{aligned}(d\theta,\ d\varphi)_{\mathfrak{J}_0} &= \int_{\mathfrak{J}_0} \left(\frac{\partial\theta}{\partial x}\frac{\partial^2\eta}{\partial x^2} + \frac{\partial\theta}{\partial y}\frac{\partial^2\eta}{\partial x\partial y}\right) dxdy \\ &= \int_{\mathfrak{J}_0} \left(\frac{\partial\theta}{\partial x}\frac{\partial^2\eta}{\partial x^2} + \frac{\partial\theta}{\partial x}\frac{\partial^2\eta}{\partial y^2}\right) dxdy = \int_{\mathfrak{J}_0} \frac{\partial\theta}{\partial x} \Delta\eta\, dxdy.\end{aligned}$$

Thus

$$\int_{\mathfrak{M}} \left(b_1 - \frac{\partial\theta}{\partial x}\right) \Delta\eta\, dxdy = 0. \tag{2.7.9}$$

By Lemma 2.5.1 we conclude that b_1 is harmonic outside $r \leqq a/2$ and that $b_1 - \partial\theta/\partial x$ is harmonic in $\mathfrak{J}_0$. A similar statement applies to b_2 outside $r \leqq a/2$ and $b_2 - \partial\theta/\partial y$ in $\mathfrak{J}_0$. In particular, we conclude that b_1 and b_2 are of class C^2 on $\mathfrak{M}$, and hence there is a single-valued function U on $\mathfrak{M}$ such that

$$(2.7.10) \qquad b_1 = \frac{\partial U}{\partial x}, \; b_2 = \frac{\partial U}{\partial y}, \quad \alpha = dU.$$

Substituting from (2.7.10) and (2.7.3) into (2.7.4), we have

$$(2.7.11) \qquad (d(\theta - U), d\eta) = 0$$

for every function η of class C^3 on $\mathfrak{M}$. The same reasoning then shows that U is harmonic outside $r \leqq a/2$ and that $U - \theta$ is harmonic in $r < a$. Hence

$$(2.7.12) \qquad u(p) = U(p) - \theta(p) + \Phi(p)$$

is harmonic on the whole surface $\mathfrak{M}$ except at the point p_0 where there is a dipole singularity. Since $U(p)$ is single-valued, so is $u(p)$.

We observe that, in terms of the particular uniformizer $z = re^{i\varphi} = x + iy$ used in defining $\Phi(p)$, we have near p_0:

$$(2.7.13) \; u(p) = \frac{x}{x^2 + y^2} + \text{regular terms} = \frac{\cos\varphi}{r} + \text{regular terms}.$$

The harmonic function $u(p)$ is unique up to an additive constant. For if u_1 and u_2 are two functions having the same local development (2.7.13), the difference $u_1 - u_2$ is an everywhere regular, single-valued harmonic function on $\mathfrak{M}$ and hence is equal to a constant.

Instead of (2.7.1), we could have taken another function, for example

$$(2.7.14) \quad \Phi = \begin{cases} \log r_1 - \log r_2 + \log r_1' - \log r_2', & 0 \leqq r \leqq a, \\ 0, & \text{elsewhere,} \end{cases}$$

where r_1, r_2 denote distances from two distinct points q_1, q_2 in $0 < r < a/2$ and r_1', r_2' are the distances from the inverse points q_1', q_2' (with respect to the circumference $r = a$).

For the function (2.7.14) has a vanishing normal derivative on $r = a$, and is harmonic throughout $r \leqq a$ except for logarithmic singularities at the points q_1, q_2. The above proof applies without modification, and we then establish the existence of a single-valued harmonic function $u_{q_1 q_2}$ which has logarithmic singularities at q_1 and

q_2. The points q_1, q_2 lie inside the same uniformizer circle. To obtain a harmonic function with logarithmic singularities at an arbitrary pair of points q_0, q on $\mathfrak{M}$, let $q_0 = q_1, q_2, \cdots, q_n = q$ be a sequence of points such that any two successive points q_{k-1}, q_k in the sequence lie inside the same uniformizer circle and form the sum

$$u_{q_0 q} = u_{q_1 q_2} + u_{q_2 q_3} + \cdots + u_{q_{n-1} q_n}. \tag{2.7.15}$$

The sum (2.7.15) is a single-valued harmonic function with logarithmic singularities at q and q_0.

If we had chosen

$$\Phi = \begin{cases} \varphi_1 - \varphi_2 - (\varphi_1' - \varphi_2'), & 0 \leqq r \leqq a, \\ 0, & \text{elsewhere}, \end{cases} \tag{2.7.16}$$

where, for example, φ_1 is the inclination of the segment joining p and q_1, we would have obtained a harmonic function $v_{q_1 q_2}$ with vortex singularities at q_1, q_2. This function is single-valued only on the surface cut along an arc joining q_1 and q_2. By forming a sum like (2.7.15), we obtain a function $v_{q_0 q}$ with vortex singularities at arbitrary points q_0, q of $\mathfrak{M}$. Let γ be a 1-cycle on $\mathfrak{M}$, and let the points $q_0 = q_1, q_2, \cdots, q_n = q_0$ be interpolated along γ such that any two successive points lie in a uniformizer circle. Since $q_1 = q_n$, the sum

$$v = v_{q_1 q_2} + v_{q_2 q_3} + \cdots + v_{q_{n-1} q_n} \tag{2.7.17}$$

is everywhere regular but not necessarily single-valued. We shall return to this function in Chapter 3, and shall calculate its periods.

2.8. Boundary-value Problems by the Method of Orthogonal Projection

We have proved, in Section 2.7, the existence of harmonic functions with singularities on a closed orientable Riemann surface. We want now to apply this result to the study of boundary-value problems on a Riemann surface $\mathfrak{M}$ with boundary C. Let $\mathfrak{F}$ be the double of $\mathfrak{M}$, and let $u_{q_0 q}$ be the harmonic function (2.7.15) on $\mathfrak{F}$. Since the logarithmic poles q and q_0 of this function can be chosen arbitrarily on $\mathfrak{F}$ and therefore, in particular, on the part $\tilde{\mathfrak{M}}$ of $\mathfrak{F}$, we can assert that there exist infinitely many single-valued regular harmonic functions u on $\mathfrak{M}$ which have finite Dirichlet integrals

$$(2.8.1)\quad D(u) = (du, du) = \int_{\mathfrak{M}} \left\{ \left(\frac{\partial u}{\partial x}\right)^2 + \left(\frac{\partial u}{\partial y}\right)^2 \right\} dxdy = \int_{\mathfrak{M}} (\text{grad } u)^2 dxdy.$$

All regular harmonic functions u on $\mathfrak{M}$ which have finite Dirichlet integrals form a Hilbert space **H** with the metric

$$(2.8.2)\quad \begin{aligned} D(u, v) &= (du, dv) \\ &= \int_{\mathfrak{M}} \left\{ \frac{\partial u}{\partial x}\frac{\partial v}{\partial x} + \frac{\partial u}{\partial y}\frac{\partial v}{\partial y} \right\} dxdy = \int_{\mathfrak{M}} \text{grad } u \cdot \text{grad } v \, dxdy. \end{aligned}$$

The Hilbert space **H** is actually a Hilbert space of differentials du. However, if we consider as identical two functions which differ by a constant, we obtain a Hilbert space of functions u.

Let $\varphi(p)$ be twice continuously differentiable on $\mathfrak{M}$ with finite Dirichlet integral and with a Laplacian $\Delta\varphi$ which is square integrable (L^2) in the neighborhood of each boundary point, the integration being expressed in terms of boundary uniformizers. If, moreover, $\varphi(p)$ is continuous in the closure of $\mathfrak{M}$, we can prove that there exists a function u of **H** which has the same boundary values on C as the function φ. We shall, therefore, be able to write

$$(2.8.3)\quad \varphi = u + \psi$$

where $\psi = \varphi - u$ is again a twice continuously differentiable function on $\mathfrak{M}$ with finite Dirichlet integral, which is continuous in the closure of $\mathfrak{M}$ and has vanishing boundary values.

If $U(p)$ is any harmonic function on $\mathfrak{M}$ which is continuously differentiable in the closure of $\mathfrak{M}$, we have by Green's identity

$$(2.8.4)\quad D(U, \psi) = -\int_C \psi \frac{\partial U}{\partial n} ds = 0$$

since $\psi = 0$ on C. From this result we readily infer that the function ψ is orthogonal to the whole space **H** in the metric (2.8.2). We are therefore led to try to prove the existence of harmonic functions with prescribed boundary values by the method of orthogonal projection.

Two possibilities arise in the application of the projection method. We may consider the Hilbert space **N** obtained by closing the linear space of all continuously differentiable functions on $\mathfrak{M} \cup C$ which have finite Dirichlet integrals and vanishing boundary values, and then project φ into **N**. This is the usual method (see [14b]), and it leads to the decomposition formula (2.8.3) where the difference term $u = \varphi - \psi$ is readily shown to be harmonic. Or we may project φ into **H** and then verify that the term $\psi = \varphi - u$ vanishes on the boundary C of $\mathfrak{M}$. Since the space **H** is simpler than the closure **N** of an incomplete linear space, we proceed by the second method. This method, which has been used by several authors (see [3], [7]), is becoming increasingly important in the theory of functions. Here we follow along the lines of a recent proof communicated to us by Dr. P. Lax.

We seek a function $u \in \mathbf{H}$ such that

$$D(\varphi - u) = \text{minimum}. \tag{2.8.5}$$

According to the general results derived earlier for a Hilbert space, an extremal function $u \in \mathbf{H}$ exists and it is characterized by the property that for every element $U \in \mathbf{H}$ we have

$$D(\varphi - u,\ U) = 0. \tag{2.8.6}$$

We shall now transform the characterizing condition (2.8.6) in order to show that the difference $\psi = \varphi - u$ vanishes on the boundary C. For this purpose, we specialize the function $U(p)$ and choose it to be of the form

$$U(p) = u_{q_0 q}(p) \tag{2.8.7}$$

where q_0 and q are points of $\tilde{\mathfrak{M}}$. We have

$$D(\varphi - u,\ u_{q_0 q}) = 0, \quad q_0,\ q \in \tilde{\mathfrak{M}}. \tag{2.8.8}$$

Let us keep q_0 fixed in $\tilde{\mathfrak{M}}$ once and for all, but let q vary over $\mathfrak{F}$. Then the improper integral

$$D(\varphi - u,\ u_{q_0 q}) = h(q) \tag{2.8.9}$$

represents a continuous function of q in $\mathfrak{M}$ and in $\tilde{\mathfrak{M}}$. We want to prove that $h(q)$ is continuous even across the boundary C of $\mathfrak{M}$, and therefore on the whole of $\mathfrak{F}$. This result will be of value in the study of the boundary behavior of the extremal function u.

We choose a point p_0 on the boundary C of $\mathfrak{M}$ where the parameter point q will cross from $\mathfrak{M}$ to $\tilde{\mathfrak{M}}$. Using a boundary uniformizer at p_0, we can map a neighborhood of p_0 on $\mathfrak{F}$ onto a disc $\mathfrak{I}$, with center at the origin of the complex plane, such that p_0 corresponds to the origin and the part of C contained in the neighborhood goes into the real axis inside $\mathfrak{J}$. Let the half $\mathfrak{J}_-$ of $\mathfrak{J}$ below the real axis correspond to points of $\mathfrak{M}$, the half $\mathfrak{J}_+$ above the real axis to points of $\tilde{\mathfrak{M}}$. Let z and ζ be the coordinates of the points p and q referred to the boundary uniformizer at p_0. It is obvious that it suffices to show that the integral

$$\int_{\mathfrak{J}_-}\left\{\frac{\partial\psi}{\partial x}\frac{\partial \log|z-\zeta|}{\partial x}+\frac{\partial\psi}{\partial y}\frac{\partial \log|z-\zeta|}{\partial y}\right\}dxdy = I(\zeta) \tag{2.8.10}$$

is a continuous function of ζ in $\mathfrak{J}$. In order to prove this, we introduce three auxiliary circles: the circle $\mathfrak{K}$ which lies in $\mathfrak{J}_-$ and touches the real axis at the origin; the circle $\tilde{\mathfrak{K}}$ obtained from $\mathfrak{K}$ by reflection on the real axis; a circle $\mathfrak{H}$ around the origin with radius ϱ. This radius will later be chosen sufficiently small to allow certain estimates.

We may assume without loss of generality that ζ lies on the imaginary axis between the center of $\mathfrak{K}$ and the origin. Let ζ_1 be the inverse point of ζ with respect to the circle $\mathfrak{K}$. It is obvious that $\operatorname{Im}\zeta_1 > 0$. We shall show that $|I(\zeta_1)-I(\zeta)|$ can be made arbitrarily small by choosing $|\operatorname{Im}\zeta|$ small; this is equivalent to the asserted continuity of $h(q)$ on $\mathfrak{F}$.

We divide the semi-circle $\mathfrak{J}_-$ into three parts: $\mathfrak{K}$, $(\mathfrak{J}_- - \mathfrak{K}) \cap \mathfrak{H}$, $\mathfrak{J}_- - \mathfrak{K} - \mathfrak{H}$. We shall show that the contribution of each region to the integral $I(\zeta) - I(\zeta_1)$ tends to zero as ζ approaches the origin. Let us choose, in fact, an arbitrarily small but fixed number $\varepsilon > 0$. We choose the radius ϱ of $\mathfrak{H}$ so small that the integral

$$\int_{\mathfrak{H}-\mathfrak{K}-\tilde{\mathfrak{K}}}\frac{1}{|z|^2}dxdy$$

is smaller than ε^2. We observe that the integral converges since $\mathfrak{H}-\mathfrak{K}-\tilde{\mathfrak{K}}$ forms a horn-angle at the origin. For $|\zeta| < \varrho$ and $\operatorname{Re}\zeta = 0$, we find by a simple geometric argument that

$$\left(\frac{\partial \log|z-\zeta|}{\partial x}\right)^2+\left(\frac{\partial \log|z-\zeta|}{\partial y}\right)^2 \leqq \frac{8}{|z|^2},$$

$$\left(\frac{\partial \log|z-\zeta_1|}{\partial x}\right)^2+\left(\frac{\partial \log|z-\zeta_1|}{\partial y}\right)^2 \leqq \frac{8}{|z|^2}.$$

Thus, we can estimate the integral

$$\int\limits_{(\mathfrak{J}_- - \mathfrak{K}) \cap \mathfrak{H}} \left\{ \frac{\partial \psi}{\partial x} \frac{\partial}{\partial x} \left(\log \left| \frac{z - \zeta}{z - \zeta_1} \right| \right) + \frac{\partial \psi}{\partial y} \frac{\partial}{\partial y} \log \left| \frac{z - \zeta}{z - \zeta_1} \right| \right) \right\} dxdy$$

by the Schwarz inequality, and we find that it is smaller than $A\varepsilon$, where A is of the order of magnitude of $\sqrt{D(\psi)}$.

Consider next the integral extended over $\mathfrak{J}_- - \mathfrak{K} - \mathfrak{H}$. Since ϱ is now kept fixed, we assert that the expression

$$\left(\frac{\partial}{\partial x} \log \left| \frac{z - \zeta}{z - \zeta_1} \right| \right)^2 + \left(\frac{\partial}{\partial y} \log \left| \frac{z - \zeta}{z - \zeta_1} \right| \right)^2$$

tends uniformly to zero in $\mathfrak{J}_- - \mathfrak{K} - \mathfrak{H}$ as ζ converges to the origin. Hence, the contribution of this domain to $I(\zeta) - I(\zeta_1)$ can be made smaller than ε by choosing $|\operatorname{Im} \zeta|$ small enough.

To estimate the remaining integral, which is extended over $\mathfrak{K}$, we integrate by parts. But first we cut out a neighborhood of the origin by means of a small circular arc c around the origin of radius δ, since we are not sure that ψ and its first partial derivatives are continuous up to the boundary C of $\mathfrak{M}$. Let the truncated domain which is obtained by removing the interior of c from $\mathfrak{K}$ be denoted by $\mathfrak{K}'$. Integrating by parts, we have

$$D_{\mathfrak{K}'}\left(\psi, \log \left| \frac{z - \zeta}{z - \zeta_1} \right| \right) = - \int\limits_{\partial \mathfrak{K}'} \left[\log \left| \frac{z - \zeta}{z - \zeta_1} \right| - k \right] \frac{\partial \psi}{\partial n} ds \tag{2.8.11}$$
$$- \int\limits_{\mathfrak{K}'} \left[\log \left| \frac{z - \zeta}{z - \zeta_1} \right| - k \right] \Delta \psi dxdy$$

where k is an arbitrary constant of integration. Since ζ and ζ_1 are inverse points with respect to $\mathfrak{K}$, the ratio $|z - \zeta|/|z - \zeta_1|$ is a constant, depending on ζ, for z on the circumference of $\mathfrak{K}$, and we choose k to be just the log of this value. Thus in the integral over $\partial \mathfrak{K}'$ only the integration over c remains.

Following a well-known argument, we can show that the integral over c tends to zero if the radius δ of c approaches zero through a suitable subsequence. In fact, suppose that there exists a positive constant A such that

$$A < \int\limits_c \left| \frac{\partial \psi}{\partial n} \right| ds = \int\limits_c \left| \frac{\partial \psi}{\partial r} \right| r d\theta \tag{2.8.12}$$

for $0 < r < r_0$, where r and θ are polar coordinates around the origin. By the Schwarz inequality we have

$$A^2 \leqq \int_c r d\theta \cdot \int_c \left(\frac{\partial \psi}{\partial r}\right)^2 r d\theta \leqq \pi r \int_c \left(\frac{\partial \psi}{\partial r}\right)^2 r d\theta \tag{2.8.13}$$

since c is less than a semicircle. Dividing by πr and integrating with respect to r from 0 to r_0, we obtain

$$\frac{A^2}{\pi} \int_0^{r_0} \frac{dr}{r} \leqq D(\psi). \tag{2.8.14}$$

The left side is infinite while the right side is finite since $D(\psi) < \infty$. This contradiction shows that a number A satisfying (2.8.12) does not exist, and hence there is a sequence of radii δ_ν convergent to zero such that for the corresponding arcs c_ν we have

$$\lim_{\nu \to \infty} \int_{c_\nu} \left| \frac{\partial \psi}{\partial n} \right| ds = 0. \tag{2.8.15}$$

If we let δ tend to zero through this sequence of values, the boundary integral in (2.8.11) disappears and we have

$$D_{\mathfrak{K}}\left(\psi, \log \left| \frac{z-\zeta}{z-\zeta_1} \right| \right) = - \int_{\mathfrak{K}} \left[\log \left| \frac{z-\zeta}{z-\zeta_1} \right| - k \right] \Delta \psi dx dy. \tag{2.8.16}$$

It is easily seen from elementary geometry that $\log(|z-\zeta|/|z-\zeta_1|) - k$ tends to zero as ζ converges toward the origin. Hence the integral

$$\int_{\mathfrak{K}} \left[\log \left| \frac{z-\zeta}{z-\zeta_1} \right| - k \right]^2 dx dy \tag{2.8.17}$$

tends to zero. Now we make use of our assumption concerning the Laplacian of φ. Since $\Delta\varphi = \Delta\psi$ by (2.8.3), we see that $\Delta\psi$ is square integrable over $\mathfrak{K}$. Hence we conclude from (2.8.16) and the inequality of Schwarz that the contribution of $\mathfrak{K}$ to $I(\zeta_1) - I(\zeta)$ can be made smaller than ε by choosing ζ near enough to the origin.

Thus, given $\varepsilon > 0$, the quantity $|h(q)|$ for q in $\mathfrak{M}$ can be made smaller than ε by choosing $|\operatorname{Im} \zeta|$ sufficiently small. In fact, $I(\zeta)$ differs from $h(q)$ by a term which is clearly continuous across the boundary and $h(q)$ vanishes in $\tilde{\mathfrak{M}}$. In view of the uniformity of the above estimates, we see that $h(q)$ is a continuous function of q on $\mathfrak{F}$, which vanishes on the boundary C of $\mathfrak{M}$.

Now let q be a point in $\mathfrak{M}$, and describe a uniformizer circle $\mathfrak{L}$ around q. We may then represent $h(q)$ as the sum of two integrals, one extended over $\mathfrak{L}$ and the other over $\mathfrak{M} - \mathfrak{L}$. On integrating by parts, we have

$$\begin{aligned} h(q) &= D_{\mathfrak{M}-\mathfrak{L}}(\psi, u_{q_0q}) - \int_{\partial\mathfrak{L}} \psi \frac{\partial u_{q_0q}(p)}{\partial n_p} ds_p + 2\pi\psi(q) \\ &= \omega(q) + 2\pi\psi(q). \end{aligned} \tag{2.8.18}$$

Since $u_{q_0q}(p)$ is harmonic in its dependence on q, $\omega(q)$ is obviously harmonic. Thus, since $\psi = \varphi - u$, we have

$$\varphi(q) = \left[u(q) - \frac{1}{2\pi}\omega(q)\right] + \frac{1}{2\pi}h(q). \tag{2.8.19}$$

The bracketed term in (2.8.19) is a harmonic function on $\mathfrak{M}$ with the same boundary values as the given function $\varphi(q)$.

We therefore have a decomposition of the form

$$\varphi = U + \Psi \tag{2.8.20}$$

where U is harmonic in $\mathfrak{M}$ and continuous in the closure of $\mathfrak{M}$, and where Ψ is continuous in the closure of $\mathfrak{M}$ with the boundary values zero. It remains to show that U and Ψ have finite Dirichlet integrals, and to establish the relationship between the harmonic function U and the extremal function u in the Hilbert space $\mathbf{H}$.

We again choose p_0 to be a boundary point of $\mathfrak{M}$ and we map a neighborhood of p_0 in $\mathfrak{M}$ onto a semicircle of the complex z-plane which is bounded above by a segment of the real axis. We decompose the function φ into a sum

$$\varphi = u_1 + \psi_1 \tag{2.8.21}$$

where u_1 and ψ_1 have finite Dirichlet integrals over the semicircle, are continuous in its closure, and where u_1 is harmonic and ψ_1 vanishes on the boundary of the semicircle. This decomposition is possible since we have only to map the semicircle onto a full circle and then apply well-known methods.

Now consider the function $U - u_1$ which is harmonic in the semicircle and zero on that part of the boundary of the semicircle which coincides with the real axis. By the reflection principle we may continue $U - u_1$ across the real axis, and we see that it is harmonic in the whole circle obtained by reflection of the semicircle on the real axis. This shows that $U - u_1$, and therefore U, has a finite Dirichlet integral over the semicircle. Since this holds for the neighborhood of each boundary point p_0 of $\mathfrak{M}$, it follows that U has a finite Dirichlet integral over $\mathfrak{M}$ and belongs, therefore, to $\mathbf{H}$. Thus $u - \omega/2\pi \in \mathbf{H}$, $\omega \in \mathbf{H}$. Furthermore, we conclude that $\Psi = h/2\pi$ has a finite Dirichlet integral. The condition which characterizes the extremal function $u \in \mathbf{H}$ can now be written in the form

$$D\left(\varphi - u, \frac{1}{2\pi}\omega\right) = D\left(\frac{1}{2\pi}(h - \omega), \frac{1}{2\pi}\omega\right) = 0. \tag{2.8.22}$$

But h vanishes on the boundary of $\mathfrak{M}$, and therefore is orthogonal to all functions of $\mathbf{H}$. Hence (2.8.22) implies that $D(\omega) = 0$, that is $\omega =$ constant. Thus we have proved that the function u which minimizes $D(\varphi - u)$ is continuous in the closure of $\mathfrak{M}$ and differs on the boundary of $\mathfrak{M}$ only by a constant from the boundary values of φ.

2.9. Harmonic Functions of a Finite Surface

Any finite Riemann surface $\mathfrak{M}$ can be completed by the doubling process to a symmetric closed orientable surface $\mathfrak{F}$. By adding together a pair of harmonic functions of $\mathfrak{F}$ whose singularities lie at conjugate points we obtain a harmonic function which is symmetric or skew-symmetric under the indirect conformal mapping of $\mathfrak{F}$ onto itself. In other words, by a procedure analogous to the Kelvin method of images, we obtain harmonic functions of the first or second kind on $\mathfrak{M}$.

Suppose that $\mathfrak{M}$ is a finite Riemann surface which has a boundary or is non-orientable, and let $\mathfrak{F}$ be its double. Let p_0 be a point of $\mathfrak{F}$ which corresponds to an interior point of $\mathfrak{M}$, and let $z = x + iy$ be a uniformizer at p_0. We denote by u_1 the single-valued harmonic function on $\mathfrak{F}$ with a dipole at p_0 such that the difference

$$u_1 - \frac{x}{x^2 + y^2} = u_1 - \mathrm{Re}\left(\frac{1}{z}\right) \tag{2.9.1}$$

is regular and vanishing at p_0. Let $\tilde{u}_1$ be the corresponding harmonic function with a dipole singularity at the conjugate point $\tilde{p}_0$ of $\mathfrak{F}$

which is defined using the uniformizer $\tilde{z} = x - iy$. Then $u_1(p) = \tilde{u}_1(\tilde{p})$ for any pair of conjugate points p, $\tilde{p}$, so $u_1 + \tilde{u}_1$ takes equal values at conjugate places of $\mathfrak{F}$. The function

$$u(p) = u_1(p) + \tilde{u}_1(p) - \tilde{u}_1(p_0) \tag{2.9.2}$$

is a harmonic function of $\mathfrak{M}$ of the second kind for which the difference

$$u - \frac{x}{x^2 + y^2} = u - \text{Re}\left(\frac{1}{z}\right) \tag{2.9.3}$$

vanishes at p_0.

Let $\mathfrak{M}_0$ be the uniformizer circle $|z| \leqq a$ with center at p_0, and let $\tilde{\mathfrak{M}}_0$ be the conjugate uniformizer circle $|\tilde{z}| \leqq a$ at $\tilde{p}_0$. Let Φ be the function (2.7.1) which vanishes outside $\mathfrak{M}_0$, and let $\tilde{\Phi}$ be its conjugate:

$$\tilde{\Phi}(p) = (\Phi(\tilde{p}))^-.$$

Write

$$V = u - \Phi - \tilde{\Phi}. \tag{2.9.4}$$

On $\mathfrak{M}$ we have

$$V = u - \Phi \tag{2.9.5}$$

since $\tilde{\Phi} \equiv 0$ there. By (2.7.11) and (2.7.12)

$$(dV, d\eta) = (d(u_1 - \Phi) + d(\tilde{u}_1 - \tilde{\Phi}), d\eta) = 0 \tag{2.9.6}$$

for every function η of class C^3 on $\mathfrak{M}$. We observe that $(dV, d\eta)$ is the classical Dirichlet integral $D(V, \eta)$:

$$(dV, d\eta) = \int_{\mathfrak{M}} \left(\frac{\partial V}{\partial x}\frac{\partial \eta}{\partial x} + \frac{\partial V}{\partial y}\frac{\partial \eta}{\partial y}\right) dxdy. \tag{2.9.7}$$

The formulas (2.9.6) and (2.9.7) will be required in the following section.

2.10. The Uniformization Principle for Finite Surfaces

We now prove a classical theorem concerning the conformal mapping of a finite Riemann surface of genus zero onto a domain of the sphere.

Assume that $\mathfrak{M}$ is a finite Riemann surface of genus zero, and let

u be the harmonic function (2.9.2) on $\mathfrak{M}$ which has a dipole singularity at the point p_0. Let $w = u + iv$ be the analytic function whose real part is u. We show first that v is single-valued on $\mathfrak{M}$; that is

$$\int dv = 0$$

where the integration is over any 1-cycle of $\mathfrak{M}$ not passing through the point p_0. Let $\mathfrak{M}$ be divided into finitely many triangles. We may then suppose that the path is a 1-cycle C^1 on $\mathfrak{M}$ which divides $\mathfrak{M}$ into two domains $\mathfrak{M}'$ and $\mathfrak{M}''$ one of which, say $\mathfrak{M}''$, contains the point p_0. The finite set of triangles of $\mathfrak{M}'$ which have points in common with C^1 forms a strip region $\mathfrak{S}$ which is bounded by C^1 and by another 1-cycle C^1_1. Let η be a function of class C^3 on $\mathfrak{M}$ which is equal to 1 in $\mathfrak{M}''$ and equal to zero in that part of $\mathfrak{M}'$ which is exterior to $\mathfrak{S}$. If we apply Green's formula to one of the triangles Δ of $\mathfrak{S}$, we obtain

$$D_\Delta(u, \eta) = -\int \eta \frac{\partial u}{\partial n} ds = \int \eta dv \tag{2.10.1}$$

where the integration is over the boundary of Δ. Adding equations (2.10.1) over all triangles of $\mathfrak{S}$ we have

$$D_{\mathfrak{M}}(u, \eta) = D_{\mathfrak{S}}(u, \eta) = \int_{C^1} dv. \tag{2.10.2}$$

But $D_{\mathfrak{M}}(u, \eta) = 0$ by (2.9.6) and (2.9.7), and this shows that v is single-valued on C^1. Since u is also single-valued, so is the analytic function $w = u + iv$.

The function $w = u + iv$ is analytic and v has a constant value on each boundary component of $\mathfrak{M}$. Let the values of v on the m boundary components of $\mathfrak{M}$ be $c_1, c_2, \cdots, c_m$, and let $a = \alpha + i\beta$ be any value for which β is different from $c_1, c_2, \cdots, c_m$. To prove that the image of $\mathfrak{M}$ by $w = u + iv$ is a schlicht domain, it is sufficient to show that $w = u + iv$ assumes the value a in $\mathfrak{M}$ once and only once. Since w assumes the value a at only finitely many points, we may divide $\mathfrak{M}$ into finitely many triangles such that w is different from a on the boundary of any triangle and such that the point p_0 is interior to one of the triangles. Let the triangles be oriented coherently and let the orientation be such that, in traversing the boundary of any triangle in the sense of the induced orientation, the area of the triangle lies to the left. We may assume that the triangle containing p_0 is so small that the value a is not taken in this triangle. We now consider the change of $\log (w - a)$ as the point describes the boundary of a triangle in the positive sense. Around the boundary of the triangle containing p_0 the change is $-2\pi i$ while around the boundary of any other triangle it is equal to $2\pi i$ times the multiplicity of the a-values inside the triangle. The sum of all these changes is equal to the sum of the changes of $\log(w-a)$ around the boundary curves and is therefore equal to zero. For the change of $\log (w - a)$ on a boundary curve is obviously equal to zero. This shows that w assumes the value a on $\mathfrak{M}$ exactly once. Thus:

THEOREM 2.10.1. *A finite Riemann surface $\mathfrak{M}$ of genus zero can be mapped onto the closed w-plane minus m rectilinear slits parallel to the real axis. The mapping function $w = u + iv$ is regular analytic at each point of $\mathfrak{M}$ except for a simple pole at one point and the corresponding differential dw has precisely two simple zeros on each of the m boundaries of $\mathfrak{M}$ and no other zeros. The zeros correspond to the ends of the rectilinear slits.*

If $\mathfrak{M}$ is a simply-connected finite Riemann surface, the function $w = u + iv$ maps $\mathfrak{M}$ either onto the closed w-plane (sphere) or onto the plane minus a single rectilinear segment parallel to the real axis. The case in which $\mathfrak{M}$ is mapped onto the plane punctured at one point is excluded since the punctured plane is an infinite complex and cannot be the image of a finite surface.

If w maps $\mathfrak{M}$ onto the plane minus a segment, we can transform the segment into

$$v = 0, \quad -1 \leqq u \leqq 1, \tag{2.10.3}$$

by a linear transformation consisting of a translation and magnification. The mapping

$$w = \frac{1}{2}\left(z + \frac{1}{z}\right)$$

then carries the w-sphere minus the segment (2.10.3) onto the interior of the unit circle $|z| < 1$.

2.11. Conformal Mapping onto Canonical Domains of Higher Genus

By Theorem 2.10.1 any finite Riemann surface $\mathfrak{M}$ of genus zero can be mapped onto the closed w-plane minus m rectilinear slits parallel to the real axis. If the finite surface $\mathfrak{M}$ is not of genus zero, then we see that it can be mapped onto a canonical domain consisting of boundary slits and, in addition, slits whose edges are identified. On each boundary curve of $\mathfrak{M}$ the differential dw arising from the mapping function $w = u + iv$ has two simple zeros corresponding to the end points of the finite boundary slits. As a consequence of a formula to be proved in the next chapter (formula 3.6.3), the total multiplicity of the zeros of dw in the interior of $\mathfrak{M}$ is equal to $2h + c$. Let us assume, for simplicity, that each of these zeros is simple and that no line $v = constant$ contains more than one crossing point. Each crossing point then gives rise to two semi-infinite slits in the w-plane with edges appropriately identified, so the total number of slits of the canonical domain is equal to $4h + 2c + m$. Each semi-infinite slit is determined by the position of its finite end point while a finite boundary slit is determined by its left end point and by its length. Since the semi-infinite slits are identified in pairs and the finite end points of a pair have the same abscissa u, we need $6h + 3c + 3m$ real numbers to describe all the slits. Two Riemann surfaces can be mapped conformally upon one another if and only if the corresponding canonical domains can be mapped upon one another by a translation and magnification. Thus two Riemann surfaces can be mapped conformally one onto the other such that a given point and direction on one goes into a given point and direction on the other provided that $6h+3c+3m-3$ real numbers have the same values. If we drop the condition that given points and directions correspond, we see that the number of real parameters needed to fix the conformal type is actually equal to $6h + 3c + 3m - 6$. However, we have neglected to take into account certain continuous groups of conformal mappings of $\mathfrak{M}$ onto itself. If ϱ denotes the number of real parameters in the continuous group of conformal mappings of $\mathfrak{M}$ onto itself and if σ is the dimension of the space of all classes of conformally equivalent finite

Riemann surfaces of the same topological type, then

$$\sigma - \varrho = 6h + 3c + 3m - 6. \tag{2.11.1}$$

This formula will be discussed further in the next chapter.

REFERENCES

1. R. COURANT, *Dirichlet's Principle, conformal mapping, and minimal surfaces*, Interscience, New York, 1950.
2. P. FATOU, *Fonctions automorphes*, Vol. II of P. Appell and E. Goursat, *Théorie des fonctions algébriques d'une variable*, Gauthier-Villars, Paris, 1930.
3. P. R. GARABEDIAN and M. SCHIFFER, "On existence theorems of potential theory and conformal mapping," *Annals of Math.*, 52 (1950), 164—187.
4. A. HURWITZ and R. COURANT, *Funktionentheorie*, Springer, Berlin, 1929.
5. F. KLEIN, *Riemann'sche Fläche I und II*, Vorlesungen Göttingen, Wintersemester 1891—1892 und Sommersemester 1892.
6. K. KODAIRA, "Harmonic fields in Riemannian manifolds (generalized potential theory)," *Annals of Math.*, 50 (1949), 587—665.
7. O. LEHTO, „Anwendung orthogonaler Systeme auf gewisse funktionentheoretische Extremal- und Abbildungsprobleme," *Ann. Acad. Sci. Fenn. A. I*, 59 (1949).
8. P. KOEBE, „Über die Uniformisierung beliebiger analytischer Kurven. Erster Teil: Das allgemeine Uniformisierungsprinzip," *Crelles Journal*, 138 (1910), 192—253.
9. E. PICARD, *Traité d'Analyse*, Vol. II, Gauthier-Villars, Paris, 1893.
10. G. DE RHAM and K. KODAIRA, "Harmonic Integrals", Institute for Advanced Study, Princeton, 1950 (mimeographed).
11. F. SCHOTTKY, „Über die conforme Abbildung mehrfach zusammenhängender ebener Flächen," *Crelles Journal*, 83 (1877), 300—351.
12. S. STOILOW, *Leçons sur les principes topologiques de la théorie des fonctions analytiques*, Gauthier-Villars, Paris, 1938.
13. O. TEICHMÜLLER, „Extremale quasikonforme Abbildungen und quadratische Differentiale," *Abh. der Preuss. Akad. der Wiss., Math.-Naturw. Kl.* 1939, 22 (1940).
14. H. WEYL, (a) *Die Idee der Riemannschen Fläche*, Teubner, Berlin, 1923. (Reprint Chelsea, New York, 1947). (b) "The method of orthogonal projection in potential theory," *Duke Math. Jour.*, 7 (1940), 411—444.
15. R. L. WILDER, *Topology of manifolds*, Colloquium Publications, Vol. 32, Amer. Math. Soc., New York, 1949.

3. Relations between Differentials

3.1. Abelian Differentials

In this chapter we bring together various classical results which will be required in the sequel, and we begin by discussing briefly the differentials of a closed orientable surface $\mathfrak{F}$. Later we shall assume that $\mathfrak{F}$ is the double of a surface $\mathfrak{M}$, in which case a subset of the differentials of $\mathfrak{F}$ forms the set of differentials of $\mathfrak{M}$.

We shall consider first the linear differentials dZ of $\mathfrak{F}$. If z is a uniformizer at the point p of $\mathfrak{F}$, then we have locally

$$\frac{dZ}{dz} = a_m z^m + a_{m+1} z^{m+1} + \cdots \tag{3.1.1}$$

where m is an integer. The number m is called the order of dZ at p, and it is plainly independent of the uniformizer chosen and is therefore a conformal invariant. The residue of dZ at p is defined to be the value of the integral

$$\frac{1}{2\pi i}\int_k dZ = \frac{1}{2\pi i}\int_k \frac{dZ}{dz}\,dz \tag{3.1.2}$$

extended in the positive sense over the circumference k of a z-circle $|z| \leqq a$ at p. This definition of residue is also independent of the particular uniformizer used. It is clear that the residue is equal to the coefficient a_{-1} of $1/z$ in the development (3.1.1).

A linear differential dZ of $\mathfrak{F}$ is said to be of the first kind if it is regular at every point of $\mathfrak{F}$, of the second kind if it is regular except for poles but never has a residue different from zero, of the third kind if it is regular except for poles and has at least one residue different from zero. The integral of a (linear) differential dZ is defined by

$$Z(p) - Z(p_0) = \int_{p_0}^{p} dZ$$

and depends on the limits of the integral and the path connecting them. The integral of a (linear) differential of the first, second or third kind is called an integral or Abelian integral of $\mathfrak{F}$ of the first, second or third kind respectively.

Let h be the genus of the closed surface $\mathfrak{F}$, and let $K_1, K_2, \cdots, K_{2h}$ be a homology basis of 1-cycles for $\mathfrak{F}$. In Section 2.6 we established the existence of a harmonic differential χ having prescribed periods

$$\int_{K_\mu} \chi.$$

Then $dw = \chi + i{*}\chi$ is a complex analytic differential, and the real parts of its periods can be prescribed arbitrarily. Moreover, since an exact harmonic differential vanishes identically, we see that dw is uniquely determined by the real parts of its periods. We therefore have the following fundamental existence theorem of Riemann:

THEOREM 3.1.1. *There exists a unique differential of the first kind with prescribed real parts for the periods.*

The function (2.7.16), which was obtained by interpolating vortices along a 1-cycle γ, also gives rise to a differential of the first kind, as we now show.

Let T denote a given triangulation of the surface $\mathfrak{F}$, and let ${*}T$ be a dual subdivision of $\mathfrak{F}$ defined as follows. First make a normal subdivision N of the given triangulation. This is accomplished by choosing an interior point of each triangle and an interior point of each edge (1-simplex), and then joining the interior point of each triangle to its three vertices and to the three interior points of its edges by six arcs. Given any vertex (0-simplex) of the original triangulation T, we associate with it the polygon formed by the triangles of the normal subdivision N which have this vertex in common. To each edge (1-simplex) of T, we associate the two new edges of N which meet at its interior point, and to each triangle (2-simplex) of T we associate the vertex of N lying in its interior. The polygons, edges and vertices associated respectively with the vertices, edges and triangles of T form a dual subdivision ${*}T$. In particular, to each edge σ_k^1 of T there is a dual edge ${*}\sigma_k^1$ of ${*}T$ which intersects it in precisely one point, the orientation of ${*}\sigma_k^1$ being such that it crosses σ_k^1 from right to left. We define the intersection number $I(\sigma_j^1, *\sigma_k^1)$ of two 1-simplexes σ_j^1, $*\sigma_k^1$ by the rule

(3.1.3) $$I(\sigma_j^1, *\sigma_k^1) = \begin{cases} 1, & j = k \\ 0, & \text{otherwise.} \end{cases}$$

The Kronecker index $I(K, *L)$ of two 1-chains

(3.1.4) $$K = \Sigma\, m_j \cdot \sigma_j^1, \quad *L = \Sigma\, n_k \cdot *\sigma_k^1$$

is then given by the sum

(3.1.5) $$I(K, *L) = \Sigma m_j n_j.$$

Let σ_j^1 be an edge of T extending from the vertex q_1 to the vertex q_2 (that is $\partial\sigma_j^1 = q_2 - q_1$), and let $Z_{\sigma_j^1}(p)$ be the analytic function whose imaginary part satisfies

(3.1.6) $$\operatorname{Im} Z_{\sigma_j^1}(p) = -\frac{1}{2\pi} v_{q_1 q_2}(p)$$

where $v_{q_1 q_2}(p)$ is the harmonic function defined at the end of Section 2.7. Let

(3.1.7) $$K = \Sigma\, m_j \sigma_j^1$$

be a cycle of T, and define

(3.1.8) $$Z_K(p) = \Sigma\, m_j Z_{\sigma_j^1}(p).$$

Then we have the formula ([5])

(3.1.9) $$\operatorname{Im} \int_{*L} dZ_K(p) = -I(K, *L).$$

To prove (3.1.9) it is sufficient to show that

(3.1.10) $$\operatorname{Im} \int_{*L} dZ_{\sigma_j^1} = -n_j;$$

formula (3.1.9) then follows from (3.1.5) and (3.1.8). Let $*\sigma_2^2$ be the polygon of $*T$ associated with the vertex q_2 of T. Then

(3.1.11) $$\int_{\partial *\sigma_2^2} dZ_{\sigma_j^1} = i$$

by the residue theorem, since $Z_{\sigma_j^1}$ has residue $1/2\pi$ at the end point

q_2 of σ_j^1. But

$$\partial{*}\sigma_2^2 = -{*}\sigma_j^1 + \cdots, \tag{3.1.12}$$

so ${*}L + n_j\partial{*}\sigma_2^2$ is a cycle of ${*}T$ which does not contain ${*}\sigma_j^1$ and therefore has no point in common with σ_j^1. Since $\operatorname{Im} Z_{\sigma_j^1}$ is single-valued on $\mathfrak{F} - \sigma_j^1$, we conclude that

$$\operatorname{Im} \int\limits_{{*}L+n_j\partial{*}\sigma_2^2} dZ_{\sigma_j^1} = 0. \tag{3.1.13}$$

Therefore

$$\operatorname{Im} \int\limits_{{*}L} dZ_{\sigma_j^1} = -n_j \operatorname{Im} \int\limits_{\partial{*}\sigma_2^2} dZ_{\sigma_j^1} = -n_j. \tag{3.1.14}$$

If dw is an arbitrary differential of the first kind, we have (compare Section 1.5)

$$(dw,\ dZ_K) = -\int V_K dw$$

where $dZ_K = dU_K + idV_K$ and where the integration is over the boundary of the domain $\mathfrak{F} - K$. Since the value of V_K on the left edge of K exceeds its value on the right edge by unity, we obtain the formula

$$(dw, dZ_K) = -\int\limits_K dw = -P(dw, K). \tag{3.1.15}$$

Let

$$dw_1, dw_2, \cdots, dw_h \tag{3.1.16}$$

be an orthonormal complex basis for the differentials of the first kind. Then

$$dZ_K = \sum_{j=1}^{h} (dZ_K, dw_j)dw_j = -\sum_{j=1}^{h} \int\limits_K d\overline{w}_j \cdot dw_j. \tag{3.1.17}$$

By (3.1.9)

$$I(K, {*}L) = \operatorname{Im}\left\{\sum_{j=1}^{h} \int\limits_K d\overline{w}_j \cdot \int\limits_{{*}L} dw_j\right\} = -I({*}L, K), \tag{3.1.18}$$

([5]). Formula (3.1.18) provides a definition of the Kronecker index of two cycles which is independent of the subdivision of $\mathfrak{F}$. In fact, for any two closed cycles we define

$$(3.1.18)' \qquad I(K_1, K_2) = \operatorname{Im}\left\{ \sum_{j=1}^{h} \int_{K_1} d\bar{w}_j \cdot \int_{K_2} dw_j \right\}.$$

We observe that the Kronecker index vanishes if either cycle bounds.

It is possible to find a homology basis

$$(3.1.19) \qquad K_1, K_2, \cdots, K_{2h-1}, K_{2h}$$

for the 1-cycles of $\mathfrak{F}$ satisfying the following conditions: (i) $K_1, K_3, \cdots, K_{2h-1}$ belong to T and $K_2, K_4, \cdots, K_{2h}$ belong to $*T$; (ii) $I(K_{2\mu-1}, K_{2\mu}) = 1$, $I(K_{2\mu-1}, K_{2\nu}) = 0$ for $\mu \neq \nu$, $I(K_{2\mu}, K_{2\nu}) = 0$, $I(K_{2\mu-1}, K_{2\nu-1}) = 0$. A basis satisfying these conditions is said to be canonical.

Let

$$(3.1.20) \qquad dZ_\mu = dZ_{K_\mu}, \quad \mu = 1, 2, \cdots, 2h,$$

where $K_1, \cdots, K_{2h}$ is a canonical basis. By (3.1.18)′, (3.1.15), and (3.1.17), we have

$$(3.1.21) \quad I(K_\mu, K_\nu) = \operatorname{Im}\{(dZ_\mu, dZ_\nu)\} = -\operatorname{Im}\left\{ \int_{K_\nu} dZ_\mu \right\}.$$

Any differential $dw = du + idv$ of the first kind may be written

$$(3.1.22) \qquad dw = \sum_{\mu=1}^{h} \left\{ dZ_{2\mu} \cdot \int_{K_{2\mu-1}} dv - dZ_{2\mu-1} \cdot \int_{K_{2\mu}} dv \right\}.$$

Let dw_1 be the differential defined by the right side of (3.1.22). Then by (3.1.21) and (3.1.22) we have

$$(3.1.23) \qquad \begin{aligned} \operatorname{Im}\{P(dw_1, K_\nu)\} &= -\operatorname{Im}\{(dw_1, dZ_\nu)\} \\ &= \sum_{\mu=1}^{h} \left\{ I(K_{2\mu-1}, K_\nu) \cdot \int_{K_{2\mu}} dv - I(K_{2\mu}, K_\nu) \cdot \int_{K_{2\mu-1}} dv \right\} \\ &= \int_{K_\nu} dv. \end{aligned}$$

Since $dw - dw_1$ is a differential whose periods have vanishing imaginary parts, we conclude that $dw = dw_1$.

From the $2h$ differentials dW_k of a real basis choose q, say $dW_1, \cdots, dW_q$, such that between $dW_1, \cdots, dW_q$ there is no linear relation with complex coefficients, and so that between any $q+1$ differentials of the basis there is a linear relation with complex coefficients. Then

$$dW_1, dW_2, \cdots, dW_q; \; idW_1, \cdots, idW_q$$

are linearly independent in the real sense, so $2q \leqq 2h$. On the other hand, every differential of the basis (3.1.16), and so every differential of the first kind, is a linear combination of $dW_1, \cdots, dW_q$ with complex coefficients and therefore a linear combination of $dW_1, \cdots, dW_q, idW_1, \cdots, idW_q$ with real coefficients. Hence $2q \geqq 2h$ so $q = h$. It follows that we can choose a complex basis $dW_1, dW_2, \cdots, dW_h$ from the differentials of a real basis, and any differential dW of the first kind is uniquely represented in the form

$$dW = C_1 dW_1 + \cdots + C_h dW_h \tag{3.1.24}$$

where $C_1, \cdots, C_h$ are complex constants.

Let dZ be any differential of $\mathfrak{F}$. Since the number of poles of dZ is finite, there is a triangulation of $\mathfrak{F}$ such that the poles all lie in the interiors of the triangles. We suppose that the triangles are given a coherent orientation. If we then integrate dZ over the boundaries of all triangles in the positive direction and add, we obtain $2\pi i$ times the sum of all the residues of dZ. Because of the coherent orientation each side of a triangle will be run over twice, once in each direction. Hence the sum of all residues is zero. That is, the sum of all the residues of a differential dZ of $\mathfrak{F}$ is equal to zero.

Let $\Omega_{q_0 q_1}(p)$ be the additive analytic function whose real part satisfies

$$\operatorname{Re} \Omega_{q_0 q_1}(p) = -u_{q_0 q_1}(p) \tag{3.1.25}$$

where $u_{q_0 q_1}$ is the function (2.7.15). The differential $d\Omega_{q_0 q_1}$ is of the third kind with simple poles of residues -1 and $+1$ at q_0, q_1 respectively.

Now let $q_1, q_2, \cdots, q_\nu$ and q_0 be any $\nu+1$ distinct points of $\mathfrak{F}$, and let $C_1, C_2, \cdots, C_\nu$ be any complex numbers, not all zero, such that

$$C_1 + C_2 + \cdots + C_\nu = 0.$$

Then

$$d\Omega = -\sum_{k=1}^{\nu} C_k d\Omega_{q_k q_0}$$

is a differential which has poles of the first order at the points q_1, $q_2, \cdots, q_\nu$ with the residues $C_1, C_2, \cdots, C_\nu$ respectively while the poles at q_0 cancel each other. It is thus possible to prescribe the residues of a differential which is regular except for poles subject to the one restriction that the sum of the residues must be zero.

If in Section 2.7 we take for Φ the function

$$\Phi = \frac{\begin{matrix}\cos\\ \sin\end{matrix}\, n\varphi}{r^n} + r^n \frac{\begin{matrix}\cos\\ \sin\end{matrix}\, n\varphi}{a^{2n}}, \tag{3.1.26}$$

we obtain a differential which is infinite at the point p_0 of $\mathfrak{F}$ like

$$-\frac{n\,dz}{z^{n+1}} \quad \text{or} \quad -i\,\frac{n\,dz}{z^{n+1}}.$$

Let dZ be a differential which near the point p has the development

$$dZ = \left(\frac{a_{-m}}{z^m} + \cdots + \frac{a_{-1}}{z} + a_0 + a_1 z + \cdots\right) dz$$

where z is a uniformizer at p. We call

$$\left(\frac{a_{-m}}{z^m} + \cdots + \frac{a_{-1}}{z}\right) dz$$

the principal part of dZ at p. Now prescribe the poles and principal parts of a differential dZ of $\mathfrak{F}$ subject to the condition that the sum of the residues is equal to zero. Moreover, prescribe the real parts of the periods

$$P(dZ, K_\mu) = \int_{K_\mu} dZ.$$

Then there is one and only one differential dZ which satisfies these conditions, and it can be constructed by addition of the elementary differentials described above. We thus have the following theorem (second fundamental existence theorem of Riemann):

THEOREM 3.1.2. *There exists a unique differential dZ which is regular analytic apart from finitely many points where dZ has principal parts assigned arbitrarily, subject to the sole condition that the sum of the residues is zero, and which has the real parts of its periods prescribed.*

3.2. The Period Matrix

Let dw_1, dw_2 be arbitrary differentials of the first kind. Writing $dw_1 = du_1 + idv_1$, $dw_2 = du_2 + idv_2$, we have by (3.1.22)

$$dw_2 = \sum_{\mu=1}^{h} \left\{ dZ_{2\mu} \cdot \int_{K_{2\mu-1}} dv_2 - dZ_{2\mu-1} \cdot \int_{K_{2\mu}} dv_2 \right\}. \tag{3.2.1}$$

Substituting from this formula into the scalar product (dw_1, dw_2) and using (3.1.15), we obtain the formula

$$\operatorname{Re}\,(dw_1, dw_2) = \sum_{\mu=1}^{h} \left\{ \int_{K_{2\mu-1}} du_1 \cdot \int_{K_{2\mu}} dv_2 - \int_{K_{2\mu-1}} dv_2 \cdot \int_{K_{2\mu}} du_1 \right\}. \tag{3.2.2}$$

Taking, in particular, $dw_1 = dw_2 = dw$, we obtain

$$\sum_{\mu=1}^{h} \left\{ \int_{K_{2\mu-1}} du \cdot \int_{K_{2\mu}} dv - \int_{K_{2\mu-1}} dv \cdot \int_{K_{2\mu}} du \right\} = \operatorname{Re} N(dw) > 0 \tag{3.2.3}$$

unless dw is identically zero.

Let

$$P_\mu = \int_{K_{2\mu-1}} dw, \qquad Q_\mu = \int_{K_{2\mu}} dw, \qquad \mu = 1, 2, \cdots, h. \tag{3.2.4}$$

From (3.2.3) we see that the periods P_μ can all vanish only if dw is identically zero, and a similar remark applies to the dual periods Q_μ. It follows that there is a unique dw with prescribed periods P_μ. For let $dw_1, dw_2, \ldots, dw_h$ be a complex basis for the differentials of $\mathfrak{F}$, and write

$$P_{\mu\nu} = \int_{K_{2\mu-1}} dw_\nu, \qquad \mu, \nu = 1, 2, \cdots, h. \tag{3.2.5}$$

The homogeneous equations

$$\sum_{\nu=1}^{h} c_\nu P_{\mu\nu} = 0, \quad \mu = 1, 2, \cdots, h, \tag{3.2.6}$$

cannot have a non-trivial solution $(c_1, \cdots, c_h)$; otherwise there would exist a non-trivial differential dw with all of its periods P_μ zero. Hence, given arbitrary complex numbers P_μ, $\mu = 1, 2, \cdots, h$, the equations

$$(3.2.7) \qquad \sum_{\nu=1}^{h} c_\nu P_{\mu\nu} = P_\mu, \quad \mu = 1, 2, \cdots, h,$$

have a unique solution. We summarize these results as follows:

THEOREM 3.2.1. *The periods P_μ determine dw uniquely and $\| P_{\mu\nu} \|$ is a non-degenerate matrix.*

Write

$$(3.2.8) \qquad \Gamma_{\mu\nu} = (dZ_\mu, dZ_\nu), \qquad \Gamma_{\mu\nu} = (\Gamma_{\nu\mu})^-,$$

where the dZ_μ are defined by (3.1.20) in terms of a canonical basis. Then by (3.1.21)

$$(3.2.9) \qquad \operatorname{Im} \Gamma_{\mu\nu} = I(K_\mu, K_\nu).$$

Let $x_1, x_2, \cdots, x_{2h}$ be arbitrary real numbers. Then by (3.2.8)

$$(3.2.9) \quad \sum_{\mu,\nu=1}^{2h} \operatorname{Re}\{\Gamma_{\mu\nu}\} x_\mu x_\nu = \sum_{\mu,\nu=1}^{2h} \Gamma_{\mu\nu} x_\mu x_\nu = N\left(\sum_{\mu=1}^{2h} x_\mu dZ_\mu\right) > 0$$

unless

$$(3.2.10) \qquad \sum_{\mu=1}^{2h} x_\mu dZ_\mu \equiv 0.$$

Computing the period of (3.2.10) around a cycle K_ν, we find by (3.1.21) that

$$(3.2.11) \quad 0 = \operatorname{Im} \sum_{\mu=1}^{2h} x_\mu \int_{K_\nu} dZ_\mu = -\sum_{\mu=1}^{2h} x_\mu I(K_\mu, K_\nu) = \sum_{\mu=1}^{2h} x_\mu I(K_\nu, K_\mu),$$

and it follows from the canonical property of the basis $K_1, \cdots, K_{2h}$ that $x_1 = x_2 = \cdots = x_{2h} = 0$. In particular, the symmetric matrix

$$(3.2.12) \qquad \| \operatorname{Re} \Gamma_{\mu\nu} \|$$

is non-singular.

3.3. NORMALIZED DIFFERENTIALS

In the sequel we shall be mainly concerned with the normalized differentials dZ_μ, $\mu = 1, 2, \cdots, 2h$. The importance of these differentials stems from the relations

$$(3.3.1) \qquad (dw, dZ_\mu) = -P(dw, K_\mu), \quad \mu = 1, 2, \cdots, 2h.$$

In addition, we shall sometimes have occasion to consider the complex basis of differentials of the first kind,

$$dw_1, \cdots, dw_h \tag{3.3.2}$$

whose periods

$$P_{\mu\nu} = \int_{K_{2\mu-1}} dw_\nu$$

satisfy $P_{\mu\nu} = \delta_{\mu\nu}$, where

$$\delta_{\mu\nu} = \begin{cases} 1, & \mu = \nu \\ 0, & \mu \neq \nu. \end{cases} \tag{3.3.3}$$

Let q be a point of $\mathfrak{F}$, $\zeta = \xi + i\eta$ a uniformizer at q. We denote by $u_{q\zeta}(p)$ the single-valued harmonic function on $\mathfrak{F}$ with a dipole at q such that the difference

$$u_{q\zeta}(p) - \frac{\xi}{\xi^2 + \eta^2} = u_{q\zeta}(p) - \operatorname{Re}\left(\frac{1}{\zeta}\right) \tag{3.3.4}$$

is regular and vanishing at q. The differential

$$2du_{q\zeta}(p) = 2\frac{\partial u_{q\zeta}(p)}{\partial p}\,dp \tag{3.3.5}$$

is of the second kind and it will be denoted by $dT_{q\zeta}(p)$. We observe that its integral has a single-valued real part. By adding linear combinations of basis differentials of the first kind to (3.3.5), we obtain further differentials of the second kind of which we distinguish one, namely the differential $dt_{q\zeta}(p)$ all of whose periods

$$P_\mu = \int_{K_{2\mu-1}} dt_{q\zeta}(p)$$

vanish. More generally, let $dT^{(r)}_{q\zeta}$ be the differential whose integral has a single-valued real part and which near q has the form

$$dT^{(r)}_{q\zeta} = \left(-\frac{r}{\zeta^{r+1}} + \text{regular terms}\right) d\zeta$$

and let $dt^{(r)}_{q\zeta}$ have the same principal part at q but with vanishing periods P_μ. In particular, $dT^{(1)}_{q\zeta} = dT_{q\zeta}$, $dt^{(1)}_{q\zeta} = dt_{q\zeta}$.

Finally, the differential $d\Omega_{q_0q_1}(p)$ defined by (3.1.25) is of the third kind and its integral has a single-valued real part. We denote by $d\omega_{q_0q_1}(p)$ the differential obtained from $d\Omega_{q_0q_1}(p)$ by adding a linear combination of basis differentials of the first kind in such a way that the periods

$$P_\mu = \int_{K_{2\mu-1}} d\omega_{q_0 q_1}(p)$$

vanish. We have two ways of normalizing differentials of the second and third kind; by requiring that their integrals have single-valued real parts or vanishing periods with respect to the cycles $K_{2\mu-1}$. The first type of normalization is called ,,real normalization", the second "complex".

3.4. Period Relations

Let $\mathfrak{J}$ be the surface obtained from $\mathfrak{F}$ by cutting it along the cycles (3.1.19). The integral of any differential df of the first or second kind is single-valued in $\mathfrak{J}$, and we may therefore consider the integral

$$\int_{\partial\mathfrak{J}} f_1 df_2$$

where df_1 and df_2 are an arbitrary pair of such differentials. By Cauchy's residue theorem,

$$\int_{\partial\mathfrak{J}} f_1 df_2 = \sum_{\mu,\nu=1}^{2h} I(K_\mu, K_\nu) \int_{K_\mu} df_1 \cdot \int_{K_\nu} df_2 \tag{3.4.1}$$

$$= \sum_{k=1}^{h} \left\{ \int_{K_{2k-1}} df_1 \cdot \int_{K_{2k}} df_2 - \int_{K_{2k-1}} df_2 \cdot \int_{K_{2k}} df_1 \right\} = 2\pi i \text{ (sum of residues of } f_1 df_2 \text{ in } \mathfrak{J}).$$

Taking first $df_1 = dZ_\mu$, $df_2 = dZ_\nu$, we see that

$$\sum_{k=1}^{h} (\Gamma_{\mu,2k-1}\Gamma_{\nu,2k} - \Gamma_{\nu,2k-1}\Gamma_{\mu,2k}) = 0.$$

The vanishing of the imaginary part of this expression gives the symmetry law

$$\operatorname{Re} \int_{K_\nu} dZ_\mu = \operatorname{Re} \int_{K_\mu} dZ_\nu, \quad \mu, \nu = 1, 2, \cdots, 2h. \tag{3.4.2}$$

We have used here the fact that $\operatorname{Im} \Gamma_{\mu\nu} = I(K_\mu, K_\nu)$. This symmetry law is also an immediate consequence of (3.2.8) and (3.1.15).

Taking $df_1 = dw_\mu$, $df_2 = dw_\nu$, we obtain

$$(3.4.2)' \qquad \int_{K_\nu} dw_\mu = \int_{K_\mu} dw_\nu, \quad \mu, \nu = 1, 2, \cdots, h.$$

Next, let $df_1 = dT_q^{(r)}(p) = dT_{q\zeta}^{(r)}(p)$, $df_2 = dZ_\mu$. In terms of the particular uniformizer ζ, we have for p near q

$$dT_{q\zeta}^{(r)}(p) = \left(-\frac{r}{\zeta^{r+1}} + \text{regular terms}\right) d\zeta.$$

Formula (3.4.1) gives

$$\sum_{k=1}^{h} \left\{ \Gamma_{\mu,\,2k-1} \cdot \int_{K_{2k}} dT_q^{(r)} - \Gamma_{\mu,\,2k} \cdot \int_{K_{2k-1}} dT_q^{(r)} \right\} = 2\pi i \frac{1}{(r-1)!} \frac{d^r Z_\mu(q)}{dq^r}.$$

Taking real parts, we have

$$(3.4.3) \qquad \int_{K_\mu} dT_q^{(r)} = -2\pi i \frac{1}{(r-1)!} \operatorname{Im}\left\{\frac{d^r Z_\mu(q)}{dq^r}\right\}, \quad \mu = 1, 2, \cdots, 2h.$$

If we choose $df_1 = dt_q^{(r)}$, $df_2 = dw_\mu$, we obtain

$$(3.4.3)' \qquad \int_{K_{2\mu}} dt_q^{(r)}(p) = -2\pi i \frac{1}{(r-1)!} \frac{d^r w_\mu(q)}{dq^r}.$$

Third, take $df_1 = d\Omega_{q_1 q_2}(p)$, $df_2 = dZ_\mu$. Since f_1 is single-valued in $\mathfrak{J}$ only if we cut $\mathfrak{J}$ along a path connecting q_1, q_2, we have in place of (3.4.1) the formula

$$(3.4.1)' \qquad \int_{\partial\mathfrak{J}} f_1 df_2 + \int_{q_1}^{q_2} \sigma f_1 df_2 = 2\pi i \text{ (sum of residues)}.$$

Here σf_1 denotes the jump of f_1 from left to right across the cut joining q_1 and q_2. We have

$$\int_{q_1}^{q_2} \sigma f_1 df_2 = 2\pi i \int_{q_1}^{q_2} dZ_\mu = 2\pi i [Z_\mu(q_2) - Z_\mu(q_1)].$$

Thus (3.4.1)′ becomes

$$(3.4.4) \qquad \int_{K_\mu} d\Omega_{q_1 q_2}(p) = 2\pi i \operatorname{Im} \int_{q_1}^{q_2} dZ_\mu.$$

In a similar way, taking $df_1 = d\omega_{q_1q_2}(p)$, $df_2 = dw_\mu$ we find that

$$\int_{K_{2\mu}} d\omega_{q_1q_2}(p) = 2\pi i \int_{q_1}^{q_2} dw_\mu. \tag{3.4.4.)'}$$

Now take $df_1 = dT_q^{(r)}$, $df_2 = dT_p^{(r)}$. The imaginary part of equation (3.4.1) then gives

$$\operatorname{Re} \frac{d^r T_q^{(r)}(p)}{dp^r} = \operatorname{Re} \frac{d^r T_p^{(r)}(q)}{dq^r}. \tag{3.4.5}$$

Similarly

$$\frac{d^r t_q^{(r)}(p)}{dp^r} = \frac{d^r t_p^{(r)}(q)}{dq^r}. \tag{3.4.5)'}$$

If we choose $df_1 = dT_q^{(r)}(p)$, $df_2 = d\Omega_{q_1q_2}$, formula (3.4.1) gives (taking imaginary parts)

$$\operatorname{Re}\{T_q^{(r)}(q_1) - T_q^{(r)}(q_2)\} = \operatorname{Re}\left\{\frac{1}{(r-1)!}\frac{d^r\Omega_{q_1q_2}(q)}{dq^r}\right\}, \tag{3.4.6}$$

while the choices $df_1 = dt_q^{(r)}(p)$, $df_2 = d\omega_{q_1q_2}$ give

$$t_q^{(r)}(q_1) - t_q^{(r)}(q_2) = \frac{1}{(r-1)!}\frac{d^r\omega_{q_1q_2}(q)}{dq^r}. \tag{3.4.6)'}$$

Finally, let $df_1 = d\Omega_{p_1p_2}$, $df_2 = d\Omega_{q_1q_2}$; we obtain

$$\operatorname{Re}\{\Omega_{q_1q_2}(p_1) - \Omega_{q_1q_2}(p_2)\} = \operatorname{Re}\{\Omega_{p_1p_2}(q_1) - \Omega_{p_1p_2}(q_2)\}, \tag{3.4.7}$$

while, if $df_1 = d\omega_{p_1p_2}$, $df_2 = d\omega_{q_1q_2}$, then

$$\omega_{q_1q_2}(p_1) - \omega_{q_1q_2}(p_2) = \omega_{p_1p_2}(q_1) - \omega_{p_1p_2}(q_2). \tag{3.4.7)'}$$

Formulas (3.4.7), (3.4.7)′ constitute the law of interchange of argument and parameter.

3.5. The Order of a Differential

The order of a differential is the difference between the sum of the orders at its zeros and the sum of the orders at its poles. Given any linear differential dZ of $\mathfrak{F}$, its order is given by the formula

$$\operatorname{ord} dZ = 2(h-1). \tag{3.5.1}$$

In order to prove (3.5.1), let z be any non-constant function of $\mathfrak{F}$. We observe first that, for each complex number a, $dz/(z-a)$ is a differential of the third kind whose residues are the orders of the function $z - a$ at its zeros and poles. Since the sum of the residues of this differential is zero, we see that $z - a$ takes the values a and ∞ equally often on $\mathfrak{F}$. Thus z takes each value on $\mathfrak{F}$ the same number of times, say n times.

Let us say that a point p of $\mathfrak{F}$ where z takes the value a lies over the point a of the z-sphere. Then $\mathfrak{F}$ is realized as an n-sheeted relatively branched covering of the z-sphere. Fewer than n points p of $\mathfrak{F}$ lie over finitely many values a; these values a correspond to points p of $\mathfrak{F}$ where the function z is of higher multiplicity than 1. Let p_0 be a point on $\mathfrak{F}$ where z takes the value a exactly m times. If t is a local uniformizer at p_0, we have

$$z = a + c_m t^m + c_{m+1} t^{m+1} + \cdots, \; c_m \neq 0.$$

Hence $(z - a)^{1/m}$ is also a uniformizer at p_0 and the point p_0 over a is a branch point of order $m - 1$. The differential dz has a zero at p_0 of order $m - 1$. If z takes the value ∞ at p_0 exactly s times, then

$$z = \frac{1}{t^s}(c_0 + c_1 t + \cdots), \; c_0 \neq 0,$$

so $(1/z)^{1/s}$ is a uniformizer at p_0. In this case dz has a pole of order $s + 1$ at p_0. The sum

$$V = \sum_0 (m - 1) + \sum_\infty (s - 1)$$

is called the branch number.

We shall now prove the formula

$$V = 2(h + n - 1). \tag{3.5.2}$$

Let the z-sphere be triangulated in such a way that $z = \infty$ and the points lying under branch points are vertices of triangles. Moreover, suppose that at most one vertex of a triangle has branch points lying over it. This triangulation can be carried over to the covering $\mathfrak{F}$. Over each triangle of the sphere there are n triangles of $\mathfrak{F}$, over each side of a triangle of the sphere there are n sides of $\mathfrak{F}$. Let α^2, α^1, α^0 be the numbers of triangles, sides and vertices of $\mathfrak{F}$, and let a^2, a^1, a^0 be the corresponding numbers for the sphere. Then

$$\alpha^2 = na^2, \; \alpha^1 = na^1. \tag{3.5.3}$$

If ν points with branch orders $r_1 - 1, \cdots, r_\nu - 1$ lie over $z = a$, then

$$\nu = n - [(r_1 - 1) + \cdots + (r_\nu - 1)]$$

and hence

$$\alpha^0 = na^0 - V. \tag{3.5.4}$$

But

$$a^0 - a^1 + a^2 = 2, \ \alpha^0 - \alpha^1 + \alpha^2 = 2 - 2h; \tag{3.5.5}$$

therefore

$$2 - 2h = \alpha^2 - \alpha^1 + \alpha^0 = n\ (a^2 - a^1 + a^0) - V = 2n - V.$$

The order of dz is equal to

$$\sum_0 (m - 1) - \sum_\infty (s + 1) = V - 2\sum_\infty s.$$

Since z takes each value (including ∞) n times, we have

$$\sum_\infty s = n.$$

Thus the order of dz is equal to $V - 2n = 2(h - 1)$. Since the ratio of two differentials is a function and the order of a function is zero (as we have seen), every linear differential of $\mathfrak{F}$ has the same order, namely $2(h - 1)$.

Since the ν-th power of a linear differential is a particular kind of differential of dimension ν, and since the ratio of two differentials of the same dimension is a function, we conclude at once from (3.5.1) that

$$\text{ord}\, dZ^\nu = 2\nu(h - 1). \tag{3.5.1$'$}$$

3.6. The Riemann-Roch Theorem for Finite Riemann Surfaces

A divisor on the double $\mathfrak{F}$ of a finite Riemann surface $\mathfrak{M}$ is a 0-cycle $D = \Sigma m_i p_i$, with integral coefficients m_i, consisting of a finite number of points p_i. We define the order of a point p_i to be 1, ord $p_i = 1$, and the order of D to be Σm_i ord $p_i = \Sigma m_i$. The set of all divisors form an Abelian group, the divisor group. To each differential dZ^ν of $\mathfrak{F}$ which is not identically zero we associate a divisor $\Sigma m_i p_i$ where m_i is the order of dZ^ν at p_i. The divisor of a function f of $\mathfrak{F}$ will be denoted by (f) and two divisors D and D' are said to be linearly equivalent if there is a function f of $\mathfrak{F}$, not identically zero, such that

$D - D' = (f)$. A class of equivalent divisors (divisor class) which contains the divisor D will be denoted by (D). All divisors in (D) plainly have the same order which is called ord (D). Clearly, the divisors of linear differentials dZ all lie in the same divisor class which will be denoted by (W). Given a divisor $D = \Sigma m_i p_i$, we write $D \geqq 0$ if all $m_i \geqq 0$. The set of all functions f of $\mathfrak{F}$ such that $(f) + D \geqq 0$ will be denoted by $F(D)$, and the *real* dimension of the linear space $F(D)$ will be called the dimension of (D): dim $(D) =$ dim $F(D)$. If $(f) + D \geqq 0$, we say that f is a multiple of $-D$.

If $D = \Sigma m_i p_i$ is a divisor of the double $\mathfrak{F}$, we call $\tilde{D} = \Sigma m_i \tilde{p}_i$ the conjugate divisor. Here $\tilde{p}$ denotes the point of $\mathfrak{F}$ which is conjugate to p. A divisor of $\mathfrak{F}$ will be called a divisor of $\mathfrak{M}$ if and only if it is equal to its conjugate. A divisor D of $\mathfrak{M}$ therefore has the form $D = \Sigma m_i P_i$ where $P_i = p_i$ if p_i is a boundary point of $\mathfrak{M}$ and $P_i = p_i + \tilde{p}_i$ if p_i is an interior point of $\mathfrak{M}$. Two divisors D and D' of $\mathfrak{M}$ are called linearly equivalent on $\mathfrak{M}$ if $D - D' = (f)$ where f is a function of $\mathfrak{M}$, and the *real* dimension of the linear space of functions f of $\mathfrak{M}$ such that $(f) + D \geqq 0$ is the dimension of (D). We have the following theorem ([4]) in which divisors, differentials and dimension refer to $\mathfrak{M}$:

THEOREM 3.6.1. *For any finite Riemann surface* $\mathfrak{M}$,

$$\dim (D) - \dim (W - D) = \operatorname{ord} (D) - G + R^0 \tag{3.6.1}$$

where

$$G = 2h + c + m + R^0 - 2, \tag{3.6.2}$$

$$\operatorname{ord} (W) = 2(G - R^0). \tag{3.6.3}$$

We remark that

$$\dim (W - D) = \dim \{ dZ \mid (dZ) \geqq D\}.$$

In fact, let dZ_0 be an arbitrary linear differential. If f is a function with $(f) \geqq D - (dZ_0)$, then $dZ = f dZ_0$ is a differential with $(dZ) \geqq D$. Conversely, if $(dZ) \geqq D$, then $f = dZ/dZ_0$ is a function with $(f) \geqq D - (dZ_0)$.

If $\mathfrak{M}$ is closed and orientable, (3.6.1) is the classical Riemann-Roch theorem. If we assume that (3.6.1) is valid for closed, orientable surfaces, the validity of (3.6.1) in general is readily proved.

For, let a divisor belonging to a function of $\mathfrak{M}$ (or $\mathfrak{F}$) be called a principal divisor of $\mathfrak{M}$ (or $\mathfrak{F}$). The group of the principal divisors of $\mathfrak{M}$ is equal to the intersection of the group of principal divisors of $\mathfrak{F}$ with the divisor group of $\mathfrak{M}$. For, let D be a divisor of $\mathfrak{M}$ which is a principal divisor of $\mathfrak{F}$. Then D defines f, where f is a function of $\mathfrak{F}$. The conjugate function $\tilde{f}$ has the divisor $\tilde{D} = D$, so $\tilde{f}/f$ is equal to a constant, $\tilde{f} = af$, say. Forming the conjugate, that is, applying the $\sim$-operation to the equation $\tilde{f} = af$, we obtain $f = \bar{a}\tilde{f}$; hence $a\bar{a} = |a|^2 = 1$. But then $g = \sqrt{a}f$ is equal to its conjugate and is therefore a function of $\mathfrak{M}$. Since g has the divisor D, we see that D is a principal divisor of $\mathfrak{M}$.

Now let D be a divisor of $\mathfrak{M}$ and let $f_1, f_2, \cdots, f_\mu$ be linearly independent functions of $\mathfrak{M}$ in the real sense such that

$$f = a_1 f_1 + \cdots + a_\mu f_\mu, \qquad a_k \text{ real},$$

is the most general function f of $\mathfrak{M}$ satisfying $(f) + D \geqq 0$. Then $\dim(D) = \mu$. Let

$$g = c_1 f_1 + \cdots + c_\mu f_\mu, \qquad c_k \text{ complex}.$$

We have to show that g is the most general function of $\mathfrak{F}$ satisfying $(g) + D \geqq 0$. It is clear that $(g) + D \geqq 0$. Let g be any function of $\mathfrak{F}$ satisfying $(g) + D \geqq 0$, and let $\tilde{g}$ be its conjugate. Then

$$g_1 = \frac{g + \tilde{g}}{2}, \quad g_2 = \frac{g - \tilde{g}}{2i}$$

are two functions of $\mathfrak{F}$ which are equal to their conjugates and are, therefore, functions of $\mathfrak{M}$. Since $(g) + D \geqq 0$ and $(\tilde{g}) + D \geqq 0$, the same is true of g_1 and g_2. Hence

$$g_1 = a_1 f_1 + \cdots + a_\mu f_\mu, \quad g_2 = b_1 f_1 + \cdots + b_\mu f_\mu,$$

where a_k and b_k are real. Thus

$$g = g_1 + ig_2 = (a_1 + ib_1)f_1 + \cdots + (a_\mu + ib_\mu)f_\mu$$

and the validity of formula (3.6.1) for closed orientable surfaces therefore implies its validity for an arbitrary finite Riemann surface.

For completeness, we indicate briefly how the formula (3.6.1) for a closed orientable surface $\mathfrak{M}$ may be deduced from the period relations of Section 3.4 (see [6]), and for simplicity we assume that

$$D = P_1 + P_2 + \cdots + P_\mu - Q_1 - Q_2 - \cdots - Q_\nu$$

where $P_i = p_i + \tilde{p}_i$, $Q_i = q_i + \tilde{q}_i$ and the points p_i, q_i are distinct points of $\mathfrak{M}$.

Let

$$dt = c_1 dt_{p_1} + \cdots + c_\mu dt_{p_\mu} \tag{3.6.4}$$

where $dt_{p_1}, dt_{p_2}, \cdots, dt_{p_\mu}$ are the normalized differentials of the second kind defined in Section 3.3 and $c_1, c_2, \cdots, c_\mu$ are μ complex constants. Then dt has double poles at $p_1, p_2, \cdots, p_\mu$ and its periods P_μ all vanish. The remaining periods will vanish provided that the following equations are satisfied:

$$\begin{aligned} c_1 \frac{dw_1}{dp_1} + c_2 \frac{dw_1}{dp_2} + \cdots + c_\mu \frac{dw_1}{dp_\mu} &= 0, \\ c_1 \frac{dw_2}{dp_1} + c_2 \frac{dw_2}{dp_2} + \cdots + c_\mu \frac{dw_2}{\partial p_\mu} &= 0, \\ \cdots\cdots\cdots\cdots\cdots\cdots & \\ c_1 \frac{dw_h}{dp_1} + c_2 \frac{dw_h}{dp_2} + \cdots + c_\mu \frac{dw_h}{dp_\mu} &= 0. \end{aligned} \tag{3.6.5}$$

This follows from equation (3.4.3)′. If equations (3.6.5) are satisfied, the integral t of dt is a function of $\mathfrak{F}$; it will vanish at $q_1, q_2, \cdots, q_\nu$ if we can choose a constant of integration c_0 such that

$$\begin{aligned} c_0 + c_1 t_{p_1}(q_1) + c_2 t_{p_2}(q_1) + \cdots + c_\mu t_{p_\mu}(q_1) &= 0, \\ c_0 + c_1 t_{p_1}(q_2) + c_2 t_{p_2}(q_2) + \cdots + c_\mu t_{p_\mu}(q_2) &= 0, \\ \cdots\cdots\cdots\cdots\cdots\cdots & \\ c_0 + c_1 t_{p_1}(q_\nu) + c_2 t_{p_2}(q_\nu) + \cdots + c_\mu t_{p_\mu}(q_\nu) &= 0. \end{aligned} \tag{3.6.6}$$

The $\mu + 1$ constants $c_0, c_1, c_2, \cdots, c_\mu$ therefore must satisfy $h + \nu$ linear homogeneous equations. If the rank of this system of $h + \nu$ equations in $\mu + 1$ unknowns is r, then there are $\mu + 1 - r$ linearly independent solutions in the complex sense; that is

$$\dim (D) = 2(\mu+1-r) = \text{ord}\,(D) - 2h+2+2(h+\nu-r). \tag{3.6.7}$$

Assume first that $\nu = 0$. Then there are no equations (3.6.6) and the transposed system corresponding to (3.6.5) is

$$(3.6.5)' \qquad \begin{aligned} &y_1 \frac{dw_1}{dp_1} + y_2 \frac{dw_2}{dp_1} + \cdots + y_h \frac{dw_h}{dp_1} = 0, \\ &y_1 \frac{dw_1}{dp_2} + y_2 \frac{dw_2}{dp_2} + \cdots + y_h \frac{dw_h}{dp_2} = 0, \\ &\cdots\cdots\cdots\cdots\cdots\cdots \\ &y_1 \frac{dw_1}{dp_\mu} + y_2 \frac{dw_2}{dp_\mu} + \cdots + y_h \frac{dw_h}{dp_\mu} = 0. \end{aligned}$$

We see that

$$(3.6.8) \qquad \dim(W - D) = 2(h - r) = 2(h + \nu - r)$$

since $\nu = 0$ by assumption. In this case formula (3.6.1) follows from (3.6.7) and (3.6.8).

Next, assume that $\nu = 1$. Then there is one equation (3.6.6) which merely determines c_0 in terms of $c_1, c_2, \cdots, c_\mu$. We therefore ignore this equation and consider the h equations (3.6.5) in the μ unknowns $c_1, c_2, \cdots, c_\mu$. If the rank is r, we have

$$(3.6.7)' \quad \dim(D) = 2(\mu - r) = \text{ord}(D) - 2h + 2 + 2(h - r).$$

But the number of linearly independent differentials which are multiples of D is equal to the number of linearly independent differentials which are multiples of $P_1 + \cdots + P_\mu = D + Q_1$. For the sum of the residues of a differential must equal zero. Hence we again have (3.6.8) and therefore (3.6.1).

Finally, suppose that $\nu > 1$. We then subtract the $\nu - 1$ last equations (3.6.6) from the first and use formula (3.4.6)'. We obtain the $\nu - 1$ equations

$$(3.6.6)' \qquad \begin{aligned} &c_1 \frac{d\omega_{q_1q_2}}{dp_1} + c_2 \frac{d\omega_{q_1q_2}}{dp_2} + \cdots + c_\mu \frac{d\omega_{q_1q_2}}{dp_\mu} = 0, \\ &c_1 \frac{d\omega_{q_1q_3}}{dp_1} + c_2 \frac{d\omega_{q_1q_3}}{dp_2} + \cdots + c_\mu \frac{d\omega_{q_1q_3}}{dp_\mu} = 0, \\ &\cdots\cdots\cdots\cdots\cdots\cdots \\ &c_1 \frac{d\omega_{q_1q_\nu}}{dp_1} + c_2 \frac{d\omega_{q_1q_\nu}}{dp_2} + \cdots + c_\mu \frac{d\omega_{q_1q_\nu}}{dp_\mu} = 0, \end{aligned}$$

plus the one equation

$$(3.6.6)'' \qquad c_0 + c_1 t_{p_1}(q_1) + c_2 t_{p_2}(q_1) + \cdots + c_\mu t_{p_\mu}(q_1) = 0.$$

The equation (3.6.6)'' determines c_0 and will again be ignored. We therefore have the system of $h + \nu - 1$ equations (3.6.5) and (3.6.6)' in the μ unknowns $c_1, c_2, \cdots, c_\mu$. If the rank is r, we have

$$(3.6.7)'' \quad \dim (D) = 2(\mu - r) = \operatorname{ord} (D) - 2h + 2 + 2(h + \nu - 1 - r).$$

The transposed system is

$$y_1 \frac{dw_1}{dp_1} + \cdots + y_h \frac{dw_h}{dp_1} + y_{h+1} \frac{d\omega_{q_1 q_2}}{dp_1} + \cdots + y_{h+\nu-1} \frac{d\omega_{q_1 q_\nu}}{dp_1} = 0,$$
$$y_1 \frac{dw_1}{dp_2} + \cdots + y_h \frac{dw_h}{dp_2} + y_{h+1} \frac{d\omega_{q_1 q_2}}{dp_2} + \cdots + y_{h+\nu-1} \frac{d\omega_{q_1 q_\nu}}{dp_2} = 0,$$
$$\cdots\cdots\cdots\cdots\cdots\cdots\cdots\cdots$$
$$y_1 \frac{dw_1}{dp_\mu} + \cdots + y_h \frac{dw_h}{dp_\mu} + y_{h+1} \frac{d\omega_{q_1 q_2}}{dp_\mu} + \cdots + y_{h+\nu-1} \frac{d\omega_{q_1 q_\nu}}{dp_\mu} = 0.$$

Since every differential dZ which has poles at the q_i may be written in the form

$$y_1 dw_1 + \cdots + y_h\, dw_h + y_{h+1} d\omega_{q_1 q_2} + \cdots + y_{h+\nu-1} d\omega_{q_1 q_\nu},$$

we have

$$(3.6.8)' \qquad \dim (W - D) = 2(h + \nu - 1 - r),$$

and formula (3.6.1) follows from (3.6.7)'' and (3.6.8)'.

3.7. Conformal Mappings of a Finite Riemann Surface onto Itself

Klein [2] classified finite Riemann surfaces by means of their algebraic genus G, where G is given by (3.6.2). There are precisely seven surfaces with $G \leq R^0$, namely:

$$\left.\begin{array}{l} h = 0,\ c = 0,\ m = 0 \text{ (sphere)} \\ h = 0,\ c = 0,\ m = 1 \text{ (disc)} \\ h = 0,\ c = 1,\ m = 0 \text{ (projective plane)} \end{array}\right\} G = 0,$$

$$\left.\begin{array}{l} h = 1,\ c = 0,\ m = 0\ \text{(torus)} \\ h = 0,\ c = 0,\ m = 2\ \text{(ring or doubly connected domain)} \\ h = 0,\ c = 1,\ m = 1\ \text{(Möbius strip)} \\ h = 0,\ c = 2,\ m = 0\ \text{(Klein bottle or non-orientable torus)} \end{array}\right\} G = R^0.$$

Let σ be the dimension of the space of all classes of conformally equivalent finite Riemann surfaces of the same topological type, and let ϱ be the number of parameters of the continuous group of conformal mappings of the surface onto itself. The formula

$$\sigma - \varrho = 6h + 3c + 3m - 6 \tag{3.7.1}$$

which was given by Klein [2], has already been discussed in Section 2.11 where, however, we made use of a fact which depends on formula (3.6.3), namely if $w = u + iv$ is the function of Section 2.11 with a simple pole at one point of $\mathfrak{M}$ (u being single-valued on $\mathfrak{F}$), then the total multiplicity of the zeros of dw in the interior of $\mathfrak{M}$ is equal to $2h + c$. As remarked in Section 2.11, the differential dw has two simple zeros on each boundary component of $\mathfrak{M}$. Since dw has order -2 at one point of $\mathfrak{M}$, we see that the number of zeros of dw in the interior of $\mathfrak{M}$ is $2h$ if $\mathfrak{M}$ is closed and orientable and $2h + c$ otherwise, and hence the statement follows.

We have the following theorem:

THEOREM 3.7.1. *The number ϱ of the parameters of the continuous group of conformal mappings of a finite Riemann surface onto itself is different from zero only in the following seven exceptional cases where*

$$\varrho = \begin{cases} 6 \text{ for } h = 0,\ c = 0,\ m = 0 \text{ (sphere)} \\ 3 \text{ for } h = 0,\ c = 0,\ m = 1 \text{ (disc)} \\ 3 \text{ for } h = 0,\ c = 1,\ m = 0 \text{ (projective plane)} \\ 2 \text{ for } h = 1,\ c = 0,\ m = 0 \text{ (torus)} \\ 1 \text{ for } h = 0,\ c = 0,\ m = 2 \text{ (doubly connected domain)} \\ 1 \text{ for } h = 0,\ c = 1,\ m = 1 \text{ (Möbius strip)} \\ 1 \text{ for } h = 0,\ c = 2,\ m = 0 \text{ (Klein bottle).} \end{cases} \tag{3.7.2}$$

This theorem is an immediate consequence of the following general theorem (due essentially to H. A. Schwarz) which we state without proof (for a proof see [6]);

THEOREM 3.7.2. *The group of the conformal mappings of an orientable*

Riemann surface onto itself is discontinuous apart from the following seven exceptional cases: sphere, simply or doubly punctured sphere, disc, punctured circle, doubly connected domain, torus.

In fact, if $\mathfrak{M}$ is orientable, Theorem 3.7.2 is immediately applicable. If $\mathfrak{M}$ is non-orientable, we apply Theorem 3.7.2 to its double $\mathfrak{F}$. If the conformal mapping of $\mathfrak{M}$ onto itself carries the point p into the point q, we extend the mapping to $\mathfrak{F}$ by sending the point $\tilde{p}$ into the point $\tilde{q}$. We therefore conclude that $\varrho = 0$ unless $\mathfrak{M}$ is one of the surfaces (3.7.2), and we then examine the cases.

The only cases (3.7.2) which need to be mentioned are the non-orientable ones. The projective plane arises from the sphere by identification of diametrically opposite points. The group of conformal mappings onto itself is therefore the group of spherical rotations. If the Möbius strip is represented in the normal form discussed in Section 2.1 we see that the only conformal mappings onto itself are given by $w' = e^{i\theta} w$ and $w' = Re^{i\theta}/w$, θ real. Finally, the Klein bottle arises from the w-plane punctured at infinity by identifying points which are equivalent with respect to the group of transformations

$$(3.7.3) \quad w \to w + m + ni\omega, \; w \to \overline{w} + \frac{1}{2} + m + ni\omega, \; \omega > 0.$$

The relatively unbranched two-sheeted orientable covering of the Klein bottle is obtained by identifying points which are equivalent with respect to the subgroup

$$w \to w + m + ni\omega.$$

3.8. Reciprocal and Quadratic Differentials

The everywhere finite reciprocal differentials dZ^{-1} of $\mathfrak{M}$,

$$dZ^{-1} = r(z)dz^{-1}, \qquad z \text{ a local uniformizer,}$$

are connected with the infinitesimal conformal mappings of $\mathfrak{M}$ onto itself. This is intuitively clear. For let z be a local uniformizer valid in the neighborhood of a point of $\mathfrak{M}$. Under the infinitesimal mapping the point with the parameter value z goes into the point with the parameter value $z + \varepsilon r(z)$ where ε is an infinitesimal real quantity. If z' is another uniformizer valid in the same neighborhood, then z' is analytic in z or in $\bar{z}$. Assume for simplicity that z' is analytic in z.

Then (neglecting infinitesimals of higher order)

$$z' + \varepsilon r_1(z') = z'(z + \varepsilon r(z)) = z' + \varepsilon \frac{dz'}{dz} r(z),$$

so

$$\frac{r_1(z')}{dz'} = \frac{r(z)}{dz}.$$

Furthermore, if z is a boundary uniformizer, then z is real at a boundary point p. Since a boundary point goes into a boundary point, $z + \varepsilon r(z)$ must also be real. Thus $r(z)/dz$ is real on the boundary and is therefore a reciprocal differential of $\mathfrak{M}$.

We show now that

$$\varrho = \dim(-W). \tag{3.8.1}$$

By (3.6.3),

$$\text{ord}(-W) = 2(R^0 - G).$$

Since the order of an everywhere finite reciprocal differential must be non-negative, there are no everywhere finite reciprocal differentials when $G > R^0$. Formula (3.8.1) is therefore proved in this case. Taking $(D) = (W)$ in (3.6.1) we see that the number of linearly independent everywhere finite linear differentials of $\mathfrak{M}$ is equal to

$$\dim(0) + 2(G - R^0) - G + R^0 = G$$

since $\dim(0) = R^0$. If $G = R^0$ we therefore have R^0 everywhere finite linear differentials (up to a constant factor) whose orders are equal to zero. Their reciprocals are the only everywhere finite reciprocal differentials and we have $\dim(-W) = R^0$. From (3.7.2) we see that (3.8.1) is true in this case. Finally assume that $G = 0$, in which case the double $\mathfrak{F}$ is the z-sphere. All reciprocal differentials of $\mathfrak{F}$ must then be equal to a rational function of z times dz^{-1}. Bearing in mind that $1/z$ and not z is a uniformizer at ∞, we see that the everywhere finite reciprocal differentials of the sphere have the form

$$\frac{a + bz + cz^2}{dz} \tag{3.8.2}$$

where a, b and c are arbitrary complex constants. Hence if $\mathfrak{M}$ is the sphere $\dim(-W) = 6$. If $\mathfrak{M}$ is the unit circle, the reciprocal differentials still have the form (3.8.2) but they must satisfy the additional restriction that they are real on $|z| = 1$. It is then seen that the everywhere finite reciprocal differentials of the circle are linear combinations (with real coefficients) of the three differentials

$$i\frac{z}{dz},\ i\frac{z}{dz}\left(z+\frac{1}{z}\right),\ \frac{z}{dz}\left(z-\frac{1}{z}\right). \tag{3.8.3}$$

Hence dim $(-W) = 3$ in this case. Finally, the projective plane is the unit circle with diametrically opposite points of its circumference identified. The three differentials

$$\frac{z}{dz},\ \frac{z}{dz}\left(z+\frac{1}{z}\right),\ i\frac{z}{dz}\left(z-\frac{1}{z}\right)$$

take the same or conjugate values at diametrically opposite points and are therefore reciprocal differentials of the projective plane. Hence dim $(-W) = 3$ in this case also. By comparison with (3.7.2) we see that formula (3.8.1) is valid when $G = 0$.

Finally, taking $(D) = (2W)$ in (3.6.1), we obtain

$$\dim (2W) - \dim (-W) = 3(G - R^0) \tag{3.8.4}$$

since ord $(2W) = 4(G - R^0)$ by (3.6.3). Thus, by (3.6.2),

$$\dim (2W) - \varrho = 6h + 3c + 3m - 6.$$

Comparing with (3.7.1) we see that

$$\dim (2W) = \sigma. \tag{3.8.5}$$

This equation shows that the quadratic differentials are connected with the moduli of the finite Riemann surface $\mathfrak{M}$.

REFERENCES

1. K. HENSEL and G. LANDSBERG, *Theorie der algebraischen Funktionen einer Variabeln*, Teubner, Leipzig, 1902.
2. F. KLEIN, *Riemann'sche Flächen I und II*, Vorlesungen Göttingen, Wintersemester 1891—1892 und Sommersemester 1892.
3. C. NEUMANN, *Vorlesungen über Riemann's Theorie der Abel'schen Integrale*, Teubner, Leipzig, 1884.
4. O. TEICHMÜLLER, „Extremale quasikonforme Abbildungen und quadratische Differentiale", *Abh. der Preuss. Akad. der Wiss., Math.-Naturw. Kl.* 1939, 22 (1940).
5. K. I. VIRTANEN, „Über Abelsche Integrale auf nullberandeten Riemannschen Flächen von unendlichem Geschlecht," *Ann. Acad. Sci. Fenn., A. I*, 56 (1949).
6. H. WEYL, *Die Idee der Riemannschen Fläche*, Teubner, Berlin, 1923. (Reprint, Chelsea, New York, 1947).

4. Bilinear Differentials

4.1. Bilinear Differentials and Reproducing Kernels

In this section we make some preliminary remarks concerning the present chapter, which centers around the bilinear differentials and, in particular, reproducing kernels. In the next section we define the Green's and Neumann's functions, and we then express as many of our functionals as possible in terms of these basic functionals. In particular, we express the differentials of the first kind in this way, as well as the mapping functions corresponding to the canonical slit domains of Chapter 2. We then define certain linear spaces of differentials on a Riemann surface, and construct the reproducing kernels for each of them.

The underlying purpose of this chapter is thus to express the various functionals in terms of one functional, namely the Green's function. In subsequent chapters we shall be concerned with the dependence of these functionals on the Riemann surface, and the results of the present chapter will enable us to confine our study of the dependence essentially to the Green's function.

We call a complex analytic expression $\lambda(p, q)$ which depends on two points p, q of a Riemann surface a bilinear differential if

$$\lambda(p, q)dzd\zeta$$

is a conformal invariant, z and ζ being local coordinates of the points p and q respectively. In other words, $\lambda(p, q)$ transforms like a linear differential of each argument and is therefore a double covariant vector.

If $\mathfrak{M}$ is a closed orientable Riemann surface, the expression

$$\frac{\partial^2 \Omega_{qq_0}(p)}{\partial p \partial q} \tag{4.1.1}$$

is a bilinear differential. Here $d\Omega_{qq_0}$ is the differential of the third kind defined in Chapter 3. This bilinear differential is symmetric in p and q, has a double pole for $p = q$, and, it should be remarked, is independent of the parameter point q_0. In order to prove these statements, we use the law of interchange of argument and parameter in the form (compare (3.4.7))

$$\begin{aligned}&\Omega_{qq_0}(p) - \Omega_{qq_0}(p_0) + (\Omega_{qq_0}(p))^- - (\Omega_{qq_0}(p_0))^- \\ &= \Omega_{pp_0}(q) - \Omega_{pp_0}(q_0) + (\Omega_{pp_0}(q))^- - (\Omega_{pp_0}(q_0))^-.\end{aligned}$$

Differentiating this identity with respect to p and q, we find that

$$\frac{\partial^2 \Omega_{qq_0}(p)}{\partial p \partial q} = \frac{\partial^2 \Omega_{pp_0}(q)}{\partial p \partial q}. \tag{4.1.2}$$

That is, (4.1.1) is a symmetric bilinear differential independent of q_0 and p_0. For p in a neighborhood of q, we have, in terms of the local uniformizer $z(p)$,

$$\Omega_{qq_0}(p) = -\log\,[z(p) - z(q)] + \text{regular terms} \tag{4.1.3}$$

and hence

$$\frac{\partial^2 \Omega_{qq_0}(p)}{\partial p \partial q} = \frac{-1}{[z(p) - z(q)]^2} + \text{regular terms.} \tag{4.1.4}$$

Let us consider now the case of an orientable finite Riemann surface $\mathfrak{M}$ with boundary C. If we complete $\mathfrak{M}$ to its double $\mathfrak{F}$, we may consider the class of differentials which are regular analytic on $\mathfrak{F}$. This class has a finite basis, namely the G basis differentials of the first kind of $\mathfrak{F}$. In order to develop a significant function-theory for the Riemann surface $\mathfrak{M}$, we introduce a wider class of differentials, namely those which are regular analytic in the interior of $\mathfrak{M}$ but which are not necessarily defined on the boundary of $\mathfrak{M}$ or elsewhere. Because of their restricted domain of definition, we call these differentials "interior differentials" on $\mathfrak{M}$. The interior differentials with finite norm over $\mathfrak{M}$ form a Hilbert space **M** which will be one of the main objects of study in the sequel.

Let $\mathfrak{F}$ be the double of $\mathfrak{M}$, and $\Omega_{qq_0}(p)$ the above integral of the third kind of $\mathfrak{F}$. In the theory of the Hilbert space **M**, the following bilinear differential plays a fundamental role:

$$L_{\mathbf{M}}(p, q) = -\frac{1}{\pi}\frac{\partial^2 \Omega_{q\tilde{q}}(p)}{\partial p \partial q}. \tag{4.1.5}$$

This bilinear differential is symmetric in p and q, and, for p near q, has the development (see (4.1.4))

$$(4.1.5)' \qquad L_{\mathbf{M}}(p, q) = \frac{1}{\pi} \frac{1}{[z(p) - z(q)]^2} + \text{regular terms.}$$

As a differential of p defined on $\mathfrak{M}$, $L_{\mathbf{M}}(p, q)$ has to be distinguished from $L_{\mathbf{M}}(p, \tilde{q})$, which is everywhere regular in $\mathfrak{M}$ because its double pole lies in $\tilde{\mathfrak{M}}$. While $L_{\mathbf{M}}(p, q)$ obviously satisfies the symmetry law

$$(4.1.6) \qquad L_{\mathbf{M}}(p, q) = L_{\mathbf{M}}(q, p),$$

we now want to prove the law of Hermitian symmetry for the differential $L_{\mathbf{M}}(p, \tilde{q})$, namely

$$(4.1.7) \qquad L_{\mathbf{M}}(p, \tilde{q}) = (L_{\mathbf{M}}(q, \tilde{p}))^-.$$

By (4.1.6), (4.1.7) is equivalent to

$$(4.1.8) \qquad L_{\mathbf{M}}(p, \tilde{q}) = (L_{\mathbf{M}}(\tilde{p}, q))^-.$$

In order to prove (4.1.8) we make use of the fact that the differential $d\Omega_{q\tilde{q}}(p)$ has imaginary periods on $\mathfrak{F}$. Hence the expression

$$\text{Re}\,\{\Omega_{\tilde{q}q}(p) - (\Omega_{q\tilde{q}}(\tilde{p}))^-\}$$

represents a single-valued harmonic function on $\mathfrak{F}$. We observe that the singularities of the two terms cancel at $\tilde{q}$ and also at q. Applying the maximum principle to this regular harmonic function on the closed Riemann surface $\mathfrak{F}$, we see that

$$(4.1.9) \qquad \Omega_{\tilde{q}q}(p) = (\Omega_{q\tilde{q}}(\tilde{p}))^- + C$$

where C is independent of p. Differentiating this identity with respect to p and $\tilde{q}$, we find that

$$(4.1.10) \qquad \frac{\partial^2 \Omega_{\tilde{q}q}(p)}{\partial p \partial \tilde{q}} = \left(\frac{\partial^2 \Omega_{q\tilde{q}}(\tilde{p})}{\partial \tilde{p} \partial q}\right)^-.$$

In this formula conjugate uniformizers must be used at conjugate places. Finally, (4.1.8) follows from (4.1.5) and (4.1.10), and this proves the Hermitian symmetry of the kernel $L_{\mathbf{M}}(p, \tilde{q})$.

Let p be a point on the boundary C of $\mathfrak{M}$. Since $p = \tilde{p}$ we have the identity

$$L_{\mathbf{M}}(p, q)dz = L_{\mathbf{M}}(\tilde{p}, q)d\bar{z}.$$

We use the law (4.1.8) and obtain

$$L_{\mathbf{M}}(p, q)dz = (L_{\mathbf{M}}(p, \tilde{q})dz)^-, \quad p \in C. \tag{4.1.11}$$

Formula (4.1.11) may be interpreted as follows: Given any fixed point q of $\mathfrak{M}$, we can find a pair of differentials taking conjugate values on C. One differential has a double pole at the point q, while the other is regular throughout $\mathfrak{M}$. It is easily seen that the differentials $L_{\mathbf{M}}(p, q)$ and $L_{\mathbf{M}}(p, \tilde{q})$ are, for fixed q, conjugate differentials of $\mathfrak{F}$ in this dependence upon p (Section 2.2).

The differential $L_{\mathbf{M}}(p, \tilde{q})$ in its dependence upon p belongs to the Hilbert space **M**. We shall show in Section 4.10 that it has the reproducing property

$$(df, L_{\mathbf{M}}(p, \tilde{q})) = -f'(q) \tag{4.1.12}$$

for any $df \in \mathbf{M}$, and on the other hand,

$$(df, L_{\mathbf{M}}(p, q)) = 0. \tag{4.1.13}$$

In formulas such as (4.1.12) and (4.1.13), we should write $(df, L_{\mathbf{M}}(p, q)dz)$ but, for simplicity, we drop dz. It will be clear which variable refers to the integration.

In the case of plane domains the bilinear differential $-L_{\mathbf{M}}(p, \tilde{q})$ has been extensively studied, and is called the reproducing kernel of the class **M** (see [2a, b, c, d], [3a], [4], [6a], [10], [13], [14b, c]). Even in this special case, it was pointed out (see [3b], [14b]) that the introduction of the kernel $L_{\mathbf{M}}(p, q)$ completes the kernel theory in a symmetric manner. However, it has been customary to regard the two kernels as essentially different, although the theory of the double of a domain shows that they are the same.

In the special case of a plane domain we may replace the abstract points p, q by their coordinates z and ζ. Consider the interior of the circle $|z| < r$; then the double may be chosen as the z-sphere and we choose

$$\tilde{z} = \frac{r^2}{\bar{z}}. \tag{4.1.14}$$

The integral of the third kind on the sphere is given by

$$\Omega_{\zeta\tilde{\zeta}}(z) = \log \frac{z - \tilde{\zeta}}{z - \zeta} + \text{constant},$$

and hence we obtain

$$L_{\mathbf{M}}(z, \zeta) = \frac{1}{\pi} \frac{1}{(z-\zeta)^2}. \tag{4.1.15}$$

Let us compute $L_{\mathbf{M}}(z, \tilde{\zeta})$ by means of (4.1.14) and (4.1.15). Using the transformation law for differentials, we have

$$L_{\mathbf{M}}(z, \tilde{\zeta})d\tilde{\zeta} = \frac{1}{\pi(z-\tilde{\zeta})^2} d\tilde{\zeta} = \frac{1}{\pi\left(z - \dfrac{r^2}{\bar{\zeta}}\right)^2} d\left(\frac{r^2}{\bar{\zeta}}\right) = -\frac{r^2}{\pi(r^2 - z\bar{\zeta})^2} d\bar{\zeta}.$$

Thus, using $\bar{\zeta}$ as uniformizer at $\tilde{\zeta}$, we find that

$$L_{\mathbf{M}}(z, \tilde{\zeta}) = \frac{-r^2}{\pi(r^2 - z\bar{\zeta})^2}. \tag{4.1.16}$$

If $\mathfrak{M}$ is an arbitrary multiply-connected domain of the z-plane, we can write

$$L(z, \zeta) = \frac{1}{\pi} \frac{1}{(z-\zeta)^2} + l(z, \zeta) \tag{4.1.17}$$

where $l(z, \zeta)$ is regular analytic in $\mathfrak{M}$. The expression $l(z, \zeta)$ plays an important role in the theory of conformal mapping of plane domains (see [3b], [6a, b], [14b, d]). The representation (4.1.17) is possible because every plane domain is, by definition, imbedded in the complex plane (Riemann sphere), and we can in this case split off the singularity of the L-differential belonging to $\mathfrak{M}$ by subtracting the L-differential of the sphere.

The theory of the logarithmic potential and of analytic functions in plane domains is essentially based on the imbedding in the complex plane. This geometrical fact has the analytical aspect that the elementary singularities are logarithmic and rational in the large as well as in the neighborhood of the singularities. Thus the principal parts may be subtracted off in the large, leaving a function everywhere regular in the interior of the domain.

In place of subdomains of the sphere, we shall consider subdomains of an abstractly given Riemann surface $\mathfrak{R}$. This more general imbedding leads to new results even in the case of plane domains (see [15a, b]). We decompose a differential on $\mathfrak{M}$ into a differential on $\mathfrak{R}$ with singularities, plus an interior differential on $\mathfrak{M}$. For example, let $L_{\mathbf{M}}(p, q)$ and $\mathscr{L}_{\mathbf{M}}(p, q)$ denote the bilinear differentials for $\mathfrak{M}$ and $\mathfrak{N}$ respectively. We can then write

$$L_{\mathbf{M}}(p, q) = \mathscr{L}_{\mathbf{M}}(p, q) + l_{\mathbf{M}}(p, q), \tag{4.1.18}$$

where $l_{\mathbf{M}}(p, q)$ is an interior bilinear differential on $\mathfrak{M}$.

In this chapter attention is confined to the domain functionals and, in particular, to the elementary singularities belonging to one given Riemann surface. The comparison of the functionals of two Riemann surfaces, one of which is imbedded in the other, will be taken up in the following chapter.

4.2. Definition of the Green's and Neumann's Functions

When $\mathfrak{M}$ is a domain of the complex z-plane, the Green's function $G(z, \zeta)$ of $\mathfrak{M}$ is defined to be the (single-valued) harmonic function on $\mathfrak{M}$ which vanishes on the boundary of $\mathfrak{M}$ and which has a logarithmic pole at the point ζ such that the difference

$$G(z, \zeta) - \log \frac{1}{|z - \zeta|}$$

remains regular at ζ. Neumann's function $N(z, \zeta)$ (Green's function of the second kind) has the same logarithmic pole at ζ as the Green's function and has on the boundary of $\mathfrak{M}$ a constant normal derivative. Green's function is uniquely determined by $\mathfrak{M}$ while Neumann's function is determined only up to an additive real constant. Green's function is a harmonic function of the first kind. Neumann's function, on the other hand, is not of the second kind and is not even a conformal invariant. However, the difference $N(z, \zeta) - N(z, \zeta_0)$ is an invariant which is a harmonic function of the second kind. We therefore redefine the difference $N(z, \zeta) - N(z, \zeta_0)$ to be the Neumann's function of $\mathfrak{M}$. This difference will be denoted by $N(z, \zeta, \zeta_0)$.

We suppose first that $\mathfrak{M}$ is an orientable finite Riemann surface of genus h (h handles) with m boundary components, $m > 0$. The double $\mathfrak{F}$ of $\mathfrak{M}$ then has genus $G = 2h + m - 1$. Let $\Omega_{qq_0}(p)$ denote the integral of the third kind of $\mathfrak{F}$ which is normalized by the condition that the real parts of its periods vanish. We define the Green's function $G(p, q)$ of $\mathfrak{M}$ by the formula ([15a, b])

$$G(p, q) = \frac{1}{2}\{\Omega_{q\tilde{q}}(p) - \Omega_{q\tilde{q}}(\tilde{p})\}. \tag{4.2.1}$$

Let us make the general remark that

$$\Omega_{qq_0}(p) + \Omega_{q_0q}(p) = C,$$

where C is a number independent of p, for arbitrary fixed points q and q_0. In fact, this combination of integrals of the third kind is regular on the whole surface since, by the definition of Ω, its poles cancel. Moreover, since the real part of each term is single-valued, the real part represents a bounded, regular, single-valued harmonic function on a closed surface, and is therefore a constant. Thus, using (4.1.9), we have

$$\Omega_{q\tilde{q}}(\tilde{p}) = -\Omega_{\tilde{q}q}(\tilde{p}) + c_1 = -(\Omega_{q\tilde{q}}(p))^{-} + c_2 \tag{4.2.2}$$

where c_1 and c_2 are numbers independent of p. Hence

$$G(p, q) = \text{Re}\,\{\Omega_{q\tilde{q}}(p)\} + c_3,$$

and Im G, as function of p, is therefore constant. On the boundary of $\mathfrak{M}$, $p = \tilde{p}$ so $G(p, q)$ vanishes when p is on the boundary C of $\mathfrak{M}$. Therefore, $\text{Im}\, G \equiv 0$ in p and q. By the law of interchange of argument and parameter

$$G(p, q) = G(q, p), \tag{4.2.3}$$

and it follows that

$$G(p, q) = 0 \tag{4.2.4}$$

whenever p or q is on the boundary C of $\mathfrak{M}$. When p is near q, q in $\mathfrak{M}$, z a uniformizer at q, we have

$$G(p, q) = \log \frac{1}{|z(p) - z(q)|} + \text{regular terms.} \tag{4.2.5}$$

Thus G has all the properties of Green's function for plane domains. We observe that

$$G(\tilde{p}, q) = -G(p, q), \quad G(p, \tilde{q}) = -G(p, q). \tag{4.2.6}$$

The Neumann's function of the surface $\mathfrak{M}$ shall be defined as

$$N(p, q, q_0) = \frac{1}{2}\{\Omega_{qq_0}(p) + \Omega_{\tilde{q}\tilde{q}_0}(\tilde{p}) + \Omega_{qq_0}(\tilde{p}) + \Omega_{\tilde{q}\tilde{q}_0}(p)\}. \tag{4.2.7}$$

As a function of p, Im N is constant by (4.1.9) and, since the function $\Omega_{qq_0}(p)$ has been defined in (3.1.25) only up to an additive constant $c(q, q_0)$, we may determine the latter such that $\operatorname{Im} N \equiv 0$. We have

$$N(\tilde{p}, q, q_0) = N(p, q, q_0), \quad N(p, \tilde{q}, \tilde{q}_0) = N(p, q, q_0). \tag{4.2.8}$$

Therefore, if $z = x + iy$ is a boundary uniformizer, we see that

$$\frac{\partial N(p, q, q_0)}{\partial y} = 0 \tag{4.2.9}$$

on the boundary C of $\mathfrak{M}$. In other words, N has a vanishing normal derivative on the boundary C. Let z be a uniformizer at q, z_0 a uniformizer at q_0. Then

$$\begin{aligned} N(p, q, q_0) &= \log \frac{1}{|z(p) - z(q)|} + \text{regular terms}, \\ N(p, q, q_0) &= \log |z_0(p) - z_0(q_0)| + \text{regular terms}, \end{aligned} \tag{4.2.10}$$

as p approaches q and q_0 respectively. Finally (law of interchange of argument and parameter)

$$N(p, q, q_0) - N(p_0, q, q_0) = N(q, p, p_0) - N(q_0, p, p_0). \tag{4.2.11}$$

It should be remarked that $N(p, q, q_0)$ is defined only up to an additive real constant $c(q, q_0)$. However, the combination (4.2.11) is a uniquely defined real-valued function on $\mathfrak{F}$.

We observe that the expressions

$$\frac{\partial^2 \{N(p, q, q_0) - N(p_0, q, q_0)\}}{\partial p \partial q} = \frac{\partial^2 N(p, q, q_0)}{\partial p \partial q}, \text{ and } \frac{\partial^2 G(p, q)}{\partial p \partial q}$$

are analytic on $\mathfrak{F}$ (apart from the obvious singularities). From (4.2.1) and (4.2.7) we have

$$\frac{\partial^2 G(p, q)}{\partial p \partial q} = \frac{1}{2} \frac{\partial^2 \Omega_{q\tilde{q}}(p)}{\partial p \partial q}, \tag{4.2.12}$$

$$\frac{\partial^2 N(p, q, q_0)}{\partial p \partial q} = \frac{1}{2} \frac{\partial^2 \Omega_{qq_0}(p)}{\partial p \partial q}. \tag{4.2.13}$$

Now the difference

$$\lambda(p, q) = \frac{\partial^2 G(p, q)}{\partial p \partial q} - \frac{\partial^2 N(p, q, q_0)}{\partial p \partial q} = \frac{1}{2} \left\{ \frac{\partial^2 \Omega_{q\tilde{q}}(p)}{\partial p \partial q} - \frac{\partial^2 \Omega_{qq_0}(p)}{\partial p \partial q} \right\} \tag{4.2.14}$$

is an everywhere finite bilinear differential on the double $\mathfrak{F}$. By

(4.1.2) $\lambda(p, q)$ is independent of q_0 and in its dependence on p, is a differential of the first kind:

$$\lambda(p, q) = \sum_{\mu=1}^{G} \gamma_\mu(q) \frac{dZ_\mu(p)}{dp} \tag{4.2.15}$$

where $dZ_1, dZ_2 \cdots, dZ_G$ are a complex basis for the differentials of the first kind on $\mathfrak{F}$. Since $\lambda(p, q) = \lambda(q, p)$ (more properly, $\lambda(p, q)dzd\zeta = \lambda(q, p)d\zeta dz$), we see that

$$\sum_{\mu=1}^{G} \gamma_\mu(q) \frac{dZ_\mu(p)}{dp} = \sum_{\mu=1}^{G} \gamma_\mu(p) \frac{dZ_\mu(q)}{dq}. \tag{4.2.16}$$

Let $p_1, p_2, \cdots, p_G$ be points of $\mathfrak{M}$ which are chosen such that the determinant

$$\left| \frac{dZ_\mu(p_k)}{dp_k} \right|$$

does not vanish. This choice is obviously possible because of the linear independence of the dZ_μ (see [17]). Solving the system of equations

$$\sum_{\mu=1}^{G} \gamma_\mu(q) \frac{dZ_\mu(p_k)}{dp_k} = \sum_{\mu=1}^{G} \gamma_\mu(p_k) \frac{dZ_\mu(q)}{dq}, \qquad k = 1, 2, \cdots, G,$$

for the $\gamma_\mu(q)$ we obtain (introducing a factor $\pi/2$ for later convenience)

$$\gamma_\mu(q) = \frac{\pi}{2} \sum_{\nu=1}^{G} c_{\mu\nu} \frac{dZ_\nu(q)}{dq}. \tag{4.2.17}$$

Substituting from (4.2.17) into (4.2.15) we find that

$$\lambda(p, q) = \frac{\pi}{2} \sum_{\mu,\nu=1}^{G} c_{\mu\nu} \frac{dZ_\mu(p)}{dp} \frac{dZ_\nu(q)}{dq}. \tag{4.2.18}$$

Since $\lambda(p, q) = \lambda(q, p)$, we have

$$c_{\mu\nu} = c_{\nu\mu}. \tag{4.2.19}$$

In view of formula (4.2.13) the quantity

$$\frac{\partial^2 N(p, q, q_0)}{\partial p \partial q}$$

does not depend on the point q_0. Therefore we may replace q_0 by $\tilde{q}_0$ in this term, and then by (4.2.6) and (4.2.8) we have

$$\frac{\partial^2 G(p, \tilde{q})}{\partial p \partial \tilde{q}} = -\frac{\partial^2 G(p, q)}{\partial p \partial \tilde{q}}, \quad \frac{\partial^2 N(p, \tilde{q}, \tilde{q}_0)}{\partial p \partial \tilde{q}} = \frac{\partial^2 N(p, q, q_0)}{\partial p \partial \tilde{q}}.$$

Thus

$$\lambda(p, \tilde{q}) = -\frac{\partial^2 G(p, q)}{\partial p \partial \tilde{q}} - \frac{\partial^2 N(p, q, q_0)}{\partial p \partial \tilde{q}}$$

$$= \frac{\pi}{2} \sum_{\mu, \nu=1}^{G} c_{\mu\nu} \frac{dZ_\mu(p)}{dp} \frac{dZ_\nu(\tilde{q})}{d\tilde{q}}. \tag{4.2.20}$$

Since G and N are real-valued functions, we have

$$\lambda(p, \tilde{q}) = (\lambda(q, \tilde{p}))^-$$

and, hence,

$$c_{\mu\nu} = \bar{c}_{\nu\mu} = c_{\nu\mu} \tag{4.2.21}$$

by (4.2.19). Therefore the $c_{\mu\nu}$ are real. By (4.2.14), (4.2.18) and (4.2.20):

$$(4.2.22) \quad \begin{cases} \dfrac{\partial^2 G(p, q)}{\partial p \partial q} = \dfrac{\partial^2 N(p, q, q_0)}{\partial p \partial q} + \dfrac{\pi}{2} \displaystyle\sum_{\mu, \nu=1}^{G} c_{\mu\nu} \frac{dZ_\mu(p)}{dp} \frac{dZ_\nu(q)}{dq}, \\ \dfrac{\partial^2 G(p, q)}{\partial p \partial \tilde{q}} = -\dfrac{\partial^2 N(p, q, q_0)}{\partial p \partial \tilde{q}} - \dfrac{\pi}{2} \displaystyle\sum_{\mu, \nu=1}^{G} c_{\mu\nu} \frac{dZ_\mu(p)}{dp} \frac{dZ_\nu(\tilde{q})}{d\tilde{q}}. \end{cases}$$

The formulas (4.2.22) show how Green's and Neumann's functions are related, and that if one of these fundamental functions is known, the other can easily be computed with the aid of the integrals of the first kind. Thus the boundary value problems of the first and second kinds for the Laplace equation appear as equivalent problems. This fact is a consequence of the Cauchy-Riemann differential equations, in virtue of which the first boundary value problem for a harmonic function is equivalent to the second problem for its harmonic conjugate.

The Green's function $G(p, q)$ of an orientable surface with boundary is symmetric with respect to its two arguments and is skew-symmetric in its dependence on conjugate points of the double. It thus has two symmetries. In the case of a closed orientable surface one of these symmetries is lacking, but the symmetry which arises from the law of interchange of argument and parameter is still present. Thus, although a Green's function in the strict sense does not exist, a function can be defined which incorporates the symmetry in argument and parameter and this function is the analogue of the Green's function. In fact, if $\mathfrak{M}$ is closed and orientable, let p_0 and q_0 be fixed points of $\mathfrak{M}$ and define

$$V(p, p_0;\ q, q_0) = \text{Re}\,\{\Omega_{qq_0}(p) - \Omega_{qq_0}(p_0)\}. \tag{4.2.23}$$

By the law of interchange of argument and parameter

$$V(p, p_0; q, q_0) = V(q, q_0; p, p_0). \tag{4.2.24}$$

Moreover, we clearly have

$$(4.2.24)' \qquad \begin{aligned} V(p, p_0; q, q_0) &= -\,V(p_0, p; q, q_0), \\ V(p, p_0; q, q_0) &= -\,V(p, p_0; q_0, q). \end{aligned}$$

The symmetry properties of the function $V(p, p_0; q, q_0)$ are therefore the same as those of the Riemann curvature tensor. We shall find it convenient for purposes of analogy to extend the definition of V if one or more of its arguments lie on the conjugate surface $\tilde{\mathfrak{M}}$. We set

$$(4.2.24)'' \qquad V(\tilde{p}, p_0; q, q_0) = -\,V(p, p_0; q, q_0);$$

similar relations for the other arguments are implied by (4.2.24) and (4.2.24)′.

In view of the formal similarity of $V(p, p_0; q, q_0)$ to a Green's function, we shall call it the Green's function of the closed orientable surface $\mathfrak{M}$. Moreover, whenever the dependence on the parameter points p_0, q_0 is not involved, we shall write

$$G(p, q) = V(p, p_0; q, q_0) \tag{4.2.25}$$

with the understanding that when p and q are interchanged, so are p_0 and q_0. Thus

$$G(p, q) = G(q, p). \tag{4.2.26}$$

By (4.2.24)″ we have also

$$(4.2.26)' \qquad G(\tilde{p}, q) = -\,G(p, q).$$

We must bear in mind the singular behavior of the Green's function for p near q_0, q near p_0. In this connection, the following electrical analogy is useful. The double $\mathfrak{F}$ of $\mathfrak{M}$ is disconnected and consists of two components $\mathfrak{M}$ and $\tilde{\mathfrak{M}}$. Let small holes be opened at the points q_0 of $\mathfrak{M}$ and $\tilde{q}_0$ of $\tilde{\mathfrak{M}}$, and let these two holes be joined by a thin tube. Imagine that an electric charge of magnitude $+ 1$ is placed at the point q, an electric charge of magnitude -1 at $\tilde{q}$. Since the lines of force issuing from q must all enter the tube, the mouth of the tube at q_0 behaves like a negative charge. If $G^*(p, q)$ is the Green's function of the surface $\mathfrak{M}^*$ obtained from $\mathfrak{M}$ by opening a hole at q_0, we shall show in Section 7.4 that

$$(4.2.27) \quad G^*(p, q) - G^*(p, q_1) \to V(p, p_0; q, q_1) - V(q_0, p_0; q, q_1)$$

as the holes shrink to the point q_0. In formula (4.2.27), q_1 is an arbitrary parameter point. Thus the choice of the function $V(p, p_0; q, q_0)$ as a Green's function is justified both on physical and on mathematical grounds.

In the case of a closed orientable surface ($m + c = 0$), we define the Neumann's function to be the Green's function when both points p, q lie in $\mathfrak{M}$ or $\tilde{\mathfrak{M}}$, and to be the negative of the Green's function when p lies in $\mathfrak{M}$, q in $\tilde{\mathfrak{M}}$ or p in $\tilde{\mathfrak{M}}$, q in $\mathfrak{M}$. Thus we define

$$(4.2.28) \qquad N(p, q, q_0) = V(p, p_0; q, q_0),$$

and we impose the symmetry condition

$$(4.2.29) \qquad N(\tilde{p}, q, q_0) = N(p, q, q_0).$$

It remains to define a Green's function for non-orientable surfaces. Since a non-orientable surface with or without boundary has a connected double which is closed and orientable, the Green's function of a non-orientable finite surface may be defined by the formula (4.2.1). However, on a non-orientable surface $\mathfrak{M}$ there are closed paths lying in its interior such that each of the two overlying paths on the double $\mathfrak{F}$ joins conjugate points, and this means that the Green's function defined by (4.2.1) reverses its sign on such closed paths (on which the orientation is reversed). On the other hand, the Neumann's function defined by (4.2.7) is single-valued.

It is in some instances desirable therefore to define a single-valued Green's function $G_s(p, q)$ for non-orientable surfaces. In the case of a closed non-orientable surface, there is no single-valued Green's function $G_s(p, q)$ with pole only at q. For such a function, $G_s(\tilde{p}, q) = G_s(p, q)$ should hold, but then $G_s(p, q)$ would have two poles, each with residue $+ 1$ on the closed orientable double. This is impossible. If the single-valued Green's function has poles at q and q_0, it must coincide with the Neumann's function (4.2.7). We therefore take this function to be the single-valued Green's function of a closed non-orientable surface.

If the non-orientable surface $\mathfrak{M}$ has a boundary, let $\mathfrak{N}$ be its two-sheeted orientable covering (see Section 2.2). The genus of $\mathfrak{N}$ is $2h + c - 1$, and the double of $\mathfrak{M}$ is formed by identifying the two boundary points of $\mathfrak{N}$ lying over each boundary point of $\mathfrak{M}$. Now the number of boundary curves of $\mathfrak{N}$ is $2m$, where m is the number of boundary components of $\mathfrak{M}$. If, instead of identifying corresponding boundaries of $\mathfrak{N}$, we form the double of $\mathfrak{N}$ in the usual way, we obtain a quadruple covering $\mathfrak{Q}$ of the surface $\mathfrak{M}$, the genus of the quadruple being $4h + 2c + 2m - 3$. Let the two points of $\mathfrak{N}$ which lie over a point of $\mathfrak{M}$ be denoted by p, $\tilde{p}$, and let their conjugate points on the double of $\mathfrak{N}$ be $\bar{p}$, $\bar{\tilde{p}}$ respectively. We observe that

$$\tilde{\bar{p}} = \bar{\tilde{p}}. \tag{4.2.30}$$

The four points of the quadruple $\mathfrak{Q}$ which lie over a point of $\mathfrak{M}$ therefore are p, $\tilde{p}$, $\tilde{\bar{p}} = \bar{\tilde{p}}$, $\bar{p}$.

In terms of the quadruple, we define a single-valued Green's function $G_s(p, q)$ by the formula

$$\begin{aligned} G_s(p, q) = \frac{1}{4}\{&\Omega_{q\bar{q}}(p) + \Omega_{\tilde{q}\bar{\tilde{q}}}(p) - \Omega_{q\bar{q}}(\bar{p}) - \Omega_{\tilde{q}\bar{\tilde{q}}}(\bar{p}) \\ &+ \Omega_{q\bar{q}}(\tilde{p}) + \Omega_{\tilde{q}\bar{\tilde{q}}}(\tilde{p}) - \Omega_{q\bar{q}}(\bar{\tilde{p}}) - \Omega_{\tilde{q}\bar{\tilde{q}}}(\bar{\tilde{p}})\}. \end{aligned} \tag{4.2.31}$$

Here the Ω's denote integrals of the third kind on the quadruple whose periods are pure imaginary. We observe that

$$G_s(\tilde{p}, q) = G_s(p, q),\ G_s(\bar{p}, q) = -G_s(p, q),\ G_s(p, q) = G_s(q, p). \tag{4.2.32}$$

Let G_1 be the Green's function for $\mathfrak{N}$. Since conjugate points of the double of $\mathfrak{N}$ are q, $\bar{q}$, we have

$$G_1(p, q) = \frac{1}{2}\{\Omega_{q\bar{q}}(p) - \Omega_{q\bar{q}}(\bar{p})\}. \tag{4.2.33}$$

Because of the symmetry,

$$G_1(\tilde{p}, q) = G_1(p, \tilde{q}),\ G_1(p, q) = G_1(\tilde{p}, \tilde{q}),\ G_1(\bar{p}, q) = -G_1(p, q). \tag{4.2.34}$$

It is clear that

$$(4.2.35)\quad \begin{aligned} G_s(p, q) &= \frac{1}{2}\{G_1(p, q) + G_1(p, \tilde{q}) + G_1(\tilde{p}, q) + G_1(\tilde{p}, \tilde{q})\} \\ &= G_1(p, q) + G_1(p, \tilde{q}). \end{aligned}$$

We observe that the Green's function (4.2.1) for $\mathfrak{M}$ has the form

$$(4.2.36)\qquad G(p, q) = G_1(p, q) - G_1(p, \tilde{q}).$$

For the difference of the left and right sides is a regular harmonic function of p (for fixed q) which vanishes on the boundary of $\mathfrak{M}$.

We define the Neumann's function of a non-orientable surface $\mathfrak{M}$ to be the function (4.2.7). Because of its symmetry property, it is single-valued on $\mathfrak{M}$.

4.3. Differentials of the First Kind Defined in Terms of the Green's Function

In Chapter 3 we defined basis differentials dZ for a closed orientable surface by interpolating vortices along the cycles of a homology basis. That is to say, we formed differentials with poles of residues $+1$, -1 at the vertices of 1-simplexes. By adding the differentials belonging to the 1-simplexes of a 1-cycle, we obtained a differential of the first kind.

If, in the case of an orientable surface with boundary, this interpolation along a cycle α is carried out using differentials from the Green's function, the symmetry of the Green's function automatically makes a corresponding interpolation along the conjugate cycle $\tilde{\alpha}$ of the double. Thus interpolation with the Green's function yields differentials which take conjugate values at conjugate places and therefore belong to the surface. This interpolation will be carried out along each cycle of a homology basis, which consists of G cycles $K_1, \cdots, K_G$, $G = 2h + m - 1$. For a closed orientable surface, the differentials obtained in this way agree with the differentials dZ_μ defined in Chapter 3.

Non-orientable surfaces differ from orientable ones in that they are not imbedded in their doubles, but the differentials of such surfaces still take conjugate values at conjugate places of the doubles. In terms of the surfaces themselves, we require (see Section 2.2) that the differentials are invariant under direct conformal transformations of the local parameter, and go over into conjugate complex values under an indirect conformal transformation. This requirement forces us to interpolate using differentials formed by means of the Green's function (4.2.33). The differentials of the first kind belonging to a non-orientable surface are defined only on the Betti group which has G generators, $G = 2h + c + m - 1$. If $m = 0$ there is one torsion coefficient of value 2, while if $m > 0$ there is no torsion coefficient (the cycle which, taken with multiplicity 2, bounds on the closed surface is now homologous to the sum of the m boundaries). We thus interpolate along the G cycles $K_1, \cdots, K_G$ of a Betti basis using the Green's function (4.2.33), the K_μ forming closed curves on the double.

In all cases we therefore define

$$(4.3.1) \qquad Z'_\mu(q) = -\frac{2}{\pi}\int_{K_\mu} \frac{\partial^2 G(p, q)}{\partial p \partial q}\, dp$$

or

$$(4.3.1)' \quad \operatorname{Im} Z_\mu(q) = \frac{1}{2\pi}\int_{K_\mu} \frac{\partial G(p, q)}{\partial n_p}\, ds_p = 2i \cdot \frac{1}{2\pi}\int_{K_\mu} \frac{\partial G(p, q)}{\partial p}\, dp$$

where $\partial/\partial n$ denotes differentiation with respect to the normal which points to the left with respect to the oriented cycle K_μ. Here we use the fact that

$$(4.3.2) \qquad \frac{\partial G(p, q)}{\partial p}\, dp = \frac{1}{2}\, dG(p, q) + \frac{1}{2i}\frac{\partial G(p, q)}{\partial n_p}\, ds_p.$$

Let $T(p, q)$ denote the analytic function of p whose real part is the Green's function $G(p, q)$. For future reference we note that, by (4.3.1)′,

$$(4.3.3) \qquad P(dT, K_\mu) = \int_{K_\mu} dT = -\, 2\pi i \cdot \operatorname{Im} Z_\mu(q).$$

Also, in the case of a non-orientable surface, by (4.2.34), (4.2.35), and (4.3.1),

$$(4.3.4) \quad Z'_\mu(q) = Z^{(1)\prime}_\mu(q) + \tilde{Z}^{(1)\prime}_\mu(q), \ \operatorname{Im} Z_\mu(q) = \operatorname{Im} Z^{(1)}_\mu(q) - \operatorname{Im} Z^{(1)}_\mu(\tilde{q}),$$

where

$$(4.3.5)\quad Z_\mu^{(1)\prime}(q) = -\frac{2}{\pi}\int\limits_{K_\mu} \frac{\partial^2 G_1(p, q)}{\partial p \partial q}\, dp,\ \tilde{Z}_\mu^{(1)\prime}(q) = (Z_\mu^{(1)\prime}(\tilde{q}))^-.$$

If $T_s(p, q)$ is the analytic function whose real part is $G_s(p, q)$, we have in a similar way

$$(4.3.6)\quad P(dT_s, K_\mu) = \int\limits_{K_\mu} dT_s = -2\pi i\,\{\operatorname{Im} Z_\mu^{(1)}(q) + \operatorname{Im} Z_\mu^{(1)}(\tilde{q})\}$$
$$= -2\pi i\,\{\operatorname{Im} Z_\mu^{(1)}(q) - \operatorname{Im} \tilde{Z}_\mu^{(1)}(q)\}.$$

Finally,

$$\tilde{Z}'_\mu(q) = (Z'_\mu(\tilde{q}))^- = -\frac{2}{\pi}\int\limits_{K_\mu}\left(\frac{\partial^2 G(p, \tilde{q})}{\partial p \partial \tilde{q}}\, dp\right)^- = \frac{2}{\pi}\int\limits_{K_\mu}\frac{\partial^2 G(p, q)}{\partial \tilde{p} \partial q}\, d\tilde{p}$$

$$= \frac{2}{\pi}\frac{d}{dq}\int\limits_{K_\mu}\left(dG - \frac{\partial G}{\partial p}\, dp\right) = -\frac{2}{\pi}\int\limits_{K_\mu}\frac{\partial^2 G(p, q)}{\partial p \partial q}\, dp = Z'_\mu(q).$$

Thus

$$(4.3.7)\qquad d\tilde{Z}_\mu(q) = dZ_\mu(q)$$

and the differentials dZ_μ belong to $\mathfrak{M}$ (Section 2.2). We observe that

$$(4.3.8)\qquad \tilde{Z}'_\mu(q) = Z'_{\tilde{\mu}}(q) = -\frac{2}{\pi}\int\limits_{\tilde{K}_\mu}\frac{\partial^2 G(p, q)}{\partial p \partial q}\, dp$$

where $\tilde{K}_\mu$ is the cycle on the double conjugate to K_μ.

For orientable surfaces with boundary we choose on $\mathfrak{M}$ a canonical basis $K_1, K_2, \cdots, K_G$ satisfying the following two conditions:

(i) $K_1, K_3, \cdots, K_{2h-1}$ belong to a subdivision S of the double, $K_2, K_4, \cdots, K_{2h}$ belong to the dual subdivision $*S$, and the remaining cycles $K_{2h+1}, \cdots, K_{2h+m-1}$ are each homologous to a boundary component;

(ii) for $1 \leqq \mu, \nu \leqq 2h$ we have $I(K_{2\mu-1}, K_{2\mu}) = 1$, $I(K_{2\mu-1}, K_{2\nu}) = 0$ $(\mu \neq \nu)$, $I(K_{2\mu-1}, K_{2\nu-1}) = 0$, $I(K_{2\mu}, K_{2\nu}) = 0$. The cycles $K_{2h+1}, \cdots, K_{2h+m-1}$ will be called boundary cycles.

For closed orientable surfaces we saw in Chapter 3 that

$$(4.3.9)\qquad (dw, dZ_\mu) = -P(dw, K_\mu) = -\int\limits_{K_\mu} dw$$

for any differential dw of the first kind. This formula carries over at once to orientable surfaces with boundary, and is valid for interior differentials with finite norms (space **M**). It is sufficient to prove (4.3.9) for interior differentials which are regular in the closure of $\mathfrak{M}$. If K_μ is not a boundary cycle, $\operatorname{Im} Z_\mu$ vanishes on the whole boundary and the argument is similar to that for a closed orientable surface while, if K_μ is a boundary cycle, it may be assumed to coincide with a boundary component of the surface. Then $\operatorname{Im} Z_\mu$ is single-valued in the interior of the surface and vanishes on each boundary component except K_μ where it has the value unity and we have

$$(dw, dZ_\mu) = -\int_{K_\mu} \operatorname{Im} Z_\mu \cdot dw = -\int_{K_\mu} dw.$$

As in Chapter 3, we write

$$(4.3.10) \qquad (dZ_\mu, dZ_\nu) = -\int_{K_\nu} dZ_\mu = \Gamma_{\mu\nu}.$$

Then $\Gamma_{\mu\nu} = (\Gamma_{\nu\mu})^-$ and

$$(4.3.11) \qquad \operatorname{Im} \Gamma_{\mu\nu} = I(K_\mu, K_\nu).$$

If $\mathfrak{M}$ is a non-orientable surface, let $\mathfrak{N}$ be its two-sheeted orientable covering which has $2m$ boundary curves, $m \geqq 0$. A canonical basis $K_1, K_2, \cdots, K_G$ for $\mathfrak{M}$ will consist of one cycle from each dual pair of cycles belonging to $\mathfrak{N}$, m of these cycles being boundary cycles of $\mathfrak{N}$.

In the case of a non-orientable surface $\mathfrak{M}$ the scalar product (dw, dZ_μ) is not defined, but

$$(4.3.12) \qquad [dw, dZ_\mu] = \operatorname{Re}\{(dw, dZ_\mu)_{\mathfrak{M}}\}$$

has a uniquely determined value. We suppose that each differential dw remains invariant under a direct conformal transformation of parameter and that it goes into the complex conjugate under an indirect conformal transformation. The differentials dw defined in the interior of $\mathfrak{M}$ are differentials in the ordinary sense in the interior of the two-sheeted orientable covering $\mathfrak{N}$ of $\mathfrak{M}$, but they have the special property that they take conjugate values at conjugate places of $\mathfrak{N}$. In particular, the space **M** of interior differentials of $\mathfrak{M}$ is just the space of differentials in $\mathfrak{R}$ having finite norms and taking conjugate values at conjugate places of $\mathfrak{R}$. We have

$$(4.3.13)\qquad [dw, dZ_\mu] = \frac{1}{2}\,\mathrm{Re}\{(dw, dZ_\mu)_{\mathfrak{R}}\} = \frac{1}{2}\,(dw, dZ_\mu)_{\mathfrak{R}},$$

where the subscript $\mathfrak{R}$ denotes that the scalar product is to be extended over $\mathfrak{R}$. If K_μ is not a boundary cycle,

$$[dw, dZ_\mu] = -\frac{1}{2}\int_{K_\mu+\tilde{K}_\mu} dw = -\,\mathrm{Re}\int_{K_\mu} dw,$$

while, if K_μ is a boundary cycle and dw is regular on the boundary, then $K_\mu = \tilde{K}_\mu$ and

$$[dw, dZ_\mu] = -\frac{1}{2}\int_{K_\mu} (dw + d\bar{w}) = -\,\mathrm{Re}\int_{K_\mu} dw.$$

Thus

$$(4.3.14)\qquad [dw, dZ_\mu] = -\,\mathrm{Re}\{P(dw, K_\mu)\} = -\,\mathrm{Re}\int_{K_\mu} dw.$$

We write

$$(4.3.15)\qquad [dZ_\mu, dZ_\nu] = \Gamma_{\mu\nu}.$$

Then $\Gamma_{\mu\nu} = \Gamma_{\nu\mu}$, and

$$(4.3.16)\qquad \mathrm{Im}\,\Gamma_{\mu\nu} = I(K_\mu, K_\nu) = 0.$$

For simplicity of notation, we shall henceforth write (dw, dZ_μ) for $[dw, dZ_\mu]$, but we shall understand by the scalar product over a non-orientable surface one-half the scalar product extended over $\mathfrak{R}$.

4.4. Differentials of the First Kind Defined in Terms of the Neumann's Function

For reasons of symmetry, we also introduce differentials based on the Neumann's function defined by (4.2.7). We proceed in a manner entirely analogous to that followed in the preceding section where the definitions were based on the Green's function. We define:

$$(4.4.1)\qquad Z_\mu^{*\prime}(q) = -\frac{2}{\pi}\int_{K_\mu} \frac{\partial^2 N(p, q, q_0)}{\partial p\,\partial q}\,dp$$

or

$$\text{(4.4.2)}\quad \begin{aligned} \operatorname{Im} Z^*_\mu(q) &= \frac{1}{2\pi}\int_{K_\mu} \frac{\partial N(p,q,q_0)}{\partial n_p}\, ds_p \\ &= 2i\cdot\frac{1}{2\pi}\int_{K_\mu}\frac{\partial N(p,q,q_0)}{\partial p}\, dp. \end{aligned}$$

From (4.4.2) we see that $Z^{*\prime}_\mu(q) \equiv 0$ if K_μ is a boundary cycle. Therefore, there are $2h$ differentials (4.4.1) if the surface is orientable, $2h + c - 1$ otherwise.

We have

$$\begin{aligned} \tilde{Z}^{*\prime}_\mu(q) &= (Z'_\mu(\tilde{q}))^- = -\frac{2}{\pi}\int_{K_\mu}\left(\frac{\partial^2 N(p,\tilde{q},q_0)}{\partial p\,\partial\tilde{q}}\, dp\right)^- = -\frac{2}{\pi}\int_{K_\mu}\frac{\partial^2 N(p,q,q_0)}{\partial\tilde{p}\,\partial q}\, d\tilde{p} \\ &= -\frac{2}{\pi}\frac{d}{dq}\int_{K_\mu}\left(dN - \frac{\partial N}{\partial p}\, dp\right) = \frac{2}{\pi}\int_{K_\mu}\frac{\partial^2 N(p,q,q_0)}{\partial p\,\partial q}\, dp = -Z^{*\prime}_\mu(q). \end{aligned}$$

The relation

$$\text{(4.4.3)}\qquad d\tilde{Z}^*_\mu(q) = -dZ^*_\mu(q)$$

shows that idZ^*_μ is a differential of $\mathfrak{M}$. We have also

$$\text{(4.4.4)}\qquad \tilde{Z}^{*\prime}_\mu(q) = Z^{*\prime}_{\tilde{\mu}}(q) = -\frac{2}{\pi}\int_{\tilde{K}_\mu}\frac{\partial^2 N(p,q,q_0)}{\partial p\,\partial q}\, dp.$$

This formula, together with (4.4.3), shows once more that $Z^{*\prime}_\mu(q) \equiv 0$ for a boundary cycle since, for a boundary cycle, we may assume that $\tilde{K}_\mu = K_\mu$.

Let

$$\text{(4.4.5)}\quad W'_\mu(q) = \frac{1}{2}[Z'_\mu(q) + Z^{*\prime}_\mu(q)] = -\frac{1}{\pi}\int_{K_\mu}\frac{\partial^2[G(p,q)+N(p,q,q_0)]}{\partial p\,\partial q}\, dp.$$

By (4.3.7), (4.3.8), (4.4.3) and (4.4.4),

$$\text{(4.4.6)}\quad W'_{\mu}(q) = \frac{1}{2}[Z'_{\tilde{\mu}}(q) - Z^{*\prime}_{\tilde{\mu}}(q)] = -\frac{1}{\pi}\int_{\tilde{K}_\mu}\frac{\partial^2[G(p,q) - N(p,q,q_0)]}{\partial p\,\partial q}\, dp$$

and, from symmetry,

$$(4.4.6)' \quad W'_{\tilde{\mu}}(q)=\frac{1}{2}[Z'_\mu(q)-Z^{*\prime}_\mu(q)]=-\frac{1}{\pi}\int_{K_\mu}\frac{\partial^2[G(p,q)-N(p,q,q_0)]}{\partial p\partial q}dp.$$

Thus

$$(4.4.7) \qquad dZ_\mu = dW_\mu + dW_{\tilde{\mu}}$$

and

$$(4.4.8) \qquad dZ^*_\mu = dW_\mu - dW_{\tilde{\mu}}.$$

The differentials $W'_\mu(q)$ are normalised differentials of the first kind for the closed orientable double. In fact, the Green's function $V(p, p_0; q, q_0)$ of the closed surface, as defined in Section 4.2, satisfies

$$2\frac{\partial^2 V}{\partial p \partial q} = \frac{\partial^2[G(p,q)+N(p,q,q_0)]}{\partial p\partial q}.$$

Thus the differentials $W'_\mu(q)$ given by (4.4.5) are exactly the differentials $Z'_\mu(q)$ defined by (4.3.1) for the double, on the cycles K_μ.

If the surface is closed and orientable, we see from (4.4.6)′ and our definitions in Section 4.2. that

$$(4.4.9) \qquad dW_{\tilde{\mu}} \equiv 0.$$

Hence by (4.4.7) and (4.4.8):

$$(4.4.10) \qquad dZ_\mu = dW_\mu,$$

$$(4.4.11) \qquad dZ^*_\mu = dW_\mu.$$

The vanishing of $dW_{\tilde{\mu}}$ is an expression of the fact that the double is disconnected.

4.5. Period Matrices

If $\mathfrak{M}$ is an orientable surface with boundary, let $x_1, x_2, \cdots, x_G$ be arbitrary complex numbers. Then by (4.3.10)

$$(4.5.1) \qquad N(\sum_{\mu=1}^{G} x_\mu dZ_\mu) = \sum_{\mu,\nu=1}^{G} \Gamma_{\mu\nu}x_\mu\bar{x}_\nu > 0$$

unless

$$(4.5.2) \qquad \sum_{\mu=1}^{G} x_\mu\, dZ_\mu \equiv 0.$$

The identity (4.5.2) implies that

$$\sum_{\mu=1}^{G} x_\mu \operatorname{Im} Z_\mu(p) = F(\tilde{p}) \tag{4.5.3}$$

where $F(\tilde{p})$ depends analytically upon $\tilde{p}$. Since $F(\tilde{p})$ is constant on each boundary component of $\mathfrak{M}$, it is identically constant on $\mathfrak{M}$. The relation (4.5.3) is then valid by continuation over the double $\mathfrak{F}$. By computing the increments of (4.5.3) around the cycles K_μ, we conclude that $x_\nu = 0$, $\nu = 1, 2, \cdots, 2h$. For $\nu = 2h+1, \cdots, G$, we have $\operatorname{Im} Z_\mu = 2 \operatorname{Im} W_\mu$ (modulo constants) by (4.4.7) and the same reasoning, applied now to the double $\mathfrak{F}$ in place of $\mathfrak{M}$, shows that $x_\nu = 0$, $\nu = 2h+1, \cdots, G$. Thus

$$\sum_{\mu, \nu=1}^{G} \Gamma_{\mu\nu} x_\mu \bar{x}_\nu > 0 \tag{4.5.4}$$

unless $x_1 = x_2 = \cdots = x_G = 0$. In particular, the matrix $\| \Gamma_{\mu\nu} \|$ is non-singular and the same is true of any principal submatrix $\| \Gamma_{i_\mu i_\nu} \|$, $\mu, \nu = 1, 2, \cdots, n$, $1 \leq n \leq G$. Hence the dZ_μ $(\mu = 1, 2, \cdots, G)$ are a real basis for the everywhere finite differentials of $\mathfrak{M}$, and a complex basis for the corresponding differentials of the double $\mathfrak{F}$.

If $\mathfrak{M}$ is non-orientable, let $x_1, x_2, \cdots, x_G$ be real. We have

$$N\Big(\sum_{\mu=1}^{G} x_\mu dZ_\mu\Big) = \sum_{\mu, \nu=1}^{G} \Gamma_{\mu\nu} x_\mu x_\nu > 0 \tag{4.5.5}$$

unless

$$\sum_{\mu=1}^{G} x_\mu dZ_\mu \equiv 0. \tag{4.5.6}$$

This last formula implies that

$$\sum_{\mu=1}^{G} x_\mu \operatorname{Im} Z_\mu \equiv \text{constant} \tag{4.5.7}$$

on the double, and we again find that $x_1 = x_2 = \cdots = x_G = 0$.

In all three cases (closed orientable, orientable with boundary, non-orientable) the matrix $\| \operatorname{Re} \Gamma_{\mu\nu} \|$ is non-singular, while if $m + c > 0$ the matrix $\|\Gamma_{\mu\nu}\|$ is non-singular. If the surfgce is non-orientable, $\operatorname{Im} \Gamma_{\mu\nu} = 0$.

4.6. Relations between the Green's and Neumann's Functions

If $m + c > 0$, the surface has a Green's function (4.2.1) and a Neumann's function (4.2.7) which are connected by the relation

$$\frac{\partial^2 G(p, q)}{\partial p \partial q} = \frac{\partial^2 N(p, q, q_0)}{\partial p \partial q} + \frac{\pi}{2} \sum_{\mu, \nu=1}^{G} c_{\mu\nu} Z'_\mu(p) Z'_\nu(q), \tag{4.6.1}$$

where the $c_{\mu\nu}$ are real coefficients, $c_{\mu\nu} = c_{\nu\mu}$. This formula was established in the case $m > 0$, $c = 0$; but the proof when $c > 0$ is the same. We now evaluate the coefficients $c_{\mu\nu}$. By (4.4.6) and (4.4.6)′

$$\begin{cases} W'_\sigma(q) = -\dfrac{1}{2} \displaystyle\sum_{\mu, \nu=1}^{G} c_{\mu\nu} P(dZ_\mu, \tilde{K}_\sigma)\, Z'_\nu(q), \\ W'_{\tilde{\sigma}}(q) = -\dfrac{1}{2} \displaystyle\sum_{\mu, \nu=1}^{G} c_{\mu\nu} P(dZ_\mu, K_\sigma)\, Z'_\nu(q). \end{cases} \tag{4.6.2}$$

Adding the equations (4.6.2) and using (4.4.7), we obtain

$$Z'_\sigma(q) = -\sum_{\mu, \nu=1}^{G} c_{\mu\nu} \operatorname{Re}\{P(dZ_\mu, K_\sigma)\} Z'_\nu(q) = \sum_{\mu, \nu=1}^{G} c_{\mu\nu} \operatorname{Re} \Gamma_{\mu\sigma} Z'_\nu(q). \tag{4.6.3}$$

Since the $Z'_\nu(q)$ are linearly independent, we conclude that

$$\sum_{\mu=1}^{G} c_{\mu\nu} \operatorname{Re} \Gamma_{\mu\sigma} = \delta_{\nu\sigma}.$$

That is,

$$\| c_{\mu\nu} \| = \| \operatorname{Re} \Gamma_{\mu\nu} \|^{-1}. \tag{4.6.4}$$

Subtracting the equations (4.6.2) and using (4.4.8), we have

$$Z^{*\prime}_\sigma(q) = i \sum_{\mu, \nu=1}^{G} c_{\mu\nu} \operatorname{Im}\{P(dZ_\mu, K_\sigma)\} Z'_\nu(q). \tag{4.6.5}$$

The differentials $W'_\varrho(q)$ are the differentials $Z'_\varrho(q)$ for the closed orientable double, and we therefore have

$$\operatorname{Im} P(dW_\varrho, K_\sigma) = I(K_\sigma, K_\varrho). \tag{4.6.6}$$

By (4.4.7)

$$\begin{aligned} \operatorname{Im} P(dZ_\mu, K_\sigma) &= \operatorname{Im} P(dW_\mu + dW_{\tilde{\mu}}, K_\sigma) \\ &= I(K_\sigma, K_\mu + \tilde{K}_\mu) = \begin{cases} I(K_\sigma, K_\mu), & \text{if } c = 0, \\ I(K_\sigma, \tilde{K}_\mu), & \text{if } c > 0. \end{cases} \end{aligned} \tag{4.6.7}$$

Substituting from (4.6.7) into (4.6.5), we see that

$$Z_\sigma^{*\prime}(q) = \begin{cases} i \sum_{\mu,\nu=1}^{2h} c_{\mu\nu} I(K_\sigma, K_\mu) Z_\nu'(q), & c = 0,\ m > 0 \\ i \sum_{\mu,\nu=1}^{2h+c-1} c_{\mu\nu} I(K_\sigma, \tilde{K}_\mu)\, Z_\nu'(q), & c > 0. \end{cases} \tag{4.6.8}$$

We observe that $Z_\sigma^{*\prime}(q) \equiv 0$ for $\sigma > 2h (c = 0)$, $\sigma > 2h + c - 1$ $(c > 0)$.

4.7. Canonical Mapping Functions

If the finite Riemann surface $\mathfrak{M}$ is of genus zero, the uniformization principle states that $\mathfrak{M}$ can be mapped conformally onto a subdomain of the closed w-plane whose boundary consists of m rectilinear segments parallel to the real w-axis (Hilbert canonical domain). If $\mathfrak{M}$ is of higher genus, a mapping of this sort onto a subdomain of the closed plane (sphere) is impossible for topological reasons. However, in Chapter 2, we pointed out that the Hilbert parallel-slit domain has an analogue for higher genus in which there appear, in addition to the boundary slits, a finite number of other slits where edges are suitably identified such that the resulting domain has the required genus.

Let us recall how such a function is constructed. Let $\mathfrak{F}$ be the double of $\mathfrak{M}$ and let $u_q = u_{q\zeta}$ be the single valued harmonic function on $\mathfrak{F}$ with a dipole singularity at the point q of $\mathfrak{M}$ which, expressed in terms of a particular uniformizer ζ at q, $\zeta(q) = 0$, is such that the difference

$$u_q - \operatorname{Re}\left(\frac{1}{\zeta}\right) \tag{4.7.1}$$

is regular harmonic in a neighborhood of q and vanishing at the point q itself. Then u_q is unique. Let $u_{\tilde{q}} = u_{\tilde{q}\bar{\zeta}}$ be the corresponding harmonic function with a dipole singularity at the conjugate point $\tilde{q}$ of $\mathfrak{F}$ which is defined in terms of the uniformizer $\bar{\zeta}$ at $\tilde{q}$. Then

$$u_q(p) = u_{\tilde{q}}(\tilde{p}). \tag{4.7.2}$$

Writing

$$u_q^*(p) = u_q(p) + u_{\tilde{q}}(p), \tag{4.7.3}$$

we see that

(4.7.4) $$u_q^*(p) = u_q^*(\tilde{p}).$$

Thus u_q^* is a harmonic function of $\mathfrak{M}$ of the second kind with a vanishing normal derivative on the boundary of $\mathfrak{M}$, and the conjugate harmonic function v has a constant value along each boundary, by the Cauchy-Riemann equations. We denote the analytic function whose real part is $u_{q\zeta}^*$ by $f_{q\zeta}$, and it is clear that

(4.7.5) $$df_q(p) = (df_q(p))^-.$$

The periods of df_q are pure imaginary. Since df_q is real on the boundary of $\mathfrak{M}$, it follows that for the boundary components C_ν,

(4.7.6) $$P(df_q, C_\nu) = 0, \qquad \nu = 1, 2, \cdots, m.$$

Any determination of f_q therefore maps C_ν onto a finite rectilinear segment parallel to the real axis in the plane of f_q. Moreover, each choice of uniformizer ζ leads to a different function with these properties. All possible functions f_q can be constructed from the two particular functions $f_{q\zeta}$ and $f_{q,i\zeta}$.

The function

(4.7.7) $$g_{q\zeta}(p) = if_{q,i\zeta}(p)$$

maps the boundary components C_ν of $\mathfrak{M}$ onto slits which are parallel to the imaginary axis, and it has a single-valued imaginary part. More generally, the function

(4.7.8) $$e^{i\theta} f_{q,e^{i\theta}\zeta}(p), \qquad \theta \text{ real},$$

defines slits parallel to the direction $e^{i\theta}$. Let

(4.7.9) $$\begin{cases} f_\theta = e^{i\theta}(f_{q\zeta}\cos\theta - ig_{q\zeta}\sin\theta), \\ g_\theta = e^{i\theta}(g_{q\zeta}\cos\theta - if_{q\zeta}\sin\theta). \end{cases}$$

The image of a boundary component C_ν by f_θ is a slit parallel to the direction $e^{i\theta}$ while the corresponding image by g_θ is parallel to the direction $ie^{i\theta}$. We observe that

(4.7.10) $$f_{q\zeta} + g_{q\zeta} = f_\theta + g_\theta.$$

Since

$$\operatorname{Re}\{f_{q,e^{i\theta}\zeta} - (f_{q\zeta}\cos\theta - ig_{q\zeta}\sin\theta)\}$$

is everywhere regular and single-valued on the double $\mathfrak{F}$, we see from the maximum principle for harmonic functions that it is a constant. Hence (apart possibly from a constant) (see [7])

$$e^{i\theta} f_{q, e^{i\theta}\zeta} = e^{i\theta}(f_{q\zeta} \cos\theta + f_{q, i\zeta} \sin\theta) = f_\theta. \tag{4.7.11}$$

Now any determination of the function

$$\frac{1}{2}\{f_{q\zeta} + g_{q\zeta}\} \tag{4.7.12}$$

maps a boundary component C_ν of $\mathfrak{M}$ onto a curve which is cut at most twice by a line parallel to the real or the imaginary axis in the image plane and, by (4.7.10), the same property is true of a line having arbitrary inclination. It follows that the image of each C_ν by the function (4.7.12) is a convex curve without double points. In the case when $\mathfrak{M}$ is of genus zero it can be shown that the mapping defined by (4.7.12) is schlicht (see [14a], [6a]). If the genus is greater than zero, the mapping cannot be schlicht on topological grounds.

Let $\zeta = \xi + i\eta$ be a uniformizer at q, $\zeta(q) = 0$. Then

$$u_q^\Delta(p) = -\frac{\partial N(p, q, q_1)}{\partial \xi}$$

is a single-valued harmonic function of p on the double $\mathfrak{F}$. When p is near q, let ζ be chosen as the local coordinate of p. Then, near q,

$$u_q^\Delta(p) = \operatorname{Re}\left(\frac{1}{\zeta}\right) + \text{regular terms},$$

and similarly near $\tilde{q}$. If $z = x + iy$ is a boundary uniformizer of $\mathfrak{M}$, it is clear that

$$\frac{\partial u_q^\Delta}{\partial y} = 0$$

on the boundary. Let

$$F_{q\zeta} = u_q^\Delta(p) + iv_q^\Delta(p)$$

be the analytic function with real part $u_q^\Delta(p)$. Since the function $\operatorname{Re}\{F_{q\zeta}(p) - f_{q\zeta}(p)\}$ is single-valued and regular on the double $\mathfrak{F}$, it is equal to a constant. Therefore (apart possibly from a constant)

$$F_{q\zeta}(p) = f_{q\zeta}(p).$$

In other words,

$$\text{Re}\, f_{q\zeta}(p) = -\frac{\partial N(p, q, q_1)}{\partial \xi}. \tag{4.7.13}$$

Hence

$$\text{Re}\, f_{q,\, i\zeta}(p) = \frac{\partial N(p, q, q_1)}{\partial \eta}. \tag{4.7.14}$$

Thus

$$\begin{aligned} \text{Re}\, f_{q\zeta}(p) + i\, \text{Re} f_{q,\, i\zeta}(p) &= -\left\{\frac{\partial N(p, q, q_1)}{\partial \xi} - i\frac{\partial N(p, q, q_1)}{\partial \eta}\right\} \\ &= -2\frac{\partial N(p, q, q_1)}{\partial \zeta} = -2\frac{\partial N(p, q, q_1)}{\partial q}, \end{aligned} \tag{4.7.15}$$

$$\text{Re}\, f_{q\zeta}(p) - i\, \text{Re}\, f_{q,\, i\zeta}(p) = -2\frac{\partial N(p, q, q_1)}{\partial \tilde{q}}; \tag{4.7.15$'$}$$

or

$$\text{Re}\, f_{q\zeta}(p) + i\, \text{Im}\, g_{q\zeta}(p) = -2\frac{\partial N(p, q, q_1)}{\partial q}, \tag{4.7.16}$$

$$\text{Re}\, f_{q\zeta}(p) - i\, \text{Im}\, g_{q\zeta}(p) = -2\frac{\partial N(p, q, q_1)}{\partial \tilde{q}}. \tag{4.7.16$'$}$$

Let $z = x + iy$ be a uniformizer at p. Differentiating (4.7.16) and (4.7.16)$'$ with respect to p and observing that

$$\begin{cases} \dfrac{\partial}{\partial x}\text{Re}\, f_{q\zeta}(p) = \text{Re}\dfrac{df_{q\zeta}(p)}{dz}, & \dfrac{\partial}{\partial x}\text{Im}\, g_{q\zeta}(p) = \text{Im}\dfrac{dg_{q\zeta}(p)}{dz}, \\ \dfrac{\partial}{\partial y}\text{Re}\, f_{q\zeta}(p) = -\text{Im}\dfrac{df_{q\zeta}(p)}{dz}, & \dfrac{\partial}{\partial y}\text{Im}\, g_{q\zeta}(p) = \text{Re}\dfrac{dg_{q\zeta}(p)}{dz}, \end{cases} \tag{4.7.17}$$

we obtain

$$\frac{df_{q\zeta}(p)}{dp} + \frac{dg_{q\zeta}(p)}{dp} = -4\frac{\partial^2 N(p, q, q_1)}{\partial p \partial q}, \tag{4.7.18}$$

$$\frac{df_{q\zeta}(p)}{dp} - \frac{dg_{q\zeta}(p)}{dp} = -4\frac{\partial^2 N(p, q, q_1)}{\partial p \partial \tilde{q}}. \tag{4.7.18$'$}$$

Formulas (4.7.13), (4.7.14), (4.7.18), (4.7.18)$'$ show the close relation between the canonical mapping functions and the Neumann's function. This relation may be explained by the fact that the canonical functions have constant imaginary part on the boundary components of $\mathfrak{M}$, and that their real parts have therefore vanishing normal derivative.

4.8. Classes of Differentials

Let K_μ, $\mu = 1, 2, \cdots, G$, be a canonical basis for the 1-cycles as described in Section 4.3, and let **M** be the Hilbert space of differentials df which are regular at each interior point of $\mathfrak{M}$ and have finite norms over $\mathfrak{M}$. If $\mathfrak{M}$ is non-orientable, the differentials are further required to assume conjugate values under an indirect conformal change of uniformizer.

Let $\mathbf{F}(i_1, i_2, \cdots, i_n)$ be the subclass of **M** composed of differentials df for which

$$(4.8.1) \qquad (df, dZ_{i_\mu}) = 0, \qquad \mu = 1, 2, \cdots, n, \ n \leqq G.$$

The smallest of these subclasses is $\mathbf{F}(1, 2, \cdots, G)$, and we denote it by the letter **S**.

We begin with the following lemma:

Lemma 4.8.1. *If any differential of the first kind of the double $\mathfrak{F}$ belongs to* **S**, *then it is identically zero.*

Proof: The lemma is immediate if $\mathfrak{M}$ is closed and orientable. If it is not, let $dw_1, dw_2, \cdots, dw_G$ be any basis for the everywhere finite differentials of $\mathfrak{F}$ (differentials of the first kind). Since $dw_1, dw_2, \cdots, dw_G$ are linearly independent, we can orthonormalize them over $\mathfrak{M}$ by the Gram-Schmidt process and we may therefore assume that

$$(4.8.2) \qquad (dw_\mu, dw_\nu) = \delta_{\mu\nu}, \qquad \mu, \nu = 1, 2, \cdots, G.$$

Every differential dw of the first kind belonging to $\mathfrak{F}$ has the form

$$dw = \sum_{\mu=1}^{G} c_\mu dw_\mu.$$

If this differential has vanishing periods, then

$$(4.8.3) \qquad (dw, dZ_\mu) = 0, \qquad \mu = 1, 2, \cdots, G.$$

Since each dw_ϱ is a linear combination

$$dw_\varrho = \sum_{\nu=1}^{G} \gamma_{\varrho\nu}\, dZ_\nu,$$

it follows that

$$c_\varrho = (dw, dw_\varrho) = 0, \qquad \varrho = 1, 2, \cdots, G.$$

Now let

(4.8.4) $$dw = \sum_{\mu=1}^{G} c_\mu dZ_\mu$$

belong to $\mathbf{F}(i_1, i_2, \cdots, i_n)$. Then

(4.8.5) $$(dw, dZ_{i_\nu}) = \sum_{\mu=1}^{G} c_\mu (dZ_\mu, dZ_{i_\nu}) = \sum_{\mu=1}^{G} c_\mu \Gamma_{\mu i_\nu} = 0\,, \quad \nu = 1, 2, \cdots, n.$$

Thus

(4.8.6) $$\sum_{\mu=1}^{n} c_{i_\mu} \Gamma_{i_\mu i_\nu} = -\Sigma\, c_\varrho \Gamma_{\varrho i_\nu}, \qquad \nu = 1, 2, \cdots, n,$$

where the sum on the right runs over the set of integers which are complementary to $i_1, i_2, \cdots, i_n$. If $\| \Gamma_{i_\mu i_\nu} \|$ $(\mu, \nu = 1, 2, \cdots, n)$ is non-singular, the equations (4.8.6) always have a non-trivial solution $(c_{i_1}, \cdots, c_{i_n})$ no matter how the values c_ϱ are prescribed, and in this case the dimension of the space of differentials of $\mathfrak{F}$ which belong to $\mathbf{F}(i_1, i_2, \cdots, i_n)$ is precisely $G - n$. Thus if $m + c > 0$, in which case $\| \Gamma_{i_\mu i_\nu} \|$ is non-singular, the basis for the differentials of $\mathfrak{F}$ of the first kind in $\mathbf{F}(i_1, \cdots, i_n)$ has the form

(4.8.7) $$dw_1, dw_2, \cdots, dw_{G-n}.$$

The differentials dZ_{i_μ}, $\mu = 1, \cdots, n$ form the orthogonal complement of the differentials (4.8.7).

In general, any differential df of $\mathfrak{F}$ of the first kind can be represented in the form

(4.8.8) $$df = dw + dW$$

where $dw \in \mathbf{F}(i_1, i_2, \cdots, i_n)$ and dW belongs to the orthogonal complement, that is $(dw, dW) = 0$. This follows by orthogonal projection, which in this case is trivial since the spaces have finite dimension.

If

(4.8.9) $$I(K_{i_\mu}, K_{i_\nu}) = 0, \qquad \mu, \nu = 1, 2, \cdots, n,$$

we say that the class $\mathbf{F}(i_1, i_2, \cdots, i_n)$ is symmetric. We see by (4.3.10) and (4.3.11) that for a symmetric class, the matrix $\| \Gamma_{i_\mu i_\nu} \|$ is real, symmetric and non-singular.

The period conditions imposed on the class $\mathbf{F}(i_1, i_2, \cdots, i_n)$ are

$$(4.8.10) \quad \begin{cases} P(dg, K_{i_\mu}) = 0, & \text{if } c = 0 \\ \operatorname{Re}\{P(dg, K_{i_\mu})\} = 0, & \text{if } c > 0 \end{cases} \quad (\mu = 1, 2, \cdots, n).$$

Suppose that $m + c > 0$ and let dg be a meromorphic differential satisfying the period conditions (4.8.10). Write

$$df = c_0 dg + \sum_{\mu=1}^{G-n} c_\mu dw_\mu.$$

The condition that the periods of df vanish on the $G - n$ cycles K_μ, $\mu \neq i_1, i_2, \cdots, i_n$, give $G - n$ conditions in $G - n + 1$ unknowns c_μ, and therefore these equations have a non-trivial solution with $c_0 \neq 0$. It follows that

$$(4.8.11) \qquad dg = df + \sum_{\mu=1}^{G-n} b_\mu dw_\mu$$

where

$$(4.8.12) \quad \begin{cases} P(df, K_\mu) = 0, & \text{if } c = 0 \\ \operatorname{Re}\{P(df, K_\mu)\} = 0, & \text{if } c > 0 \end{cases} \quad (\mu = 1, 2, \cdots, G).$$

A meromorphic differential satisfying the conditions (4.8.12) will be said to be single-valued on $\mathfrak{M}$.

On an orientable surface $\mathfrak{M}$ we may also define the class of differentials dg satisfying

$$(4.8.13) \qquad \operatorname{Re}\{P(dg, K_{i_\mu})\} = 0, \qquad \mu = 1, 2, \cdots, n.$$

This class will be denoted by $\mathbf{G}(i_1, i_2, \cdots, i_n)$. By the argument given above, we see that there are precisely $G - n$ real linearly independent basis differentials (4.8.7) in $\mathbf{G}$.

The class $\mathbf{F}(i_1, i_2, \cdots, i_n)$ on a non-orientable surface $\mathfrak{M}$ may now be interpreted as follows. Let $\mathfrak{N}$ be the two-sheeted orientable covering of $\mathfrak{M}$, and let $\mathbf{G}(i_1, i_2, \cdots, i_n)$ be the class $\mathbf{G}$ on $\mathfrak{N}$. Given any differential dg of $\mathbf{G}$, let

$$d\tilde{g}(p) = (dg(\tilde{p}))^-$$

be its conjugate. The class $\mathbf{F}$ is the subclass of $\mathbf{G}$ composed of differentials dg which satisfy $d\tilde{g} = dg$.

4.9. The Bilinear Differentials for the Class F

In the orientable case an expression $L_{\mathbf{F}}(p, \tilde{q})$ is called a reproducing kernel of a class **F** of differentials df on $\mathfrak{M}$ if:

(1) For each fixed q in the interior of $\mathfrak{M}$, $L_{\mathbf{F}}(p, \tilde{q})$ as a differential of p belongs to **F**.

(2) $L_{\mathbf{F}}(p, \tilde{q})dzd\tilde{\zeta}$ is invariant; that is, $L_{\mathbf{F}}(p, \tilde{q})$ is a bilinear differential.

(3) For each $df \in \mathbf{F}$, $L_{\mathbf{F}}(p, \tilde{q})$ has the reproducing property

$$(df, L_{\mathbf{F}}dzd\tilde{\zeta}) = -df(q). \tag{4.9.1}$$

A reproducing kernel is unique. For let $L^*_{\mathbf{F}}$ be another. Then

$$N(L_{\mathbf{F}} - L^*_{\mathbf{F}}) = (L_{\mathbf{F}} - L^*_{\mathbf{F}}, L_{\mathbf{F}}) - (L_{\mathbf{F}} - L^*_{\mathbf{F}}, L^*_{\mathbf{F}}) = 0.$$

Hence the difference $L_{\mathbf{F}} - L^*_{\mathbf{F}}$ is identically zero. We remark that property (2) follows from (1) and (3). In fact, by (4.9.1) we have

$$(L_{\mathbf{F}}(p, \tilde{r}), L_{\mathbf{F}}(p, \tilde{q})) = -L_{\mathbf{F}}(q, \tilde{r}) \tag{4.9.2}$$

and this relation implies

$$L_{\mathbf{F}}(q, \tilde{r} = (L_{\mathbf{F}}(r, \tilde{q}))^-. \tag{4.9.3}$$

We shall see that the reproducing kernel $L_{\mathbf{F}}(p, \tilde{q})$ exists for all classes **F**, but it is clear from (4.9.1) that $L_{\mathbf{F}}(p, \tilde{q})$ is different from zero if and only if the class **F** contains elements df which are not identically zero.

Heretofore the kernel has been defined mostly for plane domains, and it has been customary to call the Hermitian bilinear differential $-L(p, \tilde{q})dzd\tilde{\zeta}$ the "kernel function."

From the reproducing property of the kernel $L_{\mathbf{F}}(p, \tilde{q})$ belonging to an orientable domain $\mathfrak{M}$, we may derive an important formula for representing $L_{\mathbf{F}}(p, \tilde{q})$ in terms of any complete orthonormal system $\{d\varphi_n\}$ for the class **F**. Since $L_{\mathbf{F}}(p, \tilde{q})dzd\overline{\zeta}$, as a differential in p, belongs to **F**, we may develop it in the form

$$L_{\mathbf{F}}(p, \tilde{q})dzd\overline{\zeta} = \sum_{n=1}^{\infty} a_n(\tilde{q})d\overline{\zeta}d\varphi_n(p). \tag{4.9.4}$$

Here the Fourier coefficient $a_n(\tilde{q})d\overline{\zeta}$ is given by

$$\begin{aligned} a_n(\tilde{q})d\overline{\zeta} &= (L_{\mathbf{F}}(p, \tilde{q})dzd\overline{\zeta}, d\varphi_n(p)) \\ &= (d\varphi_n(p), L_{\mathbf{F}}(p, \tilde{q})dzd\overline{\zeta})^- = -(d\varphi_n(q))^- \end{aligned} \tag{4.9.5}$$

by virtue of the reproducing property (4.9.1) of the kernel $L_{\mathbf{F}}$.

Introducing (4.9.5) into (4.9.4) we obtain

$$L_{\mathbf{F}}(p, \tilde{q})dzd\bar{\zeta} = -\sum_{n=1}^{\infty} d\varphi_n(p)(d\varphi_n(q))^{-}. \tag{4.9.6}$$

Thus, $L_{\mathbf{F}}(p, \tilde{q})$ is indeed the kernel of the orthonormal system of differentials $\{d\varphi_n(p)\}$. The explicit formula (4.9.6) clarifies the significance of the reproducing property of $L_{\mathbf{F}}$ and gives, on the other hand, a convenient tool for constructing the kernel $L_{\mathbf{F}}$ in terms of any complete orthonormal system in $\mathbf{F}$. In the case of a closed Riemann surface, the space $\mathbf{F}$ has finite dimension while, if the surface has a boundary, there are infinitely many independent interior differentials, and $\mathbf{F}$ has infinite dimension.

When $\mathfrak{M}$ has a boundary, it turns out that, for fixed q in $\mathfrak{M}$, the bilinear differential $L_{\mathbf{F}}(p, \tilde{q})$ is regular everywhere on $\mathfrak{F}$ except at $p = \tilde{q}$ where it has a double pole:

$$L_{\mathbf{F}}(p, \tilde{q}) = \frac{1}{\pi}\frac{1}{[z(p) - z(\tilde{q})]^2} + \text{regular terms.}$$

Thus $L_{\mathbf{F}}(p, q)$ is a singular bilinear differential on $\mathfrak{M}$, with a double pole at $p = q$. Because of the singularity, $L_{\mathbf{F}}(p, q)$ is not an element of $\mathbf{F}$ although it satisfies the period relations for $\mathbf{F}$. Further, for each $df \in \mathbf{F}$, we have

$$(df,\ L_{\mathbf{F}}(p, q)) = 0, \tag{4.9.7}$$

that is, $L_{\mathbf{F}}(p, q)$ is orthogonal to the elements of $\mathbf{F}$. These properties determine $L_{\mathbf{F}}(p, q)$ uniquely, for if $L_{\mathbf{F}}^*$ is another, then $L_{\mathbf{F}} - L_{\mathbf{F}}^*$ is an element of $\mathbf{F}$ and we again have

$$N(L_{\mathbf{F}} - L_{\mathbf{F}}^*) = (L_{\mathbf{F}} - L_{\mathbf{F}}^*,\ L_{\mathbf{F}}) - (L_{\mathbf{F}} - L_{\mathbf{F}}^*,\ L_{\mathbf{F}}^*) = 0.$$

We shall take the singular bilinear differential $L_{\mathbf{F}}(p, q)$ as fundamental, rather than $L_{\mathbf{F}}(p, \tilde{q})$. When the double $\mathfrak{F}$ of $\mathfrak{M}$ is connected they are, of course, the same and the choice of the factor $1/\pi$ in normalizing the singularity is dictated by the form of the reproducing property for $L_{\mathbf{F}}(p, \tilde{q})$.

In the case of the singular bilinear differential, the scalar product in the preceding statements has not yet been defined so we begin by defining a scalar product valid for differentials df, dg whose singularities are included among a finite set of points q_i, $i = 1, \cdots, n$.

This definition must satisfy the requirement that it reduces to the ordinary one when both differentials are regular.

Assume that $\mathfrak{M}$ is orientable. Let $\mathfrak{M}_{q_i}$ be a disc of radius a_i in the plane of a local uniformizer at q_i, with center q_i, and let $\mathfrak{M}'$ denote the union of the $\mathfrak{M}_{q_i}$, $i = 1, \cdots, n$. We define

$$(4.9.8) \qquad \begin{aligned}(df, dg) &= \lim \int\limits_{\mathfrak{M}-\mathfrak{M}'} f'(p)\,(g'(p))^{-}\,dA \\ &= \lim\ (df, dg)_{\mathfrak{M}-\mathfrak{M}'}\end{aligned}$$

as $\sum_{i=1}^{n} a_i$ approaches zero, provided that this limit exists. It is easily seen that the extended definition of the scalar product retains the properties (2) (*a*) — (*e*) of Section 2.3.

We shall also interpret the integral

$$\int\limits_{\mathfrak{M}} f'(p)\,(g'(p))^{-}\,dA$$

as a limit according to the above definition whenever the integrand is singular.

The singular differentials dg to be considered below are regular on $\mathfrak{M}$ except for a double pole at a point q of $\mathfrak{M}$. In particular, dg is without residue at q. Let ζ be a uniformizer at q, $\zeta = \zeta(p)$, $\zeta(q) = 0$. Then, except for a constant factor,

$$g' = \frac{1}{\zeta^2} + \text{regular terms}$$

near q. Let $\mathfrak{M}_0$ be the disc $|\,\zeta\,| < b$, where b is sufficiently small and fixed, while $\mathfrak{M}'$ is given by $|\,\zeta\,| < a < b$. We denote the boundaries of these domains by C_0 and C' respectively. Then the integral of dg is defined up to a constant and single-valued in $\mathfrak{M}_0 - \mathfrak{M}'$, and we may verify that the scalar product (4.9.8) is defined when df is regular in $\mathfrak{M}$. In fact,

$$\begin{aligned}\lim \int\limits_{\mathfrak{M}-\mathfrak{M}'} f'(p)\,(g'(p))^{-}\,dA &= \int\limits_{\mathfrak{M}-\mathfrak{M}_0} f'(p)\,(g'(p))^{-}\,dA + \lim \int\limits_{\mathfrak{M}_0-\mathfrak{M}'} f'(p)\,(g'(p))^{-}\,dA \\ &= \int\limits_{\mathfrak{M}-\mathfrak{M}_0} f'(p)\,(g'(p))^{-}\,dA + \frac{1}{2i}\int\limits_{C_0} \bar{g}\,df - \lim \frac{1}{2i}\int\limits_{C'} \bar{g}\,df\end{aligned}$$

using (1.5.24), and it is clear that

$$\int_{C'} \bar{g}\, df = \int_{C'} (g(p))^{-} f'(p) d\zeta = o(1)$$

as a tends to zero, so the limit exists. We note that the additive constant in $\bar{g}$ does not affect this result since df has vanishing period around C_0 and C' if b is sufficiently small. Thus the scalar product in (4.9.7) is defined.

Analogously, we may verify that (df, dg) is defined when df has a singularity of the above type at a point r distinct from q. In particular

$$(L_{\mathbf{F}}(p, q),\ L_{\mathbf{F}}(p, r)), \quad r \neq q, \tag{4.9.9}$$

is defined by (4.9.8). The value of this product will be given in (4.12.3).

On the other hand, $N(dg) = (dg, dg)$ is not defined by (4.9.8) and a further extension is necessary. This extension will be based on the limit of the product (4.9.9) as r tends to q (see Section 4.12).

In extending the definition of scalar product in the case of non-orientable surfaces $\mathfrak{M}$, we note that the differentials df, dg take conjugate values at conjugate places q, $\tilde{q}$ of $\mathfrak{R}$, so we may choose conjugate regions $\mathfrak{R}'$ and $\tilde{\mathfrak{R}}'$ containing the singular points q_i, $\tilde{q}_i$. Then we define

$$\begin{aligned}(df, dg) &= \lim \frac{1}{2} \int_{\mathfrak{R}-\mathfrak{R}'-\tilde{\mathfrak{R}}'} f'(p)\, (g'(p))^{-}\, dA \\ &= \lim \frac{1}{2} (df,\ dg)_{\mathfrak{R}-\mathfrak{R}'-\tilde{\mathfrak{R}}'}\,.\end{aligned} \tag{4.9.8$'$}$$

The analogue of $L_{\mathbf{F}}(p, \tilde{q})$ in the case of a non-orientable surface $\mathfrak{M}$ is the unique kernel $L_{\mathbf{F}}(p, \bar{q}, \bar{\tilde{q}})$ satisfying

$$(df,\ L_{\mathbf{F}}) = -\operatorname{Re}\{f'(q)\} \tag{4.9.1$'$}$$

for every $df \in \mathbf{F}$. The singular bilinear differential $L_{\mathbf{F}}(p, q, \tilde{q})$ with singularities at $p = q$ and $p = \tilde{q}$ is again orthogonal to all $df \in \mathbf{F}$:

$$(df,\ L_{\mathbf{F}}(p, q, \tilde{q})) = 0. \tag{4.9.7$'$}$$

4.10. Construction of the Bilinear Differential for the Class M in Terms of the Green's Function

We shall construct the bilinear differential $L_{\mathbf{F}}$ of the class **F** in terms of the Green's function and the differentials of the first kind, and we begin with the case in which $\mathbf{F} = \mathbf{M}$. We suppose first that $\mathfrak{M}$ is orientable and has a boundary.

We set

$$L_{\mathbf{M}}(p, q) = -\frac{2}{\pi}\frac{\partial^2 G(p, q)}{\partial p \partial q} = -\frac{1}{\pi}\frac{\partial^2 \Omega_{q\tilde{q}}(p)}{\partial p \partial q}. \tag{4.10.1}$$

Then by (4.2.6) we have

$$L_{\mathbf{M}}(p, \tilde{q}) = -\frac{2}{\pi}\frac{\partial^2 G(p, \tilde{q})}{\partial p \partial \tilde{q}} = \frac{2}{\pi}\frac{\partial^2 G(p, q)}{\partial p \partial \tilde{q}} = -\frac{1}{\pi}\frac{\partial^2 \Omega_{\tilde{q}q}(p)}{\partial p \partial \tilde{q}}. \tag{4.10.2}$$

We have to show that $L_{\mathbf{M}}(p, q)$ is orthogonal to the differentials df of **M** and that $L_{\mathbf{M}}(p, \tilde{q})$ reproduces them.

When p and q lie in the same neighborhood, let both of these points be expressed by the same uniformizing variable. If $\zeta = \zeta(p)$, $\zeta(q) = 0$, we have by (4.10.1)

$$L_{\mathbf{M}}(p, q) = \frac{1}{\pi\zeta^2} + \text{regular terms}. \tag{4.10.3}$$

Hence, if $\mathfrak{M}'$ is the disc $|\zeta| < a$,

$$(df, L_{\mathbf{M}}(p, q)) = \frac{1}{2i}\int_{\mathfrak{M}-\mathfrak{M}'} f'(p)(L_{\mathbf{M}}(p, q))^{-} d\bar{z}dz + o\,(1) \tag{4.10.4}$$

as the radius a tends to zero. Now assume that df is regular analytic on the boundary C of $\mathfrak{M}$. Since $\partial G(p, q)/\partial\tilde{q}$ is single-valued on $\mathfrak{M}$ and equal to zero for p on the boundary of $\mathfrak{M}$, integration by parts gives

$$\begin{aligned}\frac{1}{2i}\int_{\mathfrak{M}-\mathfrak{M}'} f'(p)(L_{\mathbf{M}}(p, q))^{-} d\bar{z}dz &= -\frac{1}{\pi i}\int_{\mathfrak{M}-\mathfrak{M}'} f'(p)\,\frac{\partial^2 G(p, q)}{\partial\bar{p}\partial\tilde{q}}\,d\bar{z}dz \\ &= \frac{1}{\pi i}\int_{C'} f'(p)\,\frac{\partial G(p, q)}{\partial\tilde{q}}\,d\zeta.\end{aligned} \tag{4.10.5}$$

Since

(4.10.6) $$\frac{\partial G(p,q)}{\partial \tilde{q}} = \frac{1}{2\bar{\zeta}} + \text{regular terms}$$

near q, we have

(4.10.7) $$\frac{1}{\pi i}\int_{C'} f'(p)\,\frac{\partial G(p,q)}{\partial \tilde{q}}\,d\zeta = \frac{1}{2\pi i}\int_{C'} f'(p)\,\frac{d\zeta}{\bar{\zeta}} + o(1) = o(1).$$

Hence the orthogonality property

(4.10.8) $$(df, L_{\mathbf{M}}(p,q)) = 0, \qquad df \in \mathbf{M},$$

is proved subject to the restriction that f is regular on the boundary C of $\mathfrak{M}$.

If df is regular on the boundary, the reproducing property of $L_{\mathbf{M}}(p,\tilde{q})$ is proved in an analogous fashion. In fact, since $L_{\mathbf{M}}(p,\tilde{q})$ is everywhere regular in $\mathfrak{M}$, we have

(4.10.4)′ $$(df, L_{\mathbf{M}}(p,\tilde{q})) = \frac{1}{2i}\int_{\mathfrak{M}-\mathfrak{M}'} f'(p)(L_{\mathbf{M}}(p,\tilde{q}))^{-} d\bar{z}dz + o(1).$$

Integrating by parts, we obtain

(4.10.5)′ $$\begin{aligned}\frac{1}{2i}\int_{\mathfrak{M}-\mathfrak{M}'} f'(p)(L_{\mathbf{M}}(p,\tilde{q}))^{-} d\bar{z}dz &= \frac{1}{\pi i}\int_{\mathfrak{M}-\mathfrak{M}'} f'(p)\,\frac{\partial^2 G(p,q)}{\partial \bar{p}\,\partial q}\,d\bar{z}dz \\ &= -\frac{1}{\pi i}\int_{C'} f'(p)\,\frac{\partial G(p,q)}{\partial q}\,d\zeta.\end{aligned}$$

Near q we have instead of (4.10.6)

(4.10.6)′ $$\frac{\partial G(p,q)}{\partial q} = \frac{1}{2\zeta} + \text{regular terms},$$

so

(4.10.7)′ $$\frac{1}{\pi i}\int_{C'} f'(p)\,\frac{\partial G(p,q)}{\partial q}\,d\zeta = f'(q).$$

Letting a tend to zero, we obtain the reproducing formula

(4.10.8)′ $$(df, L_{\mathbf{M}}(p,\tilde{q})) = -f'(q)$$

subject to the restriction that f' is regular on the boundary C of $\mathfrak{M}$.

We now remove this restriction. Given the point q interior to $\mathfrak{M}$, we construct a sequence of finite Riemann surfaces $\mathfrak{M}_\mu$ satisfying the following conditions:

(1) $\mathfrak{M}_\mu$ together with its boundary lies in the interior of $\mathfrak{M}$.

(2) Each $\mathfrak{M}_\mu$ contains the point q in its interior.

(3) $\mathfrak{M}_\mu \subset \mathfrak{M}_{\mu+1}$ and, given any point p interior to $\mathfrak{M}$, there is a positive integer $\mu_0 = \mu_0(p)$ such that $p \in \mathfrak{M}_\mu$ for $\mu \geqq \mu_0$.

(4) At each point of $\mathfrak{M}_\mu$ the uniformizers are admissible uniformizers of $\mathfrak{M}$.

The sequence $\mathfrak{M}_\mu$ is easily constructed using the level lines of the Green's function $G(p, q)$ where q is the given interior point of $\mathfrak{M}$. In fact, we define $\mathfrak{M}_\mu$ to be the subdomain of points p of $\mathfrak{M}$ for which

$$G(p, q) \geqq \frac{\varepsilon}{\mu}$$

where ε is a positive number. If ε is small enough, then for each positive integer μ the level curves $G(p, q) = \varepsilon/\mu$ consist of precisely m analytic Jordan curves each of which is homotopic to precisely one boundary curve of $\mathfrak{M}$, and as μ tends to infinity these curves approach the boundary curves of $\mathfrak{M}$. Let p be a boundary point of $\mathfrak{M}_\mu$, z a uniformizer of $\mathfrak{M}$ at p. In the plane of z a subarc of the boundary of $\mathfrak{M}_\mu$ through p appears as an analytic arc bounding a half-neighborhood of points belonging to $\mathfrak{M}_\mu$. By the Riemann mapping theorem the half-neighborhood can be mapped onto a subdomain of the upper half-plane of a variable t such that the analytic boundary arc goes into a segment of the real t-axis. It is then clear that t is a boundary uniformizer of $\mathfrak{M}_\mu$ at p, and that t is also a uniformizer of $\mathfrak{M}$ at p.

We shall prove the validity of (4.10.8) for a general $df \in \mathbf{M}$. The corresponding proof for (4.10.8)′ is similar. Let $\mathfrak{M}'$ be a fixed uniformizer disc $|\zeta| < a$ at q with boundary C'. Let $\mathfrak{M}_0$ be a compact subdomain lying in the interior of $\mathfrak{M}$ and containing $\mathfrak{M}'$ in its interior. There is a μ_1 such that $\mathfrak{M}_0$ is a compact subdomain of $\mathfrak{M}_\mu$ for $\mu \geqq \mu_1$. In $\mathfrak{M}_0 - \mathfrak{M}'$ the Green's function $G_\mu(p, q)$ of $\mathfrak{M}_\mu$, together with its derivatives, converges uniformly to the Green's function $G(p, q)$ of $\mathfrak{M}$ and its corresponding derivatives. It follows that the regular sequence $\{L_{\mathbf{M}}(p, q) - L_{\mathbf{M}}^{(\mu)}(p, q)\}$ converges uniformly to zero in

$\mathfrak{M}_0$ and that $\{L_{\mathbf{M}}^{(\mu)}(p,q)\,\partial G_\mu(p,q)/\partial\bar{q}\}$ converges uniformly on C'. Further

$$(df,\ L_{\mathbf{M}}^{(\mu)}(p,q))_{\mathfrak{M}_\mu} = 0 \tag{4.10.9}$$

by (4.10.8), since df is regular on the boundary of $\mathfrak{M}_\mu$. Then

$$\begin{aligned}(df, L_{\mathbf{M}}(p,q)) &= (df, L_{\mathbf{M}}(p,q))_{\mathfrak{M}-\mathfrak{M}_0} \\ &+ (df, L_{\mathbf{M}}(p,q) - L_{\mathbf{M}}^{(\mu)}(p,q))_{\mathfrak{M}_0} - (df, L_{\mathbf{M}}^{(\mu)}(p,q))_{\mathfrak{M}_\mu-\mathfrak{M}_0}.\end{aligned} \tag{4.10.10}$$

By the Schwarz inequality

$$\left|\int\limits_{\mathfrak{M}_\mu-\mathfrak{M}_0} f'(p)(L_{\mathbf{M}}^{(\mu)}(p,q))^- dA\right|^2 \leqq \int\limits_{\mathfrak{M}_\mu-\mathfrak{M}_0} |L_{\mathbf{M}}^{(\mu)}(p,q)|^2 dA \cdot \int\limits_{\mathfrak{M}-\mathfrak{M}_0} |f'(p)|^2 dA. \tag{4.10.11}$$

Here

$$\int\limits_{\mathfrak{M}_\mu-\mathfrak{M}_0} |L_{\mathbf{M}}^{(\mu)}(p,q)|^2 dA \leqq \int\limits_{\mathfrak{M}_\mu-\mathfrak{M}'} |L_{\mathbf{M}}^{(\mu)}(p,q)|^2 dA = \frac{1}{\pi i}\int\limits_{C'} L_{\mathbf{M}}^{(\mu)}(p,q)\,\frac{\partial G_\mu(p,q)}{\partial\bar{q}}\,d\zeta \leqq K$$

where K is independent of μ and $\mathfrak{M}_0$, by the uniform convergence on C'. Hence

$$\left|\int\limits_{\mathfrak{M}_\mu-\mathfrak{M}_0} f'(p)(L_{\mathbf{M}}^{(\mu)}(p,q))^- dA\right|^2 \leqq K \cdot \int\limits_{\mathfrak{M}-\mathfrak{M}_0} |f'(p)|^2 dA. \tag{4.10.12}$$

Analogously,

$$\left|\int\limits_{\mathfrak{M}-\mathfrak{M}_0} f'(p)\,(L_{\mathbf{M}}(p,q))^- dA\right|^2 \leqq k \cdot \int\limits_{\mathfrak{M}-\mathfrak{M}_0} |f'(p)|^2 dA \tag{4.10.13}$$

where

$$k = \int\limits_{\mathfrak{M}-\mathfrak{M}'} |L_{\mathbf{M}}(p,q)|^2 dA.$$

Letting μ tend to infinity, we obtain from (4.10.10)

$$|(df, L_{\mathbf{M}}(p,q))|^2 \leqq K' \cdot \int\limits_{\mathfrak{M}-\mathfrak{M}_0} |f'(p)|^2 dA, \tag{4.10.14}$$

where K' is independent of $\mathfrak{M}_0$. Letting $\mathfrak{M}_0$ approach $\mathfrak{M}$, we obtain (4.10.8) for any df of class $\mathbf{M}$ on $\mathfrak{M}$.

We have at once from (4.3.1), (4.10.1)

$$P(L_{\mathbf{M}}(p,q), K_\mu) = Z'_\mu(q). \tag{4.10.15}$$

This relation remains valid if we replace q by $\tilde{q}$. We remark that $L_{\mathbf{M}}(p, q)$ is symmetric in p and q and that $L_{\mathbf{M}}(p, \tilde{q})$ satisfies the law of Hermitian symmetry.

If $\mathfrak{M}$ is closed and orientable, we again define $L_{\mathbf{M}}(p, q)$ in terms of the Green's function for $\mathfrak{M}$, using the first equality of (4.10.1). The appropriate definition for $L_{\mathbf{M}}(p, \tilde{q})$ is then (4.10.2) in view of (4.2.26)'. Since the essential properties of the Green's function are the same in this case, all the properties of these bilinear differentials follow as before, but without the necessity of considering the behavior of df on the boundary.

If the surface $\mathfrak{M}$ is non-orientable, let

$$L_{\mathbf{M}}^{(1)}(p, q) = -\frac{2}{\pi}\frac{\partial^2 G_1(p, q)}{\partial p \partial q} \tag{4.10.16}$$

be the bilinear differential for the orientable covering $\mathfrak{N}$. Then it is clear that

$$\begin{aligned} L_{\mathbf{M}}(p, q, \tilde{q}) &= L_{\mathbf{M}}^{(1)}(p, q) + L_{\mathbf{M}}^{(1)}(p, \tilde{q}) = \\ &-\frac{2}{\pi}\left\{\frac{\partial^2 G_1(p, q)}{\partial p \partial q} + \frac{\partial^2 G_1(p, \tilde{q})}{\partial p \partial \tilde{q}}\right\}. \end{aligned} \tag{4.10.17}$$

By (4.2.34)

$$L_{\mathbf{M}}^{(1)}(\tilde{p}, q) = (L_{\mathbf{M}}^{(1)}(p, \tilde{q}))^-$$

and hence

$$L_{\mathbf{M}}(\tilde{p}, q, \tilde{q}) = (L_{\mathbf{M}}(p, q, \tilde{q}))^-. \tag{4.10.18}$$

Thus $L_{\mathbf{M}}(p, q, \tilde{q})$ is a differential on $\mathfrak{N}$ in its dependence on p which takes conjugate values at conjugate places. In its dependence on p the differential

$$L_{\mathbf{M}}(p, \bar{q}, \bar{\tilde{q}}) = L_{\mathbf{M}}^{(1)}(p, \bar{q}) + L_{\mathbf{M}}^{(1)}(p, \bar{\tilde{q}}) \tag{4.10.19}$$

is everywhere regular on $\mathfrak{N}$.

It is obvious that

$$(df, L_{\mathbf{M}}(p, q, \tilde{q}))_{\mathfrak{M}} = \frac{1}{2}(df, L_{\mathbf{M}}(p, q, \tilde{q}))_{\mathfrak{N}} = 0, \tag{4.10.20}$$

$$(df, L_{\mathbf{M}}(p, \bar{q}, \bar{\tilde{q}}))_{\mathfrak{M}} = \frac{1}{2}(df, L_{\mathbf{M}}(p, \bar{q}, \bar{\tilde{q}}))_{\mathfrak{N}} = -\operatorname{Re}\{f'(q)\} \tag{4.10.21}$$

for every differential of class **M**.

By (4.3.5)

(4.10.22) $$P(L_{\mathbf{M}}^{(1)}(p, q), K_\mu) = Z_\mu^{(1)\prime}(q).$$

Therefore by (4.3.4)

(4.10.23) $$P(L_{\mathbf{M}}(p, q, \tilde{q}), K_\mu) = Z_\mu^{(1)\prime}(q) + Z_\mu^{(1)\prime}(\tilde{q}).$$

4.11. CONSTRUCTION OF THE BILINEAR DIFFERENTIAL FOR THE CLASS F

Suppose first that the surface is orientable. Let

(4.11.1) $$L_{\mathbf{F}}(p, q) = L_{\mathbf{M}}(p, q) + \sum_{\mu, \nu=1}^{n} \gamma_{\mu\nu} Z'_{i_\mu}(p) Z'_{i_\nu}(q)$$

where the coefficients are to be determined such that

(4.11.2) $$P(L_{\mathbf{F}}(p, q), K_{i_\varrho}) = 0, \qquad \varrho = 1, 2, \cdots, n.$$

By (4.10.15) and (4.3.10) the period conditions (4.11.2) give

$$Z'_{i_\varrho}(q) - \sum_{\mu, \nu=1}^{n} \gamma_{\mu\nu} \Gamma_{i_\mu i_\varrho} Z'_{i_\nu}(q) = 0, \qquad \varrho = 1, 2, \cdots, n,$$

and, because of the linear independence of the $Z'_{i_\nu}(q)$, we therefore have

(4.11.3) $$\sum_{\mu=1}^{n} \gamma_{\mu\nu} \Gamma_{i_\mu i_\varrho} = \delta_{\nu\varrho}, \qquad \nu, \varrho = 1, 2, \cdots, n.$$

Thus $L_{\mathbf{F}}(p, q)$ exists if and only if the matrix $\| \Gamma_{i_\mu i_\varrho} \|$ is non-singular, and then

(4.11.4) $$\| \gamma_{\mu\nu} \| = \| \bar{\Gamma}_{i_\mu i_\nu} \|^{-1}.$$

We observe that

(4.11.5) $$\gamma_{\mu\nu} = (\gamma_{\nu\mu})^{-},$$

which implies Hermitian symmetry for $L_{\mathbf{F}}(p, \tilde{q})$.

If it exists, the expression $L_{\mathbf{F}}(p, q)$ has the desired properties. In fact, if $df \in \mathbf{F}$, then

(4.11.6) $$(df, L_{\mathbf{F}}(p, q)) = (df, L_{\mathbf{F}}(q, p)) = (df, L_{\mathbf{M}}(p, q)) = 0$$

by (4.10.8) since, by hypothesis,

$$(df, dZ_{i_\mu}) = 0, \qquad \mu = 1, 2, \cdots, n.$$

Similarly

$$(4.11.7) \qquad (df, L_{\mathbf{F}}(p, \tilde{q})) = -f'(q).$$

The bilinear differential $L_{\mathbf{F}}$ always exists if there is a boundary, for in that case the matrix $\| \Gamma_{i_\mu i_\varrho} \|$ is non-singular. But by (4.11.5) we shall have

$$(4.11.8) \qquad L_{\mathbf{F}}(p, q) = L_{\mathbf{F}}(q, p)$$

if and only if the class **F** is symmetric (see Section 4.8).

If the Riemann surface is closed, the situation is quite different for $\| \Gamma_{i_\mu i_\varrho} \|$ may be singular. However, the bilinear differentials for the symmetric classes always exist since $\| \operatorname{Re} \Gamma_{\mu\nu} \|$ is non-singular. Of the symmetric classes on the closed surface, two deserve to be singled out, namely the classes **F** $(1, 3, \cdots, G-1)$ and **F** $(2, 4, \cdots, G)$, $G = 2h$ (where h is the genus). It is sufficient to consider **F** $(1, 3, \cdots, G-1)$. Since, for fixed q, $L_{\mathbf{F}}(p, \tilde{q})$ is a differential of the first kind whose periods P_μ around the cycles $K_{2\mu-1}$ all vanish, we conclude from (3.2.3) that $L_{\mathbf{F}}(p, \tilde{q}) \equiv 0$. Thus by (4.11.1) (with q replaced by $\tilde{q}$)

$$(4.11.9) \qquad L_{\mathbf{M}}(p, \tilde{q}) = -\sum_{\mu, \nu=1}^{h} \gamma_{\mu\nu} Z'_{2\mu-1}(p) Z'_{2\nu-1}(\tilde{q})$$

where

$$(4.11.10) \qquad \| \gamma_{\mu\nu} \| = \| \Gamma_{2\mu-1, 2\nu-1} \|^{-1}.$$

On the other hand, $L_{\mathbf{F}}(p, q)$ for fixed q is a differential of the second kind all of whose periods P_μ vanish, and for p near q we have, provided p and q are expressed in terms of the same local coordinate z, $z = z(p)$, $z(q) = 0$:

$$(4.11.11) \qquad L_{\mathbf{F}}(p, q) = \frac{1}{\pi z^2} + \text{regular terms}.$$

Thus

$$(4.11.12) \qquad L_{\mathbf{F}}(p, q) = -\frac{1}{\pi} \frac{dt_q(p)}{dp},$$

and by (4.11.1)

$$(4.11.13) \quad L_{\mathbf{M}}(p, q) = -\frac{1}{\pi} \frac{dt_q(p)}{dp} - \sum_{\mu, \nu=1}^{h} \gamma_{\mu\nu} Z'_{2\mu-1}(p) Z'_{2\nu-1}(q).$$

Computing periods in (4.11.9) and (4.11.13) around a cycle $K_{2\varrho}$

we find using (3.4.3)′ and (4.3.10) that

$$(4.11.14) \qquad Z'_{2\varrho}(\tilde{q}) = \sum_{\mu,\nu=1}^{h} \gamma_{\mu\nu}\Gamma_{2\mu-1,2\varrho}Z'_{2\nu-1}(\tilde{q}),$$

$$(4.11.15) \qquad 2iw'_\varrho(q) = \sum_{\mu,\nu=1}^{h} \gamma_{\mu\nu}\Gamma_{2\mu-1,\,2\varrho}Z'_{2\nu-1}(q) - Z'_{2\varrho}(q).$$

If the double were connected, formula (4.11.13) could be obtained from (4.11.9) by analytic continuation, and similarly (4.11.15) could be obtained from (4.11.14). The fact that the double is disconnected makes the formulas with $\tilde{q}$ in place of q quite different.

Now suppose that the surface $\mathfrak{M}$ is non-orientable and set

$$(4.11.16) \quad L_{\mathbf{F}}(p, q, \tilde{q}) = L_{\mathbf{M}}(p, q, \tilde{q}) + \sum_{\mu,\nu=1}^{n} \gamma_{\mu\nu}Z'_{i_\mu}(p)\,\mathrm{Re}\{Z'_{i_\nu}(q)\}$$

where the real coefficients $\gamma_{\mu\nu}$ are to be determined by the conditions

$$(4.11.17) \quad \mathrm{Re}\{P(L_{\mathbf{F}}(p, q, \tilde{q}), K_{i_\varrho})\} = 0, \qquad \varrho = 1, 2, \cdots, n.$$

By (4.3.14) and (4.10.23)

$$\mathrm{Re}\{Z^{(1)\prime}_{i_\varrho}(q) + Z^{(1)\prime}_{i_\varrho}(\tilde{q})\} - \sum_{\mu,\nu=1}^{n} \gamma_{\mu\nu}\Gamma_{i_\mu i_\varrho}\,\mathrm{Re}\{Z'_{i_\nu}(q)\} = 0, \quad \varrho = 1, \cdots, n.$$

That is, by (4.3.4),

$$(4.11.18) \quad \mathrm{Re}\{Z'_{i_\varrho}(q)\} - \sum_{\mu,\nu=1}^{n} \gamma_{\mu\nu}\Gamma_{i_\mu i_\varrho}\,\mathrm{Re}\{Z'_{i_\nu}(q)\} = 0, \qquad \varrho = 1, \cdots, n.$$

Because of the linear independence, we have

$$\sum_{\mu=1}^{n} \gamma_{\mu\nu}\Gamma_{i_\mu i_\varrho} = \delta_{\nu\varrho}, \qquad \nu,\ \varrho = 1, 2, \cdots, n,$$

and therefore, since $\|\Gamma_{i_\mu i_\varrho}\|$ is always non-singular,

$$(4.11.19) \qquad \|\gamma_{\mu\nu}\| = \|\Gamma_{i_\mu i_\nu}\|^{-1}.$$

In view of the fact that the bilinear differential for a non-orientable $\mathfrak{M}$ is a simple linear combination of the differentials for its orientable covering, we shall henceforth assume that the surface $\mathfrak{M}$ is orientable when the contrary is not explicitly stated.

4.12. Properties of the Bilinear Differentials

If the Riemann surface has a boundary, let p be a boundary point. Then $p = \tilde{p}$ and

$$L_{\mathbf{F}}(p, q)dzd\zeta = L_{\mathbf{F}}(\tilde{p}, q)d\bar{z}d\zeta = L_{\mathbf{F}}(p, q)d\bar{z}d\zeta. \tag{4.12.1}$$

If q is on the boundary, we have in the same way

$$L_{\mathbf{F}}(p, q)dzd\zeta = L_{\mathbf{F}}(p, \tilde{q})dzd\bar{\zeta} = L_{\mathbf{F}}(p, q)dzd\bar{\zeta}. \tag{4.12.1$'$}$$

Formulas (4.12.1) and (4.12.1)′ are consequences of the fact that $L_{\mathbf{F}}(p, q)dzd\zeta$ is invariant to changes of the uniformizer and, at a point of the boundary, either the variable or its complex conjugate can be used.

Using conjugate uniformizers at conjugate places, we define

$$\tilde{L}_{\mathbf{F}}(p, q) = (L_{\mathbf{F}}(\tilde{p}, \tilde{q}))^{-}.$$

Since

$$(L_{\mathbf{F}}(p, \tilde{q}))^{-} = L_{\mathbf{F}}(q, \tilde{p})$$

we have, replacing p by $\tilde{p}$,

$$(L_{\mathbf{F}}(\tilde{p}, \tilde{q}))^{-} = L_{\mathbf{F}}(q, p);$$

that is

$$\tilde{L}_{\mathbf{F}}(p, q) = L_{\mathbf{F}}(q, p).$$

If, in particular, the class $\mathbf{F}$ is symmetric, then $\tilde{L}_{\mathbf{F}}(p, q)$ is the same as $L_{\mathbf{F}}(p, q)$, and in this case $L_{\mathbf{F}}(p, q)$ is a bilinear differential of $\mathfrak{M}$ which is real when both p and q are on the boundary.

Since the regular differential is an element of $\mathbf{F}$, we have by (4.11.6) and (4.11.7),

$$(L_{\mathbf{F}}(p, \tilde{r}),\ L_{\mathbf{F}}(p, q)) = 0$$

and

$$(L_{\mathbf{F}}(p, \tilde{r}),\ L_{\mathbf{F}}(p, \tilde{q})) = -L_{\mathbf{F}}(q, \tilde{r}).$$

We observe that these integrals are not obtained from each other by replacing q by $\tilde{q}$, even if $\mathfrak{M}$ has a boundary. The reason is that the integrals are discontinuous because of the singular behavior of the integrands at the boundary.

In particular,

$$N(L_{\mathbf{F}}(p, \tilde{q})) = -L_{\mathbf{F}}(q, \tilde{q}) \geqq 0. \tag{4.12.2}$$

We now consider the scalar product of two singular differentials.

Let q and r be distinct points in the interior of $\mathfrak{M}$. We shall prove that

$$L_{\mathbf{F}}(\bar{r}, q) = (L_{\mathbf{F}}(p, q), L_{\mathbf{F}}(p, r)). \tag{4.12.3}$$

The scalar product on the right is to be interpreted in the sense of (4.9.8).

We prove (4.12.3) first for the case in which $\mathbf{F} = \mathbf{M}$. Let ζ be a uniformizer at q, η a uniformizer at r, and let $\mathfrak{M}_q$, $\mathfrak{M}_r$ be the uniformizer circles $|\zeta| \leqq a$, $|\eta| \leqq a$ at q and r respectively. Denoting the boundaries of $\mathfrak{M}_q$, $\mathfrak{M}_r$ by C_q, C_r respectively, we have

$$(L_{\mathbf{M}}(p, q), L_{\mathbf{M}}(p, r)) = -\frac{2}{\pi}\int\limits_{\mathfrak{M}-\mathfrak{M}_q-\mathfrak{M}_r} \frac{\partial^2 G(p, q)}{\partial p \partial q} (L_{\mathbf{M}}(p, r))^- dA_z + o(1)$$

$$= \frac{1}{\pi i}\int\limits_C \frac{\partial G(p, q)}{\partial q}(L_{\mathbf{M}}(p, r))^- d\bar{z} - \frac{1}{\pi i}\int\limits_{C_q} \frac{\partial G(p, q)}{\partial q}(L_{\mathbf{M}}(p, r))^- d\bar{\zeta}$$

$$-\frac{1}{\pi i}\int\limits_{C_r} \frac{\partial G(p, q)}{\partial q}(L_{\mathbf{M}}(p, r))^- d\bar{\eta} + o(1). \tag{4.12.4}$$

Since $\partial G(p, q)/\partial q$ vanishes when p is on the boundary C of $\mathfrak{M}$, we have only to consider the remaining two integrals. For p on C_q,

$$\frac{\partial G(p, q)}{\partial q} = \frac{1}{2\zeta} + \text{regular terms}, \tag{4.12.5}$$

$$(L_{\mathbf{M}}(p, r))^- = \bar{b}_0 + \bar{b}_1\bar{\zeta} + \cdots. \tag{4.12.6}$$

Therefore

$$\frac{1}{\pi i}\int\limits_{C_q} \frac{\partial G(p, q)}{\partial q}(L_{\mathbf{M}}(p, r))^- d\bar{\zeta} = o(1) \tag{4.12.7}$$

as a tends to zero. Finally

$$\frac{1}{\pi i}\int\limits_{C_r} \frac{\partial G(p, q)}{\partial q}(L_{\mathbf{M}}(p, r))^- d\bar{\eta} \tag{4.12.8}$$

$$= \frac{1}{\pi i}\int\limits_{C_r}\left\{\frac{\partial G(r, q)}{\partial q} + \frac{\partial^2 G(r, q)}{\partial r \partial q}\eta + \frac{\partial^2 G(r, q)}{\partial \bar{r}\partial q}\bar{\eta} + \cdots\right\}\cdot$$

$$\left\{\frac{1}{\pi\bar{\eta}^2} + \bar{c}_0 + \bar{c}_1\bar{\eta} + \cdots\right\} d\bar{\eta} = -\frac{2}{\pi}\frac{\partial^2 G(r, q)}{\partial \bar{r}\partial q} + o(1)$$

$$= -L_{\mathbf{M}}(\bar{r}, q) + o(1)$$

as a approaches zero. Substituting from (4.12.7) and (4.12.8) into (4.12.4), we obtain (4.12.3) in the case $\mathbf{F} = \mathbf{M}$.

Consider now the general case. Writing

$$L_{\mathbf{F}}(p, q) = L_{\mathbf{M}}(p, q) + \Lambda_{\mathbf{F}}(p, q) \tag{4.12.9}$$

where $\Lambda_{\mathbf{F}}(p, q)$ is the bilinear sum of differentials occurring in (4.11.1), we have

$$(L_{\mathbf{F}}(p, q), L_{\mathbf{F}}(p, r)) = (L_{\mathbf{M}}(p, q), L_{\mathbf{M}}(p, r)) + (\Lambda_{\mathbf{F}}(p, q), \Lambda_{\mathbf{F}}(p, r))$$

since the cross terms vanish by (4.11.6). By the case already proved, the first term is simply $L_{\mathbf{M}}(\tilde{r}, q)$. For the second, we have

$$\begin{aligned}(\Lambda_{\mathbf{F}}(p, q), \Lambda_{\mathbf{F}}(p, r)) &= \sum_{\mu, \nu, \sigma, \varrho} \gamma_{\mu\nu}\gamma_{\sigma\varrho} Z'_{i_\nu}(q) Z'_{i_\sigma}(\tilde{r}) \cdot (Z'_{i_\mu}(p), Z'_{i_\varrho}(p)) \\ &= \sum_{\mu, \nu, \sigma, \varrho} \gamma_{\mu\nu} \Gamma_{i_\mu i_\varrho} \gamma_{\sigma\varrho} Z'_{i_\nu}(q) Z'_{i_\sigma}(\tilde{r}) = \sum_{\nu, \sigma} \gamma_{\sigma\nu} Z'_{i_\nu}(q) Z'_{i_\sigma}(\tilde{r}) \\ &= \Lambda_{\mathbf{F}}(\tilde{r}, q).\end{aligned} \tag{4.12.10}$$

Here we have used the fact that the matrices $\| \gamma_{\mu\nu} \|$ and $\| \bar{\Gamma}_{i_\mu i_\nu} \|$ are inverse.

We now consider the definition of a norm for differentials dg which are regular on $\mathfrak{M}$ except for a double pole without residue at a fixed point q of $\mathfrak{M}$.

The scalar product in (4.12.3) is not defined by (4.9.8) if $r = q$, and we now define

$$N(L_{\mathbf{M}}(p, q)) = \lim_{r \to q} (L_{\mathbf{M}}(p, q), L_{\mathbf{M}}(p, r)) = L_{\mathbf{M}}(\tilde{q}, q). \tag{4.12.11}$$

We note that $L_{\mathbf{M}}(\tilde{q}, q) = L_{\mathbf{M}}(q, \tilde{q}) \leqq 0$ by (4.12.2). That is, the norm defined by (4.12.11) is real. However, the extended norm may be negative in the case of singular differentials.

For any $df \in \mathbf{M}$, we have

$$(df + L_{\mathbf{M}}(p, q), df + L_{\mathbf{M}}(p, r)) = (df, df) + (L_{\mathbf{M}}(p, q), L_{\mathbf{M}}(p, r)) \tag{4.12.12}$$

using (4.10.8), so we define

$$N(df + L_{\mathbf{M}}(p, q)) = N(df) + N(L_{\mathbf{M}}(p, q)). \tag{4.12.13}$$

If dg is regular on $\mathfrak{M}$ except at q, where it has the same singularity as $L_{\mathbf{M}}(p, q)$, we may take $df = dg - L_{\mathbf{M}}(p, q)$. Then (4.12.13) gives

$$N(dg) = N(dg - L_{\mathbf{M}}(p, q)) + N(L_{\mathbf{M}}(p, q)). \tag{4.12.14}$$

In particular, we have defined

$$(4.12.15)\quad N(L_{\mathbf{F}}(p,q)) = N(L_{\mathbf{F}}(p,q) - L_{\mathbf{M}}(p,q)) + N(L_{\mathbf{M}}(p,q)).$$

On the other hand, the identity (4.12.3) implies that we should have

$$(4.12.15)'\quad N(L_{\mathbf{F}}(p,q)) = \lim_{r \to q} (L_{\mathbf{F}}(p,q),\ L_{\mathbf{F}}(p,r)) = L_{\mathbf{F}}(\tilde{q},q).$$

To verify that the result (4.12.15)′ agrees with (4.12.15), we note that (4.12.10), in which we may take $r = q$, may be rewritten as

$$(4.12.16)\quad N(L_{\mathbf{F}} - L_{\mathbf{M}}) = L_{\mathbf{F}}(\tilde{q},q) - L_{\mathbf{M}}(\tilde{q},q) = L_{\mathbf{F}}(\tilde{q},q) - N(L_{\mathbf{M}})$$

using the definition of $\Lambda_{\mathbf{F}}$ in (4.12.9). Similarly, for $df \in \mathbf{F}$ we have

$$(4.12.12)'\quad (df + L_{\mathbf{F}}(p,q),\ df + L_{\mathbf{F}}(p,r)) = (df, df) + (L_{\mathbf{F}}(p,q), L_{\mathbf{F}}(p,r)),$$

so we should have

$$(4.12.14)'\quad N(dg) = N(dg - L_{\mathbf{F}}(p,q)) + N(L_{\mathbf{F}}(p,q))$$

for singular differentials satisfying the period conditions for the class $\mathbf{F}$. Again the definitions are consistent. In fact, for the regular differential $dg - L_{\mathbf{M}}(p,q)$ we have

$$\begin{aligned} N(dg - L_{\mathbf{M}}) &= N(dg - L_{\mathbf{F}} + L_{\mathbf{F}} - L_{\mathbf{M}}) = N(dg - L_{\mathbf{F}}) + N(L_{\mathbf{F}} - L_{\mathbf{M}}) \\ &= N(dg - L_{\mathbf{F}}) + N(L_{\mathbf{F}}) - N(L_{\mathbf{M}}) \end{aligned}$$

using the orthogonality properties of $L_{\mathbf{F}}$ and $L_{\mathbf{M}}$, and (4.12.16). Further, since $N(dg - L_{\mathbf{F}}) \geqq 0$, we derive from (4.12.14)′ an important characterization of the singular bilinear differential for a class $\mathbf{F}$: the singular bilinear differential $L_{\mathbf{F}}(p,q)$ minimizes the norm among all differentials dg having the same singularity at q and satisfying the period conditions for the class $\mathbf{F}$. Further, the norms of all normalized singular differentials are bounded from below by $L_{\mathbf{M}}(\tilde{q},q)$, by (4.12.14).

If $df \in \mathbf{S}$ is regular on the boundary C of $\mathfrak{M}$, the scalar product on the left in (4.12.12)′, with $\mathbf{F} = \mathbf{S}$, may be integrated by parts. If C_q and C_r are uniformizer circles of radius a about q and r, respectively, and if $\Psi_{\mathbf{S}}(p,q)$ is the integral of $L_{\mathbf{S}}(p,q)$:

$$\frac{\partial \Psi_{\mathbf{S}}(p,q)}{\partial p} = L_{\mathbf{S}}(p,q),$$

then

$$(df + L_{\mathbf{S}}(p, q),\ df + L_{\mathbf{S}}(p, r))$$

$$= -\frac{1}{2i}\int_C (f(p) + \Psi_{\mathbf{S}}(p, q))\,(f'(p) + L_{\mathbf{S}}(p, r))^- d\bar{z}$$

$$+\frac{1}{2i}\int_{C_q} (f(p) + \Psi_{\mathbf{S}}(p, q))\,(f'(p) + L_{\mathbf{S}}(p, r))^- d\bar{\zeta}$$

$$(4.12.17)\qquad +\frac{1}{2i}\int_{C_r} (f(p) + \Psi_{\mathbf{S}}(p, q))\,(f'(p) + L_{\mathbf{S}}(p, r))^- d\bar{\eta} + o(1)$$

$$= -\frac{1}{2i}\int_C (f(p) + \Psi_{\mathbf{S}}(p, q))\,(f'(p) + L_{\mathbf{S}}(p, r))^- d\bar{z},$$

letting a tend to zero. Letting r tend to q and choosing $df = dg - L_{\mathbf{S}}(p, q)$, we obtain the simple formula

$$(4.12.18)\qquad N(dg) = -\frac{1}{2i}\int_C g\, d\bar{g},$$

valid for normalized singular differentials dg which have vanishing periods and are regular on the boundary C of $\mathfrak{M}$.

The reproducing kernel $L_{\mathbf{F}}(p, \tilde{q})$ also gives the solution of certain minimum problems. For arbitrary $df \in \mathbf{F}$, we have by (4.11.7), (2.3.3), and (4.12.2)

$$(4.12.19)\qquad \left|\left(\frac{df}{d\zeta}\right)_q\right|^2 \leqq |(df, L_{\mathbf{F}}(p, \tilde{q}))|^2 \leqq N(df)N(L_{\mathbf{F}}(p, \tilde{q})) = -L_{\mathbf{F}}(q, \tilde{q})N(df)$$

and there is equality in (4.12.19) if and only if $f'(p)$ is proportional to $L_{\mathbf{F}}(p, \tilde{q})$. Thus the best value for $k(p_0)$ in (2.3.16)′ is $-L_{\mathbf{M}}(p_0, \tilde{p}_0)$. Analogously, the best value when df is restricted to be an element of a class $\mathbf{F}$ is $-L_{\mathbf{F}}(p_0, \tilde{p}_0)$.

When $L_{\mathbf{F}}(p, \tilde{q})$ is not identically zero, it can be defined in terms of a minimum problem suggested by (4.12.19). Let df_0 be the differential which minimizes $N(df)$ over all $df \in \mathbf{F}$ with $df(q) \neq 0$, normalized in terms of a particular uniformizer at q so that $f'(q) = 1$.

The minimizing property implies that df_0 is orthogonal to $dh(p) = df(p) - f'(q) \, df_0(p)$, for arbitrary $df \in \mathbf{F}$, since $dh(q) = 0$. Thus $-df_0(p)/N(df_0)$ has the reproducing property and must coincide with $L_{\mathbf{F}}(p, \tilde{q})dz$. In particular, the minimizing differential df_0 may be expressed in the form

$$(4.12.20) \qquad df_0 = \frac{L_{\mathbf{F}}(p, \tilde{q})}{L_{\mathbf{F}}(q, \tilde{q})} \, dp,$$

using (4.12.2).

Let $\mathbf{F}_1$ and $\mathbf{F}_2$ be two classes of differentials on $\mathfrak{M}$ such that

$$(4.12.21) \qquad\qquad \mathbf{F}_1 \subseteq \mathbf{F}_2$$

and the bilinear differential exists for each class. In particular, $L_{\mathbf{F}_1}(p, \tilde{q}) \in \mathbf{F}_2$.

By (4.12.19) and (4.12.2) we have

$$(4.12.22) \qquad\qquad 0 \geq L_{\mathbf{F}_1}(q, \tilde{q}) \geq L_{\mathbf{F}_2}(q, \tilde{q}).$$

Since $\mathbf{S}$ is a subclass of every $\mathbf{F}$ while $\mathbf{M}$ contains every class $\mathbf{F}$, we have

$$(4.12.22)' \qquad 0 \geq L_{\mathbf{S}}(q, \tilde{q}) \geq L_{\mathbf{F}}(q, \tilde{q}) \geq L_{\mathbf{M}}(q, \tilde{q}).$$

On the other hand, $L_{\mathbf{F}_2}(p, q)$ minimizes the norm among all differentials dg having the same singularity at q and satisfying the period relations for $\mathbf{F}_2$. Since $L_{\mathbf{F}_1}(p, q)$ is among the dg we have

$$(4.12.23) \qquad\qquad L_{\mathbf{F}_2}(\tilde{q}, q) \leq L_{\mathbf{F}_1}(\tilde{q}, q)$$

and

$$(4.12.23)' \qquad L_{\mathbf{M}}(\tilde{q}, q) \leq L_{\mathbf{F}}(\tilde{q}, q) \leq L_{\mathbf{S}}(\tilde{q}, q).$$

The quantities $L_{\mathbf{F}}(q, \tilde{q})$ and $L_{\mathbf{F}}(\tilde{q}, q)$ need not be equal unless $\mathbf{F}$ is a symmetric class.

These inequalities may be generalized at once to the Hermitian quadratic forms

$$(4.12.24) \qquad \sum_{\mu, \nu=1}^{\varrho} L_{\mathbf{F}}(q_\mu, \tilde{q}_\nu) x_\mu \bar{x}_\nu \quad \text{and} \quad \sum_{\mu, \nu=1}^{\varrho} L_{\mathbf{F}}(\tilde{q}_\nu, q_\mu) x_\mu \bar{x}_\nu.$$

For example, we have from (4.12.23)$'$

$$(4.12.25) \quad \sum_{\mu, \nu=1}^{\varrho} L_{\mathbf{M}}(\tilde{q}_\nu, q_\mu) x_\mu \bar{x}_\nu \leq \sum_{\mu, \nu=1}^{\varrho} L_{\mathbf{F}}(\tilde{q}_\nu, q_\mu) x_\mu \bar{x}_\nu \leq \sum_{\mu, \nu=1}^{\varrho} L_{\mathbf{S}}(\tilde{q}_\nu, q_\mu) x_\mu \bar{x}_\nu.$$

Now, by (4.12.2),

$$(4.12.26) \qquad -N\Big(\sum_{\nu=1}^{\varrho} \bar{x}_\nu L_{\mathbf{F}}(p, \tilde{q}_\nu)\Big) = \sum_{\mu,\nu=1}^{\varrho} L_{\mathbf{F}}(q_\mu, \tilde{q}_\nu) x_\mu \bar{x}_\nu \leqq 0.$$

Thus the first Hermitian form (4.12.24) is non-positive and, if $\mathbf{F}$ is a symmetric class, the same is true of the second.

When $\mathfrak{M}$ is orientable and has a boundary, the bilinear differentials for the class $\mathbf{S}$ may be given a geometric interpretation. In fact, the integral f of each differential $df \in \mathbf{S}$ is single-valued on $\mathfrak{M}$ and, unless df vanishes identically, gives a mapping of $\mathfrak{M}$ onto a Riemann surface spread over the complex plane. Conversely, each such mapping f corresponds to a differential df in $\mathbf{S}$. In particular, since such mappings exist, the class $\mathbf{S}$ contains elements df which do not vanish identically and $L_{\mathbf{S}}(p, \tilde{q})$ is non-trivial. For each mapping the internal area of the image domain is given by $N(df)$. Let q be a fixed point interior to $\mathfrak{M}$. Then if we consider the set of all mappings with $df(q) \neq 0$, normalized in terms of a particular uniformizer at q so that $f'(q) = 1$, the mapping which minimizes the internal area is given by

$$df = \frac{L_{\mathbf{S}}(p, \tilde{q})}{L_{\mathbf{S}}(q, \tilde{q})}\, dp$$

and the value of the minimum area is

$$(4.12.27) \qquad A = -\frac{1}{L_{\mathbf{S}}(q, \tilde{q})}.$$

The singular differentials dg whose periods vanish also correspond to mapping functions g. If $dg = du + idv$ is regular on the boundary C of $\mathfrak{M}$,

$$(4.12.28) \qquad \begin{aligned} N(dg) &= -\frac{1}{2i}\int_C g\,d\bar{g} = -\frac{1}{4i}\int_C d\,|g|^2 + \frac{1}{2}\int_C (udv - vdu) \\ &= \frac{1}{2}\int_C (udv - vdu) \end{aligned}$$

since g is single-valued.

Let Σ_ν be the image of C_ν by g, and let Σ_ν be oriented such that it is described positively by $w = g(p)$ as p describes C_ν in the negative sense. We define $\Omega_\nu(w_0)$ to be the order of Σ_ν with respect to the point w_0. By (4.12.28)

$$N(dg) = -\sum_{\nu=1}^{m} \int \Omega_\nu(w_0) dA_{w_0} \tag{4.12.29}$$

where each integration in the sum is over the whole w_0-plane. We therefore say that $-N(dg)$ is the area external to the image of $\mathfrak{M}$ by dg.

In the class of all mappings dg, normalized to have the same singularity at q as $L_{\mathbf{S}}(p, q)\, dp$, the norm is minimized by $L_{\mathbf{S}}(p, q)$. That is, the mapping which gives the maximum external area is given by

$$dg = L_{\mathbf{S}}(p, q)\, dp,$$

and the maximum external area is

$$E = -L_{\mathbf{S}}(\tilde{q}, q), \tag{4.12.30}$$

using (4.12.15)'.

Multiplying (4.12.27) and (4.12.30) we obtain

$$AE = \frac{L_{\mathbf{S}}(\tilde{q}, q)}{L_{\mathbf{S}}(q, \tilde{q})} \tag{4.12.31}$$

which is a generalization of a known identity for multiply-connected domains of the plane (see [6a], [14a]). In fact, the class $\mathbf{S}$ is symmetric whenever the surface $\mathfrak{M}$ is of genus zero, in which case $L_{\mathbf{S}}(\tilde{q}, q) = L_{\mathbf{S}}(q, \tilde{q})$ and $AE = 1$. We remark that, in the case of genus zero, the function g corresponding to the maximum external area is schlicht and maps $\mathfrak{M}$ onto a subdomain of the sphere.

As another application of the bilinear differentials we give a generalization of the Poisson formula which permits the representation of an analytic function inside a circle by means of the boundary values of its real part. Let $\mathfrak{M}$ be an orientable Riemann surface with boundary C and q an interior point of $\mathfrak{M}$. If $f(p)$ is single-valued on $\mathfrak{M}$ and regular on C, then we may integrate by parts in the regular scalar product in (4.11.7), with $\mathbf{F} = \mathbf{S}$, to obtain

$$\begin{aligned} f'(q) &= -(df, L_{\mathbf{S}}(p, \tilde{q})) = \frac{1}{2i} \int_{\mathfrak{M}} f'(p)\, (L_{\mathbf{S}}(p, \tilde{q}))^{-} dz\, d\bar{z} \\ &= \frac{1}{2i} \int_{C} f(p)\, L_{\mathbf{S}}(q, \bar{p})\, d\bar{z}. \end{aligned} \tag{4.12.32}$$

Similarly, if $\mathfrak{M}'$ is the disc $|\zeta| < a$, where ζ is a uniformizer at q, $\zeta = \zeta(p)$, $\zeta(q) = 0$, the formula (4.11.6) yields

$$0 = (L_{\mathbf{S}}(q, p), df) = \frac{1}{2i} \int\limits_{\mathfrak{M}-\mathfrak{M}'} L_{\mathbf{S}}(q, p)\ (f'(p))^{-} d\bar{z} dz + o(1)$$

$$(4.12.33) \qquad = \frac{1}{2i} \int\limits_{C} L_{\mathbf{S}}(q, p)(f(p))^{-} dz - \frac{1}{2i} \int\limits_{C'} L_{\mathbf{S}}(q, p)(f(p))^{-} d\zeta + o(1)$$

$$= \frac{1}{2i} \int\limits_{C} L_{\mathbf{S}}(q, p)(f(p))^{-} dz,$$

letting a tend to zero. Since $L_{\mathbf{S}}(q, \tilde{p}) d\bar{z} = L_{\mathbf{S}}(q, p) dz$ on C, we obtain by adding (4.12.32) and (4.12.33)

$$(4.12.34) \qquad f'(q) = \frac{1}{i} \int\limits_{C} L_{\mathbf{S}}(q, p)\ \operatorname{Re}\{f(p)\}\, dz.$$

This is a generalization of Poisson's formula, for the case of an arbitrary surface with boundary.

4.13. Approximation of Differentials

In the following chapter certain formulas will be established for differentials regular on the boundary, and their validity for any differential with finite norm will require an approximation argument which we now supply. In Section 4.10 we approximated the bilinear differential by the corresponding differentials of domains $\mathfrak{M}_\mu$ lying in the interior of the given surface. Now, given any differential df of class **F**, we define a sequence of differentials df_μ of **F** each of which is analytic on the boundary of $\mathfrak{M}$ and such that $N(df - df_\mu)$ tends to zero.

Given a differential df of class **F**, let $\mathfrak{M}_\mu$ be a compact subdomain lying in the interior of $\mathfrak{M}$ which converges to $\mathfrak{M}$ as μ tends to infinity. We do not need to suppose that $\mathfrak{M}_\mu$ is a finite Riemann surface. Define

$$(4.13.1) \qquad f'_\mu(q) = -(df,\ L_{\mathbf{F}}(p, \tilde{q}))_{\mathfrak{M}_\mu}.$$

It is clear that $df_\mu(q)$ is regular up to and including the boundary of $\mathfrak{M}$ and that it belongs to **F**. We have

$$| f'_\mu(q) - f'(q) |^2 = \int_{\mathfrak{M}-\mathfrak{M}_\mu} \int_{\mathfrak{M}-\mathfrak{M}_\mu} L_{\mathbf{F}}(p_1, \bar{q})(L_{\mathbf{F}}(p_2, \bar{q}))^-(f'(p_1))^- f'(p_2) dA_1 dA_2,$$

and therefore

(4.13.2)

$$\int_{\mathfrak{M}} | f'_\mu(q) - f'(q) |^2 dA = -\int_{\mathfrak{M}-\mathfrak{M}_\mu} \int_{\mathfrak{M}-\mathfrak{M}_\mu} L_{\mathbf{F}}(p_1, \bar{p}_2)(f'(p_1))^- f'(p_2) dA_1 dA_2$$

since

$$\int_{\mathfrak{M}} L_{\mathbf{F}}(p_1, \bar{q})(L_{\mathbf{F}}(p_2, \bar{q}))^- dA = \int_{\mathfrak{M}} L_{\mathbf{F}}(q, \bar{p}_2)(L_{\mathbf{F}}(q, \bar{p}_1))^- dA = -L_{\mathbf{F}}(p_1, \bar{p}_2).$$

In view of the fact that

$$-\int_{\mathfrak{M}} \int_{\mathfrak{M}} L_{\mathbf{F}}(p_1, \bar{p}_2)(f'(p_1))^- f'(p_2) dA_1 dA_2 = \int_{\mathfrak{M}} | f'(p_1)|^2 dA_1,$$

we see that the right side of (4.13.2) tends to zero as μ approaches infinity; that is

$$N(df - df_\mu) \to 0. \tag{4.13.3}$$

4.14. A SPECIAL COMPLETE ORTHONORMAL SYSTEM

In the sequel we require a complete orthonormal system of differentials in **F** with special properties, and we now construct this system by a method which closely parallels that applied in the case of plane domains ([2a]).

Assume that $\mathfrak{M}$ is orientable, let q be a point of $\mathfrak{M}$, ζ a uniformizer at q, and let $\mathbf{F}_n(q) = \mathbf{F}_n(q, \zeta)$ be the subclass of **F** which is composed of differentials df which satisfy the conditions

$$f^{(\nu)}(q) = 0, \quad \nu = 1, 2, \cdots, n-1; \quad f^{(n)}(q) = 1 \tag{4.14.1}$$

where

$$f^{(\nu)}(q) = \left.\frac{d^\nu f}{d\zeta^\nu}\right|_q.$$

Let us consider first the case that $\mathfrak{M}$ possesses a boundary. In this case, the class $\mathbf{F}_n(q)$ is non-empty for every $n \geqq 1$. In fact, let r_i, $i = 1, 2, \cdots, n$, be n arbitrary points on $\mathfrak{M}$ and consider the differential of the class $\mathbf{F}$

$$dT(p) = \sum_{i=1}^{n} x_i \, L_{\mathbf{F}}(p, \tilde{r}_i) dp. \tag{4.14.2}$$

The numbers x_i can be chosen so that dT belongs to the class $\mathbf{F}_n(q)$ provided that the determinant

$$\det \left\| \frac{\partial^{k-1}}{\partial \zeta^{k-1}} \left(L_{\mathbf{F}}(q, \tilde{r}_i) \frac{dq}{d\zeta} \right) \right\| \tag{4.14.3}$$

does not vanish. We want to show that we can choose the r_i so that this determinant is not zero. Now, if this determinant vanishes for every choice of the $r_i \in \mathfrak{M}$, we would obviously have

$$\sum_{k=0}^{n-1} a_k \frac{\partial^k}{\partial \zeta^k} \left(L_{\mathbf{F}}(q, \tilde{r}) \frac{dq}{d\zeta} \right) \equiv 0 \tag{4.14.4}$$

where the coefficients a_k do not depend on r. Continuing this identity over into $\tilde{\mathfrak{M}}$, we would then obtain

$$\sum_{k=0}^{n-1} a_k \frac{\partial^k}{\partial \zeta^k} \left(L_{\mathbf{F}}(q, r) \frac{\partial q}{\partial \zeta} \right) \equiv 0. \tag{4.14.5}$$

But this is clearly impossible since for $r = q$ the left-hand side becomes infinite.

We have thus proved that the determinant (4.14.3) does not vanish identically for arbitrary choice of the r_i and, hence, we can always construct differentials of the class $\mathbf{F}$ and of the form (4.14.2) which belong to $\mathbf{F}_n(q)$.

Thus, the class $\mathbf{F}_n(q)$ is non-empty for every $n \geqq 1$. Let d be the greatest lower bound of $N(df)$ for differentials of $\mathbf{F}_n(q)$, and let $\{df_\mu\}$ be a sequence of differentials of $\mathbf{F}_n(q)$ for which

$$\lim_{\mu \to \infty} N(df_\mu) = d.$$

The sequence $\{df_\mu\}$ plainly converges uniformly in any compact subregion to a differential $df_{(n)} \in \mathbf{F}_n(q)$. It then follows from the usual reasoning that $N(df_{(n)}) = d$. Moreover, if df is any differential of $\mathbf{F}$ satisfying

$$f^{(\nu)}(q) = 0, \qquad \nu = 1, 2, \cdots, n,$$

then by the minimizing property

$$(df, df_{(n)}) = 0. \tag{4.14.6}$$

It follows from (4.14.6) that the minimizing differential $df_{(n)}$ is unique. Write

$$d\varphi_n = \frac{df_{(n)}}{\sqrt{N(df_{(n)})}}, \quad n = 1, 2, \cdots. \tag{4.14.7}$$

Then $\{d\varphi_n\}$ is an orthonormal system, and we now show that it is complete.

Let df be an arbitrary differential of **F**, and let

$$a_\mu = (df, d\varphi_\mu), \qquad \mu = 1, 2, \cdots. \tag{4.14.8}$$

From Section 2.3 we know that the Fourier series

$$dg = \sum_{\mu=1}^{\infty} a_\mu d\varphi_\mu$$

is a differential of **F**. To show that the system is complete, we have to prove that $df = dg$. Let numbers $a_\nu^{(s)}$ be determined in such a way that the differential

$$df_s = \sum_{\nu=1}^{s-1} a_\nu^{(s)}\, d\varphi_\nu \tag{4.14.9}$$

satisfies the conditions

$$f_s^{(\mu)}(q) = f^{(\mu)}(q), \quad \mu = 1, 2, \cdots, s-1, \quad s \geqq 2. \tag{4.14.9$'$}$$

Since

$$f_s^{(\mu)}(q) = \sum_{\nu=1}^{\mu} a_\nu^{(s)} \varphi_\nu^{(\mu)}(q), \quad \varphi_\mu^{(\mu)}(q) \neq 0, \tag{4.14.10}$$

the $a_\nu^{(s)}$ are uniquely determined. By (4.14.6), (4.14.8) and (4.14.9),

$$(d(f_s - f), d\varphi_\nu) = a_\nu^{(s)} - a_\nu = 0, \quad \nu = 1, 2, \cdots, s-1. \tag{4.14.11}$$

By (4.14.9) and (4.14.9)′

$$f^{(\mu)}(q) - f_s^{(\mu)}(q) = f^{(\mu)}(q) - \sum_{\nu=1}^{s-1} a_\nu \varphi_\nu^{(\mu)}(q) = 0, \quad \mu = 1, 2, \cdots, s-1.$$

Letting s tend to infinity we obtain, since a uniformly convergent series of analytic functions can be differentiated term-by-term,

$$f^{(\mu)}(q) = g^{(\mu)}(q), \qquad \mu = 1, 2, \cdots.$$

Thus $f \equiv g$, and the existence of a complete orthonormal system is established.

In the case of a closed surface $\mathfrak{M}$, we have only finitely many independent differentials and the selection of a complete orthonormal system becomes a problem of elementary algebra and can always be solved. There are only finitely many classes $\mathbf{F}_n(q)$ which are non-empty; but in each one we may solve the preceding minimum problem for the norm and construct a normalized differential $d\varphi_n$ which is orthogonal to all differentials of the classes $\mathbf{F}_m(q)$ with $m > n$. By this procedure, we can construct a particular complete orthonormal system of differentials which is useful in various applications.

REFERENCES

1. N. Aronszajn, „La théorie des noyaux reproduisants et ses applications, I," *Proc. Cambridge Phil. Soc.* 39 (1943), 133—153.
2. S. Bergman, (a) "Partial differential equations, advanced topics," Brown University, Providence. R.I., 1941 (mimeographed). (b) "Sur les fonctions orthogonales de plusieurs variables complexes avec les applications à la théorie des fonctions analytiques," Mém. des Sci. Math., Vol. 106, Gauthier-Villars, Paris, 1947. (c) *The kernel function and conformal mapping*, Math. Surveys, No. 5, Amer. Math. Soc., New York, 1950. (d) „Ueber die Entwicklung der harmonischen Funktionen der Ebene und des Raumes nach Orthogonalfunktionen," *Math. Annalen*, 96 (1922), 237—271.
3. S. Bergman and M. Schiffer, (a) "A representation of Green's and Neumann's function in the theory of partial differential equations of second order," *Duke Math. Journ.*, 14 (1947), 609—638. (b) "Kernel functions and conformal mapping," *Compositio Math.* 8 (1951), 205—249.
4. S. Bochner, „Ueber orthogonale Systeme analytischer Funktionen," *Math. Zeit.*, 14 (1922), 180—207.
5. R. Courant and D. Hilbert, *Methoden der mathematischen Physik*, Vol. II, Springer, Berlin, 1937. (Reprint, Interscience, New York, 1943).
6. P. R. Garabedian and M. Schiffer, (a) "Identities in the theory of conformal mapping," *Trans. Amer. Math. Soc.*, 65 (1949), 187—238. (b) "On existence theorems of potential theory and conformal mapping," *Annals of Math.*, 52 (1950), 164—187.
7. H. Grötzsch, „Ueber das Parallelschlitztheorem der konformen Abbildung schlichter Bereiche", *Ber. über die Verh. der Sächs. Akad. der Wiss.*, Leipzig, Math.-Phys. Kl., 84 (1932), 15—36.
8. K. Hensel and G. Landsberg, *Theorie der algebraischen Funktionen einer Variabeln*, Teubner, Leipzig, 1902.
9. O. D. Kellogg, *Foundations of potential theory*, Springer, Berlin, 1929. (Reprint, Murray Publishing Co.).

10. O. LEHTO, „Anwendung orthogonaler Systeme auf gewisse funktionentheoretische Extremal- und Abbildungsprobleme," *Ann. Acad. Sci, Fenn, A. I*, 59 (1949).

11. J. E. LITTLEWOOD, *Theory of functions*, Oxford Univ. Press, 1944.

12. J. MARCINKIEWICZ and A. ZYGMUND, "A theorem of Lusin," *Duke Math. Jour.*, 4 (1938), 473—485.

13. Z. NEHARI, "The kernel function and canonical conformal maps," *Duke Math. Jour.*, 16 (1949), 165—178.

14. M. SCHIFFER, (a) "The span of multiply connected domains," *Duke Math. Jour.*, 10 (1943), 209—216. (b) "The kernel function of an orthonormal system," *Duke Math. Jour.*, 13 (1946), 529—540. (c) "An application of orthonormal functions in the theory of conformal mapping," *Amer. Jour. of Math.*, 70 (1948), 147—156. (d) "Various types of orthogonalization," *Duke Math. Jour.*, 17 (1950), 329—366.

15. M. SCHIFFER and D. C. SPENCER, (a) "The coefficient problem for multiply-connected domains," *Annals of Math.*, 52 (1950), 362—402. (b) "Lectures on conformal mapping and extremal methods," Princeton University, 1949—1950 (mimeographed).

16. J. L. WALSH, *Interpolation and approximation by rational functions in the complex domain*, Colloquium Publications, Vol. 20, Amer. Math. Soc., New York, 1935.

17. H. WEYL, *Die Idee der Riemannschen Fläche*, Teubner, Berlin, 1923. (Reprint, Chelsea, New York, 1947).

5. Surfaces Imbedded in a Given Surface

5.1. One Surface Imbedded in Another

Suppose that $\mathfrak{M}$ is imbedded in another finite Riemann surface $\mathfrak{R}$. In other words, suppose that $\mathfrak{M}$ is a subdomain of $\mathfrak{R}$. Our purpose is to find identities and inequalities which connect the domain functionals of $\mathfrak{M}$ (bilinear differentials, differentials of the first kind and their periods) with the corresponding functionals of $\mathfrak{R}$.

The orientable covering of a non-orientable surface is an orientable surface with one symmetry. Therefore, as we saw in Chapter 4, the theory of the functionals of a non-orientable surface is merely that of an orientable surface in which the functionals satisfy a symmetry condition. Moreover, if a non-orientable surface is imbedded in another, then its orientable covering is imbedded in that of the larger surface. Hence the imbedding of non-orientable surfaces can always be reduced to the imbedding of symmetrical orientable surfaces, and we may therefore suppose that the surfaces are orientable. Henceforth, unless it is explicitly stated to the contrary, we assume that all surfaces are orientable.

If $\mathfrak{M}$ is imbedded in $\mathfrak{R}$, we shall assume unless it is stated to the contrary that the boundary components of $\mathfrak{M}$ are analytic curves on $\mathfrak{R}$. Then the boundary uniformizers of $\mathfrak{M}$ are admissible uniformizers at the corresponding points of $\mathfrak{R}$. This hypothesis is made for convenience in proving identities; it can afterwards be removed by a limiting process. We remark that, if $\mathfrak{M}$ is closed, $\mathfrak{R}$ is closed and $\mathfrak{M}$ coincides with $\mathfrak{R}$. We exclude this trivial case; then $\mathfrak{M}$ will always have a boundary.

We shall say that the imbedding of $\mathfrak{M}$ in $\mathfrak{R}$ is essential if each boundary component of $\mathfrak{M}$ which bounds on $\mathfrak{R}$ also bounds on $\mathfrak{M}$, and that it is proper if each boundary point of $\mathfrak{M}$ is an interior point of $\mathfrak{R}$.

Let h be the genus of $\mathfrak{M}$, m the number of boundaries, and let

$$K_1, K_2, \cdots, K_G, \quad G = 2h + m - 1,$$

be a canonical homology basis for $\mathfrak{M}$ in which $K_{2h+1}, \cdots, K_{2h+m-1}$ are the boundary cycles, $K_{2h+\mu}$ being homologous to the boundary component C_μ of $\mathfrak{M}$. The cycles $K_1, \cdots, K_{2h}$ are homologously independent on $\mathfrak{R}$, but this is not necessarily true of the cycles $K_{2h+1}, \cdots, K_{2h+m-1}$, some of which may even bound on $\mathfrak{R}$. Let

$$\mathscr{K}_1, \mathscr{K}_2, \cdots, \mathscr{K}_{G_0}$$

be a canonical basis for $\mathfrak{R}$. Then, on $\mathfrak{R}$,

$$K_\mu = \sum_{\nu=1}^{G_0} a_{\mu\nu}\mathscr{K}_\nu, \quad \mu = 1, 2, \cdots, G. \tag{5.1.1}$$

If $\mathscr{G}(p, q)$ is the Green's function of $\mathfrak{R}$, and if

$$\mathscr{L}_{\mathbf{M}}(p, q) = -\frac{2}{\pi}\frac{\partial^2 \mathscr{G}(p, q)}{\partial p \partial q}, \tag{5.1.2}$$

then, as in Chapter 4, we have the canonical basis differentials

$$\mathscr{Z}'_\mu(q) = \int_{\mathscr{K}_\mu} \mathscr{L}_{\mathbf{M}}(p, q)dz, \quad \mu = 1, 2, \cdots, G_0. \tag{5.1.3}$$

Let

$$P_{\mu\nu} = (d\mathscr{Z}_\mu, d\mathscr{Z}_\nu)_{\mathfrak{R}} = -P(d\mathscr{Z}_\mu, \mathscr{K}_\nu), \quad \mu, \nu = 1, 2, \cdots, G_0. \tag{5.1.4}$$

Corresponding to the cycles K_μ, $\mu = 1, 2, \cdots, G$, we define

$$\mathscr{Y}'_\mu(q) = \int_{K_\mu} \mathscr{L}_{\mathbf{M}}(p, q)dz = \sum_{\nu=1}^{G_0} a_{\mu\nu}\mathscr{Z}'_\nu(q), \tag{5.1.5}$$

and we write

$$\Pi_{\mu\nu} = (d\mathscr{Y}_\mu, d\mathscr{Y}_\nu)_{\mathfrak{R}}, \quad \mu, \nu = 1, 2, \cdots, G. \tag{5.1.6}$$

If dZ_μ is the differential of $\mathfrak{M}$ corresponding to the cycle K_μ, and df is a differential of class $\mathbf{M}$ on $\mathfrak{R}$, then

$$\begin{aligned} (df, dZ_\mu)_{\mathfrak{M}} &= -P(df, K_\mu) = -\sum_{\nu=1}^{G_0} a_{\mu\nu} P(df, \mathscr{K}_\nu) \\ &= \sum_{\nu=1}^{G_0} a_{\mu\nu}(df, d\mathscr{Z}_\nu)_{\mathfrak{R}} = (df, \sum_{\nu=1}^{G_0} a_{\mu\nu} d\mathscr{Z}_\nu)_{\mathfrak{R}} = (df, d\mathscr{Y}_\mu)_{\mathfrak{R}}; \end{aligned}$$

that is,

$$(df, dZ_\mu)_{\mathfrak{M}} = (df, d\mathscr{Y}_\mu)_{\mathfrak{R}}, \quad \mu = 1, 2, \cdots, G. \tag{5.1.7}$$

In particular,

(5.1.8) $$\Pi_{\mu\nu} = - P(d\mathscr{Y}_\mu, K_\nu).$$

Let

(5.1.9) $$\varepsilon_{\mu\nu} = (d\mathscr{X}_\mu, dZ_\nu)_{\mathfrak{M}} = -P(d\mathscr{X}_\mu, K_\nu),\ \mu = 1, 2, \cdots, G_0; \nu = 1, 2, \cdots, G.$$

Then, using (5.1.7),

(5.1.10) $$\varepsilon_{\mu\nu} = \sum_{\varrho=1}^{G_0} a_{\nu\varrho} P_{\mu\varrho}.$$

Also, starting from (5.1.6), we find

(5.1.11) $$\Pi_{\mu\nu} = \sum_{\varrho,\sigma=1}^{G_0} a_{\mu\varrho} a_{\nu\sigma} P_{\varrho\sigma}.$$

Thus the periods of $\mathscr{X}'_\mu$ and $\mathscr{Y}'_\mu$ over the cycles K_ν of $\mathfrak{M}$ can be expressed in terms of the periods of the basis differentials on $\mathfrak{R}$ and the coefficients of the transformation for the homology basis. Further, by (5.1.1)

(5.1.12) $$I(K_\mu, K_\nu) = \sum_{\varrho,\sigma=1}^{G_0} a_{\mu\varrho} a_{\nu\sigma} I(\mathscr{K}_\varrho, \mathscr{K}_\sigma).$$

Now let $\mathbf{F}_{\mathfrak{R}}$ be any class of differentials on $\mathfrak{R}$. A class $\mathbf{F}_{\mathfrak{M}}$ of differentials on $\mathfrak{M}$ will be called a corresponding class if each differential of $\mathbf{F}_{\mathfrak{R}}$ satisfies the period relations of $\mathbf{F}_{\mathfrak{M}}$. Specifically, if $\mathbf{F}_{\mathfrak{R}} = \mathbf{F}_{\mathfrak{R}}(j_1, \cdots, j_k)$, so that

(5.1.13) $$(df, d\mathscr{X}_\varrho)_{\mathfrak{R}} = 0, \quad \varrho = j_1, \cdots, j_k,$$

for each $df \in \mathbf{F}_{\mathfrak{R}}$, then $\mathbf{F}_{\mathfrak{M}} = \mathbf{F}_{\mathfrak{M}}(i_1, \cdots, i_l)$ is a corresponding class if df, considered as a differential on $\mathfrak{M}$, satisfies the period relations for $\mathbf{F}_{\mathfrak{M}}$, namely

(5.1.14) $$(df, dZ_\varrho)_{\mathfrak{M}} = 0, \quad \varrho = i_1, \cdots, i_l.$$

By (5.1.12), a corresponding class $\mathbf{F}_{\mathfrak{M}}$ is symmetric whenever $\mathbf{F}_{\mathfrak{R}}$ is symmetric.

Given a class $\mathbf{F}_{\mathfrak{R}}$ in $\mathfrak{R}$, let $\mathscr{L}_{\mathbf{F}}(p, q)$ be the bilinear differential for the class $\mathbf{F}_{\mathfrak{R}}$ in $\mathfrak{R}$, and let $L_{\mathbf{F}}(p, q)$ be the bilinear differential on $\mathfrak{M}$ for some corresponding class $\mathbf{F}_{\mathfrak{M}}$. We define

(5.1.15) $$l_{\mathbf{F}}(p, q) = L_{\mathbf{F}}(p, q) - \mathscr{L}_{\mathbf{F}}(p, q).$$

The double of $\mathfrak{M}$ is composed of two surfaces $\mathfrak{M}$ and $\tilde{\mathfrak{M}}$, and the double of $\mathfrak{R}$ of two surfaces $\mathfrak{R}$ and $\tilde{\mathfrak{R}}$. It is convenient to take $\tilde{\mathfrak{M}}$ to be the subdomain of $\tilde{\mathfrak{R}}$ which is conjugate to $\mathfrak{M}$ on the double of $\mathfrak{R}$. The double of $\mathfrak{M}$ is then obtained by identifying the boundaries of $\tilde{\mathfrak{M}}$ and $\mathfrak{M}$. With this convention, a point $\tilde{p}$ of $\tilde{\mathfrak{M}}$ is uniquely defined and may be regarded either as a point of $\tilde{\mathfrak{M}}$ or of $\tilde{\mathfrak{R}}$. Meaning is thus attached to the symbols $l_{\mathbf{F}}(\tilde{p}, q)$, $l_{\mathbf{F}}(p, \tilde{q})$.

In order to simplify the notation, we set

$$L_{\mathbf{F}}(p, q) = L_{\mathbf{M}}(p, q) + B_{\mathbf{F}}(p, q) \tag{5.1.16}$$

and

$$\mathscr{L}_{\mathbf{F}}(p, q) = \mathscr{L}_{\mathbf{M}}(p, q) + \mathscr{B}_{\mathbf{F}}(p, q) \tag{5.1.17}$$

where

$$B_{\mathbf{F}}(p, q) = \sum_{\mu, \nu=1}^{G} \alpha_{\mu\nu} Z'_{\mu}(p) Z'_{\nu}(q) \tag{5.1.16$'$}$$

and

$$\mathscr{B}_{\mathbf{F}}(p, q) = \sum_{\mu, \nu=1}^{G_0} \beta_{\mu\nu} \mathscr{Z}'_{\mu}(p) \mathscr{Z}'_{\nu}(q). \tag{5.1.17$'$}$$

Comparing these formulas with (4.11.1), we see that $\alpha_{\mu\nu}$ and $\beta_{\mu\nu}$ vanish if either μ or ν has a value different from i_τ, $\tau = 1, 2, \cdots, l$, or j_τ, $\tau = 1, 2, \cdots, k$ respectively. Also, by (4.11.5),

$$\alpha_{\mu\nu} = (\alpha_{\nu\mu})^-, \quad \beta_{\mu\nu} = (\beta_{\nu\mu})^-. \tag{5.1.18}$$

From (4.11.3) we have

$$\sum_{\mu=1}^{G} \alpha_{\mu\nu} \Gamma_{\mu\varrho} = \delta_{\nu\varrho}, \quad \varrho = i_1, i_2, \cdots, i_l, \tag{5.1.19}$$

where $\Gamma_{\mu\varrho} = (dZ_\mu, dZ_\varrho)_{\mathfrak{M}}$, and

$$\sum_{\mu=1}^{G_0} \beta_{\mu\nu} P_{\mu\varrho} = \delta_{\nu\varrho}, \quad \varrho = j_1, j_2, \cdots, j_k. \tag{5.1.20}$$

Since $\mathscr{L}_{\mathbf{F}}(p, \tilde{q})$ is of class $\mathbf{F}_{\mathfrak{R}}$ on $\mathfrak{R}$, it is also of class $\mathbf{F}_{\mathfrak{M}}$ and we have

$$(\mathscr{L}_{\mathbf{F}}(p, \tilde{q}), Z'_\varrho(p))_{\mathfrak{M}} = 0, \quad \varrho = i_1, i_2, \cdots, i_l. \tag{5.1.14$'$}$$

Substituting from (5.1.17) and (5.1.17)$'$, this becomes

$$\mathscr{Y}'_\varrho(q) = \sum_{\mu, \nu=1}^{G_0} \beta_{\mu\nu} \varepsilon_{\mu\varrho} \mathscr{Z}'_\nu(q), \quad \varrho = i_1, \cdots, i_l. \tag{5.1.21}$$

Comparing (5.1.21) with (5.1.5) we see that

$$a_{\varrho\nu} = \sum_{\mu=1}^{G_0} \beta_{\mu\nu}\,\varepsilon_{\mu\varrho}, \qquad \varrho = i_1, \cdots, i_l. \tag{5.1.22}$$

5.2. Several Surfaces Imbedded in a Given Surface

We now consider the case in which the set $\mathfrak{M}$ imbedded in $\mathfrak{R}$ consists of several components each of which is a finite Riemann surface. Assume, then, that $\mathfrak{M}$ is the union of a finite number of domains $\mathfrak{M}_\nu$, $\nu = 1, \cdots, k$, each of which is a finite Riemann surface imbedded in $\mathfrak{R}$. We suppose that no two domains $\mathfrak{M}_\mu$ have points in common.

If the component $\mathfrak{M}_\nu$ of $\mathfrak{M}$ has genus h_ν and m_ν boundaries, we write

$$G_\nu = 2h_\nu + m_\nu - 1,$$

and we define the algebraic genus of $\mathfrak{M}$ to be

$$G = \sum_{\nu=1}^{k} G_\nu.$$

Let

$$K_1^{(\nu)}, \cdots, K_{G_\nu}^{(\nu)}$$

be a canonical homology basis for $\mathfrak{M}_\nu$, and let

$$K_1^{(1)}, \cdots, K_{G_1}^{(1)}, \cdots, K_1^{(k)}, \cdots, K_{G_k}^{(k)}$$

be denoted in order by

$$K_1, \cdots, K_G. \tag{5.2.1}$$

We call (5.2.1) a canonical basis for $\mathfrak{M}$. Let $dZ_\mu^{(\nu)}$ be the differential corresponding to the cycle $K_\mu^{(\nu)}$ on $\mathfrak{M}_\nu$, and define $dZ_\mu^{(\nu)}$ to be identically zero in $\mathfrak{R} - \mathfrak{M}_\nu$. Then

$$dZ_1^{(1)}, \cdots, dZ_{G_1}^{(1)}, \cdots, dZ_1^{(k)}, \cdots, dZ_{G_k}^{(k)}$$

is a basis for the differentials of $\mathfrak{M}$ which we denote by

$$dZ_1, \cdots, dZ_G. \tag{5.2.2}$$

The formulas (5.1.1)—(5.1.12) are clearly valid under these more general circumstances.

The class $\mathbf{F}_\mathfrak{R}$ on $\mathfrak{R}$ is assumed given, and we suppose that $\mathbf{F}_\nu$ is a corresponding class on $\mathfrak{M}_\nu$. We extend the definition of a differential df_ν of $\mathfrak{M}_\nu$ over $\mathfrak{R}$ by setting it equal to zero outside the component $\mathfrak{M}_\nu$, and we define $\mathbf{F}_{\mathfrak{M}}$ to be the class of differentials

$$df = df_1 + df_2 + \cdots + df_k, \quad df_\nu \in \mathbf{F}_\nu. \tag{5.2.3}$$

In other words, df is identically zero in $\mathfrak{R} - \mathfrak{M}$ and is equal to some differential of $\mathbf{F}_\nu$ on $\mathfrak{M}_\nu$. We say that $\mathbf{F}_{\mathfrak{M}}$ corresponds to the class $\mathbf{F}_{\mathfrak{R}}$.

If $\mathbf{F} = \mathbf{F}_{\mathfrak{M}}$ corresponds to the given class $\mathbf{F}_{\mathfrak{R}}$, we define the bilinear differential $L_{\mathbf{F}}(p, q)$ of this class on $\mathfrak{M}$ to have the value zero for any two points p, q of $\mathfrak{R}$ unless both points lie in the closure of the same domain $\mathfrak{M}_\nu$ in which case $L_{\mathbf{F}}(p, q)$ is equal to the bilinear differential $L^{(\nu)}_{\mathbf{F}_\nu}$ of the class $\mathbf{F}_\nu$ on $\mathfrak{M}_\nu$.

The definition of $L_{\mathbf{F}}(p, q)$ which has been given is the natural one, in view of the minimizing property of the singular bilinear differential in the case of a single finite Riemann surface. The norm of a differential on $\mathfrak{M}$ is the sum of the norms on the component surfaces. If q is a point of $\mathfrak{M}_\nu$, the differential which minimizes the norm among all differentials with the given singularity at q must clearly vanish on $\mathfrak{M}_\mu$, $\mu \neq \nu$, where it is regular, and coincide with $L^{(\nu)}_{\mathbf{F}_\nu}$ on $\mathfrak{M}_\nu$.

If $df \in \mathbf{F}_{\mathfrak{M}}$, we have the formulas

$$(df, L_{\mathbf{F}}(p, q))_{\mathfrak{R}} = 0, \tag{5.2.4}$$

$$(df, L_{\mathbf{F}}(p, \bar{q}))_{\mathfrak{R}} = -f'(q). \tag{5.2.5}$$

Writing

$$\Gamma_{\mu\nu} = (dZ_\mu, dZ_\nu), \tag{5.2.6}$$

we see that the period matrix $\| \Gamma_{\mu\nu} \|$ is non-singular. For its determinant is simply the product of the corresponding period determinants for each component of $\mathfrak{M}$. Therefore the formulas (5.1.16)—(5.1.20) are valid. In fact all formalism carries over immediately to the case where $\mathfrak{M}$ is the union of a finite number of disjoint subdomains of $\mathfrak{R}$.

5.3. FUNDAMENTAL IDENTITIES

We assume that $\mathfrak{M}$ is the union of a finite number of disjoint surfaces imbedded in $\mathfrak{R}$, and we find formulas for the differences

$$\Gamma_{\mu\nu} - \Pi_{\mu\nu},\quad Z'_{\mu} - \mathscr{Y}'_{\mu},\quad l_{\mathbf{F}} = L_{\mathbf{F}} - \mathscr{L}_{\mathbf{F}}.$$

We begin by showing that for fixed $q \in \mathfrak{R}$, $l_{\mathbf{F}}(p, q)$ is a differential of class $\mathbf{F}_{\mathfrak{M}}$ on $\mathfrak{M}$. Clearly $l_{\mathbf{F}}(p, q)$ is regular, and therefore of class $\mathbf{M}$, on $\mathfrak{M}$. It remains to show that $l_{\mathbf{F}}(p, q)$ satisfies the period relations of $\mathbf{F}_{\mathfrak{M}}$. Now

$$(5.3.1)\qquad (l_{\mathbf{M}}(p, q),\ Z'_{\mu}(p))_{\mathfrak{M}} = -P(l_{\mathbf{M}}(p, q), K_{\mu})$$
$$= -P(L_{\mathbf{M}}(p, q), K_{\mu}) + P(\mathscr{L}_{\mathbf{M}}(p, q), K_{\mu}) = -Z'_{\mu}(q) + \mathscr{Y}'_{\mu}(q)$$

by (4.10.15) and (5.1.5). We note that (5.3.1) remains true if q is replaced by $\tilde{q}$. For $\varrho = i_{\tau}$, $\tau = 1, 2, \cdots, l$,

$$(l_{\mathbf{F}}(p, q), Z'_{\varrho}(p))_{\mathfrak{M}} = (l_{\mathbf{M}}(p, q), Z'_{\varrho}(p))_{\mathfrak{M}} + (B_{\mathbf{F}}(p, q), Z'_{\varrho}(p))_{\mathfrak{M}}$$
$$- (\mathscr{B}_{\mathbf{F}}(p, q), Z'_{\varrho}(p))_{\mathfrak{M}}$$

using (5.1.16) and (5.1.17);

$$= -Z'_{\varrho}(q) + \mathscr{Y}'_{\varrho}(q) + \sum_{\mu,\nu=1}^{G} \alpha_{\mu\nu}\, \Gamma_{\mu\varrho} Z'_{\nu}(q) - \sum_{\mu,\nu=1}^{G_0} \beta_{\mu\nu}\, \varepsilon_{\mu\varrho}\, \mathscr{X}'_{\nu}(q)$$

by (5.1.16)′, (5.1.17)′ and (5.3.1);

$$= -Z'_{\varrho}(q) + \mathscr{Y}'_{\varrho}(q) + Z'_{\varrho}(q) - \mathscr{Y}'_{\varrho}(q) = 0$$

using (5.1.19) and (5.1.22). Thus

$$(5.3.2)\qquad (l_{\mathbf{F}}(p, q),\ Z'_{\varrho}(p))_{\mathfrak{M}} = 0,\quad \varrho = i_1, \cdots, i_l,$$

and $l_{\mathbf{F}}(p, q)$ is in $\mathbf{F}_{\mathfrak{M}}$. The same argument can be applied to $l_{\mathbf{F}}(p, \tilde{q})$ but this is unnecessary since $L_{\mathbf{F}}(p, \tilde{q})$ and $\mathscr{L}_{\mathbf{F}}(p, \tilde{q})$ are in $\mathbf{F}_{\mathfrak{M}}$.

The fundamental identity is

$$(5.3.3)\qquad (l_{\mathbf{F}}(p, q),\ l_{\mathbf{F}}(p, r))_{\mathfrak{R}} = -l_{\mathbf{F}}(\tilde{r}, q).$$

To prove this, we note first that

$$(5.3.4)\qquad (L_{\mathbf{F}}(p, q),\ L_{\mathbf{F}}(p, r))_{\mathfrak{R}} = L_{\mathbf{F}}(\tilde{r}, q)$$

for all q and r in $\mathfrak{R}$. For if q and r do not lie in the same component of $\mathfrak{M}$, then both sides of (5.3.4) vanish, while if q and r lie in $\mathfrak{M}_{\mu}$, we have

$$(L_{\mathbf{F}}(p, q),\ L_{\mathbf{F}}(p, r))_{\mathfrak{R}} = (L_{\mathbf{F}}(p, q),\ L_{\mathbf{F}}(p, r)_{\mathfrak{M}_{\mu}}$$
$$= (L_{\mathbf{F}}^{(\mu)}(p, q),\ L_{\mathbf{F}}^{(\mu)}(p, r))_{\mathfrak{M}_{\mu}} = L_{\mathbf{F}}^{(\mu)}(\tilde{r}, q) = L_{\mathbf{F}}(\tilde{r}, q)$$

by (4.12.3) if q is distinct from r and by taking the limit as r tends to q if $q = r$. Also

$$(5.3.4)' \qquad (\mathscr{L}_{\mathbf{F}}(p, q), \mathscr{L}_{\mathbf{F}}(p, r))_{\mathfrak{R}} = \mathscr{L}_{\mathbf{F}}(\tilde{r}, q).$$

By (5.2.4),

$$(l_{\mathbf{F}}(p, q), L_{\mathbf{F}}(p, r))_{\mathfrak{R}} = 0$$

so that

$$(5.3.5) \quad (\mathscr{L}_{\mathbf{F}}(p, q), L_{\mathbf{F}}(p, r))_{\mathfrak{R}} = (L_{\mathbf{F}}(p, q), L_{\mathbf{F}}(p, r))_{\mathfrak{R}} = L_{\mathbf{F}}(\tilde{r}, q).$$

Finally

$$\begin{aligned}&(l_{\mathbf{F}}(p, q), l_{\mathbf{F}}(p, r))_{\mathfrak{R}}\\ &= (L_{\mathbf{F}}(p, q), l_{\mathbf{F}}(p, r))_{\mathfrak{R}} - (\mathscr{L}_{\mathbf{F}}(p, q), L_{\mathbf{F}}(p, r))_{\mathfrak{R}} + (\mathscr{L}_{\mathbf{F}}(p, q), \mathscr{L}_{\mathbf{F}}(p, r))_{\mathfrak{R}}\\ &= - L_{\mathbf{F}}(\tilde{r}, q) + \mathscr{L}_{\mathbf{F}}(\tilde{r}, q) = - l_{\mathbf{F}}(\tilde{r}, q)\end{aligned}$$

and this is (5.3.3).

The identity (5.3.3) is also true if we replace r by $\tilde{r}$:

$$(5.3.6) \qquad (l_{\mathbf{F}}(p, q), l_{\mathbf{F}}(p, \tilde{r}))_{\mathfrak{R}} = - l_{\mathbf{F}}(r, q),$$

or if we replace q by $\tilde{q}$ and r by $\tilde{r}$:

$$(5.3.7) \qquad (l_{\mathbf{F}}(p, \tilde{q}), l_{\mathbf{F}}(p, \tilde{r}))_{\mathfrak{R}} = - l_{\mathbf{F}}(r, \tilde{q}).$$

However, these identities are not immediate consequences of (5.3.3) by analytic continuation. Using (5.2.4) and (5.2.5), we have

$$\begin{aligned}&(l_{\mathbf{F}}(p, q), l_{\mathbf{F}}(p, \tilde{r}))_{\mathfrak{R}}\\ &= (l_{\mathbf{F}}(p, q), L_{\mathbf{F}}(p, \tilde{r}))_{\mathfrak{R}} - (L_{\mathbf{F}}(p, q), \mathscr{L}_{\mathbf{F}}(p, \tilde{r}))_{\mathfrak{R}} + (\mathscr{L}_{\mathbf{F}}(p, q), \mathscr{L}_{\mathbf{F}}(p, \tilde{r}))_{\mathfrak{R}}\\ &= - l_{\mathbf{F}}(r, q).\end{aligned}$$

Similarly, by (5.2.5),

$$\begin{aligned}&(l_{\mathbf{F}}(p, \tilde{q}), l_{\mathbf{F}}(p, \tilde{r}))_{\mathfrak{R}}\\ &= (l_{\mathbf{F}}(p, \tilde{q}), L_{\mathbf{F}}(p, \tilde{r}))_{\mathfrak{R}} - (L_{\mathbf{F}}(p, \tilde{q}), \mathscr{L}_{\mathbf{F}}(p, \tilde{r}))_{\mathfrak{R}} + (\mathscr{L}_{\mathbf{F}}(p, \tilde{q}), \mathscr{L}_{\mathbf{F}}(p, \tilde{r}))_{\mathfrak{R}}\\ &= - l_{\mathbf{F}}(r, \tilde{q}) + (\mathscr{L}_{\mathbf{F}}(q, \tilde{r}))^{-} - \mathscr{L}_{\mathbf{F}}(r, \tilde{q}) = - l_{\mathbf{F}}(r, \tilde{q}).\end{aligned}$$

Let

$$(5.3.8) \qquad \mathfrak{z}'_{\mu}(q) = Z'_{\mu}(q) - \mathscr{Y}'_{\mu}(q).$$

In (5.3.3) and (5.3.7), take $\mathbf{F} = \mathbf{M}$ and suppose that q is fixed. If we compute the periods with respect to r around a cycle K_{μ}, we find by (5.3.1)

$$(l_{\mathbf{M}}(p, q),\ \mathfrak{z}'_\mu(p))_{\mathfrak{R}} = -\,\mathfrak{z}'_\mu(q) \tag{5.3.9}$$

and

$$(l_{\mathbf{M}}(p, \tilde{q}),\ \mathfrak{z}'_\mu(p))_{\mathfrak{R}} = -\,\mathfrak{z}'_\mu(\tilde{q}). \tag{5.3.10}$$

Finally, writing

$$\gamma_{\mu\nu} = \Pi_{\mu\nu} - \Gamma_{\mu\nu}, \tag{5.3.11}$$

we compute the period of (5.3.9) with respect to q around a cycle K_ν and obtain (since $\gamma_{\mu\nu}$ is real and therefore symmetric)

$$(d\mathfrak{z}_\mu,\ d\mathfrak{z}_\nu)_{\mathfrak{R}} = -\,\gamma_{\mu\nu}. \tag{5.3.12}$$

We have also

$$(\mathscr{L}_{\mathbf{F}}(p, q), L_{\mathbf{F}}(p, \tilde{r}))_{\mathfrak{R}} = -\,(l_{\mathbf{F}}(p, q),\ L_{\mathbf{F}}(p, \tilde{r}))_{\mathfrak{R}} = l_{\mathbf{F}}(r, q). \tag{5.3.13}$$

The identity (5.3.13) shows that $l_{\mathbf{F}}(r, q)$ is the projection of the singular differential $\mathscr{L}_{\mathbf{F}}(p, q)$ into the space of the differentials of class $\mathbf{F}_{\mathfrak{M}}$ on $\mathfrak{M}$.

Other identities can easily be derived. For example, using (5.1.19), (5.1.20) and (5.1.22), respectively, we obtain

$$(B_{\mathbf{F}}(p, q),\ B_{\mathbf{F}}(p, r))_{\mathfrak{R}} = B_{\mathbf{F}}(\tilde{r}, q), \tag{5.3.14}$$

$$(\mathscr{B}_{\mathbf{F}}(p, q),\ \mathscr{B}_{\mathbf{F}}(p, r))_{\mathfrak{R}} = \mathscr{B}_{\mathbf{F}}(\tilde{r}, q), \tag{5.3.15}$$

$$\begin{aligned}(B_{\mathbf{F}}(p, q),\ \mathscr{B}_{\mathbf{F}}(p, r))_{\mathfrak{R}} &= \sum_{\mu,\nu=1}^{G} \sum_{\varrho,\sigma=1}^{G_0} \alpha_{\mu\nu}(\beta_{\varrho\sigma})^-(\varepsilon_{\varrho\mu})^- Z'_\nu(q)\,\mathscr{Z}'_\sigma(\tilde{r}) \\ &= \sum_{\mu,\nu=1}^{G} \alpha_{\mu\nu}\,\mathscr{Y}'_\mu(\tilde{r})\, Z'_\nu(q).\end{aligned} \tag{5.3.16}$$

From (5.3.1) it follows that

$$(l_{\mathbf{M}}(p, q),\ B_{\mathbf{F}}(p, r))_{\mathfrak{R}} = \sum_{\mu,\nu=1}^{G} \alpha_{\mu\nu} Z'_\mu(\tilde{r})\mathscr{Y}'_\nu(q) - B_{\mathbf{F}}(\tilde{r}\ q) \tag{5.3.17}$$

while (5.3.2) leads to

$$(l_{\mathbf{F}}(p, q),\ B_{\mathbf{F}}(p, r))_{\mathfrak{R}} = 0. \tag{5.3.18}$$

Finally

$$\begin{aligned}&(l_{\mathbf{M}}(p, q),\ \mathscr{B}_{\mathbf{F}}(p, r))_{\mathfrak{R}} \\ &= (L_{\mathbf{M}}(p, q),\ \mathscr{B}_{\mathbf{F}}(p, r))_{\mathfrak{M}} - (\mathscr{L}_{\mathbf{M}}(p, q),\ \mathscr{B}_{\mathbf{F}}(p, r))_{\mathfrak{R}} = 0.\end{aligned} \tag{5.3.19}$$

Taking $r = q,\ q \in \mathfrak{M}$, in (5.3.3) we obtain

$$(5.3.20)\quad l_{\mathbf{F}}(\tilde{q}, q) + (\mathscr{L}_{\mathbf{F}}(p, q), \mathscr{L}_{\mathbf{F}}(p, q))_{\mathfrak{R}-\mathfrak{M}} = -(l_{\mathbf{F}}(p, q), l_{\mathbf{F}}(p, q))_{\mathfrak{M}} \leqq 0.$$

This inequality has a geometrical interpretation when $\mathbf{F} = \mathbf{S}$, where $\mathbf{S}$ is the class of differentials with single-valued integrals. In fact, let

$$(5.3.21)\qquad \mathscr{E} = -\mathscr{L}_{\mathbf{S}}(\tilde{q}, q).$$

From Chapter 4 we know that $\mathscr{E}$ is the maximum external area of image domains of $\mathfrak{R}$ under mappings by single-valued functions $\mathfrak{f}$ which have a pole at q such that $d\mathfrak{f} - \mathscr{L}_{\mathbf{S}}(p, q)$ is regular there. If E denotes the corresponding maximum external area for $\mathfrak{M}$ then

$$(5.3.22)\qquad E \geqq \mathscr{E} + (\mathscr{L}_{\mathbf{S}}(p, q), \mathscr{L}_{\mathbf{S}}(p, q))_{\mathfrak{R}-\mathfrak{M}},$$

since the maximum external area for $\mathfrak{M}$ is not less than the corresponding external area of $\mathfrak{M}$ defined by the mapping $\mathfrak{f}$, $d\mathfrak{f} = \mathscr{L}_{\mathbf{S}}(p, q)$. But (5.3.22) is just the inequality (5.3.20).

An analogous inequality is obtained by comparing internal areas. Suppose that q is a point of $\mathfrak{M}$, and let

$$df_{\mathbf{S}}(p) = \frac{L_{\mathbf{S}}(p, \tilde{q})}{L_{\mathbf{S}}(q, \tilde{q})}, \quad d\mathfrak{f}_{\mathbf{S}}(p) = \frac{\mathscr{L}_{\mathbf{S}}(p, \tilde{q})}{\mathscr{L}_{\mathbf{S}}(q, \tilde{q})}.$$

Since we know that $N_{\mathfrak{M}}(df_{\mathbf{S}})$ is minimal, we have

$$N_{\mathfrak{M}}(d\mathfrak{f}_{\mathbf{S}}) \geqq N_{\mathfrak{M}}(df_{\mathbf{S}}) = \frac{-1}{L_{\mathbf{S}}(q, \tilde{q})}.$$

But

$$N_{\mathfrak{M}}(d\mathfrak{f}_{\mathbf{S}}) = N_{\mathfrak{R}}(d\mathfrak{f}_{\mathbf{S}}) - N_{\mathfrak{R}-\mathfrak{M}}(d\mathfrak{f}_{\mathbf{S}}) = \frac{-1}{\mathscr{L}_{\mathbf{S}}(q, \tilde{q})} - N_{\mathfrak{R}-\mathfrak{M}}(d\mathfrak{f}_{\mathbf{S}}).$$

Writing

$$A = \frac{-1}{L_{\mathbf{S}}(q, \tilde{q})}, \quad \mathscr{A} = \frac{-1}{\mathscr{L}_{\mathbf{S}}(q, \tilde{q})},$$

we therefore have

$$(5.3.23)\qquad A \leqq \mathscr{A} - \frac{1}{(\mathscr{L}_{\mathbf{S}}(q, \tilde{q}))^2}(\mathscr{L}_{\mathbf{S}}(p, \tilde{q}), \mathscr{L}_{\mathbf{S}}(p, \tilde{q}))_{\mathfrak{R}-\mathfrak{M}},$$

in analogy with (5.3.22).

The inequality (5.3.23) may also be obtained from an identity. In corresponding notation, let

$$(5.3.24)\qquad k_{\mathbf{F}}(p, \tilde{q}) = df_{\mathbf{F}}(p) - d\mathfrak{f}_{\mathbf{F}}(p).$$

If $q, r \in \mathfrak{M}$, we have

$$(5.3.25)\quad \begin{aligned}(k_{\mathbf{F}}(p, \tilde{r}), k_{\mathbf{F}}(p, \tilde{q}))_{\mathfrak{R}} = &-\frac{L_{\mathbf{F}}(q, \tilde{r})}{L_{\mathbf{F}}(q, \tilde{q})L_{\mathbf{F}}(r, \tilde{r})} + \frac{\mathscr{L}_{\mathbf{F}}(q, \tilde{r})}{L_{\mathbf{F}}(q, \tilde{q})\mathscr{L}_{\mathbf{F}}(r, \tilde{r})} \\ &+\frac{\mathscr{L}_{\mathbf{F}}(q, \tilde{r})}{\mathscr{L}_{\mathbf{F}}(q, \tilde{q})L_{\mathbf{F}}(r, \tilde{r})} - \frac{\mathscr{L}_{\mathbf{F}}(q, \tilde{r})}{\mathscr{L}_{\mathbf{F}}(q, \tilde{q})\mathscr{L}_{\mathbf{F}}(r, \tilde{r})}.\end{aligned}$$

Taking $r = q$, we obtain in particular

$$(5.3.26)\quad \begin{aligned}&\frac{1}{L_{\mathbf{F}}(q, \tilde{q})} - \frac{1}{\mathscr{L}_{\mathbf{F}}(q, \tilde{q})} - \frac{1}{(\mathscr{L}_{\mathbf{F}}(q, \tilde{q}))^2}(\mathscr{L}_{\mathbf{F}}(p, \tilde{q}), \mathscr{L}_{\mathbf{F}}(p, \tilde{q}))_{\mathfrak{R}-\mathfrak{M}} \\ &\qquad = (k_{\mathbf{F}}(p, \tilde{q}), k_{\mathbf{F}}(p, \tilde{q}))_{\mathfrak{M}} \geqq 0,\end{aligned}$$

and this gives a generalization of (5.3.23).

We observe that (5.3.3) may be put into the form

$$(5.3.27)\quad l_{\mathbf{F}}(\tilde{r}, q) = -(l_{\mathbf{F}}(p, q), l_{\mathbf{F}}(p, r))_{\mathfrak{M}} - (\mathscr{L}_{\mathbf{F}}(p, q), \mathscr{L}_{\mathbf{F}}(p, r))_{\mathfrak{R}-\mathfrak{M}}.$$

This formula will play an important role in the theory of variations of domains $\mathfrak{M}$. In fact, if $\mathfrak{M}$ is near to $\mathfrak{R}$ (in a sense to be specified later), then $l_{\mathbf{F}}(\tilde{r}, q) = L_{\mathbf{F}}(\tilde{r}, q) - \mathscr{L}_{\mathbf{F}}(\tilde{r}, q)$ will be small and the first term in (5.3.27) will be of still higher order. Thus, we shall find

$$(5.3.27)'\quad L_{\mathbf{F}}(\tilde{r}, q) - \mathscr{L}_{\mathbf{F}}(\tilde{r}, q) \sim -(\mathscr{L}_{\mathbf{F}}(p, q), \mathscr{L}_{\mathbf{F}}(p, r))_{\mathfrak{R}-\mathfrak{M}}$$

and the right-hand side can be evaluated if the $\mathscr{L}$-differential of $\mathfrak{R}$ is known. Thus, $L_{\mathbf{F}}(p, q)$ can be determined approximately in terms of $\mathscr{L}_{\mathbf{F}}(p, q)$.

5.4. Inequalities for Quadratic and Hermitian Forms

We now apply the identities of Section 5.3 to obtain inequalities for quadratic and Hermitian forms with coefficient matrices

$$\| l_{\mathbf{F}}(q_\mu, q_\nu) \|, \quad \| L_{\mathbf{F}}(q_\mu, \tilde{q}_\nu) \|,$$

respectively. In this way we obtain necessary conditions expressed in the form of inequalities between quadratic and Hermitian forms in order that one or more surfaces can be imbedded in a given surface $\mathfrak{R}$. In the following section we derive from these inequalities necessary and sufficient conditions in order that a given surface $\mathfrak{M}$ can be conformally imbedded in another given surface $\mathfrak{R}$. A special case of these inequalities was derived by Grunsky [2] in connection with the conformal mapping of multiply-connected plane domains (see also [1]).

We define

$$\Gamma_{\mathbf{F}}(\tilde{r}, q) = (\mathscr{L}_{\mathbf{F}}(p, q), \mathscr{L}_{\mathbf{F}}(p, r))_{\mathfrak{R}-\mathfrak{M}}, \tag{5.4.1}$$

and correspondingly if r or q is replaced by $\tilde{r}$ or $\tilde{q}$. For example,

$$\Gamma_{\mathbf{F}}(r, q) = (\mathscr{L}_{\mathbf{F}}(p, q), \mathscr{L}_{\mathbf{F}}(p, \tilde{r}))_{\mathfrak{R}-\mathfrak{M}}.$$

Let λ be a complex number and consider the norm over $\mathfrak{M}$

$$N_{\mathfrak{M}}\left(\sum_{\mu=1}^{N} x_\mu l_{\mathbf{F}}(p, q_\mu) - \lambda \sum_{\mu=1}^{N} \bar{x}_\mu L_{\mathbf{F}}(p, \tilde{q}_\mu)\right), \tag{5.4.2}$$

where the x_μ are arbitrary complex numbers and the q_μ are points of $\mathfrak{M}$. By (4.11.7), (4.9.2) and (5.3.27) we find that (5.4.2) is equal to

$$\begin{aligned} &-|\lambda|^2 \sum_{\mu,\nu=1}^{N} L_{\mathbf{F}}(q_\mu, \tilde{q}_\nu) x_\mu \bar{x}_\nu + 2 \operatorname{Re}\left\{\bar{\lambda} \sum_{\mu,\nu=1}^{N} l_{\mathbf{F}}(q_\mu, q_\nu) x_\mu x_\nu\right\} \\ &\qquad - \sum_{\mu,\nu=1}^{N} [l_{\mathbf{F}}(\tilde{q}_\nu, q_\mu) + \Gamma_{\mathbf{F}}(\tilde{q}_\nu, q_\mu)] x_\mu \bar{x}_\nu. \end{aligned} \tag{5.4.3}$$

Since the expression (5.4.3) is non-negative for all choices of the complex number λ, we obtain the inequality

$$\begin{aligned} &\left|\sum_{\mu,\nu=1}^{N} l_{\mathbf{F}}(q_\mu, q_\nu) x_\mu x_\nu\right|^2 \\ &\leqq \sum_{\mu,\nu=1}^{N} L_{\mathbf{F}}(q_\mu, \tilde{q}_\nu) x_\mu \bar{x}_\nu \cdot \sum_{\mu,\nu=1}^{N} [l_{\mathbf{F}}(\tilde{q}_\nu, q_\mu) + \Gamma_{\mathbf{F}}(\tilde{q}_\nu, q_\mu)] x_\mu \bar{x}_\nu. \end{aligned} \tag{5.4.4}$$

We remark that, by the reproducing property,

$$-\sum_{\mu,\nu=1}^{N} L_{\mathbf{F}}(q_\mu, \tilde{q}_\nu) x_\mu \bar{x}_\nu = N_{\mathfrak{M}}\left(\sum_{\mu=1}^{N} \bar{x}_\mu L_{\mathbf{F}}(p, \tilde{q}_\mu)\right) \geqq 0. \tag{5.4.5}$$

Similarly

$$\sum_{\mu,\nu=1}^{N} \Gamma_{\mathbf{F}}(\tilde{q}_\nu, q_\mu) x_\mu \bar{x}_\nu = N_{\mathfrak{R}-\mathfrak{M}}\left(\sum_{\mu=1}^{N} x_\mu \mathscr{L}_{\mathbf{F}}(p, q_\mu)\right) \geqq 0. \tag{5.4.6}$$

Hence

$$\sum_{\mu,\nu=1}^{N} L_{\mathbf{F}}(q_\mu, \tilde{q}_\nu) x_\mu \bar{x}_\nu \cdot \sum_{\mu,\nu=1}^{N} \Gamma_{\mathbf{F}}(\tilde{q}_\nu, q_\mu) x_\mu \bar{x}_\nu \leqq 0, \tag{5.4.7}$$

and it follows from (5.4.4) that the inequality

$$(5.4.8)\qquad \left|\sum_{\mu,\nu=1}^{N} l_{\mathbf{F}}(q_\mu, q_\nu)x_\mu x_\nu\right|^2 \leqq \sum_{\mu,\nu=1}^{N} L_{\mathbf{F}}(q_\mu, \tilde{q}_\nu)x_\mu \bar{x}_\nu \cdot \sum_{\mu,\nu=1}^{N} l_{\mathbf{F}}(\tilde{q}_\nu, q_\mu)x_\mu \bar{x}_\nu$$

is true a fortiori. In particular, we observe from (5.4.5) and (5.4.8) that

$$(5.4.9)\qquad \sum_{\mu,\nu=1}^{N} l_{\mathbf{F}}(\tilde{q}_\nu, q_\mu)x_\mu \bar{x}_\nu \leqq 0.$$

Instead of the norm (5.4.2), we can take

$$(5.4.10)\qquad N_{\mathfrak{M}}\left(\sum_{\mu=1}^{N} \bar{x}_\mu l_{\mathbf{F}}(p, \tilde{q}_\mu) - \lambda \sum_{\mu=1}^{N} \bar{x}_\mu L_{\mathbf{F}}(p, \tilde{q}_\mu)\right).$$

Using the formula (5.3.7) we find that the value of this integral is equal to

$$(5.4.11)\qquad \begin{aligned} &-|\lambda|^2 \sum_{\mu,\nu=1}^{N} L_{\mathbf{F}}(q_\mu, \tilde{q}_\nu)x_\mu \bar{x}_\nu + 2\,\mathrm{Re}\left\{\bar{\lambda} \sum_{\mu,\nu=1}^{N} l_{\mathbf{F}}(q_\mu, \tilde{q}_\nu)x_\mu \bar{x}_\nu\right\} \\ &\qquad - \sum_{\mu,\nu=1}^{N} [l_{\mathbf{F}}(q_\mu, \tilde{q}_\nu) + \Gamma_{\mathbf{F}}(q_\mu, \tilde{q}_\nu)]x_\mu \bar{x}_\nu. \end{aligned}$$

Since the expression (5.4.11) is non-negative for all complex λ, we have

$$(5.4.12)\qquad \begin{aligned} &\left|\sum_{\mu,\nu=1}^{N} l_{\mathbf{F}}(q_\mu, \tilde{q}_\nu)x_\mu \bar{x}_\nu\right|^2 \\ &\leqq \sum_{\mu,\nu=1}^{N} L_{\mathbf{F}}(q_\mu, \tilde{q}_\nu)x_\mu \bar{x}_\nu \cdot \sum_{\mu,\nu=1}^{N} [l_{\mathbf{F}}(q_\mu, \tilde{q}_\nu) + \Gamma_{\mathbf{F}}(q_\mu, \tilde{q}_\nu)]x_\mu \bar{x}_\nu. \end{aligned}$$

Now

$$(5.4.13)\qquad \sum_{\mu,\nu=1}^{N} \Gamma_{\mathbf{F}}(q_\mu, \tilde{q}_\nu)x_\mu \bar{x}_\nu = N_{\mathfrak{R}-\mathfrak{M}}\left(\sum_{\mu=1}^{N} \bar{x}_\mu \mathscr{L}_{\mathbf{F}}(p, \tilde{q}_\mu)\right) \geqq 0,$$

and it follows from (5.4.5), (5.4.12) and (5.4.13) that

$$(5.4.14)\qquad \sum_{\mu,\nu=1}^{N} l_{\mathbf{F}}(q_\mu, \tilde{q}_\nu)x_\mu \bar{x}_\nu \leqq 0.$$

For simplicity, write

$$(5.4.15)\qquad L = \sum_{\mu,\nu=1}^{N} L_{\mathbf{F}}(q_\mu, \tilde{q}_\nu)x_\mu \bar{x}_\nu,\quad \mathscr{L} = \sum_{\mu,\nu=1}^{N} \mathscr{L}_{\mathbf{F}}(q_\mu, \tilde{q}_\nu)x_\mu \bar{x}_\nu,$$

$$\Gamma = \sum_{\mu,\nu=1}^{N} \Gamma_{\mathbf{F}}(q_\mu, \tilde{q}_\nu)x_\mu \bar{x}_\nu,$$

and let

$$l = \sum_{\mu,\nu=1}^{N} l_F(q_\mu, \tilde{q}_\nu) x_\mu \bar{x}_\nu = L - \mathscr{L}. \tag{5.4.16}$$

By (5.4.14)

$$L \leqq \mathscr{L} \leqq 0, \tag{5.4.17}$$

$$|\mathscr{L}| \leqq |L|, \tag{5.4.18}$$

and therefore

$$|l| = |L| - |\mathscr{L}|. \tag{5.4.19}$$

The inequality (5.4.12) becomes

$$\begin{aligned} l^2 - |L|\,|l| + \Gamma|L| &= |l|\,\{|l| - |L|\} + \Gamma|L| \\ &= -|\mathscr{L}|\,\{|L| - |\mathscr{L}|\} + \Gamma|L| \leqq 0, \end{aligned}$$

and therefore

$$|\mathscr{L}|^2 \leqq (|\mathscr{L}| - \Gamma)\,|L|. \tag{5.4.20}$$

Thus $|\mathscr{L}| \geqq \Gamma$ and, if $|\mathscr{L}| > \Gamma$, this formula provides an estimate for $|L|$. From (5.4.13) and a formula analogous to (4.9.2) with respect to $\mathfrak{R}$, we have

$$\begin{aligned} \Gamma &= N_{\mathfrak{R}}\left(\sum_{\mu=1}^{N} \bar{x}_\mu \mathscr{L}_F(p, \tilde{q}_\mu)\right) - N_{\mathfrak{M}}\left(\sum_{\mu=1}^{N} \bar{x}_\mu \mathscr{L}_F(p, \tilde{q}_\mu)\right) \\ &= |\mathscr{L}| - N_{\mathfrak{M}}\left(\sum_{\mu=1}^{N} \bar{x}_\mu \mathscr{L}_F(p, \tilde{q}_\mu)\right). \end{aligned} \tag{5.4.21}$$

Hence (5.4.20) may be written

$$|L| \geqq \frac{|\mathscr{L}|^2}{N_{\mathfrak{M}}\left(\sum_{\mu=1}^{N} \bar{x}_\mu \mathscr{L}_F(p, \tilde{q}_\mu)\right)} \geqq |\mathscr{L}|. \tag{5.4.22}$$

Inequality (5.4.22) is thus a strengthened version of (5.4.18). Taking $N = 1$, $x_1 = 1$, and dropping the subscript 1 on q, we have

$$L = L_F(q, \tilde{q}), \ \mathscr{L} = \mathscr{L}_F(q, \tilde{q}), \ \Gamma = \Gamma_F(q, \tilde{q}). \tag{5.4.23}$$

In this particular case inequality (5.4.22) takes the form

$$|L_F(q, \tilde{q})| \geqq \frac{(\mathscr{L}_F(q, \tilde{q}))^2}{N_{\mathfrak{M}}(\mathscr{L}_F(p, \tilde{q}))}. \tag{5.4.24}$$

We recognize that (5.4.24) is simply the inequality (5.3.26) written in a slightly different form.

An inequality may also be obtained by use of the identity (5.3.12). Write

$$(5.4.25) \qquad \Lambda_{\mu\nu} = (d\mathfrak{z}_\mu, d\mathfrak{z}_\nu)_{\mathfrak{R}-\mathfrak{M}},$$

and consider the norm

$$(5.4.26) \qquad N_{\mathfrak{M}}\left(\sum_{\mu=1}^{N} x_\mu \mathfrak{z}'_\mu(p) - \lambda \sum_{\mu=1}^{N} \bar{x}_\mu L_{\mathrm{M}}(p, \bar{q}_\mu)\right).$$

Using formula (5.3.12) we see that this integral is equal to

$$(5.4.27) \qquad \begin{aligned} -|\lambda|^2 \sum_{\mu,\nu=1}^{N} L_{\mathrm{M}}(q_\mu, \bar{q}_\nu) x_\mu \bar{x}_\nu + 2 \operatorname{Re}\left\{\bar{\lambda} \sum_{\mu,\nu=1}^{N} \mathfrak{z}'_\nu(q_\mu) x_\mu x_\nu\right\} \\ - \sum_{\mu,\nu=1}^{N} [\gamma_{\mu\nu} + \Lambda_{\mu\nu}] x_\mu \bar{x}_\nu. \end{aligned}$$

Since this expression is non-negative for all choices of λ, we obtain the inequality

$$(5.4.28) \qquad \begin{aligned} &\left|\sum_{\mu,\nu=1}^{N} \mathfrak{z}'_\nu(q_\mu) x_\mu x_\nu\right|^2 \\ &\leq \sum_{\mu,\nu=1}^{N} L_{\mathrm{M}}(q_\mu, \bar{q}_\nu) x_\mu \bar{x}_\nu \cdot \sum_{\mu,\nu=1}^{N} [\gamma_{\mu\nu} + \Lambda_{\mu\nu}] x_\mu \bar{x}_\nu, \end{aligned}$$

Since

$$\sum_{\mu,\nu=1}^{N} \Lambda_{\mu\nu} x_\mu \bar{x}_\nu = N_{\mathfrak{R}-\mathfrak{M}}\left(\sum_{\mu=1}^{N} x_\mu \mathfrak{z}'_\mu(p)\right) \geq 0,$$

we have a fortiori

$$(5.4.29) \qquad \left|\sum_{\mu,\nu=1}^{N} \mathfrak{z}'_\nu(q_\mu) x_\mu x_\nu\right|^2 \leq \sum_{\mu,\nu=1}^{N} L_{\mathrm{M}}(q_\mu, \bar{q}_\nu) x_\mu \bar{x}_\nu \cdot \sum_{\mu,\nu=1}^{N} \gamma_{\mu\nu} x_\mu \bar{x}_\nu.$$

We conclude in particular that

$$\sum_{\mu,\nu=1}^{N} \gamma_{\mu\nu} x_\mu \bar{x}_\nu = \sum_{\mu,\nu=1}^{N} (\Pi_{\mu\nu} - \Gamma_{\mu\nu}) x_\mu \bar{x}_\nu \leq 0,$$

that is,

$$(5.4.30) \qquad \sum_{\mu,\nu=1}^{N} \Pi_{\mu\nu} x_\mu \bar{x}_\nu \leq \sum_{\mu,\nu=1}^{N} \Gamma_{\mu\nu} x_\mu \bar{x}_\nu.$$

Here both sums are non-negative.

We can even derive from (5.4.28) the better estimate

$$\sum_{\mu,\nu=1}^{N} (\gamma_{\mu\nu} + \Lambda_{\mu\nu}) x_\mu \bar{x}_\nu \leqq 0;$$

that is,

$$(5.4.31) \quad -\sum_{\mu,\nu=1}^{N} \gamma_{\mu\nu} x_\mu \bar{x}_\nu \geqq \sum_{\mu,\nu=1}^{N} \Lambda_{\mu\nu} x_\mu \bar{x}_\nu = N_{\mathfrak{R}-\mathfrak{M}}\left(\sum_{\mu=1}^{N} x_\mu \mathfrak{z}'_\mu(p)\right).$$

5.5. Extension of a Local Complex Analytic Imbedding of One Surface in Another

We now apply the inequality (5.4.8) to obtain necessary and sufficient conditions in order that a local analytic mapping of a neighborhood in a given surface $\mathfrak{M}^*$ onto a neighborhood in another surface $\mathfrak{R}$ can be extended over the whole of $\mathfrak{M}^*$ to give a one- one conformal mapping of $\mathfrak{M}^*$ onto a subdomain of $\mathfrak{R}$. In other words, we derive necessary and sufficient conditions in order that a local complex analytic imbedding be a complex analytic imbedding in the large. For plane domains these conditions have been stated when the surface $\mathfrak{R}$ is the sphere (see [1]).

We shall assume throughout this and the following sections of this chapter that the class $\mathbf{F}_{\mathfrak{R}}$ is symmetric, in which case we may assume in the preceding section the symmetry relations

$$(5.5.1) \qquad L_{\mathbf{F}}(p, q) = L_{\mathbf{F}}(q, p), \ \mathscr{L}_{\mathbf{F}}(p, q) = \mathscr{L}_{\mathbf{F}}(q, p).$$

The inequality (5.4.8) may then be written

$$(5.5.2) \quad \left|\sum_{\mu,\nu=1}^{N} l_{\mathbf{F}}(q_\mu, q_\nu) x_\mu x_\nu\right|^2 \leqq \sum_{\mu,\nu=1}^{N} L_{\mathbf{F}}(q_\mu, \tilde{q}_\nu) x_\mu \bar{x}_\nu \cdot \sum_{\mu,\nu=1}^{N} l_{\mathbf{F}}(q_\mu, \tilde{q}_\nu) x_\mu \bar{x}_\nu.$$

Since

$$(5.5.3) \qquad \sum_{\mu,\nu=1}^{N} L_{\mathbf{F}}(q_\mu, \tilde{q}_\nu) x_\mu \bar{x}_\nu \leqq 0, \ \sum_{\mu,\nu=1}^{N} \mathscr{L}_{\mathbf{F}}(q_\mu, \tilde{q}_\nu) x_\mu \bar{x}_\nu \leqq 0,$$

we have a fortiori

$$(5.5.4) \qquad \left|\sum_{\mu,\nu=1}^{N} l_{\mathbf{F}}(q_\mu, q_\nu) x_\mu x_\nu\right|^2 \leqq \left(\sum_{\mu,\nu=1}^{N} L_{\mathbf{F}}(q_\mu, \tilde{q}_\nu) x_\mu \bar{x}_\nu\right)^2,$$

or

$$(5.5.5) \qquad \left|\sum_{\mu,\nu=1}^{N} l_{\mathbf{F}}(q_\mu, q_\nu) x_\mu x_\nu\right| \leqq -\sum_{\mu,\nu=1}^{N} L_{\mathbf{F}}(q_\mu, \tilde{q}_\nu) x_\mu \bar{x}_\nu.$$

Given an inequality such as (5.5.5) for a quadratic form, we can pass immediately to an inequality for the corresponding bilinear form. In fact, if $y_1, y_2, \cdots, y_N$ are an independent set of complex numbers, we have

$$4 \sum_{\mu,\nu=1}^{N} l_{\mathbf{F}}(q_\mu, q_\nu)x_\mu y_\nu = \sum_{\mu,\nu=1}^{N} l_{\mathbf{F}}(q_\mu, q_\nu)(x_\mu + y_\mu)(x_\nu + y_\nu)$$
$$- \sum_{\mu,\nu=1}^{N} l_{\mathbf{F}}(q_\mu, q_\nu)(x_\mu - y_\mu)(x_\nu - y_\nu).$$

Therefore by (5.5.5)

$$4\left| \sum_{\mu,\nu=1}^{N} l_{\mathbf{F}}(q_\mu, q_\nu)x_\mu y_\nu \right|$$
$$\leqq - \sum_{\mu,\nu=1}^{N} L_{\mathbf{F}}(q_\mu, \tilde{q}_\nu)\left[(x_\mu + y_\mu)(\bar{x}_\nu + \bar{y}_\nu) + (x_\mu - y_\mu)(\bar{x}_\nu - \bar{y}_\nu)\right];$$

that is

$$\left| \sum_{\mu,\nu=1}^{N} l_{\mathbf{F}}(q_\mu, q_\nu)x_\mu y_\nu \right| \leqq -\frac{1}{2} \sum_{\mu,\nu=1}^{N} L_{\mathbf{F}}(q_\mu, \tilde{q}_\nu)(x_\mu \bar{x}_\nu + y_\mu \bar{y}_\nu). \tag{5.5.6}$$

By a limiting process it is possible to pass from the inequalities (5.5.5) to inequalities between integrals. Let p_0 be a point in the interior of $\mathfrak{M}$, and let z_0 be a uniformizer at p_0. We denote the circle $|z_0| \leqq a$, $a > 0$, by $\mathfrak{M}_0$ and, if p, q lie in $\mathfrak{M}_0$, we express the coordinates of p by z, those of q by ζ; that is $z = z_0(p)$, $\zeta = z_0(q)$. Now let $\varrho(z_0)$ be a continuous complex-valued function of z_0 on the boundary $\partial\mathfrak{M}_0$ of $\mathfrak{M}_0$. By a limiting process we obtain from (5.5.5) the inequality

$$\left| \int_{\partial\mathfrak{M}_0} \int_{\partial\mathfrak{M}_0} l_{\mathbf{F}}(p, q)\varrho(z)\varrho(\zeta)dz d\zeta \right| \leqq - \int_{\partial\mathfrak{M}_0} \int_{\partial\mathfrak{M}_0} L_{\mathbf{F}}(p, \tilde{q})\varrho(z)(\varrho(\zeta))^{-} dz d\bar{\zeta}. \tag{5.5.7}$$

In $\mathfrak{M}_0$ we have the series developments

$$l_{\mathbf{F}}(p, q) = \sum_{\mu,\nu=0}^{\infty} a_{\mu\nu} z^\mu \zeta^\nu, \quad -L_{\mathbf{F}}(p, \tilde{q}) = \sum_{\mu,\nu=0}^{\infty} b_{\mu\nu} z^\mu \bar{\zeta}^\nu, \tag{5.5.8}$$

where, because of the symmetry assumptions (5.5.1),

$$a_{\mu\nu} = a_{\nu\mu}, \quad b_{\mu\nu} = \bar{b}_{\nu\mu}. \tag{5.5.9}$$

Taking

$$\varrho(z) = \frac{1}{2\pi i} \sum_{\mu=0}^{N} \alpha_\mu z^{-(\mu+1)} \tag{5.5.10}$$

in (5.5.7), we obtain

$$\left| \sum_{\mu,\nu=0}^{N} a_{\mu\nu}\alpha_\mu\alpha_\nu \right| \leqq \sum_{\mu,\nu=0}^{N} b_{\mu\nu}\alpha_\mu\bar{\alpha}_\nu. \tag{5.5.11}$$

The inequalities (5.5.11) are valid for arbitrary choice of the finite set $\alpha_0, \alpha_1, \cdots, \alpha_N$. Thus inequalities for the bilinear differentials are easily transformed into similar inequalitites for the coefficients in their local developments.

Let us consider now an orientable surface $\mathfrak{N}$ with boundary C. In saying that the surface $\mathfrak{N}$ is complex analytically imbedded in the surface $\mathfrak{R}$, we mean that there is a one-one conformal mapping of $\mathfrak{N}$ onto a subdomain $\mathfrak{M}$ of $\mathfrak{R}$.

In order that such an imbedding be possible at all, we must assume that a topological imbedding of $\mathfrak{N}$ into $\mathfrak{R}$ exists. If $\mathfrak{M}'$ is the image of $\mathfrak{N}$ in $\mathfrak{R}$ under the topological mapping, there is a correspondence between the cycles of $\mathfrak{N}$ and $\mathfrak{M}'$. By introducing on $\mathfrak{M}'$ as uniformizers the parameters at the corresponding points of $\mathfrak{R}$, we make $\mathfrak{M}'$ into a Riemann surface analytically imbedded in $\mathfrak{R}$. Given a class $\mathbf{F}_{\mathfrak{R}}$, let $\mathbf{F}_{\mathfrak{M}'}$ be a corresponding class on $\mathfrak{M}'$ in the sense of Section 5.1. We then choose $\mathbf{F}_{\mathfrak{N}}$ to be that class of differentials on $\mathfrak{N}$ which have vanishing periods on those cycles of $\mathfrak{N}$ which correspond to the cycles associated with the class $\mathbf{F}_{\mathfrak{M}'}$. The question now arises whether this topological imbedding can be realized analytically.

Of course, different topological imbeddings give rise to different analytic imbeddings, provided the latter exist. Consider, for example, the case of two circular rings $\mathfrak{N}$ and $\mathfrak{R}$ in the complex plane. The ring $\mathfrak{N}$ may be topologically imbedded in $\mathfrak{R}$ in two different ways according as the essential cycle of $\mathfrak{N}$ bounds in $\mathfrak{R}$ or not. The first type of topological imbedding is always realizable analytically while the second is possible only if the moduli of $\mathfrak{N}$ and $\mathfrak{R}$ satisfy a suitable inequality.

The choice of a corresponding class $\mathbf{F}_{\mathfrak{M}'}$ depends on the topological type of the imbedding. For example, in the case of the rings $\mathfrak{N}$ and $\mathfrak{R}$, we have two classes of differentials in each domain — namely the classes **S** and **M**. In the first type of imbedding, the classes **M** and **S** on $\mathfrak{N}$ correspond both to the class **M** and to the class **S** on $\mathfrak{R}$ while, in the second type of imbedding, the class **S** on $\mathfrak{N}$ corresponds only to the class **S** on $\mathfrak{R}$.

If there is a conformal mapping of $\mathfrak{N}$ into $\mathfrak{R}$, let the points π, $\varkappa$ of $\mathfrak{N}$ correspond to the points p, q respectively of $\mathfrak{R}$. Then, if $\Lambda_{\mathbf{F}}(\pi, \varkappa)$ is the bilinear differential of $\mathfrak{N}$, we have

$$\Lambda_{\mathbf{F}}(\pi, \varkappa)d\pi d\varkappa = L_{\mathbf{F}}(p, q)dpdq \tag{5.5.12}$$

where $L_{\mathbf{F}}(p, q)$ is the bilinear differential of $\mathfrak{M}$. Again let

$$\begin{aligned} l_{\mathbf{F}}(p, q)dpdq &= [L_{\mathbf{F}}(p, q) - \mathscr{L}_{\mathbf{F}}(p, q)]dpdq \\ &= \Lambda_{\mathbf{F}}(\pi, \varkappa)d\pi d\varkappa - \mathscr{L}_{\mathbf{F}}(p, q)dpdq. \end{aligned} \tag{5.5.13}$$

Writing

$$d_{\mathbf{F}}(\pi, \varkappa) = \Lambda_{\mathbf{F}}(\pi, \varkappa) - \mathscr{L}_{\mathbf{F}}(p, q)\frac{dp}{d\pi}\frac{dq}{d\varkappa}, \tag{5.5.14}$$

the inequality (5.5.5) becomes

$$\left| \sum_{\mu, \nu=1}^{N} d_{\mathbf{F}}(\pi_\mu, \pi_\nu)x_\mu x_\nu \right| \leqq - \sum_{\mu, \nu=1}^{N} \Lambda_{\mathbf{F}}(\pi_\mu, \tilde{\pi}_\nu)x_\mu \bar{x}_\nu. \tag{5.5.15}$$

The inequalities (5.5.15) are obtained from (5.5.5), the latter being derived from (5.5.2) by neglecting certain non-negative terms. If we do not neglect these terms, we obtain instead of (5.5.15)

$$\left| \sum_{\mu, \nu=1}^{N} d_{\mathbf{F}}(\pi_\mu, \pi_\nu)x_\mu x_\nu \right|^2 \leqq \sum_{\mu, \nu=1}^{N} \Lambda_{\mathbf{F}}(\pi_\mu, \tilde{\pi}_\nu)x_\mu \bar{x}_\nu \cdot \sum_{\mu, \nu=1}^{N} d_{\mathbf{F}}(\pi_\mu, \tilde{\pi}_\nu)x_\mu \bar{x}_\nu, \tag{5.5.16}$$

where

$$d_{\mathbf{F}}(\pi, \tilde{\varkappa}) = \Lambda_{\mathbf{F}}(\pi, \tilde{\varkappa}) - \mathscr{L}_{\mathbf{F}}(p, \tilde{q})\frac{dp}{d\pi}\frac{d\tilde{q}}{d\tilde{\varkappa}}. \tag{5.5.17}$$

The inequalities (5.5.16) are necessary conditions in order that $\mathfrak{N}$ can be complex analytically imbedded in $\mathfrak{R}$.

Now assume that the surface $\mathfrak{N}$ is locally imbedded complex analytically in $\mathfrak{R}$. That is to say, assume that a neighborhood of a point π_0 of $\mathfrak{N}$ is mapped one-one and conformally onto a neighborhood in $\mathfrak{R}$. Then for π, $\varkappa$ in a neighborhood of π_0 we have the local series developments with $z = z(\pi)$, $\zeta = \zeta(\varkappa)$:

$$\begin{aligned} d_{\mathbf{F}}(\pi, \varkappa) &= \sum_{\mu, \nu=0}^{\infty} c_{\mu\nu} z^\mu \zeta^\nu, \quad -d_{\mathbf{F}}(\pi, \tilde{\varkappa}) = \sum_{\mu, \nu=0}^{\infty} C_{\mu\nu} z^\mu \bar{\zeta}^\nu, \\ -\Lambda_{\mathbf{F}}(\pi, \tilde{\varkappa}) &= \sum_{\mu, \nu=0}^{\infty} b_{\mu\nu} z^\mu \bar{\zeta}^\nu, \end{aligned} \tag{5.5.18}$$

and (5.5.15), (5.5.16) may be transformed respectively into the inequalitites

$$\left| \sum_{\mu,\nu=0}^{N} c_{\mu\nu} x_\mu x_\nu \right| \leq \sum_{\mu,\nu=0}^{N} b_{\mu\nu} x_\mu \bar{x}_\nu, \tag{5.5.19}$$

$$\left| \sum_{\mu,\nu=0}^{N} c_{\mu\nu} x_\mu x_\nu \right|^2 \leq \sum_{\mu,\nu=0}^{N} b_{\mu\nu} x_\mu \bar{x}_\nu \cdot \sum_{\mu,\nu=0}^{N} C_{\mu\nu} x_\mu \bar{x}_\nu. \tag{5.5.20}$$

We investigate the circumstances under which this local imbedding can be extended to a complex analytic imbedding in the large, and we base our considerations on the following theorem:

THEOREM 5.5.1. *Let*

$$V(\pi, \varkappa) = \sum_{\mu,\nu=0}^{\infty} v_{\mu\nu} z^\mu \zeta^\nu, \quad v_{\mu\nu} = v_{\nu\mu}, \tag{5.5.21}$$

be an analytic bilinear differential defined for π, $\varkappa$ in a neighborhood of the point π_0 of $\mathfrak{R}$. If the kernel $\Lambda_{\mathbf{F}}(\pi, \tilde{\varkappa})$ of $\mathfrak{R}$ has the local development

$$-\Lambda_{\mathbf{F}}(\pi, \tilde{\varkappa}) = \sum_{\mu,\nu=0}^{\infty} b_{\mu\nu} z^\mu \bar{\zeta}^\nu \tag{5.5.22}$$

and if for every finite set of complex numbers $x_0, x_1, \cdots, x_N$ the inequalities

$$\left| \sum_{\mu,\nu=0}^{N} v_{\mu\nu} x_\mu x_\nu \right| \leq \sum_{\mu,\nu=0}^{N} b_{\mu\nu} x_\mu \bar{x}_\nu \tag{5.5.23}$$

are satisfied, then (5.5.21) *is the local development of a symmetric analytic bilinear differential $V(\pi, \varkappa)$, of the class* **F** *on $\mathfrak{R}$, which is everywhere regular.*

Similarly, if

$$W(\pi, \tilde{\varkappa}) = \sum_{\mu,\nu=0}^{\infty} w_{\mu\nu} z^\mu \bar{\zeta}^\nu, \quad w_{\mu\nu} = \bar{w}_{\nu\mu}, \tag{5.5.24}$$

is defined for π, $\varkappa$ near π_0 and if

$$0 \leq \sum_{\mu,\nu=0}^{N} w_{\mu\nu} x_\mu \bar{x}_\nu \leq \sum_{\mu,\nu=0}^{N} b_{\mu\nu} x_\mu \bar{x}_\nu \tag{5.5.25}$$

for every finite set of complex numbers $x_0, x_1, \cdots, x_N$, then W can be extended over $\mathfrak{R}$ and defines an analytic Hermitian bilinear differential of the class **F** *in its dependence on π.*

PROOF: Let $\{d\psi_\nu\}$ be a complete orthonormal system for the class **F** of differentials on $\mathfrak{R}$ and, for π near π_0, let

$$\psi'_\mu(\pi) = \sum_{\nu=\mu}^{\infty} \beta_{\mu\nu} z^\nu, \qquad \beta_{\mu\mu} \neq 0. \tag{5.5.26}$$

The existence of such a system has been proved in Section 4.14. For formal reasons it is convenient to define $\beta_{\mu\nu} = 0$ for $\mu > \nu$. The $\beta_{\mu\nu}$ then form an infinite triangular matrix with zeros under the principal diagonal.

We have by (4.9.6)

$$-\Lambda_{\mathbf{F}}(\pi, \tilde{\varkappa}) = \sum_{\mu=0}^{\infty} \psi'_\mu(\pi)\,(\psi'_\mu(\varkappa))^-, \tag{5.5.27}$$

and hence by (5.5.22), (5.5.26) and (5.5.27)

$$b_{\mu\nu} = \sum_{\varrho=0}^{\infty} \beta_{\varrho\mu}\,(\beta_{\varrho\nu})^-. \tag{5.5.28}$$

Since $\beta_{\varrho\mu}$ vanishes for $\varrho > \mu$, the sum in (5.5.28) involves only finitely many terms. We remark next that the infinite triangular matrix $\| \beta_{\mu\nu} \|$ has an inverse matrix $\| \gamma_{\mu\nu} \|$ of the same form and that each element of the inverse matrix can be calculated by finite algebraic operations. Let

$$t_{\mu\nu} = \sum_{\varrho,\,\sigma=0}^{\infty} v_{\varrho\sigma}\gamma_{\varrho\mu}\gamma_{\sigma\nu}, \quad t_{\mu\nu} = t_{\nu\mu}, \tag{5.5.29}$$

define a matrix associated with the coefficient matrix $\| v_{\mu\nu} \|$ of the given local development (5.5.21). Using (5.5.26) we obtain by formal calculations with power series

$$V(\pi, \varkappa) = \sum_{\varrho,\,\sigma=0}^{\infty} v_{\varrho\sigma} z^\varrho \zeta^\sigma = \sum_{\mu,\,\nu=0}^{\infty} t_{\mu\nu}\psi'_\mu(\pi)\psi'_\nu(\varkappa). \tag{5.5.30}$$

We know nothing about the convergence of the last sum and the equality is to be understood in the sense that formal operations on both sides of the equation lead to the same coefficient for each term $z^\varrho\zeta^\sigma$. On the other hand, it should be observed that the numbers $t_{\mu\nu}$ exist since the defining sums (5.5.29) are finite.

Let a_ν, $\nu = 1, 2, \cdots, N$, be arbitrary complex numbers and set

$$\alpha_\mu = \sum_{\nu=0}^{N} \gamma_{\mu\nu} a_\nu. \tag{5.5.31}$$

Then by (5.5.29)

$$\sum_{\mu,\,\nu=0}^{N} t_{\mu\nu} a_\mu a_\nu = \sum_{\varrho,\,\sigma=0}^{N} v_{\varrho\sigma}\alpha_\varrho\alpha_\sigma. \tag{5.5.32}$$

Using the inequality (5.5.23) and the identity (5.5.28), we obtain

$$(5.5.33)\qquad \left|\sum_{\mu,\nu=0}^{N} t_{\mu\nu}a_\mu a_\nu\right| \leqq \sum_{\varrho=0}^{N}\left|\sum_{\mu=0}^{N}\beta_{\varrho\mu}\alpha_\mu\right|^2 = \sum_{\varrho=0}^{N}|a_\varrho|^2$$

for arbitrary choice of the a_μ.

We now extend (5.5.33) from quadratic forms to bilinear forms. We clearly have

$$(5.5.34)\qquad \begin{aligned}&\left|\sum_{\mu,\nu=0}^{N} t_{\mu\nu}a_\mu a_\nu + \sum_{\mu,\nu=0}^{N} t_{\mu\nu}b_\mu b_\nu \pm 2\sum_{\mu,\nu=0}^{N} t_{\mu\nu}a_\mu b_\nu\right|\\ &\qquad\leqq \sum_{\varrho=0}^{N}(|a_\varrho|^2 + |b_\varrho|^2 \pm 2\,\mathrm{Re}\,\{a_\varrho b_\varrho\})\end{aligned}$$

from which we infer that

$$(5.5.35)\qquad \left|\sum_{\mu,\nu=0}^{N} t_{\mu\nu}a_\mu b_\nu\right| \leqq \frac{1}{2}\Big(\sum_{\varrho=0}^{N}|a_\varrho|^2 + \sum_{\varrho=0}^{N}|b_\varrho|^2\Big)$$

for any choice of the finite sets $a_1, a_2, \cdots, a_N$ and $b_1, b_2, \cdots, b_N$.

Choose in particular

$$(5.5.36)\qquad a_\mu = \psi'_\mu(\pi), \quad b_\mu = \psi'_\mu(\varkappa).$$

Then (5.5.35) becomes

$$(5.5.37)\qquad \left|\sum_{\mu,\nu=0}^{N} t_{\mu\nu}\psi'_\mu(\pi)\psi'_\nu(\varkappa)\right| \leqq \frac{1}{2}\left\{\sum_{\mu=0}^{N}|\psi'_\mu(\pi)|^2 + \sum_{\mu=0}^{N}|\psi'_\mu(\varkappa)|^2\right\}.$$

If π and $\varkappa$ are interior points of $\mathfrak{R}$ we may let N tend to infinity and we obtain by (5.5.27)

$$(5.5.38)\qquad \left|\sum_{\mu,\nu=0}^{\infty} t_{\mu\nu}\psi'_\mu(\pi)\psi'_\nu(\varkappa)\right| \leqq -\frac{1}{2}\{\Lambda_{\mathbf{F}}(\pi, \tilde{\pi}) + \Lambda_{\mathbf{F}}(\varkappa, \tilde{\varkappa})\}.$$

Let us return to the formal identity (5.5.30) which, by (5.5.38), has now become a valid equation connecting two series uniformly convergent in a neighborhood of the point π_0. Moreover, the second sum in (5.5.30) is regular analytic over the whole of $\mathfrak{R}$ and is obviously of the class **F**. Thus the extension of the local development (5.5.21) over $\mathfrak{R}$ has been effected.

The second part of the theorem is proved in an analogous way. Let

$$(5.5.39)\qquad s_{\mu\nu} = \sum_{\varrho,\sigma=0}^{\infty} w_{\varrho\sigma}\gamma_{\varrho\mu}(\gamma_{\sigma\nu})^-, \qquad s_{\mu\nu} = (s_{\nu\mu})^-,$$

be the matrix associated with the coefficients $w_{\varrho\sigma}$. Again introducing numbers a_ν and α_ν related by (5.5.31), we find

$$\sum_{\mu,\nu=0}^{N} s_{\mu\nu} a_\mu \bar{a}_\nu = \sum_{\varrho,\sigma=0}^{N} w_{\varrho\sigma} \alpha_\varrho \overline{\alpha}_\sigma. \tag{5.5.40}$$

By (5.5.25) and (5.5.28) we obtain

$$0 \leqq \sum_{\mu,\nu=0}^{N} s_{\mu\nu} a_\mu \bar{a}_\nu \leqq \sum_{\varrho=0}^{N} | a_\varrho |^2. \tag{5.5.41}$$

Hence

$$\left| \sum_{\mu,\nu=0}^{N} s_{\mu\nu} a_\mu \bar{b}_\nu \right| \leqq \frac{1}{2} \left(\sum_{\varrho=0}^{N} | a_\varrho |^2 + \sum_{\varrho=0}^{N} | b_\varrho |^2 \right). \tag{5.5.42}$$

Thus

$$\left| \sum_{\mu,\nu=0}^{\infty} s_{\mu\nu} \psi'_\mu(\pi) (\psi'_\nu(\varkappa))^- \right| \leqq -\frac{1}{2} \{ \Lambda_{\mathbf{F}}(\pi, \tilde{\pi}) + \Lambda_{\mathbf{F}}(\varkappa, \tilde{\varkappa}) \}. \tag{5.5.43}$$

For π, $\varkappa$ in a neighborhood of π_0 we have by construction

$$\sum_{\mu,\nu=0}^{\infty} s_{\mu\nu} \psi'_\mu(\pi) (\psi'_\nu(\varkappa))^- = \sum_{\varrho,\sigma=0}^{\infty} w_{\varrho\sigma} z^\varrho \bar{\zeta}^\sigma = W(\pi, \tilde{\varkappa}). \tag{5.5.44}$$

By means of this identity we are able to extend $W(\pi, \tilde{\varkappa})$ over the whole of $\mathfrak{N}$ and it defines a Hermitian bilinear differential which, in its dependence on π, is of the class **F**. This completes the proof of our theorem.

Let us apply the above theorem to a given local complex analytic imbedding of the surface $\mathfrak{N}$ in the surface $\mathfrak{R}$. Assume that a neighborhood $\mathfrak{J}$ of the point π_0 of $\mathfrak{N}$ is mapped one-one and analytically onto a neighborhood of a point p_0 of $\mathfrak{R}$. In other words, if z is a local coordinate near the point π_0 of $\mathfrak{N}$ and if w is a local coordinate near p_0 on $\mathfrak{R}$, we suppose that w is an analytic function of z with a non-vanishing derivative in $\mathfrak{J}$. If π and $\varkappa$ are two points of $\mathfrak{N}$ lying in this neighborhood of π_0 which have coordinates z and ζ respectively, we then have the local developments (5.5.18). Now suppose that the inequalities (5.5.19) are fulfilled for arbitrary choice of the finite set $x_1, x_2, \cdots, x_N$. From Theorem 5.5.1 we conclude that the bilinear differential $d_{\mathbf{F}}(\pi, \varkappa)$ given by (5.5.14) is regular everywhere on $\mathfrak{N}$.

We now extend the local mapping by analytic continuation, and we assume first that the point $\varkappa$ is fixed, $\varkappa \in \mathfrak{J}$. From (5.5.14)

$$dW(p) = dQ(\pi) \tag{5.5.45}$$

where

$$dW(p) = \frac{dq}{d\varkappa}\mathscr{L}_{\mathbf{F}}(p, q)dp \tag{5.5.46}$$

and where $dQ(\pi)$ is everywhere regular on $\mathfrak{R}$. Let $\mathfrak{J}_0$ be a subdomain of $\mathfrak{J}$ containing $\varkappa$ in its interior but with π_0 in its exterior. Integrating (5.5.45) along a path beginning at the point π_0 and avoiding the domain $\mathfrak{J}_0$, we obtain

$$W(p) - W(p_0) = Q(\pi) - Q(\pi_0). \tag{5.5.47}$$

This relation may be used to define $p(\pi)$ outside $\mathfrak{J}$. Moreover, $p(\pi)$ defined in this way will be regular analytic except at points where $W'(p)$ vanishes. But if

$$W'(p) = \frac{dq}{d\varkappa}\mathscr{L}_{\mathbf{F}}(p, q) \tag{5.5.48}$$

vanishes at a point p of $\mathfrak{R}$, then by varying the point $\varkappa$ inside $\mathfrak{J}_0$ we can make $\mathscr{L}_{\mathbf{F}}(p, q)$ different from zero at the point in question. Thus the mapping $p(\pi)$ defined by continuation will be everywhere regular so long as the point p remains in the closure of $\mathfrak{R}$.

In this mapping two distinct points π, $\varkappa$ of $\mathfrak{R}$ go into distinct points p, q of $\mathfrak{R}$. For assume that $p(\pi) = q(\varkappa)$, $\pi \neq \varkappa$. Then $\mathscr{L}_{\mathbf{F}}(p, q)$ becomes infinite in (5.5.14), which contradicts the regularity of $d_{\mathbf{F}}(\pi, \varkappa)$, unless $\dfrac{dp}{d\pi}$ or $\dfrac{dq}{d\varkappa}$ vanishes. But if $\dfrac{dp}{d\pi}$ vanishes, say at a point π_1, then there are two points π', π'' in a neighborhood of π_1 where $\dfrac{dp}{d\pi}$ is different from zero and which map into the same point of $\mathfrak{R}$, contradicting the regularity of $d_{\mathbf{F}}(\pi, \varkappa)$. Thus distinct points of $\mathfrak{R}$ go into distinct points of $\mathfrak{R}$ and $\dfrac{dp}{d\pi}$ is regular analytic and non-vanishing. But then we conclude from (5.5.14) and the regularity of $d_{\mathbf{F}}(\pi, \varkappa)$ that distinct points of $\mathfrak{R}$ correspond to distinct points of $\mathfrak{R}$, and therefore the mapping $p(\pi)$ is single-valued on $\mathfrak{R}$. Hence the correspondence $p(\pi)$ defined by continuation is one-one and analytic so long as the point p remains in the closure of $\mathfrak{R}$.

In order to prove that the point p remains on $\mathfrak{R}$ we must suppose that the stronger inequalities (5.5.20) are valid for arbitrary finite sets $x_1, \cdots, x_N$. Since

$$\sum_{\mu,\nu=0}^{N} b_{\mu\nu} x_\mu \bar{x}_\nu \geqq 0, \tag{5.5.49}$$

by (5.5.19), we have from (5.5.20)

$$\sum_{\mu,\nu=0}^{N} C_{\mu\nu} x_\mu \bar{x}_\nu \geqq 0. \tag{5.5.50}$$

Let

$$D_{\mathbf{F}}(\pi, \tilde{\varkappa}) = d_{\mathbf{F}}(\pi, \tilde{\varkappa}) - \Lambda_{\mathbf{F}}(\pi, \tilde{\varkappa}) = -\mathscr{L}_{\mathbf{F}}(p, \tilde{q}) \frac{dp}{d\pi} \frac{d\tilde{q}}{d\tilde{\varkappa}}. \tag{5.5.51}$$

By (5.5.18)

$$D_{\mathbf{F}}(\pi, \tilde{\varkappa}) = \sum_{\mu,\nu=0}^{\infty} (b_{\mu\nu} - C_{\mu\nu}) z^\mu \bar{\zeta}^\nu. \tag{5.5.52}$$

Moreover

$$0 \leqq \sum_{\mu,\nu=0}^{N} (b_{\mu\nu} - C_{\mu\nu}) x_\mu \bar{x}_\nu \leqq \sum_{\mu,\nu=0}^{N} b_{\mu\nu} x_\mu \bar{x}_\nu. \tag{5.5.53}$$

Thus the condition (5.5.25) is satisfied for $D_{\mathbf{F}}(\pi, \tilde{\varkappa})$, and we have from Theorem 5.5.1 that $D_{\mathbf{F}}(\pi, \tilde{\varkappa})$ can be extended analytically over the whole of $\mathfrak{N}$. If an interior point π maps into a boundary point $p = p(\pi)$ of $\mathfrak{R}$, we see from (5.5.51) and the non-vanishing of $\dfrac{dp}{d\pi}$ that $D_{\mathbf{F}}(\pi, \tilde{\pi})$ is infinite, a contradiction. Thus the mapping $p(\pi)$ carries $\mathfrak{N}$ onto a subdomain of $\mathfrak{R}$, and we have the following theorem:

THEOREM 5.5.2. *A local complex analytic imbedding $p(\pi)$ of $\mathfrak{N}$ into $\mathfrak{R}$ defines an analytic imbedding of the whole of $\mathfrak{N}$ into $\mathfrak{R}$ if and only if*

$$\left| \sum_{\mu,\nu=1}^{N} d_{\mathbf{F}}(\varkappa_\mu, \varkappa_\nu) x_\mu x_\nu \right|^2 \leqq \sum_{\mu,\nu=1}^{N} \Lambda_{\mathbf{F}}(\varkappa_\mu, \tilde{\varkappa}_\nu) x_\mu \bar{x}_\nu \cdot \sum_{\mu,\nu=1}^{N} d_{\mathbf{F}}(\varkappa_\mu, \tilde{\varkappa}_\nu) x_\mu \bar{x}_\nu \tag{5.5.54}$$

for every N-tuple of points $\varkappa_1, \varkappa_2, \cdots, \varkappa_N$ lying in some neighborhood where $p(\pi)$ is defined and for arbitrary choice of the finite set $x_1, x_2, \cdots, x_N$ (x_μ complex). Here

$$d_{\mathbf{F}}(\pi, \varkappa) = \Lambda_{\mathbf{F}}(\pi, \varkappa) - \mathscr{L}_{\mathbf{F}}(p, q) \frac{dp}{d\pi} \frac{dq}{d\varkappa}. \tag{5.5.55}$$

We may also express a necessary and sufficient condition for an analytic imbedding of $\mathfrak{N}$ into $\mathfrak{R}$ in terms of the coefficients of the series development of the map function $p(\pi)$ at a given point $\pi_0 \in \mathfrak{N}$ and in terms of a given local parameter. We have

THEOREM 5.5.3. *Let $p(\pi)$ map a neighborhood of $\pi_0 \in \mathfrak{N}$ into a neighborhood of $p_0 = p(\pi_0) \in \mathfrak{R}$. Using a local parameter $z(\pi)$ around π_0 we may calculate the series developments* (5.5.18). *Then, a necessary and sufficient condition that $p(\pi)$ can be extended over $\mathfrak{N}$ to give an analytic imbedding of $\mathfrak{N}$ into $\mathfrak{R}$ is the validity of the inequalities* (5.5.20) *for every N-tuple of complex numbers x_ν.*

Let us proceed next to the case that $\mathfrak{N}$ is a closed surface. For the sake of simplicity, we restrict ourselves to the consideration of the class **M**. We are sure that the kernels $\Lambda_{\mathbf{M}}(\pi, \varkappa)$ and $\Lambda_{\mathbf{M}}(\pi, \tilde{\varkappa})$ exist in $\mathfrak{N}$ and obtain again (5.5.20) as a necessary condition for an analytic imbedding. But now we can make even a more definite statement. $\mathfrak{N}$ must be mapped into $\mathfrak{R}$ by the function $p(\pi)$ and we must necessarily have the equality

$$\Lambda_{\mathbf{M}}(\pi, \tilde{\varkappa})d\pi d\tilde{\varkappa} = \mathscr{L}_{\mathbf{M}}(p, \tilde{q})dpd\tilde{q} \tag{5.5.56}$$

since there is only one differential $\mathscr{L}_{\mathbf{M}}$ on any surface. Similarly

$$\Lambda_{\mathbf{M}}(\pi, \varkappa)d\pi d\varkappa = \mathscr{L}_{\mathbf{M}}(p, q)dpdq. \tag{5.5.57}$$

Thus, we have identically on $\mathfrak{N}$

$$d_{\mathbf{M}}(\pi, \varkappa) = d_{\mathbf{M}}(\pi, \tilde{\varkappa}) = 0 \tag{5.5.58}$$

and instead of the inequalities (5.5.20), we may write the equalities

$$c_{\mu\nu} = 0, \qquad C_{\mu\nu} = 0. \tag{5.5.59}$$

These conditions are, conversely, sufficient in order that $\mathfrak{N}$ be analytically imbedded into $\mathfrak{R}$ as can be derived from (5.5.57) in just the same way as before.

5.6. APPLICATIONS TO SCHLICHT FUNCTIONS

If $\mathfrak{R}$ is the w-sphere and $\mathfrak{N}$ is a subdomain thereof, a function $p(\pi)$ which gives an imbedding of $\mathfrak{N}$ in $\mathfrak{R}$ is called "schlicht." In particular, $p(\pi)$ may be the identity mapping. Or we may take $\mathfrak{R}$ to be a disc $|w| < a$, in which case the functions $p(\pi)$ become bounded schlicht functions.

We show first how the coefficient problem for schlicht functions in multiply-connected domains of the complex plane can be treated. In this case, we choose $\mathfrak{R}$ to be the whole z-sphere and then the complex variable z is a valid uniformizer at each point of $\mathfrak{R}$ except ∞.

The $\mathscr{L}$-kernels have the simple forms

$$\mathscr{L}(z, \zeta) = \frac{1}{\pi(z-\zeta)^2}, \quad \mathscr{L}(z, \bar{\zeta}) = 0. \tag{5.6.1}$$

Let $\mathfrak{R}$ be a domain in the z-plane which contains the origin and has finite connectivity. Let $w = \varphi(z)$ be regular analytic around the point $z = 0$. Let us assume further that $\varphi(0) = 0$. We want to derive from Theorem 5.5.2 the conditions that w can be extended to a schlicht mapping of $\mathfrak{R}$ into $\mathfrak{R}$. We assume that the class **F** is **S**. In the particular case under consideration, **S** is a symmetric class.

By (5.5.14) and (5.6.1) we have

$$d(z, \zeta) = \Lambda(z, \zeta) - \frac{\varphi'(z)\varphi'(\zeta)}{\pi[\varphi(z)-\varphi(\zeta)]^2}. \tag{5.6.2}$$

We have to develop this expression into a power series around the origin and then insert the coefficients of the development into the inequalities (5.5.19). For this it is convenient to introduce certain sets of auxiliary functions which will provide an interpretation of the inequalities obtained.

Let $A'_\nu(z)$ and $B'_\nu(z)$ be defined by the generating functions

$$\Lambda(z, \zeta) = \sum_{\nu=0}^{\infty} (\nu+1)A'_{\nu+1}(z)\zeta^\nu, \quad \Lambda(z, \bar{\zeta}) = \sum_{\nu=0}^{\infty} (\nu+1)B'_{\nu+1}(z)\bar{\zeta}^\nu. \tag{5.6.3}$$

Clearly, the differentials $A'_\nu(z)$ and $B'_\nu(z)$ are defined over the whole of $\mathfrak{R}$. In fact, denoting by C the boundary of $\mathfrak{R}$, we have

$$\begin{cases} (\nu+1)A'_{\nu+1}(z) = \dfrac{1}{2\pi i}\displaystyle\int_C \frac{\Lambda(z, \zeta)}{\zeta^{\nu+1}}\,d\zeta + \frac{\nu+1}{\pi z^{\nu+2}}, \\[2ex] (\nu+1)(B'_{\nu+1}(z))^- = \dfrac{1}{2\pi i}\displaystyle\int_C \frac{(\Lambda(z, \bar{\zeta}))^-}{\zeta^{\nu+1}}\,d\zeta. \end{cases} \tag{5.6.3)'$$

We see from (5.6.3)' that $B'_\nu(z)$ is regular analytic in $\mathfrak{R}$ while $A'_\nu(z)$ has a pole of order $\nu + 1$ at the origin. Set, therefore,

$$\begin{aligned} A'_\nu(z) &= \frac{1}{\pi}\left[\frac{1}{z^{\nu+1}} + \sum_{\mu=0}^{\infty} (\mu+1)\alpha_{\nu,\mu+1} z^\mu\right], \\ B'_\nu(z) &= -\frac{1}{\pi}\sum_{\mu=0}^{\infty} (\mu+1)\beta_{\nu,\mu+1} z^\mu. \end{aligned} \tag{5.6.4}$$

If z tends to the boundary C of $\mathfrak{N}$, we have

$$\Lambda(z, \zeta)dz = (\Lambda(z, \bar{\zeta})dz)^-$$

and hence, comparing coefficients in (5.6.3) we obtain

$$A'_{\nu+1}(z)dz = (B'_{\nu+1}(z)dz)^- \tag{5.6.5}$$

for z on the boundary C. Thus, with any domain $\mathfrak{N}$ in the complex plane we may associate a sequence of analytic functions A_ν, B_ν which are related by (5.6.5). Since we are dealing with the class **S**, the functions A_ν and B_ν are single-valued on $\mathfrak{N}$. On C we have

$$A_{\nu+1}(z) = (B_{\nu+1}(z))^- + \text{constant}, \tag{5.6.6}$$

and the functions A_ν and B_ν are easily seen to be determined except for an additive constant. By (5.6.3) and (5.6.4) we have the series developments

$$\begin{cases} \Lambda(z, \zeta) = \dfrac{1}{\pi}\left[\dfrac{1}{(z-\zeta)^2} + \displaystyle\sum_{\mu,\nu=1}^{\infty} \mu\nu\alpha_{\mu\nu}z^{\mu-1}\zeta^{\nu-1}\right], \\ \Lambda(z, \bar{\zeta}) = -\dfrac{1}{\pi}\displaystyle\sum_{\mu,\nu=1}^{\infty} \mu\nu\beta_{\mu\nu}z^{\mu-1}\bar{\zeta}^{\nu-1}, \end{cases} \tag{5.6.7}$$

where

$$\alpha_{\mu\nu} = \alpha_{\nu\mu}, \quad \beta_{\mu\nu} = (\beta_{\nu\mu})^-, \tag{5.6.7$'$}$$

by the symmetry property of the Λ-kernels.

Having defined sequences of functions related to a given domain $\mathfrak{N}$, let us now define sequences of functions related to a given function $\varphi(z)$. We consider the generating function

$$\log\frac{\varphi(z) - \varphi(\zeta)}{z - \zeta} = \sum_{\mu,\nu=0}^{\infty} \gamma_{\mu\nu}z^\mu\zeta^\nu. \tag{5.6.8}$$

We have, consequently,

$$\log\left[1 - \frac{\varphi(\zeta)}{\varphi(z)}\right] + \log\frac{\varphi(z)}{z} = -\sum_{\mu=1}^{\infty}\frac{\zeta^\mu}{\mu z^\mu} + \sum_{\mu,\nu=0}^{\infty}\gamma_{\mu\nu}z^\mu\zeta^\nu. \tag{5.6.8$'$}$$

On the other hand, for fixed $z \neq 0$ the function $\log\left[1 - \dfrac{\varphi(\zeta)}{\varphi(z)}\right]$ is analytic in ζ around the origin and may, therefore, be developed into a power series

$$\log\left[1-\frac{\varphi(\zeta)}{\varphi(z)}\right]=\sum_{\mu=1}^{\infty}F_{\mu}\left(\frac{1}{\varphi(z)}\right)\zeta^{\mu}. \tag{5.6.9}$$

It is easily seen that $F_{\mu}(t)$ is a polynomial of degree μ in its argument t. Comparing (5.6.8)′ and (5.6.9) we obtain

$$F_{\nu}\left(\frac{1}{\varphi(z)}\right)=-\frac{1}{\nu z^{\nu}}+\sum_{\mu=0}^{\infty}\gamma_{\mu\nu}z^{\mu}. \tag{5.6.10}$$

For $\nu=0$, we define F_0 identically zero. Thus $-\nu F_{\nu}$ is a polynomial of degree ν in $1/\varphi(z)$ whose principal part at the origin is exactly $1/z^{\nu}$. It is uniquely determined by $\varphi(z)$ except for an additive constant. It is called the ν-th Faber polynomial of the function $\varphi(z)$, and it plays an important role in conformal mapping (see [5]). The above definition of F_{ν} also provides a simple construction for it.

Differentiating (5.6.8) with respect to z and ζ, we obtain

$$\frac{\varphi'(z)\,\varphi'(\zeta)}{\pi[\varphi(z)-\varphi(\zeta)]^2}-\frac{1}{\pi(z-\zeta)^2}=\frac{1}{\pi}\sum_{\mu,\nu=1}^{\infty}\mu\nu\gamma_{\mu\nu}z^{\mu-1}\zeta^{\nu-1}. \tag{5.6.11}$$

From (5.6.2), (5.6.7) and (5.6.11) we obtain

$$d(z,\zeta)=\frac{1}{\pi}\sum_{\mu,\nu=1}^{\infty}\mu\nu(\alpha_{\mu\nu}-\gamma_{\mu\nu})z^{\mu-1}\zeta^{\nu-1}. \tag{5.6.12}$$

Now we are in a position to give necessary and sufficient conditions that $\varphi(z)$ be schlicht, in terms of inequalities which depend on the coefficients $\alpha_{\mu\nu}$ and $\beta_{\mu\nu}$ determined by the domain $\mathfrak{N}$. From (5.5.19) we have

$$\left|\sum_{\mu,\nu=1}^{N}(\alpha_{\mu\nu}-\gamma_{\mu\nu})x_{\mu}x_{\nu}\right|\leqq\sum_{\mu,\nu=1}^{N}\beta_{\mu\nu}x_{\mu}\bar{x}_{\nu}. \tag{5.6.13}$$

The inequalities (5.6.13) are Grunsky's necessary and sufficient conditions that $\varphi(z)$ be schlicht ([2]). Since the $\gamma_{\mu\nu}$ are easily expressed in terms of the coefficients in the Taylor's expansion of $\varphi(z)$ around the origin, the inequalities are conditions on the coefficients of $\varphi(z)$. Unfortunately, these conditions appear in such an implicit form that it is difficult to extract much information from them concerning the possible values of these coefficients.

As another example, we consider bounded schlicht functions φ in a plane domain $\mathfrak{N}$ which satisfy, say, the inequality $|\varphi|\leqq 1$.

In this case $\mathfrak{R}$ is the unit disc and we have

$$(5.6.14)\quad \mathscr{L}(z, \zeta) = \frac{1}{\pi(z-\zeta)^2}, \quad -\mathscr{L}(z, \bar{\zeta}) = \frac{1}{\pi(1-z\bar{\zeta})^2}.$$

Since $\mathscr{L}(z, \bar{\zeta})$ does not enter into the inequalities (5.5.19), we obtain the same conditions (5.6.13) as before if we apply (5.5.19) alone. We therefore apply (5.5.20).

We have to introduce a new sequence of functions connected with $\varphi(z)$. Starting from (5.6.9), we find

$$(5.6.15)\qquad \log\,[1-\varphi(z)(\varphi(\zeta))^-] = \sum_{\mu=0}^{\infty} F_\mu(\overline{\varphi(\zeta)})z^\mu.$$

Set

$$(5.6.15)'\qquad F_\mu(\overline{\varphi(\zeta)}) = \sum_{\nu=0}^{\infty} \lambda_{\mu\nu}\bar{\zeta}^\nu.$$

Then

$$(5.6.16)\qquad \log\,[1-\varphi(z)(\varphi(\zeta))^-] = \sum_{\mu,\nu=0}^{\infty} \lambda_{\mu\nu}z^\mu\bar{\zeta}^\nu.$$

Differentiating this identity with respect to z and $\bar{\zeta}$, we obtain

$$(5.6.17)\qquad \frac{\varphi'(z)(\varphi'(\zeta))^-}{[1-\varphi(z)(\varphi(\zeta))^-]^2} = -\sum_{\mu,\nu=1}^{\infty} \mu\nu\lambda_{\mu\nu}z^{\mu-1}\bar{\zeta}^{\nu-1}.$$

From (5.5.17), (5.5.18) and (5.6.17), we compute

$$(5.6.18)\qquad C_{\mu-1,\,\nu-1} = \frac{\nu\mu}{\pi}(\beta_{\mu\nu} + \lambda_{\mu\nu}).$$

Thus we obtain finally the following necessary and sufficient conditions for the coefficients of bounded schlicht functions:

$$(5.6.19)\quad \left|\sum_{\mu,\nu=1}^{N} (\alpha_{\mu\nu}-\gamma_{\mu\nu})x_\mu x_\nu\right|^2 \leqq \sum_{\mu,\nu=1}^{N} \beta_{\mu\nu}x_\mu\bar{x}_\nu \cdot \sum_{\mu,\nu=1}^{N} (\beta_{\mu\nu}+\lambda_{\mu\nu})x_\mu\bar{x}_\nu.$$

In particular, let $\mathfrak{R}$ be the unit disc with center at the origin. In this case we find easily $\alpha_{\mu\nu} = 0$, $\beta_{\mu\nu} = \delta_{\mu\nu}/\nu$ (where $\delta_{\mu\nu} = 1$ if $\mu = \nu$, 0 otherwise). Thus (5.6.19) simplifies to

$$(5.6.19)'\quad \left|\sum_{\mu,\nu=1}^{N} \gamma_{\mu\nu}x_\mu x_\nu\right|^2 \leqq \sum_{\mu=1}^{N} \frac{1}{\mu}|x_\mu|^2 \cdot \left(\sum_{\mu=1}^{N} \frac{1}{\mu}|x_\mu|^2 + \sum_{\mu,\nu=1}^{N} \lambda_{\mu\nu}x_\mu\bar{x}_\nu\right).$$

Taking, for example, $N = 1$ we have the condition

$$(5.6.19)'' \qquad | a_2^2 - a_3 a_1 |^2 \leqq | a_1 |^4 (1 - | a_1 |^2)$$

for the first three coefficients in the development

$$\varphi(z) = \sum_{\mu=1}^{\infty} a_\mu z^\mu .$$

It is worthwhile to remark that the inequalities (5.5.15) can be applied in a different way to obtain information about schlicht functions in plane domains. For example let $\mathfrak{N}$ be the unit disc; then $\Lambda(z, \zeta)$ and $\Lambda(z, \bar{\zeta})$ are given by (5.6.14). We let

$$(5.6.20) \qquad U(z, \zeta) = \frac{1}{\pi} \left\{ \frac{\varphi'(z)\varphi'(\zeta)}{[\varphi(z) - \varphi(\zeta)]^2} - \frac{1}{[z - \zeta]^2} \right\}.$$

Then $U(z, \zeta)$ is regular in $\mathfrak{N}$ if $\varphi(z)$ is schlicht there. For $z = \zeta$ we find

$$(5.6.21) \quad U(z, z) = \frac{1}{6\pi} \left[\frac{\varphi'''(z)}{\varphi'(z)} - \frac{3}{2} \left(\frac{\varphi''(z)}{\varphi'(z)} \right)^2 \right] = \frac{1}{6\pi} \{\varphi, z\}$$

where $\{\varphi, z\}$ is the Schwarz differential parameter.

In view of (5.6.2) and the fact that $\Lambda(z, \zeta) = 1/\pi(z - \zeta)^2$, we may write (5.5.15) in the form

$$(5.6.22) \qquad \left| \sum_{\mu, \nu=1}^{N} U(z_\mu, z_\nu) x_\mu x_\nu \right| \leqq \frac{1}{\pi} \sum_{\mu, \nu=1}^{N} \frac{x_\mu \bar{x}_\nu}{(1 - z_\mu \bar{z}_\nu)^2}.$$

In the special case $N = 1$ we derive from (5.6.21) and (5.6.22) the estimate

$$(5.6.23) \qquad | \{\varphi, z\} | \leqq \frac{6}{(1 - | z |^2)^2}.$$

The example of the function $z/(1 - z)^2$ shows that the estimate (5.6.23) is best possible.

5.7. Extremal Mappings

In Section 5.5 we established sets of inequalities for mappings of a domain $\mathfrak{N}$ into a domain $\mathfrak{R}$. We now study the class of extremal mappings for which some inequality becomes an equality, and we are thus led to examine the process by which we passed from identities to inequalities.

The inequality (5.5.5) was obtained from (5.5.2) which, in turn, was obtained from (5.4.4). In passing from (5.5.2) to (5.5.5) we neglected the term

$$\sum_{\mu,\nu=1}^{N} \mathscr{L}_{\mathbf{F}}(q_\mu, \tilde{q}_\nu)x_\mu\bar{x}_\nu$$

while, in passing from (5.4.4) to (5.5.2), we neglected

$$\sum_{\mu,\nu=1}^{N} \Gamma_{\mathbf{F}}(q_\mu, \tilde{q}_\nu)x_\mu\bar{x}_\nu.$$

Here

$$\Gamma_{\mathbf{F}}(q_\mu, \tilde{q}_\nu) = (\mathscr{L}_{\mathbf{F}}(p, q_\mu), \mathscr{L}_{\mathbf{F}}(p, q_\nu))_{\mathfrak{R}-\mathfrak{M}}. \tag{5.7.1}$$

If there is equality in (5.5.5), then necessarily

$$\sum_{\mu,\nu=1}^{N} \mathscr{L}_{\mathbf{F}}(q_\mu, \tilde{q}_\nu)x_\mu\bar{x}_\nu = \sum_{\mu,\nu=1}^{N} \Gamma_{\mathbf{F}}(q_\mu, \tilde{q}_\nu)x_\mu\bar{x}_\nu. \tag{5.7.2}$$

But then, since the left side of (5.7.2) is non-positive by (5.5.3), the right non-negative by (5.4.6), each side must be zero. This implies, in particular, because the right side of (5.7.2) is zero,

$$\sum_{\mu=1}^{N} \mathscr{L}_{\mathbf{F}}(p, q_\mu)x_\mu = 0 \tag{5.7.3}$$

at each interior point p of $\mathfrak{R} - \mathfrak{M}$. By analytic continuation the sum in (5.7.3) is therefore identically zero in $\mathfrak{R}$ if there exists an interior point of $\mathfrak{R} - \mathfrak{M}$. On the other hand, $\mathscr{L}_{\mathbf{F}}(p, q_\mu)$ becomes infinite if $p = q_\mu$ and hence it is impossible that (5.7.3) should hold throughout $\mathfrak{R}$. We have therefore proved that, in the case of an extremal mapping, the set $\mathfrak{R} - \mathfrak{M}$ has no interior points. In particular, we conclude from (5.7.1) that

$$\Gamma_{\mathbf{F}}(p, q) = 0 \tag{5.7.4}$$

for an extremal image domain $\mathfrak{M}$.

From the identity

$$-\sum_{\mu,\nu=1}^{N} \mathscr{L}_{\mathbf{F}}(q_\mu, \tilde{q}_\nu)x_\mu\bar{x}_\nu = N_{\mathfrak{R}}\left(\sum_{\mu=1}^{N} \bar{x}_\mu \mathscr{L}_{\mathbf{F}}(p, \tilde{q}_\mu)\right) \tag{5.7.5}$$

and the fact that the left side vanishes for an extremal mapping,

we have

$$\sum_{\mu=1}^{N} \mathcal{L}_{\mathbf{F}}(p, \tilde{q}_\mu)\bar{x}_\mu = 0 \tag{5.7.6}$$

for each p of $\mathfrak{R}$.

If there is equality in (5.5.5) for some particular set $x_1, x_2, \cdots, x_N$, we see moreover that the integral (5.4.2) must vanish for some value of the complex number λ. For this particular value we therefore have

$$\sum_{\mu=1}^{N} x_\mu l_{\mathbf{F}}(p, q_\mu) = \lambda \sum_{\mu=1}^{N} \bar{x}_\mu L_{\mathbf{F}}(p, \tilde{q}_\mu). \tag{5.7.7}$$

Take the scalar product of (5.7.7) and $l_{\mathbf{F}}(p, q)$ with respect to p over $\mathfrak{M}$. By (5.3.3), (5.4.1) and the reproducing property of $L_{\mathbf{F}}(p, \tilde{q})$ we obtain

$$\sum_{\mu=1}^{N} x_\mu[l_{\mathbf{F}}(\tilde{q}, q_\mu) + \Gamma_{\mathbf{F}}(\tilde{q}, q_\mu)] = \lambda \sum_{\mu=1}^{N} \bar{x}_\mu(l_{\mathbf{F}}(q, q_\mu))^-. \tag{5.7.8}$$

In the extremal case now being considered, (5.7.4) and (5.7.6) hold. Hence

$$\sum_{\mu=1}^{N} \bar{x}_\mu L_{\mathbf{F}}(q, \tilde{q}_\mu) = \bar{\lambda} \sum_{\mu=1}^{N} x_\mu l_{\mathbf{F}}(q, q_\mu) \tag{5.7.9}$$

for $q \in \mathfrak{M}$. Comparing (5.7.7) with (5.7.9), we see that $|\lambda| = 1$, and we therefore write (5.7.7) and (5.7.9) in the common form

$$\sum_{\mu=1}^{N} \lambda^{\frac{1}{2}} \bar{x}_\mu L_{\mathbf{F}}(q, \tilde{q}_\mu) = \sum_{\mu=1}^{N} (\lambda^{\frac{1}{2}})^- x_\mu l_{\mathbf{F}}(q, q_\mu). \tag{5.7.10}$$

That is, setting $(\lambda^{\frac{1}{2}})^- x_\mu = y_\mu$ and replacing q by p, we have

$$\sum_{\mu=1}^{N} \bar{y}_\mu L_{\mathbf{F}}(p, \tilde{q}_\mu) = \sum_{\mu=1}^{N} y_\mu l_{\mathbf{F}}(p, q_\mu). \tag{5.7.11}$$

Using (5.5.12) and (5.5.13) we may express (5.7.11) as follows:

$$\sum_{\mu=1}^{N} y_\mu \mathcal{L}_{\mathbf{F}}(p, q_\mu)\frac{dp}{d\pi} = \sum_{\mu=1}^{N} \left[y_\mu \Lambda_{\mathbf{F}}(\pi, \varkappa_\mu)\left(\frac{d\varkappa}{dq}\right)_{q_\mu} - \bar{y}_\mu \Lambda_{\mathbf{F}}(\pi, \tilde{\varkappa}_\mu)\left(\frac{d\tilde{\varkappa}}{d\tilde{q}}\right)_{q_\mu} \right]. \tag{5.7.12}$$

Let π tend to a boundary point of $\mathfrak{N}$, and use a boundary uniformizer at π. Then $\Lambda_{\mathbf{F}}(\pi, \varkappa_\mu) = (\Lambda_{\mathbf{F}}(\pi, \tilde{\varkappa}_\mu))^-$ and hence, in view of (5.7.12),

$$\text{Re}\left\{ \sum_{\mu=1}^{N} y_\mu \mathscr{L}_{\mathbf{F}}(p, q_\mu) \frac{dp}{d\pi} \right\} = 0 \tag{5.7.13}$$

for π on the boundary of $\mathfrak{N}$. This equation determines at each point $\pi \in C$ the direction of the tangent vector dp of the image in $\mathfrak{R}$ of the boundary C of $\mathfrak{N}$. Thus (5.7.13) is a differential equation for the boundary of $\mathfrak{M}$ imbedded in $\mathfrak{R}$. Using a boundary uniformizer, we may integrate (5.7.13) and we obtain

$$\text{Re}\left\{ \sum_{\mu=1}^{N} y_\mu \Xi_{\mathbf{F}}(p, q_\mu) \right\} = \text{constant} \tag{5.7.14}$$

where

$$\frac{\partial}{\partial p} \Xi_{\mathbf{F}}(p, q) = \mathscr{L}_{\mathbf{F}}(p, q). \tag{5.7.14$'$}$$

The boundary curves of $\mathfrak{M}$ are, therefore, the level curves of a harmonic function on $\mathfrak{R}$ which has poles at the points q_μ. These level curves are generalizations of the isothermal curves in the plane which are defined as level curves of rational functions.

Since $\mathfrak{R} - \mathfrak{M}$ has no interior points, the boundary curves (5.7.14) of $\mathfrak{M}$ are piecewise analytic slits lying on $\mathfrak{R}$. We assumed at the beginning of this chapter that the boundary uniformizers of a domain $\mathfrak{N}$ imbedded in $\mathfrak{R}$ are admissible uniformizers at the corresponding points of $\mathfrak{R}$. This restriction may now be removed in the main formulas and identities by a limiting process. For if $\mathfrak{M}$ is any finite Riemann surface imbedded in $\mathfrak{R}$, it can be approximated by a sequence of finite Riemann surfaces $\mathfrak{M}_n$, $\mathfrak{M}_n \subset \mathfrak{M}_{n+1}$, where $\mathfrak{M}_n$ tends to $\mathfrak{M}$ as n becomes infinite. Since all functionals considered are derived from the Green's function and since the Green's function of $\mathfrak{M}_n$ together with its derivatives tends uniformly to the Green's function of $\mathfrak{M}$ in any compact region interior to $\mathfrak{M}$, we see that the functionals of $\mathfrak{M}_n$ converge to the corresponding functionals of $\mathfrak{M}$ in any compact subdomain interior to $\mathfrak{M}$. We shall not carry out the details of the approximation process, but we observe in particular that the inequality (5.4.8) is valid for any finite Riemann surface imbedded in $\mathfrak{R}$. This remark is necessary since we found that the extremal domains $\mathfrak{M}$

are slit domains in $\mathfrak{R}$ with boundary uniformizers which must be singular uniformizers of $\mathfrak{R}$ at a finite number of points.

We have shown that every domain $\mathfrak{M} \subset \mathfrak{R}$ which is obtained from $\mathfrak{R}$ by an extremal mapping into $\mathfrak{R}$ is a slit domain whose boundary slits satisfy a differential equation (5.7.13). We now remark that a partial converse of this statement is valid; namely if $\mathfrak{R}$ is mapped on a slit subdomain $\mathfrak{M}$ of $\mathfrak{R}$ with differential equation (5.7.13) for the boundary slits, where $\mathbf{F}_{\mathfrak{R}}$ is such that $\mathbf{F}_{\mathfrak{M}} = \mathbf{S}$ is a corresponding class, then the mapping solves somes extremum problem for the inequalities (5.5.16). In fact, suppose that $\mathfrak{M}$ is a slit domain whose boundary satisfies (5.7.13). Consider the expression

$$(5.7.15) \quad \sum_{\mu=1}^{N} y_\mu[\mathscr{L}_{\mathbf{F}}(p, q_\mu) - L_{\mathbf{F}}(p, q_\mu)] + \sum_{\mu=1}^{N} \bar{y}_\mu L_{\mathbf{F}}(p, \tilde{q}_\mu) = \Delta(p).$$

It represents a differential on $\mathfrak{M}$, and for p on the boundary of $\mathfrak{M}$ we have, in view of (5.7.13) and the behavior of the $L_{\mathbf{F}}$-kernels there,

$$(5.7.15)' \qquad \operatorname{Re}\{\Delta(p)dp\} = 0.$$

Moreover, $\Delta(p)$ is obviously regular on $\mathfrak{M}$. Thus, since $\mathbf{F}_{\mathfrak{M}} = \mathbf{S}$, $\Delta(p)$ vanishes identically. Then

$$(5.7.16) \qquad -\sum_{\mu=1}^{N} y_\mu l_{\mathbf{F}}(p, q_\mu) + \sum_{\mu=1}^{N} \bar{y}_\mu L_{\mathbf{F}}(p, \tilde{q}_\mu) = 0.$$

Taking the norm over $\mathfrak{M}$ of (5.7.16) we obtain by (5.3.3)

$$(5.7.17) \quad \sum_{\mu,\nu=1}^{N} y_\mu \bar{y}_\nu [l_{\mathbf{F}}(\tilde{q}_\nu, q_\mu) + L_{\mathbf{F}}(q_\mu, \tilde{q}_\nu)] - 2\operatorname{Re}\left\{\sum_{\mu,\nu=1}^{N} y_\mu y_\nu l_{\mathbf{F}}(q_\mu, q_\nu)\right\} = 0.$$

On the other hand, the norm over $\mathfrak{M}$ of the first sum in (5.7.16) must equal that of the second, and hence we have

$$(5.7.17)' \qquad \sum_{\mu,\nu=1}^{N} y_\mu \bar{y}_\nu l_{\mathbf{F}}(\tilde{q}_\nu, q_\mu) = \sum_{\mu,\nu=1}^{N} y_\mu \bar{y}_\nu L_{\mathbf{F}}(q_\mu, \tilde{q}_\nu).$$

Thus, instead of (5.7.17), we may write

$$(5.7.18) \qquad \left[\operatorname{Re}\left\{\sum_{\mu,\nu=1}^{N} y_\mu y_\nu l_{\mathbf{F}}(q_\mu, q_\nu)\right\}\right]^2 = \sum_{\mu,\nu=1}^{N} y_\mu \bar{y}_\nu L_{\mathbf{F}}(q_\mu, \tilde{q}_\nu) \cdot \sum_{\mu,\nu=1}^{N} y_\mu \bar{y}_\nu l_{\mathbf{F}}(\tilde{q}_\nu, q_\mu).$$

Comparing this result with (5.4.8), we see that the domain $\mathfrak{M}$ has extremal character in $\mathfrak{R}$ with respect to that inequality. Referring back to our domain $\mathfrak{N}$, we can express this property as an extremal property with respect to the inequality (5.5.16). Thus the converse of the above result is proved; every domain bounded by slits satisfying the differential equation (5.7.13) for suitable **F** results from an extremal mapping.

5.8. NON-SCHLICHT MAPPINGS

So far we have dealt with the problem of the schlicht imbedding of a domain $\mathfrak{N}$ into another domain $\mathfrak{R}$. The question arises whether similar criteria might be developed for the case of a non-schlicht mapping of $\mathfrak{N}$ into $\mathfrak{R}$. In order to show the use which can be made of our L-differentials let us consider the following special problem.

We consider an orientable surface $\mathfrak{N}$ with boundary and distinguish on it a point $\varkappa_0$. We choose a fixed uniformizer z at $\varkappa_0$ and consider all functions on $\mathfrak{N}$ which have near $\varkappa_0$ a series development

$$(5.8.1) \qquad f(z) = 1 + a_1 z + a_2 z^2 + \cdots + a_n z^n + \cdots$$

and have positive real parts everywhere on $\mathfrak{N}$. We want to give necessary conditions for such functions in terms of their coefficients a_ν. Clearly, this is a special case of the general subordination problem with $\mathfrak{R}$ the half-plane $\mathrm{Re}\, w > 0$.

We derive our conditions on $f(\pi)$ in the following form. We select N points $\varkappa_i$ on $\mathfrak{N}$ and N complex numbers x_i. We form the differential

$$(5.8.2) \qquad H'(\pi) = \sum_{i=1}^{N} \bar{x}_i \Lambda_{\mathbf{M}}(\pi, \tilde{\varkappa}_i)$$

and the integral

$$(5.8.3) \qquad \int_{\mathfrak{N}} | H'(\pi) |^2 \,\mathrm{Re}\, \{f(\pi)\} dA_\pi = J_f.$$

Because of our assumption on $f(\pi)$, the number J_f will be non-negative. On the other hand, we can easily evaluate the integral. Using (4.10.8)' and (5.8.2), we have

$$J_f = \sum_{i,k=1}^{N} \bar{x}_i x_k \int_{\mathfrak{N}} \Lambda_{\mathbf{M}}(\pi, \tilde{\varkappa}_i)(\Lambda_{\mathbf{M}}(\pi, \tilde{\varkappa}_k))^- \cdot \frac{1}{2}[f(\pi)+(f(\pi))^-]dA_\pi \tag{5.8.4}$$
$$= -\frac{1}{2} \sum_{i,k=1}^{N} \bar{x}_i x_k \Lambda_{\mathbf{M}}(\varkappa_k, \tilde{\varkappa}_i)[f(\varkappa_k) + (f(\varkappa_i))^-] \geqq 0.$$

We thus derive the result:

THEOREM 5.8.1. *If $f(\pi)$ is an analytic function on $\mathfrak{N}$ with positive real part, the kernel $-\Lambda_{\mathbf{M}}(\pi, \tilde{\varkappa})[f(\pi) + (f(\varkappa))^-]$ is positive-definite.*

Here $\Lambda_{\mathbf{M}}(\pi, \tilde{\varkappa})$ may be replaced by $\Lambda_{\mathbf{F}}(\pi, \tilde{\varkappa})$ for an arbitrary class $\mathbf{F}$, since $f(\pi)$ is single-valued on $\mathfrak{N}$.

Let

$$-\Lambda_{\mathbf{M}}(\pi, \tilde{\varkappa})[f(\pi) + (f(\varkappa))^-] = \sum_{\mu,\nu=0}^{\infty} c_{\mu\nu} z^\mu \bar{\zeta}^\nu$$

be the series development of the above kernel near $\varkappa_0$ in terms of the local uniformizer $z(\pi) = z$, $z(\varkappa) = \zeta$. If we assume that $\Lambda_{\mathbf{M}}(\pi, \tilde{\varkappa})$ is known, the coefficients $c_{\mu\nu}$ can easily be calculated in terms of the coefficients a_ν of the function $f(\pi)$ considered. From the inequalities (5.8.4) we derive, by the method of Section 5.5, the following inequality in terms of the $c_{\mu\nu}$:

$$\sum_{\mu,\nu=0}^{N} x_\mu \bar{x}_\nu c_{\mu\nu} \geqq 0 \tag{5.8.5}$$

for every choice of the complex N-vector x_μ, $\mu = 1, 2, \cdots, N$.

The inequalities (5.8.5) impose an infinity of conditions on the coefficients of a function $f(\pi)$ on $\mathfrak{N}$ with positive real part. They are closely related to analogous inequalities established by Carathéodory and Toeplitz in the case when the domain $\mathfrak{N}$ is a circle.

It is easy to generalize the above method to a wider class of domains $\mathfrak{R}$. Suppose there exists on $\mathfrak{R}$ a function $\Phi(p)$ which has on $\mathfrak{R}$ a positive real part. In this case, we may give necessary conditions for a function $f(\pi)$ on $\mathfrak{N}$ to map $\mathfrak{N}$ (non-schlicht) into $\mathfrak{R}$. For this purpose we would consider the integral

$$\int_{\mathfrak{N}} |H'(\pi)|^2 \operatorname{Re}\{\Phi(f(\pi))\} dA_\pi = J_{f,\Phi}. \tag{5.8.6}$$

This integral must again be positive and treating it in the above way, we can again derive an infinity of inequalities for the coefficients a_ν of the function $f(\pi)$.

REFERENCES

1. S. BERGMAN and M. SCHIFFER, "Kernel functions and conformal mapping," *Compositio Math.*, 8 (1951), 205—249.

2. H. GRUNSKY, ,,Koeffizientenbedingungen für schlicht abbildende meromorphe Funktionen," *Math. Zeit.*, 45 (1939), 29—61.

3. J. A. JENKINS and D. C. SPENCER, "Hyperelliptic trajectories," *Annals of Math.*, 53 (1951), 4—35.

4. A. C. SCHAEFFER and D. C. SPENCER, *Coefficient regions for schlicht functions*, Colloquium Publications, Vol. 35, Amer. Math. Soc., New York, 1950.

5. M. SCHIFFER, "Faber polynomials in the theory of univalent functions," *Bull. Amer. Math. Soc.*, 54 (1948), 503—517.

6. M. SCHIFFER and D. C. SPENCER, "On the conformal mapping of one Riemann surface into another," *Ann. Sci. Fenn.*, *A. I*, 94 (1951).

6. Integral Operators

6.1. DEFINITION OF THE OPERATORS T, $\tilde{T}$ AND S

We continue to suppose that $\mathfrak{M}$ is a subdomain of a finite Riemann surface $\mathfrak{R}$, and that the classes $\mathbf{F}_{\mathfrak{M}}$ and $\mathbf{F}_{\mathfrak{R}}$ of differentials are related as described at the beginning of the preceding chapter. Moreover, for simplicity, we suppose that $\mathfrak{R}$ is orientable and has a boundary. In this chapter we shall be concerned with integral operators T which transform a differential f' of $\mathbf{F}_{\mathfrak{M}}$ into a differential Tf' on $\mathfrak{R}$.

We shall again denote the bilinear differentials of $\mathfrak{M}$ and $\mathfrak{R}$ by $L_{\mathbf{F}}(p, q)$ and $\mathscr{L}_{\mathbf{F}}(p, q)$, respectively, and we introduce the integrals $\Psi_{\mathbf{F}}(p, q)$, $\Xi_{\mathbf{F}}(p, q)$, $\Xi^*_{\mathbf{F}}(q, p)$:

$$\frac{\partial \Psi_{\mathbf{F}}(p, q)}{\partial p} = L_{\mathbf{F}}(p, q), \qquad \frac{\partial \Xi_{\mathbf{F}}(p, q)}{\partial p} = \mathscr{L}_{\mathbf{F}}(p, q), \tag{6.1.1}$$

$$\frac{\partial \Xi^*_{\mathbf{F}}(q, p)}{\partial p} = \mathscr{L}_{\mathbf{F}}(q, p).$$

We write

$$l_{\mathbf{F}}(p, q) = L_{\mathbf{F}}(p, q) - \mathscr{L}_{\mathbf{F}}(p, q) \tag{6.1.2}$$

(as in Chapter 5). Scalar products and norms over $\mathfrak{M}$, $\mathfrak{R}$ will be distinguished, wherever necessary, by attaching subscripts $\mathfrak{M}$, $\mathfrak{R}$.

From the point of view of the comparison of the two domains $\mathfrak{M}$ and $\mathfrak{R}$, one is led quite naturally to the following question. Scalar multiplication of any differential of $\mathbf{F}_{\mathfrak{M}}$ with the kernels $-L_{\mathbf{F}}(p, \tilde{q})$ and $L_{\mathbf{F}}(p, q)$ leads on the one hand to reproduction of the differential and zero on the other. What then will be the outcome of the same multiplication with the kernels $-\mathscr{L}_{\mathbf{F}}(p, \tilde{q})$ and $\mathscr{L}_{\mathbf{F}}(p, q)$? In each case the result will be a differential on $\mathfrak{R}$ which is a linear functional of the given differential on $\mathfrak{M}$.

To form these linear functionals, define

$$T_f(q) = (\mathscr{L}_{\mathbf{F}}(q, p), f'(p))_{\mathfrak{M}}, \tag{6.1.3}$$

$$(6.1.4)\quad \tilde{T}_f(q) = (f'(p), \mathscr{L}_{\mathbf{F}}(p, \tilde{q}))_{\mathfrak{M}} = (\mathscr{L}_{\mathbf{F}}(q, \tilde{p}), (f'(p))^-)_{\mathfrak{M}},$$

$$(6.1.5)\quad S_f(q) = T_f(q) + \tilde{T}_f(q).$$

The differential $\tilde{T}$ is regular analytic on the whole of $\mathfrak{R}$, while T and S are regular analytic on $\mathfrak{M}$ and on $\mathfrak{R} - \mathfrak{M}$ but, as we shall see, are discontinuous across the boundary of $\mathfrak{M}$. If $\mathscr{L}_{\mathbf{F}}(q, p)$ is symmetric, that is if $\mathscr{L}_{\mathbf{F}}(q, p) = \mathscr{L}_{\mathbf{F}}(p, q)$, then

$$(6.1.6)\quad \mathscr{L}_{\mathbf{F}}(q, \tilde{p}) = \mathscr{L}_{\mathbf{F}}(\tilde{p}, q) = (\mathscr{L}_{\mathbf{F}}(\tilde{q}, p))^-.$$

In this important case we see that, formally at least,

$$(6.1.7)\quad \tilde{T}_f(q) = (T_f(\tilde{q}))^-,$$

so $\tilde{T}$ is obtained from T by applying the "$\sim$-operation" described in Chapter 2. It will often be convenient to extend the definition of a differential of $\mathbf{F}_{\mathfrak{M}}$ over the double $\mathfrak{D}$ of $\mathfrak{R}$ by defining f' to be identically zero in $\mathfrak{R} - \mathfrak{M}$ and setting $f'(\tilde{p}) = (f'(p))^-$. Then

$$(6.1.8)\quad S_f(q) = (\mathscr{L}_{\mathbf{F}}(q, p), f'(p))_{\mathfrak{D}}.$$

Computing periods with respect to q in (6.1.4), we see that $\tilde{T}_f$ satisfies the period conditions for a differential of class $\mathbf{F}_{\mathfrak{R}}$. By (6.2.10) below, $N_{\mathfrak{R}}(\tilde{T}_f) < \infty$, so $\tilde{T}_f$ is in $\mathbf{F}_{\mathfrak{R}}$, and we may therefore think of $\tilde{T}$ as an operator which projects each differential of $\mathbf{F}_{\mathfrak{M}}$ into a corresponding differential of $\mathbf{F}_{\mathfrak{R}}$. If q is a point of $\mathfrak{M}$, the integral on the right of (6.1.3) is to be interpreted in the sense of (4.9.8). However by (4.11.6)

$$(6.1.9)\quad (L_{\mathbf{F}}(q, p), f'(p))_{\mathfrak{M}} = (L_{\mathbf{F}}(p, q), f'(p))_{\mathfrak{M}} = 0$$

since f' belongs to the class $\mathbf{F}_{\mathfrak{M}}$. Thus, if q is a point of $\mathfrak{M}$,

$$(6.1.3)'\quad T_f(q) = -(l_{\mathbf{F}}(q, p), f'(p))_{\mathfrak{M}}$$

and it follows that T_f is regular analytic throughout the closure of $\mathfrak{M}$ and by (5.3.2) is in $\mathbf{F}_{\mathfrak{M}}$ when considered as a differential on $\mathfrak{M}$. From (6.1.3) we see that T_f satisfies the period conditions for a differential of $\mathbf{F}_{\mathfrak{R}}$ on $\mathfrak{R} - \mathfrak{M}$.

Let q be an interior point of $\mathfrak{R}$ lying on the boundary C of $\mathfrak{M}$, and suppose that f' is analytic on an arc of C containing q. Then the difference between the values of $T_f(q)$ on the two edges of C is equal to $(f'(q))^-$. Let ζ be a uniformizer at q which is also a boundary uniformizer for $\mathfrak{M}$. Let $\mathfrak{k}$ in $\mathfrak{R}$ correspond to the uniformizer circle $|\zeta| < a$ about q and write $\mathfrak{M}' = \mathfrak{M} \cap \mathfrak{k}$. Choose q_i and q_e to be points in $\mathfrak{k}$, in $\mathfrak{M}$ and $\mathfrak{R} - \mathfrak{M}$ respectively. We wish to consider the difference $T_f(q_i) - T_f(q_e)$ as q_i and q_e approach q. Let $t_i = \zeta(q_i)$ and $t_e = \zeta(q_e)$ and, for $p \in \mathfrak{M}'$, $z = \zeta(p)$. By (4.11.1) and (4.1.5)′, we have for $p \in \mathfrak{M}'$, $q' = q_i$ or q_e,

$$\mathscr{L}_{\mathbf{F}}(q', p) = \mathscr{L}_{\mathbf{M}}(q', p) + \text{regular terms}$$

$$= \frac{1}{\pi[\zeta(q') - z]^2} + \text{regular terms}.$$

Thus

$$\Xi^*_{\mathbf{F}}(q', p) = \frac{-1}{\pi(z - \zeta(q'))} + \mathscr{E}(q', p)$$

where $\mathscr{E}(q', p)$ is continuous at q. Also, $\mathscr{L}_{\mathbf{F}}(q', p)$ is continuous at q for $p \in \mathfrak{M} - \mathfrak{M}'$. Then

$$T_f(q_i) - T_f(q_e) = (\mathscr{L}_{\mathbf{F}}(q_i, p) - \mathscr{L}_{\mathbf{F}}(q_e, p), f'(p))_{\mathfrak{M}-\mathfrak{M}'}$$

$$- \frac{1}{2i} \int_{\partial \mathfrak{M}'} [\mathscr{E}(q_i, p) - \mathscr{E}(q_e, p)] \; (f'(p))^- d\bar{z}$$

$$+ \frac{1}{2\pi i} \int_{\partial \mathfrak{M}'} \left[\frac{1}{z - t_i} - \frac{1}{z - t_e}\right] (f'(p))^- d\bar{z}$$

where the scalar product over $\mathfrak{M}'$ has been evaluated according to (4.9.8). As q_i and q_e tend to q, the first two terms tend to zero since the integrands are continuous at q. The discontinuity of the last integral can be evaluated by Cauchy's theorem to give $(f'(q))^-$. In fact, the boundary of $\mathfrak{M}'$, considered in the ζ-plane, consists of a semicircle s and a segment γ of the real axis. The integral along s also vanishes at q. In the integral along γ, $d\bar{z}$ may be replaced by dz. The conjugate of this integral is

$$-\frac{1}{2\pi i} \int_{\gamma} \left[\frac{1}{z - \bar{t}_i} - \frac{1}{z - \bar{t}_e}\right] f'(z) dz.$$

If we complete the path along s, we add an integral which vanishes at q, and the result may be evaluated by Cauchy's theorem, giving

f' at the point $\bar{t}_e$. As q_i and q_e tend to q, this yields $f'(q)$ where $f'(q) = df/d\zeta$, ζ a boundary uniformizer at q. Thus for $q_i = q_e = q$, we have

$$T_f(q_i) - T_f(q_e) = (f'(q))^-. \tag{6.1.10}$$

We observe that

$$S_{f_1+f_2} = S_{f_1} + S_{f_2}. \tag{6.1.11}$$

On the other hand,

$$S_{\lambda f} = \bar{\lambda} T_f + \lambda \tilde{T}_f. \tag{6.1.12}$$

Taking in particular $\lambda = -i$, we have

$$S_{-if} = i(T_f - \tilde{T}_f). \tag{6.1.13}$$

Also, assuming symmetry of $\mathbf{F}_{\mathfrak{R}}$,

$$\begin{aligned}(T_{f_1}, f_2') &= \int_{\mathfrak{M}} \left(\int_{\mathfrak{M}} \mathscr{L}_{\mathbf{F}}(q, p)(f_2'(q))^- dA_\zeta \right) (f_1'(p))^- dA_z \\ &= (T_{f_2}, f_1').\end{aligned} \tag{6.1.14}$$

The interchange may be justified if we replace $\mathscr{L}_{\mathbf{F}}$ by $l_{\mathbf{F}}$. Similarly (but without requiring symmetry)

$$(\tilde{T}_{f_1}, f_2') = \int_{\mathfrak{M}} \left(\int_{\mathfrak{M}} \mathscr{L}_{\mathbf{F}}(q, \bar{p})(f_2'(q))^- dA_\zeta \right) f_1'(p) dA_z = (f_1', \tilde{T}_{f_2}). \tag{6.1.15}$$

Finally (with symmetry), we find, by adding (6.1.14) and (6.1.15) and taking real parts,

$$\operatorname{Re}\{(S_{f_1}, f_2')\} = \operatorname{Re}\{(S_{f_2}, f_1')\}. \tag{6.1.16}$$

Thus $\tilde{T}$ is a self-adjoint operator in the Hilbert space with Dirichlet metric. On the other hand, T, and hence S, are self-adjoint operators only if we consider a Hilbert space in which the metric is based upon the scalar product

$$[f', g'] = \operatorname{Re}\{(f', g')\}. \tag{6.1.17}$$

6.2. Scalar Products of Transforms

The operators defined in Section 6.1 transform the Hilbert space $\mathbf{F}_{\mathfrak{M}}$, which consists of differentials analytic in $\mathfrak{M}$, into linear spaces of piecewise analytic differentials on $\mathfrak{R}$. In these spaces, the metric is again defined by Dirichlet products over $\mathfrak{R}$. In the case of differentials which are discontinuous across the boundary of $\mathfrak{M}$, the scalar product over $\mathfrak{R}$ is the sum of the scalar products over $\mathfrak{M}$ and the components of $\mathfrak{R}-\mathfrak{M}$. In this section we shall investigate two preliminary questions: first, to find certain metric relations among these spaces, such as orthogonality; and second, to express the metrics over $\mathfrak{R}$ of the transformed differentials in terms of the metrics over $\mathfrak{M}$ of the original differentials.

We show first that

$$(6.2.1) \qquad (T_{f_1}, \tilde{T}_{f_2})_{\mathfrak{R}} = 0$$

where f_1', f_2' are any two differentials of the class $\mathbf{F}_{\mathfrak{M}}$ on $\mathfrak{M}$. In fact, by (6.1.3)′ and (6.1.4),

$$(T_{f_1}, \tilde{T}_{f_2})_{\mathfrak{R}} = -\int_{\mathfrak{M}}\int_{\mathfrak{M}}\int_{\mathfrak{M}} l_{\mathbf{F}}(q, p_1)(f_1'(p_1))^{-}\mathscr{L}_{\mathbf{F}}(p_2, \tilde{q})(f_2'(p_2))^{-} dA_{z_1} dA_{z_2} dA_{\zeta}$$

$$+\int_{\mathfrak{R}-\mathfrak{M}}\int_{\mathfrak{M}}\int_{\mathfrak{M}} \mathscr{L}_{\mathbf{F}}(q, p_1)(f_1'(p_1))^{-}\mathscr{L}_{\mathbf{F}}(p_2, \tilde{q})(f_2'(p_2))^{-} dA_{z_1} dA_{z_2} dA_{\zeta}.$$

Since

$$-(l_{\mathbf{F}}(q, p_1), \mathscr{L}_{\mathbf{F}}(q, \tilde{p}_2))_{\mathfrak{M}} = (\mathscr{L}_{\mathbf{F}}(q, p_1), \mathscr{L}_{\mathbf{F}}(q, \tilde{p}_2))_{\mathfrak{M}}$$

by (6.1.9), we have

$$(T_{f_1}, \tilde{T}_{f_2})_{\mathfrak{R}} = \int_{\mathfrak{M}}\int_{\mathfrak{M}}\left\{\int_{\mathfrak{R}} \mathscr{L}_{\mathbf{F}}(q, p_1)(\mathscr{L}_{\mathbf{F}}(q, \tilde{p}_2))^{-} dA_{\zeta}\right\}(f_1'(p_1))^{-}(f_2'(p_2))^{-} dA_{z_1} dA_{z_2}.$$

But $\mathscr{L}_{\mathbf{F}}(q, p)$ is orthogonal to all differentials of the class $\mathbf{F}$ on $\mathfrak{R}$; hence

$$(\mathscr{L}_{\mathbf{F}}(q, p_1), \mathscr{L}_{\mathbf{F}}(q, \tilde{p}_2))_{\mathfrak{R}} = 0,$$

and this proves (6.2.1). Thus we have proved the following lemma:

Lemma 6.2.1. *The T-transforms and $\tilde{T}$-transforms of any two differentials on $\mathfrak{M}$ are orthogonal with respect to the scalar product over $\mathfrak{R}$.*

Next, by (6.1.3)′, (5.3.27) and (5.4.1):

$$(T_{f_1}, T_{f_2})_{\mathfrak{R}} = \int\limits_{\mathfrak{M}} \left\{ \int\limits_{\mathfrak{M}} \int\limits_{\mathfrak{M}} l_{\mathbf{F}}(q, p_1)(l_{\mathbf{F}}(q, p_2))^{-}(f_1'(p_1))^{-} f_2'(p_2) dA_{z_1} dA_{z_2} \right\} dA_{\zeta}$$

$$+ \int\limits_{\mathfrak{R}-\mathfrak{M}} \left\{ \int\limits_{\mathfrak{M}} \int\limits_{\mathfrak{M}} \mathscr{L}_{\mathbf{F}}(q, p_1)(\mathscr{L}_{\mathbf{F}}(q, p_2))^{-}(f_1'(p_1))^{-} f_2'(p_2) dA_{z_1} dA_{z_2} \right\} dA_{\zeta}$$

$$= -\int\limits_{\mathfrak{M}} \int\limits_{\mathfrak{M}} [l_{\mathbf{F}}(\tilde{p}_2, p_1) + \Gamma_{\mathbf{F}}(\tilde{p}_2, p_1)] (f_1'(p_1))^{-} f_2'(p_2) dA_{z_1} dA_{z_2}$$

$$+ \int\limits_{\mathfrak{M}} \int\limits_{\mathfrak{M}} \Gamma_{\mathbf{F}}(\tilde{p}_2, p_1)(f_1'(p_1))^{-} f_2'(p_2) dA_{z_1} dA_{z_2}$$

$$= -\int\limits_{\mathfrak{M}} \int\limits_{\mathfrak{M}} l_{\mathbf{F}}(\tilde{p}_2, p_1)(f_1'(p_1))^{-} f_2'(p_2) dA_{z_1} dA_{z_2}.$$

This gives:

LEMMA 6.2.2. *The scalar product over $\mathfrak{R}$ of two T-transforms can be expressed as a Hermitian form over $\mathfrak{M}$ with kernel l:*

$$(6.2.2) \quad (T_{f_1}, T_{f_2})_{\mathfrak{R}} = -\int\limits_{\mathfrak{M}} \int\limits_{\mathfrak{M}} l_{\mathbf{F}}(\tilde{p}_2, p_1)(f_1'(p_1)))^{-} f_2'(p_2) dA_{z_1} dA_{z_2}.$$

Even more simply we have, for the $\tilde{T}$-transforms,

$$(\tilde{T}_{f_1}, \tilde{T}_{f_2})_{\mathfrak{R}} = \int\limits_{\mathfrak{R}} \left\{ \int\limits_{\mathfrak{M}} \int\limits_{\mathfrak{M}} \mathscr{L}_{\mathbf{F}}(q, \tilde{p}_1)(\mathscr{L}_{\mathbf{F}}(q, \tilde{p}_2))^{-} f_1'(p_1)(f_2'(p_2))^{-} dA_{z_1} dA_{z_2} \right\} dA_{\zeta}$$

(6.2.3)

$$= -\int\limits_{\mathfrak{M}} \int\limits_{\mathfrak{M}} \mathscr{L}_{\mathbf{F}}(p_2, \tilde{p}_1)\, f_1'(p_1)(f_2'(p_2))^{-} dA_{z_1} dA_{z_2}.$$

We are now able to express the product of two S-transforms by integrals extended over $\mathfrak{M}$. By (6.2.1), (6.2.2) and (6.2.3),

$$(S_{f_1}, S_{f_2})_{\mathfrak{R}} = (T_{f_1}, T_{f_2})_{\mathfrak{R}} + (\tilde{T}_{f_1}, \tilde{T}_{f_2})_{\mathfrak{R}}$$

$$= -\int\limits_{\mathfrak{M}} \int\limits_{\mathfrak{M}} l_{\mathbf{F}}(\tilde{p}_2, p_1)(f_1'(p_1))^{-} f_2'(p_2) dA_{z_1} dA_{z_2}$$

(6.2.4)

$$-\int\limits_{\mathfrak{M}} \int\limits_{\mathfrak{M}} \mathscr{L}_{\mathbf{F}}(p_2, \tilde{p}_1)\, f_1'(p_1)(f_2'(p_2))^{-} dA_{z_1} dA_{z_2}$$

$$
\begin{aligned}
&= -\int_{\mathfrak{M}}\int_{\mathfrak{M}} L_{\mathbf{F}}(p_1, \tilde{p}_2)(f_1'(p_1))^- f_2'(p_2)\,dA_{z_1}dA_{z_2} \\
&\quad + \int_{\mathfrak{M}}\int_{\mathfrak{M}} [L_{\mathbf{F}}(p_1, \tilde{p}_2) - L_{\mathbf{F}}(\tilde{p}_2, p_1)](f_1'(p_1))^- f_2'(p_2)\,dA_{z_1}dA_{z_2} \\
&\quad - \int_{\mathfrak{M}}\int_{\mathfrak{M}} [\mathscr{L}_{\mathbf{F}}(p_1, \tilde{p}_2) - \mathscr{L}_{\mathbf{F}}(\tilde{p}_2, p_1)](f_1'(p_1))^- f_2'(p_2)\,dA_{z_1}dA_{z_2} \qquad (6.2.4)\\
&\quad + \int_{\mathfrak{M}}\int_{\mathfrak{M}} \mathscr{L}_{\mathbf{F}}(p_1, \tilde{p}_2)(f_1'(p_1))^- f_2'(p_2)\,dA_{z_1}dA_{z_2} \\
&\quad - \int_{\mathfrak{M}}\int_{\mathfrak{M}} (\mathscr{L}_{\mathbf{F}}(p_1, \tilde{p}_2))^- f_1'(p_1)(f_2'(p_2))^-\,dA_{z_1}dA_{z_2}.
\end{aligned}
$$

Now

$$
(6.2.5)\qquad -\int_{\mathfrak{M}}\int_{\mathfrak{M}} L_{\mathbf{F}}(p_1, \tilde{p}_2)(f_1'(p_1))^- f_2'(p_2)\,dA_{z_1}dA_{z_2} = (f_2', f_1')_{\mathfrak{M}}.
$$

Further, by (4.1.6), (5.1.16), (5.1.16)' and (5.1.18),

$$
(6.2.6)\qquad
\begin{aligned}
&\int_{\mathfrak{M}}\int_{\mathfrak{M}} [L_{\mathbf{F}}(p_1, \tilde{p}_2) - L_{\mathbf{F}}(\tilde{p}_2, p_1)](f_1'(p_1))^- f_2'(p_2)\,dA_{z_1}dA_{z_2} \\
&= 2i \sum_{\mu,\nu=1}^{G} \operatorname{Im}\{\alpha_{\mu\nu}\}(dZ_\mu, df_1)_{\mathfrak{M}}((dZ_\nu, df_2)_{\mathfrak{M}})^- = 0
\end{aligned}
$$

since all terms $(dZ_\mu, df_i)_{\mathfrak{M}}$ occurring in the above sum vanish for differentials of class $\mathbf{F}_{\mathfrak{M}}$ on $\mathfrak{M}$. Similarly

$$
(6.2.7)\qquad
\begin{aligned}
&\int_{\mathfrak{M}}\int_{\mathfrak{M}} [\mathscr{L}_{\mathbf{F}}(p_1, \tilde{p}_2) - \mathscr{L}_{\mathbf{F}}(\tilde{p}_2, p_1)](f_1'(p_1))^- f_2'(p_2)\,dA_{z_1}dA_{z_2} \\
&= 2i \sum_{\mu,\nu=1}^{G_0} \operatorname{Im}\{\beta_{\mu\nu}\}\,(d\mathscr{Z}_\mu, df_1)_{\mathfrak{M}}\,((d\mathscr{Z}_\nu, df_2)_{\mathfrak{M}})^-.
\end{aligned}
$$

The last two terms in (6.2.4) combine to give

$$
2i \operatorname{Im}\left\{\int_{\mathfrak{M}}\int_{\mathfrak{M}} \mathscr{L}_{\mathbf{F}}(p_1, \tilde{p}_2)(f_1'(p_1))^- f_2'(p_2)\,dA_{z_1}dA_{z_2}\right\}.
$$

Substituting these results into (6.2.4), we obtain finally

$$(S_{f_1}, S_{f_2})_{\mathfrak{R}} = (T_{f_1}, T_{f_2})_{\mathfrak{R}} + (\tilde{T}_{f_1}, \tilde{T}_{f_2})_{\mathfrak{R}}$$

$$(6.2.8)\qquad = (f_2', f_1')_{\mathfrak{M}} - 2i \sum_{\mu,\nu=1}^{G_0} \operatorname{Im}\{\beta_{\mu\nu}\} (d\mathscr{Z}_\mu, df_1)_{\mathfrak{M}}((d\mathscr{Z}_\nu, df_2)_{\mathfrak{M}})^- + 2i \operatorname{Im}\left\{\int_{\mathfrak{M}}\int_{\mathfrak{M}} \mathscr{L}_{\mathbf{F}}(p_1, \tilde{p}_2)(f_1'(p_1))^- f_2'(p_2) dA_{z_1} dA_{z_2}\right\}.$$

In particular, taking $f_1 = f_2 = f$, we have

$$N_{\mathfrak{R}}(S_f) = N_{\mathfrak{R}}(T_f) + N_{\mathfrak{R}}(\tilde{T}_f)$$

$$(6.2.9)\qquad = N_{\mathfrak{M}}(df) - 2i \sum_{\mu,\nu=1}^{G_0} \operatorname{Im}\{\beta_{\mu\nu}\} (d\mathscr{Z}_\mu, df)_{\mathfrak{M}}((d\mathscr{Z}_\nu, df)_{\mathfrak{M}})^-.$$

This formula shows that the norm of the S-transform is, up to a bilinear combination of terms $(d\mathscr{Z}_\mu, df)_{\mathfrak{M}}$ equal to the norm of df. We observe in particular that

$$(6.2.10)\qquad N_{\mathfrak{R}}(\tilde{T}_f) < \infty.$$

Formula (6.2.9) becomes particularly elegant if $\mathscr{L}_{\mathbf{F}}(p, q) = \mathscr{L}_{\mathbf{F}}(q, p)$. In this case of symmetry, we have $\operatorname{Im} \beta_{\mu\nu} = 0$, so

$$(6.2.11)\qquad N_{\mathfrak{R}}(S_f) = N_{\mathfrak{M}}(df) = N_{\mathfrak{R}}(df)$$

since $df \equiv 0$ in $\mathfrak{R} - \mathfrak{M}$. Thus:

LEMMA 6.2.3. *If the class $\mathbf{F}_{\mathfrak{R}}$ is symmetric, S_f is a norm-preserving transformation.*

Let dh be any differential of the class $\mathbf{F}_{\mathfrak{R}}$, df any differential of $\mathbf{F}_{\mathfrak{M}}$. We shall prove the following lemma:

LEMMA 6.2.4. *The discontinuous T-transforms in $\mathfrak{R}$ are orthogonal to all analytic differentials of $\mathbf{F}_{\mathfrak{R}}$, and the scalar product of a $\tilde{T}$-transform in the metric of $\mathfrak{R}$ is a scalar product over $\mathfrak{M}$:*

$$(6.2.12)\qquad (T_f, dh)_{\mathfrak{R}} = 0, \quad (\tilde{T}_f, dh)_{\mathfrak{R}} = -(df, dh)_{\mathfrak{M}}.$$

In fact,

$$(T_f, dh)_{\mathfrak{R}} = \int_{\mathfrak{M}}\left\{\int_{\mathfrak{R}} \mathscr{L}_{\mathbf{F}}(q, p)(h'(q))^- dA_\zeta\right\}(f'(p))^- dA_z = 0$$

and

$$(\tilde{T}_f, dh)_{\mathfrak{R}} = \int_{\mathfrak{M}} \left\{ \int_{\mathfrak{R}} \mathscr{L}_{\mathbf{F}}(q, \bar{p})(h'(q))^{-} dA_{\zeta} \right\} f'(p) dA_z$$

$$= -\int_{\mathfrak{M}} f'(p)(h'(p))^{-} dA_z = -(df, dh)_{\mathfrak{M}}.$$

We point out that from the point of view of the Dirichlet metric in $\mathfrak{R}$ the operators $S_f = T_f + \tilde{T}_f$ and $S_{-if} = i(T_f - \tilde{T}_f)$ are indistinguishable. We clearly have by (6.2.1)

(6.2.13) $$(S_{f_1}, S_{f_2}) = (S_{-if_1}, S_{-if_2}).$$

We might equally well have defined $S_f = T_f - \tilde{T}_f$. Merely for definiteness we adopt the definition (6.1.5).

6.3. The Iterated Operators

In Section 6.1, various operators were defined for the classes $\mathbf{F}_{\mathfrak{M}}$ which are one of our main interests in this book. However, since these operators carry differentials of $\mathfrak{M}$ into differentials analytic on $\mathfrak{M}$ and on $\mathfrak{R} - \mathfrak{M}$, it appears desirable to extend the domain of the operators T, $\tilde{T}$, and S to such differentials on $\mathfrak{R}$. We extend each differential in $\mathbf{F}_{\mathfrak{M}}$ into a piecewise analytic differential on $\mathfrak{R}$ by defining it to be identically zero in $\mathfrak{R} - \mathfrak{M}$. Instead of formulas (6.1.3) and (6.1.4) we shall now use the following extended definitions, viz.

(6.3.1) $$T_f(q) = (\mathscr{L}_{\mathbf{F}}(q, p), f'(p))_{\mathfrak{R}},$$

and

(6.3.2) $$\tilde{T}_f(q) = (f'(p), \mathscr{L}_{\mathbf{F}}(p, \bar{q}))_{\mathfrak{R}},$$

valid for all differentials piecewise analytic on $\mathfrak{R}$.

Let $\mathbf{E}_{\mathfrak{R}}$ be the class of differentials which are regular in the interior of $\mathfrak{M}$ and in the interior of each component of $\mathfrak{R} - \mathfrak{M}$, but which are not necessarily continuous across the boundary C of $\mathfrak{M}$. We suppose that the differentials of $\mathbf{E}_{\mathfrak{R}}$ have finite norms over $\mathfrak{R}$. Particular subclasses $\mathbf{D}_{\mathfrak{R}}$ of $\mathbf{E}_{\mathfrak{R}}$ may be distinguished by supposing that the differentials of $\mathbf{D}_{\mathfrak{R}}$ form a corresponding class on $\mathfrak{M}$ and on each component of $\mathfrak{R} - \mathfrak{M}$. In particular, if all the corresponding classes are the class $\mathbf{M}$, then $\mathbf{D}_{\mathfrak{R}} = \mathbf{E}_{\mathfrak{R}}$. We observe that all the results of Sections 6.1 and 6.2 remain valid under these more general circumstances.

We remark that the class $\mathbf{F}_{\mathfrak{R}}$ is contained in every $\mathbf{D}_{\mathfrak{R}}$; in particular the $\tilde{T}$-transform of an arbitrary $df \in \mathbf{E}_{\mathfrak{R}}$ also lies in every $\mathbf{D}_{\mathfrak{R}}$. However, a class $\mathbf{F}_{\mathfrak{M}}$ will not lie in very $\mathbf{D}_{\mathfrak{R}}$ if $\mathbf{F}_{\mathfrak{M}}$ contains a proper subclass which is also a corresponding class. The operators T and $\tilde{T}$ transform a class $\mathbf{D}_{\mathfrak{R}}$ into itself, hence the importance of these classes in the theory of these operators.

Now we can iterate our operations, and form operator products on differentials of $\mathbf{D}_{\mathfrak{R}}$:

$$(T^2(q))_f = (\mathscr{L}_{\mathbf{F}}(q, p),\ T_f(p))_{\mathfrak{R}}, \tag{6.3.3}$$

$$(\tilde{T}^2(q))_f = (\mathscr{L}_{\mathbf{F}}(q, \bar{p}),\ (\tilde{T}_f(p))^-)_{\mathfrak{R}}, \tag{6.3.4}$$

$$(T\tilde{T}(q))_f = (\mathscr{L}_{\mathbf{F}}(q, p),\ \tilde{T}_f(p))_{\mathfrak{R}}, \tag{6.3.5}$$

$$(\tilde{T}T(q))_f = (\mathscr{L}_{\mathbf{F}}(q, \bar{p}),\ (T_f(p))^-)_{\mathfrak{R}} \tag{6.3.6}$$

and, therefore, we have also defined:

$$S^2 = (T + \tilde{T})^2 = T^2 + T\tilde{T} + \tilde{T}T + \tilde{T}^2. \tag{6.3.7}$$

We shall prove, with no assumption of symmetry, that

$$(S^2(q))_f = f'(q), \tag{6.3.8}$$

or, symbolically,

$$S^2 = I, \tag{6.3.8$'$}$$

where I stands for the identity transformation. Thus we have:

LEMMA 6.3.1. *S is an involutory transformation of $\mathbf{D}_{\mathfrak{R}}$ into itself.*

In particular, if $f \in \mathbf{F}_{\mathfrak{M}}$, this means that $(S^2(q))_f$ is equal to $f'(q)$ for q in $\mathfrak{M}$ and equal to zero for q in $\mathfrak{R} - \mathfrak{M}$. This elegant result justifies our extension of the operators to the $\mathfrak{R}$-domain, and forms the basis of significant further developments.

Proof of (6.3.8)′ turns essentially upon the formula

$$T^2 = I + \tilde{T} \tag{6.3.9}$$

which we now proceed to establish, by rather lengthy calculation. We assume at first that df is a regular analytic differential in the closure of $\mathfrak{M}$ and also in the closure of $\mathfrak{R} - \mathfrak{M}$. Then df can be written as the sum of differentials df_i, each of which is regular analytic in

$\mathfrak{M}$ and in each component of $\mathfrak{R}-\mathfrak{M}$, and different from zero in one only of these domains. Since all the operators are linear it is sufficient to prove all necessary identities for differentials of the type df_i. Finally we shall be able to remove the restriction of analyticity in the closed regions by a limiting process of the usual form. We prove first:

LEMMA 6.3.2. *For each differential $\mathscr{Z}'_\mu(p)$ on $\mathfrak{R}$ such that the class $\mathbf{F}_\mathfrak{R}$ is single-valued on the cycle $\mathscr{K}_\mu$ belonging to $\mathscr{Z}'_\mu$, and for each $df \in \mathbf{F}_\mathfrak{M}$, we have*

$$(\mathscr{Z}'_\mu, T_f)_\mathfrak{R} = (f', \mathscr{Z}'_\mu)_\mathfrak{R}. \tag{6.3.10}$$

In fact, by (5.1.17), (5.1.17)′ and (5.1.4) we have

$$(\mathscr{L}_\mathbf{F}(p, q), \mathscr{Z}'_\mu(p))_\mathfrak{R} = (\mathscr{L}_\mathbf{M}(p, q), \mathscr{Z}'_\mu(p))_\mathfrak{R} + \sum_{\varrho, \sigma=1}^{G_0} \beta_{\varrho\sigma} P_{\varrho\mu} \mathscr{Z}'_\sigma(q). \tag{6.3.11}$$

Using (5.1.20) and the fact that $\mathscr{L}_\mathbf{M}$ is orthogonal to all differentials on $\mathfrak{R}$, we thus obtain

$$(\mathscr{L}_\mathbf{F}(p, q), \mathscr{Z}'_\mu(p))_\mathfrak{R} = \mathscr{Z}'_\mu(q). \tag{6.3.12}$$

Now we apply formula (6.3.12) and we derive the relation

$$\begin{aligned} (\mathscr{Z}'_\mu, T_f) &= \int_\mathfrak{R} \mathscr{Z}'_\mu(p_1) \left\{ \int_\mathfrak{R} (\mathscr{L}_\mathbf{F}(p_1, p_2))^- f'(p_2) dA_{z_2} \right\} dA_{z_1} \\ &= \int_\mathfrak{R} f'(p_2) \left(\int_\mathfrak{R} \mathscr{L}_\mathbf{F}(p_1, p_2) (\mathscr{Z}'_\mu(p_1))^- dA_{z_1} \right)^- dA_{z_2} \\ &= \int_\mathfrak{R} f'(p_2) (\mathscr{Z}'_\mu(p_2))^- dA_{z_2} = (f', \mathscr{Z}'_\mu). \end{aligned} \tag{6.3.13}$$

This proves the lemma.

Next,

$$\begin{aligned} (T^2(q))_f &= (\mathscr{L}_\mathbf{F}(q, p_1), T_f(p_1))_\mathfrak{R} \\ &= (\mathscr{L}_\mathbf{F}(p_1, q), T_f(p_1))_\mathfrak{R} - 2i \sum_{\mu, \nu=1}^{G_0} \operatorname{Im}\{\beta_{\mu\nu}\} (df, d\mathscr{Z}_\mu)_\mathfrak{R} \mathscr{Z}'_\nu(q) \end{aligned} \tag{6.3.14}$$

using (4.11.1) and Lemma 6.3.2. Further, $\mathscr{L}_\mathbf{F}(p_1, q) - \mathscr{L}_\mathbf{S}(p_1, q)$ is a differential of class $\mathbf{F}_\mathfrak{R}$, so by (6.2.12) we have

$$(\mathscr{L}_\mathbf{F}(p_1, q) - \mathscr{L}_\mathbf{S}(p_1, q), T_f(p_1))_\mathfrak{R} = 0$$

or

$$(6.3.15)\qquad (\mathscr{L}_F(p_1, q),\ T_f(p_1))_{\mathfrak{R}} = (\mathscr{L}_S(p_1, q),\ T_f(p_1))_{\mathfrak{R}}.$$

Suppose first that f' vanishes outside $\mathfrak{M}$ and is regular in the closure of $\mathfrak{M}$, and that $\mathfrak{M}$ is properly imbedded in $\mathfrak{R}$. Let C be the boundary of $\mathfrak{M}$ oriented in the positive sense with respect to $\mathfrak{M}$, and let $-C$ denote the oppositely oriented boundary. When the integration is over C we assume that the boundary values of the integrand are derived from $\mathfrak{M}$; when the integration is over $-C$, we assume that the boundary values are derived from $\mathfrak{R} - \mathfrak{M}$. The boundary of $\mathfrak{R}$ will be denoted by B. Let q be an interior point of $\mathfrak{M}$ or of $\mathfrak{R} - \mathfrak{M}$ and let $\mathfrak{k}$ correspond to the circle $|\zeta| < a$ at q. Integrating by parts we obtain

$$(\mathscr{L}_S(p_1,q),\ T_f(p_1))_{\mathfrak{R}} = (\mathscr{L}_S(p_1, q),\ T_f(p_1))_{\mathfrak{M}} + (\mathscr{L}_S(p_1,q),\ T_f(p_1))_{\mathfrak{R}-\mathfrak{M}}$$

$$(6.3.16)\qquad \begin{aligned} &= -\frac{1}{2i}\int_C \Xi_S(p_1, q)\ (T_f(p_1))^- dz_1 \\ &\quad -\frac{1}{2i}\int_{-C+B} \Xi_S(p_1, q)\ (T_f(p_1))^- dz_1 \\ &\quad +\frac{1}{2i}\int_{\partial\mathfrak{k}} \Xi_S(p_1, q)\ (T_f(p_1))^- d\bar{z}_1 \end{aligned}$$

where the variable z_1 in the boundary integrals is a boundary uniformizer. Now

$$(6.3.17)\qquad \frac{1}{2i}\int_{\partial\mathfrak{k}} \Xi_S(p_1, q)\ (T_f(p_1))^- d\bar{z}_1 = o(1) \text{ as } a \to 0,$$

since $T_f(p_1)$ is regular near q. By (6.1.10), we have

$$(6.3.18)\quad -\frac{1}{2i}\int_{C-C} \Xi_S(p_1, q)(T_f(p_1))^- dz_1 = -\frac{1}{2i}\int_C \Xi_S(p_1, q) f'(p_1) dz_1 = f'(q)$$

by Cauchy's theorem, using the fact that $f'(q) = 0$ if q is in $\mathfrak{R} - \mathfrak{M}$. In the integral over B, p_1 is a boundary point of $\mathfrak{R}$, so that $p_1 = \tilde{p}_1$. Then

$$\begin{aligned}(T_f(p_1))^- = (f'(p_2),\ \mathscr{L}_F(p_1, p_2))_{\mathfrak{R}} &= (f'(p_2),\ \mathscr{L}_F(\tilde{p}_1, p_2))_{\mathfrak{R}} \\ &= (f'(p_2),\ (\mathscr{L}_F(\tilde{p}_2, p_1))^-)_{\mathfrak{R}}.\end{aligned}$$

Thus

$$-\frac{1}{2i}\int_B \Xi_{\mathbf{S}}(p_1,q)(T_f(p_1))^- dz_1 = -\frac{1}{2i}\int_B \Xi_{\mathbf{S}}(p_1,q)\left[\int_{\mathfrak{R}} f'(p_2)\mathscr{L}_{\mathbf{F}}(\bar{p}_2,p_1)dA_{z_2}\right]dz_1$$

$$= \int_{\mathfrak{R}} f'(p_2)\left[-\frac{1}{2i}\int_B \Xi_{\mathbf{S}}(p_1,q)\mathscr{L}_{\mathbf{F}}(\bar{p}_2,p_1)dz_1\right]dA_{z_2}$$

$$= \int_{\mathfrak{R}} f'(p_2)\,\mathscr{L}_{\mathbf{F}}(\bar{p}_2,q)dA_{z_2}$$

by Cauchy's theorem;

$$= \int_{\mathfrak{R}} \mathscr{L}_{\mathbf{F}}(q,\bar{p}_2)f'(p_2)dA_{z_2} + 2i\sum_{\mu,\nu=1}^{G_0} \operatorname{Im}\{\beta_{\mu\nu}\}(df, d\mathscr{Z}_\mu)_{\mathfrak{R}}\mathscr{Z}'_\nu(q)$$

using (5.1.17) and (5.1.17)′. That is,

$$-\frac{1}{2i}\int_B \Xi_{\mathbf{S}}(p_1,q)(T_f(p_1))^- dz_1 \tag{6.3.19}$$
$$= \tilde{T}_f(q) + 2i\sum_{\mu,\nu=1}^{G} \operatorname{Im}\{\beta_{\mu\nu}\}\,(df, d\mathscr{Z}_\mu)_{\mathfrak{R}}\mathscr{Z}'_\nu(q).$$

Combining (6.3.14) through (6.3.19) we obtain (6.3.9).

If $\mathfrak{M}$ is not properly imbedded in $\mathfrak{R}$, we let $\mathfrak{M}_\mu$ be a sequence of surfaces which approach $\mathfrak{M}$ as μ tends to infinity, $\mathfrak{M}_\mu \subset \mathfrak{M}_{\mu+1} \subset \mathfrak{M}$. If q is a fixed interior point of $\mathfrak{M}$ or of $\mathfrak{R} - \mathfrak{M}$, it is clear that the functionals T_μ, $\tilde{T}_\mu$ for the surface $\mathfrak{M}_\mu$ approach the corresponding functionals T, $\tilde{T}$ for the surface $\mathfrak{M}$ and therefore (6.3.9) is valid for $\mathfrak{M}$.

To remove the restriction that df is regular in the closure of $\mathfrak{M}$, we use the fact, proved in Section 4.13, that we can approximate df, in the sense of the norm over $\mathfrak{M}$, by differentials df_n regular in the closure of $\mathfrak{M}$. If q is a point of $\mathfrak{M}$, we have by (6.1.3)′

$$\begin{cases} T_f(q) = -(l_{\mathbf{F}}(q,p),\ f'(p))_{\mathfrak{M}}, \\ T_n(q) = T_{f_n}(q) = -\,(l_{\mathbf{F}}(q,p),\ f'_n(p))_{\mathfrak{M}}. \end{cases} \tag{6.3.20}$$

Hence, by the inequality of Schwarz,

$$| T_f(q) - T_n(q) |^2 \leqq \left\{ \int_{\mathfrak{M}} | l_{\mathrm{F}}(q, p) | \, | f'(p) - f'_n(p) | \, dA_z \right\}^2$$

$$\leqq \int_{\mathfrak{M}} | l_{\mathrm{F}}(q, p) |^2 dA_z \cdot \int_{\mathfrak{M}} | f'(p) - f'_n(p) |^2 dA_z.$$

Since the right side tends to zero by (4.13.3), we see that

$$T_n(q) \to T_f(q) \tag{6.3.21}$$

uniformly in the closure of $\mathfrak{M}$. Hence, in particular,

$$\int_{\mathfrak{M}} | T_f(q) - T_n(q) |^2 dA_\zeta \to 0 \tag{6.3.22}$$

as n tends to infinity.

If q is in the interior of $\mathfrak{R} - \mathfrak{M}$, then

$$\begin{cases} T_f(q) = (\mathscr{L}_{\mathrm{F}}(q, p), f'(p))_{\mathfrak{M}}, \\ T_n(q) = (\mathscr{L}_{\mathrm{F}}(q, p), f'_n(p))_{\mathfrak{M}}, \end{cases} \tag{6.3.23}$$

where $\mathscr{L}_{\mathrm{F}}(q, p)$ as a function of p is bounded. Hence (6.3.21) holds uniformly for q in any compact domain interior to $\mathfrak{R} - \mathfrak{M}$.

Suppose that q is an interior point of $\mathfrak{M}$, in which case

$$(T^2(q))_f - (T^2(q))_n = - \int_{\mathfrak{M}} l_{\mathrm{F}}(q, p)(T_f(p) - T_n(p))^- dA_z$$

$$+ \int_{\mathfrak{R}-\mathfrak{M}} \mathscr{L}_{\mathrm{F}}(q, p)(T_f(p) - T_n(p))^- dA_z.$$

By the Schwarz inequality

$$\left| \int_{\mathfrak{M}} l_{\mathrm{F}}(q, p)(T_f(p) - T_n(p))^- dA_z \right|$$

$$\leqq \left\{ \int_{\mathfrak{M}} | l_{\mathrm{F}}(q, p) |^2 \, dA_z \right\}^{\frac{1}{2}} \left\{ \int_{\mathfrak{M}} | T_f(p) - T_n(p) |^2 \, dA_z \right\}^{\frac{1}{2}},$$

and the right side tends to zero as n approaches infinity by (6.3.22) Let $\mathfrak{J}$ denote a compact region interior to $\mathfrak{R} - \mathfrak{M}$. Given $\varepsilon > 0$, we can choose $\mathfrak{J}$ such that

$$\int\limits_{\mathfrak{R}-\mathfrak{M}-\mathfrak{J}} |\mathscr{L}_{\mathbf{F}}(q, p)|^2 dz < \varepsilon^2.$$

By the Schwarz inequality

$$\left|\int\limits_{\mathfrak{R}-\mathfrak{M}-\mathfrak{J}} \mathscr{L}_{\mathbf{F}}(q, p)(T_f(p) - T_n(p))^- dA_z\right|$$

$$\leqq \left\{\int\limits_{\mathfrak{R}-\mathfrak{M}-\mathfrak{J}} |\mathscr{L}_{\mathbf{F}}(q, p)|^2 dA_z\right\}^{\frac{1}{2}} \left\{\int\limits_{\mathfrak{R}-\mathfrak{M}-\mathfrak{J}} |T_f(p) - T_n(p)|^2 dA_z\right\}^{\frac{1}{2}}$$

where, by the triangle inequality,

$$\left\{\int\limits_{\mathfrak{R}-\mathfrak{M}-\mathfrak{J}} |T_f(p) - T_n(p)|^2 dA_z\right\}^{\frac{1}{2}} \leqq \left\{\int\limits_{\mathfrak{R}-\mathfrak{M}-\mathfrak{J}} |T_f(p)|^2 dA_z\right\}^{\frac{1}{2}} + \left\{\int\limits_{\mathfrak{R}-\mathfrak{M}-\mathfrak{J}} |T_n(p)|^2 dA_z\right\}^{\frac{1}{2}}$$

$$\leqq \{N_{\mathfrak{R}}(T_f)\}^{\frac{1}{2}} + \{N_{\mathfrak{R}}(T_n)\}^{\frac{1}{2}}.$$

By (6.2.9)

$$N_{\mathfrak{R}}(T_f) \leqq N_{\mathfrak{M}}(df) + 2 \sum_{\mu, \nu=1}^{G_0} |\operatorname{Im} \beta_{\mu\nu}|\, |(d\mathscr{Z}_\mu, df)_{\mathfrak{M}}|\, |(d\mathscr{Z}_\nu, df)_{\mathfrak{M}}|.$$

Since

$$|(d\mathscr{Z}_\mu, df)_{\mathfrak{M}}| \leqq \{N_{\mathfrak{M}}(d\mathscr{Z}_\mu)\}^{\frac{1}{2}} \{N_{\mathfrak{M}}(df)\}^{\frac{1}{2}},$$

we have

$$N_{\mathfrak{R}}(T_f) \leqq A N_{\mathfrak{M}}(df),$$

where

$$A = 1 + 2 \sum_{\mu, \nu=1}^{G_0} |\operatorname{Im} \beta_{\mu\nu}|\, \{N_{\mathfrak{M}}(d\mathscr{Z}_\mu)\}^{\frac{1}{2}} \{N_{\mathfrak{M}}(d\mathscr{Z}_\nu)\}^{\frac{1}{2}}$$

is a number depending only on $\mathfrak{R}$. Similarly

$$N_{\mathfrak{R}}(T_n) \leqq A N_{\mathfrak{M}}(df_n).$$

Since $N_{\mathfrak{M}}(df_n)$ tends to $N_{\mathfrak{M}}(df)$, we have

$$\left\{\int\limits_{\mathfrak{R}-\mathfrak{M}-\mathfrak{J}} |T_f(p) - T_n(p)|^2 dA_z\right\}^2 \leqq B,$$

where B is independent of n. Thus

$$\left|\int\limits_{\mathfrak{R}-\mathfrak{M}-\mathfrak{J}} \mathscr{L}_{\mathbf{F}}(q, p)(T_f(p) - T_n(p))^- dA_z\right| < B\varepsilon.$$

For fixed $\mathfrak{J}$,

$$\int_{\mathfrak{J}} \mathscr{L}_{\mathbf{F}}(q, p)(T_f(p) - T_n(p))^- dA_z \to 0$$

by the uniform convergence. Since ε is arbitrary, it follows that

$$(T^2(q))_n \to (T^2(q))_f,$$

for each q in the interior of $\mathfrak{M}$.

Suppose that q is a fixed interior point of $\mathfrak{R} - \mathfrak{M}$. Then

$$\int_{\mathfrak{M}} \mathscr{L}_{\mathbf{F}}(q, p)(T_f(p) - T_n(p))^- dA_z \to 0$$

as n approaches infinity, and it remains to show that

$$\int_{\mathfrak{R}-\mathfrak{M}} \mathscr{L}_{\mathbf{F}}(q, p)(T_f(p) - T_n(p))^- dA_z \to 0.$$

Let $\mathfrak{J}$ be a compact subdomain lying in the interior of $\mathfrak{R} - \mathfrak{M}$ and containing the point q in its interior. By choice of $\mathfrak{J}$ we again have

$$\left| \int_{\mathfrak{R}-\mathfrak{M}-\mathfrak{J}} \mathscr{L}_{\mathbf{F}}(q, p)(T_f(p) - T_n(p))^- dA_z \right| < B\varepsilon.$$

Let $\mathfrak{k}_b$ be the circle $|\zeta| < b$ at q. Then

$$\int_{\mathfrak{J}} \mathscr{L}_{\mathbf{F}}(q, p)(T_f(p) - T_n(p))^- dA_z$$

$$= \int_{\mathfrak{J}-\mathfrak{k}_b} \mathscr{L}_{\mathbf{F}}(q, p)(T_f(p) - T_n(p))^- dA_z$$

$$- \frac{1}{2i} \int_{\partial\mathfrak{k}_b} \Xi^*_{\mathbf{F}}(q, p)(T_f(p) - T_n(p))^- d\bar{z},$$

where the integral over $\mathfrak{k}_b$ has been evaluated according to (4.9.8). Here the right side tends to zero by the uniform convergence. It follows that

$$(T^2(q))_n \to (T^2(q))_f,$$

for each q in the interior of $\mathfrak{R} - \mathfrak{M}$.

It is obvious that

$$\tilde{T}_n(q) \to \tilde{T}_f(q)$$

for each q in the interior of $\mathfrak{R}$. Further, $f'_n(q)$ tends to $f'(q)$ for each q in $\mathfrak{R}$ which is not on the boundary C of $\mathfrak{M}$. Since (6.3.9) is valid for f_n, it remains valid in the limit for each q in the interior of $\mathfrak{M}$ or in the interior of $\mathfrak{R} - \mathfrak{M}$.

This removes the restriction that f' is regular in the closure of $\mathfrak{M}$. Finally, the same argument applies to the case where f' is different from zero in one of the components of $\mathfrak{R} - \mathfrak{M}$, and (6.3.9) is therefore proved in general.

Next, we have the much easier identity

$$(\tilde{T}^2(q))_f = -\tilde{T}_f(q), \text{ or } \tilde{T}^2 = -\tilde{T}. \tag{6.3.24}$$

This follows from (4.11.7), since $\tilde{T}_f \in \mathbf{F}_{\mathfrak{R}}$. In fact,

$$(\tilde{T}^2(q))_f = (\tilde{T}_f(p), \mathscr{L}_{\mathbf{F}}(p, \tilde{q}))_{\mathfrak{R}}$$

Finally, it can be shown that

$$T\tilde{T} = \tilde{T}T = 0. \tag{6.3.25}$$

The first relation follows from (4.11.6). For the second we have

$$\begin{aligned}(\tilde{T}T(q))_f &= (\mathscr{L}_{\mathbf{F}}(q, \tilde{p}_2), (T_f(p_2)^-)_{\mathfrak{R}} \\ &= \int_{\mathfrak{R}} \mathscr{L}_{\mathbf{F}}(q, \tilde{p}_2)\left[\int_{\mathfrak{R}} (\mathscr{L}_{\mathbf{F}}(p_2, p_1))^- f'(p_1)\, dA_{z_1}\right] dA_{z_2} \\ &= \int_{\mathfrak{R}} f'(p_1) \left[\int_{\mathfrak{R}} \mathscr{L}_{\mathbf{F}}(q, \tilde{p}_2)(\mathscr{L}_{\mathbf{F}}(p_2, p_1))^- dA_{z_2}\right] dA_{z_1} = 0,\end{aligned}$$

since $\mathscr{L}_{\mathbf{F}}(p_2, p_1)$ is orthogonal to $\mathscr{L}_{\mathbf{F}}(q, \tilde{p}_2)$ by (4.11.6).

Collecting these results, namely (6.3.7), (6.3.9), (6.3.24) and (6.3.25), we have

$$\begin{aligned}S^2 &= (T + \tilde{T})^2 = T^2 + T\tilde{T} + \tilde{T}T + \tilde{T}^2 \\ &= I + \tilde{T} + 0 + 0 - \tilde{T} = I.\end{aligned}$$

We summarize the relations between the operators T, $\tilde{T}$ and S in the following theorem:

THEOREM 6.3.1. *The operators T, $\tilde{T}$ and S satisfy*

$$T^2 = I + \tilde{T},\ \tilde{T}^2 = -\tilde{T},\ T\tilde{T} = \tilde{T}T = 0,\ S^2 = I.$$

6.4. Spaces of Piecewise Analytic Differentials

The subclass of $\mathbf{D}_{\mathfrak{R}}$ composed of differentials which are orthogonal to the differentials of $\mathbf{F}_{\mathfrak{R}}$ will be denoted by $\mathbf{O}_{\mathfrak{R}}$. By (6.2.12) we see that T_f belongs to $\mathbf{O}_{\mathfrak{R}}$.

We show first that

$$\mathbf{D}_{\mathfrak{R}} = \mathbf{F}_{\mathfrak{R}} + \mathbf{O}_{\mathfrak{R}}. \tag{6.4.1}$$

In fact, let a differential dh of $\mathbf{D}_{\mathfrak{R}}$ be given. We determine the orthogonal projection of dh into the space $\mathbf{F}_{\mathfrak{R}}$, that is, we determine the differential df of $\mathbf{F}_{\mathfrak{R}}$ for which $N_{\mathfrak{R}}(dh - df)$ is a minimum. The existence of such a differential df is guaranteed by the completeness of the Hilbert space. The difference $dh - df$ is clearly orthogonal to each differential dg in $\mathbf{F}_{\mathfrak{R}}$. This is geometrically obvious, and also follows from the usual analytic argument, depending upon the minimum property of df. Thus, if $dg \in \mathbf{F}_{\mathfrak{R}}$, and ε is an arbitrary complex number,

$$N_{\mathfrak{R}}(dh - df - \varepsilon dg) = N_{\mathfrak{R}}(dh - df) - 2 \operatorname{Re} \{\bar{\varepsilon}(dh - df, dg)_{\mathfrak{R}}\}$$
$$+ |\varepsilon|^2 N_{\mathfrak{R}}(dg) \geqq N_{\mathfrak{R}}(dh - df),$$

whence

$$(dh - df, dg)_{\mathfrak{R}} = 0.$$

Thus $dh - df$ belongs to $\mathbf{O}_{\mathfrak{R}}$. It follows that $\mathbf{D}_{\mathfrak{R}} \subset \mathbf{O}_{\mathfrak{R}} + \mathbf{F}_{\mathfrak{R}}$. But it is obvious that $\mathbf{D}_{\mathfrak{R}} \supset \mathbf{O}_{\mathfrak{R}} + \mathbf{F}_{\mathfrak{R}}$, so (6.4.1) is true.

By $T(\mathbf{F}_{\mathfrak{R}})$ we mean the class of differentials T_f, $df \in \mathbf{F}_{\mathfrak{R}}$, and $T(\mathbf{F}_{\mathfrak{R}}) = 0$ means that each differential of $T(\mathbf{F}_{\mathfrak{R}})$ is zero. We now prove that

$$T(\mathbf{F}_{\mathfrak{R}}) = 0,\ \tilde{T}(\mathbf{F}_{\mathfrak{R}}) = \mathbf{F}_{\mathfrak{R}},\ T(\mathbf{O}_{\mathfrak{R}}) \subset \mathbf{O}_{\mathfrak{R}},\ \tilde{T}(\mathbf{O}_{\mathfrak{R}}) = 0. \tag{6.4.2}$$

The first relation is a consequence of (4.11.6). By (4.11.7), we have

$$\tilde{T}_f(q) = (f'(p), \mathscr{L}_{\mathbf{F}}(p, \tilde{q}))_{\mathfrak{R}} = -f'(q),$$

which proves the second relation. To prove the third, suppose that $dg \in \mathbf{O}_{\mathfrak{R}}$, $df \in \mathbf{F}_{\mathfrak{R}}$. Then

$$(f', T_g) = \int_{\mathfrak{R}} f'(p_2)\Big[\int_{\mathfrak{R}} (\mathscr{L}_{\mathbf{F}}(p_2, p_1))^- g'(p_1) dA]_{z_1}\Big] dA_{z_2}$$
$$= \int_{\mathfrak{R}} g'(p_1)\Big[\int_{\mathfrak{R}} (\mathscr{L}_{\mathbf{F}}(p_2, p_1))^- f'(p_2) dA_{z_2}\Big] dA_{z_1} = 0,$$

again by (4.11.6). Finally, for $dg \in \mathbf{O}_{\mathfrak{R}}$, we have

$$\tilde{T}_g(q) = (g'(p), \mathscr{L}_{\mathbf{F}}(p, \tilde{q}))_{\mathfrak{R}} = 0$$

since $\mathscr{L}_{\mathbf{F}}(p, \tilde{q})$, as a differential in p, belongs to $\mathbf{F}_{\mathfrak{R}}$. This proves the fourth relation.

Combining (6.4.1) and (6.4.2), we have

$$T(\mathbf{D}_{\mathfrak{R}}) \subset \mathbf{O}_{\mathfrak{R}}, \; \tilde{T}(\mathbf{D}_{\mathfrak{R}}) = \mathbf{F}_{\mathfrak{R}}, \; T^2(\mathbf{D}_{\mathfrak{R}}) \subset T(\mathbf{O}_{\mathfrak{R}}) \subset \mathbf{O}_{\mathfrak{R}}. \tag{6.4.3}$$

Now we are able to characterize the operators T and $\tilde{T}$ by their projection properties in the space $\mathbf{D}_{\mathfrak{R}}$. In fact, let df be an arbitrary differential of $\mathbf{D}_{\mathfrak{R}}$; from (6.3.9), we have

$$df = T_f^2 - \tilde{T}_f. \tag{6.4.4}$$

Using the second and third relations (6.4.3), we see that (6.4.4) is just an explicit representation of the orthogonal projections of df into $\mathbf{F}_{\mathfrak{R}}$ and $\mathbf{O}_{\mathfrak{R}}$. Thus T^2 and $-\tilde{T}$ appear as the projection operators of an element of $\mathbf{D}_{\mathfrak{R}}$ onto the spaces $\mathbf{O}_{\mathfrak{R}}$ and $\mathbf{F}_{\mathfrak{R}}$ respectively.

In particular, if $df \in \mathbf{O}_{\mathfrak{R}}$, then $\tilde{T}_f = 0$ so (6.4.4) becomes

$$T_f^2 = df$$

or

$$T^2(\mathbf{O}_{\mathfrak{R}}) = \mathbf{O}_{\mathfrak{R}}.$$

But by (6.4.2),

$$T^2(\mathbf{O}_{\mathfrak{R}}) = T(T(\mathbf{O}_{\mathfrak{R}})) \subset T(\mathbf{O}_{\mathfrak{R}}) \subset \mathbf{O}_{\mathfrak{R}}.$$

Thus

$$T(\mathbf{O}_{\mathfrak{R}}) = \mathbf{O}_{\mathfrak{R}} \tag{6.4.2$'$}$$

and hence

$$T(\mathbf{D}_{\mathfrak{R}}) = \mathbf{O}_{\mathfrak{R}}. \tag{6.4.3$'$}$$

6.5. Conditions for the Vanishing of a Differential

We now consider the subclass $\mathbf{O}_{\mathfrak{M}}$ of $\mathbf{F}_{\mathfrak{M}}$ consisting of elements which are orthogonal to all elements of $\mathbf{F}_{\mathfrak{R}}$. We remark that the class $\mathbf{F}_{\mathfrak{R}}$ is non-trivial, since $\mathfrak{R}$ has a boundary. Since $\mathscr{L}_{\mathbf{F}}(p, \tilde{q})$ belongs to $\mathbf{F}_{\mathfrak{R}}$, every differential df of the class $\mathbf{O}_{\mathfrak{M}}$ satisfies

$$\tilde{T}_f(q) \equiv 0, \; q \in \mathfrak{R}. \tag{6.5.1}$$

In fact, by (6.3.2),

$$\tilde{T}_f(q) = (f'(p), \mathscr{L}_{\mathbf{F}}(p, \tilde{q}))_{\mathfrak{R}} = 0. \tag{6.5.1$'$}$$

For $df \in \mathbf{F}_{\mathfrak{M}}$ we write

$$J_f(q) = (\mathscr{L}_{\mathbf{F}}(p, q), f'(p))_{\mathfrak{R}}. \tag{6.5.2}$$

It is clear that $J_f(q)$ is an analytic function of q in $\mathfrak{R} - \mathfrak{M}$. For $q \in \mathfrak{M}$,

$$J_f(q) = -(l_{\mathbf{F}}(p, q), f'(p))_{\mathfrak{R}}$$

so that $J_f(q)$ is also regular in the closure of $\mathfrak{M}$. For $df \in \mathbf{O}_{\mathfrak{M}}$ and q on the boundary of $\mathfrak{R}$, so that $q = \tilde{q}$, we have $J_f(q) = 0$ by (6.5.1)$'$. By analytic continuation, we then have

$$J_f(q) \equiv 0, \quad q \in \mathfrak{R} - \mathfrak{M}, \quad df \in \mathbf{O}_{\mathfrak{M}}. \tag{6.5.2$'$}$$

Thus the elements of $\mathbf{O}_{\mathfrak{M}}$ satisfy the conditions (6.5.1) and (6.5.2)$'$.

We prove next:

LEMMA 6.5.1. *If df is in $\mathbf{O}_{\mathfrak{M}}$, then $(T^2(q))_f$ is regular in the closure of $\mathfrak{M}$.*

For every $df \in \mathbf{F}_{\mathfrak{M}}$ we have

$$(\mathscr{L}_{\mathbf{F}}(q, p), f'(p))_{\mathfrak{M}} = -(l_{\mathbf{F}}(q, p), f'(p))_{\mathfrak{M}} \tag{6.5.3}$$

and hence this expression is regular in q throughout the closure of $\mathfrak{M}$. In particular $T_f(q)$, considered for q in $\mathfrak{M}$, is in $\mathbf{F}_{\mathfrak{M}}$; thus

$$(\mathscr{L}_{\mathbf{F}}(q, p), T_f(p))_{\mathfrak{M}}$$

is regular in q in the closure of $\mathfrak{M}$. If $df \in \mathbf{O}_{\mathfrak{M}}$, then for $q \in \mathfrak{R} - \mathfrak{M}$ we have by (6.5.2)$'$

$$\begin{aligned} T_f(q) &= (\mathscr{L}_{\mathbf{F}}(q, p), f'(p))_{\mathfrak{R}} = (\mathscr{L}_{\mathbf{F}}(q, p) - \mathscr{L}_{\mathbf{F}}(p, q), f'(p))_{\mathfrak{R}} \\ &= 2i \sum_{\mu, \nu=1}^{G_0} \operatorname{Im}\{\beta_{\mu\nu}\} \mathscr{Z}'_\mu(q) (d\mathscr{Z}_\nu, df)_{\mathfrak{R}} \end{aligned} \tag{6.5.4}$$

using (5.1.17), (5.1.17)$'$ and (5.1.18). Moreover

$$\begin{aligned} (\mathscr{L}_{\mathbf{F}}(q, p), \mathscr{Z}'_\varrho(p))_{\mathfrak{R}-\mathfrak{M}} &= (\mathscr{L}_{\mathbf{F}}(q, p), \mathscr{Z}'_\varrho(p))_{\mathfrak{R}} \\ &\quad - (L_{\mathbf{F}}(q, p), \mathscr{Z}'_\varrho(p))_{\mathfrak{M}} + (l_{\mathbf{F}}(q, p), \mathscr{Z}'_\varrho(p))_{\mathfrak{M}} \end{aligned} \tag{6.5.5}$$

and these terms are regular in the closure of $\mathfrak{M}$. This is obvious in the case of the last term. Using (4.10.8) and the formulas of

Section 5.1, the first term is

$$\sum_{\mu,\nu=1}^{G_0} \beta_{\mu\nu} P_{\nu\varrho} \mathscr{Z}'_{\mu}(q)$$

while the second is

$$-\sum_{\mu,\nu=1}^{G} \alpha_{\mu\nu}(\varepsilon_{\varrho\nu})^{-} Z'_{\mu}(q)$$

and these expressions are regular in the closure of $\mathfrak{M}$. Using (6.5.4) and (6.5.5), together with our remarks concerning regularity, we see that

$$(\mathscr{L}_{\mathbf{F}}(q,p),\ T_f(p))_{\mathfrak{R}-\mathfrak{M}}$$

is regular in the closure of $\mathfrak{M}$. Thus

$$(T^2(q))_f = (\mathscr{L}_{\mathbf{F}}(q,p),\ T_f(p))_{\mathfrak{M}} + (\mathscr{L}_{\mathbf{F}}(q,p),\ T_f(p))_{\mathfrak{R}-\mathfrak{M}}$$

is regular in the closure of $\mathfrak{M}$. This proves the lemma.

By (6.3.9), we have

$$(T^2(q))_f = f'(q) + \tilde{T}_f(q). \tag{6.5.6}$$

However, for $df \in \mathbf{O}_{\mathfrak{M}}$, $\tilde{T}_f \equiv 0$ by (6.5.1), so (6.5.6) implies

$$(T^2(q))_f = f'(q). \tag{6.5.6$'$}$$

Combining (6.5.6)′ with the preceding lemma, we have

LEMMA 6.5.2. *Each differential of the class $\mathbf{O}_{\mathfrak{M}}$ is regular analytic in the closure of $\mathfrak{M}$.*

Suppose that $\mathfrak{M}$ is properly imbedded in $\mathfrak{R}$. Since

$$J_f(q) - T_f(q) = 2i \sum_{\mu,\nu=1}^{G_0} \operatorname{Im}\{\beta_{\mu\nu}\}(d\mathscr{Z}_{\mu}, df)_{\mathfrak{R}} \mathscr{Z}'_{\nu}(q)$$

is continuous as q crosses the boundary of $\mathfrak{M}$, and df is regular on the boundary of $\mathfrak{M}$, it follows from (6.1.10) that

$$J_f(q_i) - J_f(q_e) = T_f(q_i) - T_f(q_e) = (f'(q))^{-},$$

where $f'(q) = df/d\zeta$, ζ a boundary uniformizer at q. But by (6.5.2)′, $J_f(q_e) = 0$. Thus $J_f(q)$ has in $\mathfrak{M}$ the boundary values $(f'(q))^{-}$. Therefore the differentials $f'(p) + J_f(p)$ and $(f'(p) - J_f(p))/i$, which are regular everywhere in the closure of $\mathfrak{M}$, are real on the boundary and we have

$$(6.5.7)\qquad f'(p) + J_f(p) = \sum_{\mu=1}^{G} a_\mu Z'_\mu(p), \quad a_\mu \text{ real},$$

$$(6.5.8)\qquad (f'(p) - J_f(p))/i = \sum_{\mu=1}^{G} b_\mu Z'_\mu(p), \quad b_\mu \text{ real}.$$

Combining these equations we obtain

$$(6.5.9)\qquad f'(p) = \sum_{\mu=1}^{G} c_\mu Z'_\mu(p), \quad c_\mu = \frac{1}{2}(a_\mu + ib_\mu).$$

If $\mathbf{F}_{\mathfrak{M}}$ is the class $\mathbf{S}$, we conclude immediately that $f'(p) \equiv 0$ by Lemma 4.8.1.

If $\mathbf{F}_{\mathfrak{M}}$ is not the class $\mathbf{S}$, then we can conclude that $f'(p) \equiv 0$ if the corresponding classes and the imbedding are such that it is possible for a differential of $\mathbf{F}_{\mathfrak{R}}$ to have non-vanishing periods on any cycle of $\mathfrak{M}$ not included in the period restrictions for $\mathbf{F}_{\mathfrak{M}}$. In fact, let $\mathbf{F}_{\mathfrak{M}} = \mathbf{F}_{\mathfrak{M}}(i_1, i_2, \cdots, i_l)$ where $0 \leq l < G$. Given any cycle K_ν of $\mathfrak{M}$, $\nu \neq i_1, \cdots, i_l$, there is an integral W_ν of the first kind such that $dW_\nu \in \mathbf{F}_{\mathfrak{R}}$ and

$$(6.5.10)\qquad (dW_\nu, dZ_\mu) = -P(dW_\nu, K_\mu) = \delta_{\nu\mu}.$$

On the other hand, each dW_ν is in $\mathbf{F}_{\mathfrak{R}}$ and therefore orthogonal to $df \in \mathbf{O}_{\mathfrak{M}}$. By (6.5.9) and (6.5.10) we have

$$(6.5.11)\qquad (dW_\nu, df)_{\mathfrak{M}} = \sum_{\mu=1}^{G} \bar{c}_\mu \delta_{\nu\mu} = \bar{c}_\nu = 0, \quad \nu \neq i_1, i_2, \cdots, i_l,$$

so

$$(6.5.12)\qquad f'(p) = \sum_{\tau=1}^{l} c_{i_\tau} Z'_{i_\tau}(p).$$

But $df \in \mathbf{F}_{\mathfrak{M}}$, so we have

$$(6.5.13)\qquad (df, dZ_{i_\nu})_{\mathfrak{M}} = \sum_{\tau=1}^{l} c_{i_\tau} \Gamma_{i_\tau i_\nu} = 0, \quad \nu = 1, 2, \cdots, l.$$

Since the matrix $\| \Gamma_{i_\tau i_\nu} \|$ is non-singular because of the proper imbedding hypothesis, it follows that $c_{i_\tau} = 0$, $\tau = 1, 2, \cdots, l$. Hence $f'(p) \equiv 0$.

For the existence of the differentials $dW_\nu \in \mathbf{F}_{\mathfrak{R}}$ we must suppose that the class $\mathbf{F}_{\mathfrak{M}}$ is the minimal corresponding class on $\mathfrak{M}$, i.e., that $\mathbf{F}_{\mathfrak{M}}$ contains no proper subclass (with period restrictions on additional cycles K_ν) which is also a corresponding class. In addition we suppose that the imbedding is essential (see Section 5.1). Then we have the following

LEMMA 6.5.3. *Let $\mathfrak{M}$ be properly and essentially imbedded in $\mathfrak{R}$, and let $\mathbf{F}_{\mathfrak{M}}$ be the minimal corresponding class on $\mathfrak{M}$. Then the only element of $\mathbf{F}_{\mathfrak{M}}$ which lies in $\mathbf{O}_{\mathfrak{M}}$ is the zero differential.*

We remark that the conclusion $f'(p) \equiv 0$ remains valid under the weaker assumption that any boundary component C_μ of $\mathfrak{M}$ which is homologous to zero on $\mathfrak{R}$ but not on $\mathfrak{M}$ be a cycle K_{i_ν} associated with the class $\mathbf{F}_{\mathfrak{M}}$. However, some hypothesis of this nature is essential. For consider the case in which $\mathbf{F}_{\mathfrak{M}} = \mathbf{M}$, and suppose that there is a boundary component C_μ of $\mathfrak{M}$, which is homologous to zero on $\mathfrak{R}$ but not on $\mathfrak{M}$. Let $Z'_{2h+\mu}$ be the differential of the first kind on $\mathfrak{M}$ corresponding to the boundary cycle C_μ. Then

$$(6.5.14)\quad \tilde{T}(q) = (Z'_{2h+\mu}(p), \mathscr{L}_{\mathbf{M}}(p, \bar{q}))_{\mathfrak{M}} = -(P(\mathscr{L}_{\mathbf{M}}(p, \bar{q}), C_\mu))^- = 0$$

since C_μ is homologous to zero on $\mathfrak{R}$. Similarly, for $q \in \mathfrak{R} - \mathfrak{M}$,

$$(6.5.14)'\quad J(q) = (\mathscr{L}_{\mathbf{M}}(p, q), Z'_{2h+\mu}(p))_{\mathfrak{M}} = -P(\mathscr{L}_{\mathbf{M}}(p, q), C_\mu) = 0.$$

Hence, for the particular differential $Z'_{2h+\mu}(p)$ of class $\mathbf{M}$ on $\mathfrak{M}$, $\tilde{T}(q)$ is zero for $q \in \mathfrak{R}$ and $J(q)$ is zero for $q \in \mathfrak{R} - \mathfrak{M}$, but $Z'_{2h+\mu}(p)$ is not identically zero.

We observe that the equation (6.5.1) characterizes the class $\mathbf{O}_{\mathfrak{M}}$, that is, if $\tilde{T}_f(q)$ vanishes identically for $df \in \mathbf{F}_{\mathfrak{M}}$, then $df \in \mathbf{O}_{\mathfrak{M}}$. For let dh be any differential of class $\mathbf{F}_{\mathfrak{R}}$. Then by (6.2.12),

$$(6.5.15)\qquad (df, dh)_{\mathfrak{M}} = -(\tilde{T}_f, dh)_{\mathfrak{R}} = 0.$$

The same conclusion follows if we know merely that $\tilde{T}_f$ vanishes in $\mathfrak{R} - \mathfrak{M}$, since $\tilde{T}_f$ is analytic in $\mathfrak{R}$. Thus by Lemma 6.5.3 we have

LEMMA 6.5.4. *Under the hypotheses of the preceding lemma, if $\tilde{T}_f$ vanishes in $\mathfrak{R} - \mathfrak{M}$ and $df \in \mathbf{F}_{\mathfrak{M}}$, then df vanishes identically.*

Let us make an application of this lemma to properties of the norm of an operator. If $\mathbf{F}_{\mathfrak{R}}$ is symmetric, we have by (6.2.11)

$$(6.5.16)\qquad N_{\mathfrak{R}}(S_f) = N_{\mathfrak{R}}(T_f) + N_{\mathfrak{R}}(\tilde{T}_f) = N_{\mathfrak{M}}(df).$$

From this identity we conclude that

$$(6.5.17)\qquad N_{\mathfrak{M}}(df) \geqq N_{\mathfrak{M}}(T_f),\ N_{\mathfrak{M}}(df) \geqq N_{\mathfrak{M}}(\tilde{T}_f).$$

If equality occurs in either relation, then $\tilde{T}_f = 0$ in $\mathfrak{R} - \mathfrak{M}$. This implies

LEMMA 6.5.5. *Let* $\mathfrak{M}$ *and* $\mathbf{F}_{\mathfrak{M}}$ *satisfy the conditions of Lemma* 6.5.3, *and let* $\mathbf{F}_{\mathfrak{R}}$ *be symmetric. Then for any non-zero differential of* $\mathbf{F}_{\mathfrak{M}}$, *we have the proper inequalities*

$$N_{\mathfrak{M}}(df) > N_{\mathfrak{M}}(T_f), \quad N_{\mathfrak{M}}(df) > N_{\mathfrak{M}}(\tilde{T}_f). \tag{6.5.18}$$

Let us derive a similar result for the S-operator. We make the same assumptions that were made in the preceding lemma. By (6.2.11) we have

$$N_{\mathfrak{M}}(S_f) < N_{\mathfrak{M}}(df), \tag{6.5.19}$$

unless $S_f(q) \equiv 0$ in $\mathfrak{R} - \mathfrak{M}$. If $S_f(q) \equiv 0$ in $\mathfrak{R} - \mathfrak{M}$, we have

$$S_f(q) = \begin{cases} g'(q), & q \in \mathfrak{M}, \\ 0, & q \in \mathfrak{R} - \mathfrak{M}, \end{cases}$$

where g' is regular analytic in the closure of $\mathfrak{M}$ since S_f is regular analytic in the closure of $\mathfrak{M}$ for every differential of $\mathbf{F}_{\mathfrak{M}}$. By (6.3.8),

$$(S^2(q))_f = S_g(q) = f'(q).$$

Since $S_g(q)$ is regular analytic in the closure of $\mathfrak{M}$, so is $f'(q)$. By (6.1.10),

$$S_f(q_i) = S_f(q_i) - S_f(q_e) = T_f(q_i) - T_f(q_e) = (f'(q))^-,$$

where $f'(q) = df/d\zeta$, ζ a boundary uniformizer at q. Therefore $f'(q) + S_f(q)$ and $(f'(q) - S_f(q))/i$ are differentials of $\mathfrak{M}$, so

$$\begin{cases} f'(q) + S_f(q) = \sum_{\mu=1}^{G} a_\mu Z'_\mu(q), & a_\mu \text{ real}, \\ (f'(q) - S_f(q))/i = \sum_{\mu=1}^{G} b_\mu Z'_\mu(q), & b_\mu \text{ real}. \end{cases} \tag{6.5.20}$$

Hence

$$f'(q) = \frac{1}{2} \sum_{\mu=1}^{G} (a_\mu + ib_\mu) Z'_\mu(q). \tag{6.5.21}$$

By (6.2.12), we have for any differential dh of $\mathbf{F}_{\mathfrak{R}}$,

$$(S_f, dh)_{\mathfrak{R}} = (S_f, dh)_{\mathfrak{M}} = -(df, dh)_{\mathfrak{M}},$$

that is,

$$(S_f + df,\ dh)_{\mathfrak{M}} = 0. \tag{6.5.22}$$

Equation (6.5.22) shows that $S_f + df$ lies in $\mathbf{O}_{\mathfrak{M}}$. By Lemma 6.5.3 we therefore conclude that

$$S_f(q) + f'(q) \equiv 0,\ a_\mu = 0,\ \mu = 1, 2, \cdots, G. \tag{6.5.23}$$

Thus

$$f'(q) = \frac{i}{2} \sum_{\mu=1}^{G} b_\mu Z'_\mu(q). \tag{6.5.24}$$

Conversely, it is readily verified that each differential of the form (6.5.24) satisfies (6.5.23). For a general differential $df \in \mathbf{F}_{\mathfrak{M}}$,

$$\begin{aligned} S_f(q) &= -(l_{\mathbf{F}}(q, p),\ f'(p))_{\mathfrak{M}} + (\mathscr{L}_{\mathbf{F}}(q, \tilde{p}),\ (f'(p))^-)_{\mathfrak{M}} \\ &= -(l_{\mathbf{F}}(q, p),\ f'(p))_{\mathfrak{M}} - (l_{\mathbf{F}}(q, \tilde{p}),\ (f'(p))^-)_{\mathfrak{M}} - f'(q), \end{aligned}$$

that is,

$$\begin{aligned} S_f(q) + f'(q) = &-\int_{\mathfrak{M}} l_{\mathbf{F}}(q, p)(f'(p))^- dA_z \\ &-\int_{\mathfrak{M}} l_{\mathbf{F}}(q, \tilde{p}) f'(p) dA_z. \end{aligned} \tag{6.5.25}$$

Now substitute the special differential (6.5.24) into the general identity (6.5.25); we obtain

$$\begin{aligned} S_f(q) + f'(q) &= \frac{i}{2}\left\{ \int_{\mathfrak{M}} l_{\mathbf{F}}(q, p) \sum_{\mu=1}^{G} b_\mu (Z'_\mu(p))^- dA_z \right. \\ &\qquad \left. - \int_{\mathfrak{M}} l_{\mathbf{F}}(q, \tilde{p}) \sum_{\mu=1}^{G} b_\mu Z'_\mu(p) dA_z \right\} \\ &= -\frac{i}{2}\left\{ \sum_{\mu=1}^{G} b_\mu [P(l_{\mathbf{F}}(p, q), K_\mu) - (P(l_{\mathbf{F}}(p, \tilde{q}), K_\mu))^-] \right\} = 0 \end{aligned}$$

since

$$(P(l_{\mathbf{F}}(p, \tilde{q}), K_\mu))^- = P(l_{\mathbf{F}}(p, q), K_\mu) \tag{6.5.26}$$

because of the symmetry of the class $\mathbf{F}_{\mathfrak{R}}$ (and therefore also of $\mathbf{F}_{\mathfrak{M}}$).

We have therefore proved that (6.5.19) is satisfied unless df is of the special form (6.5.24). In this case there is equality of norms.

Under the hypotheses of Lemma 6.5.5, we have the proper inequality

(6.5.27) $$N_{\mathfrak{M}}(\tilde{T}_f) > 0$$

for each non-zero $df \in \mathbf{F}_{\mathfrak{M}}$. We now consider the possibility that

$$N_{\mathfrak{M}}(T_f) = 0$$

for every $df \in \mathbf{F}_{\mathfrak{M}}$. This can occur if and only if

(6.5.28) $$l_{\mathbf{F}}(p, q) \equiv 0 \text{ in } \mathfrak{M}.$$

In fact, for $df = l_{\mathbf{F}}(p, q) \in \mathbf{F}_{\mathfrak{M}}$ we have, for $q \in \mathfrak{M}$,

$$T_f(q) = -N_{\mathfrak{M}}(l_{\mathbf{F}}(p, q)).$$

An example in which (6.5.28) holds is provided by the case in which $\mathfrak{M}$ is a circular disc concentrically imbedded in another circular disc $\mathfrak{R}$.

In the case (6.5.28) we have $S_f = \tilde{T}_f$ in $\mathfrak{M}$ since $T_f \equiv 0$. Thus, for example, the set of possible exceptions (6.5.24) to (6.5.19) must be empty, by Lemma 6.5.5. On the whole, the effect of (6.5.28) is only to exclude certain possibilities which may occur in the general case. If either $\mathbf{F}_{\mathfrak{R}} = \mathbf{S}$ or $\mathbf{F}_{\mathfrak{M}} = \mathbf{S}$ we can show that (6.5.28) is possible only if $\mathfrak{M}$ is simply connected.

Suppose that $\mathbf{F}_{\mathfrak{R}} = \mathbf{S}$. Then, since the class $\mathbf{F}_{\mathfrak{R}}$ is assumed to be symmetric, $\mathfrak{R}$ must be of genus zero. In this case the function

$$\Xi_{\mathbf{S}}(p, q) = \int \mathscr{L}_{\mathbf{S}}(p, q)dp$$

is schlicht and maps $\mathfrak{R}$ onto a plane surface with maximal external area (cf. Section 4.12). It follows that $\mathscr{L}_{\mathbf{S}}(p, q) \neq 0$ at each point $p \in \mathfrak{R}$. Now let C be the boundary of $\mathfrak{M}$. In the case (6.5.28) we have for $p,\ q \in C$

$$\mathscr{L}_{\mathbf{S}}(p, q)dp\,dq = L_{\mathbf{S}}(p, q)dp\,dq = \text{real}.$$

Let q be a fixed point of C; then

$$\operatorname{Im}\{\Xi_{\mathbf{S}}(p, q)\} = \text{constant}$$

on each component of C. Therefore $\Xi_{\mathbf{S}}(p, q)$, which has a simple pole at $p = q$, maps $\mathfrak{M}$ onto a parallel slit domain where each boundary component of C goes into a straight line parallel to the real axis. The boundary component of C which contains q is mapped into a line containing the point at infinity, while the other components are mapped into finite segments. But $\mathscr{L}_{\mathbf{S}}(p, q)$ must vanish at those points $p \in C$ which correspond to finite endpoints of these segments. Since $\mathscr{L}_{\mathbf{S}}(p, q) \neq 0$ in $\mathfrak{R}$ and $\mathfrak{M}$ is properly imbedded in $\mathfrak{R}$, the only possibility is that $\mathfrak{M}$ is mapped onto a half-plane. Thus $\mathfrak{M}$ is simply-connected.

After a linear transformation we may suppose that $\mathfrak{R}$ is the exterior of m_0 Jordan curves $B_1, \cdots, B_{m_0}$ on the z-sphere and that $\mathfrak{M}$ is the exterior of a large circle which encloses the boundary curves of $\mathfrak{R}$ in its interior. Under these circumstances we have

$$L_{\mathbf{S}}(p, q) = \frac{1}{\pi(z - \zeta)^2},$$

and therefore

$$\mathscr{L}_{\mathbf{S}}(p, q) = \frac{1}{\pi(z - \zeta)^2}.$$

The boundary curves B_μ of $\mathfrak{R}$ are characterized by the differential equations

$$\frac{dz\, d\zeta}{(z - \zeta)^2} = \text{real}, \qquad z, \zeta \in B.$$

Integrating with respect to z and ζ we find that

$$\frac{z_1 - \zeta_1}{z_1 - \zeta_2} \cdot \frac{z_2 - \zeta_2}{z_2 - \zeta_1} = \text{real} \tag{6.5.29}$$

provided that z_1, z_2 lie on the same component B_μ and ζ_1, ζ_2 on the same component B_ν. From (6.5.29) we concluded that each boundary B_ν must be a circle through z_1 and z_2, i.e. that $\mathfrak{R}$ is bounded by a single circle. Thus, in the case $\mathbf{F}_{\mathfrak{R}} = \mathbf{S}$, the hypothesis (6.5.28) implies that the imbedding of $\mathfrak{M}$ into $\mathfrak{R}$ is conformally equivalent to the concentric imbedding of one disc in a larger one.

Assume next that $\mathbf{F}_{\mathfrak{M}}$ is the class $\mathbf{S}$. Then $\mathfrak{M}$ must be of genus zero. In this case the function

$$\Psi_{\mathbf{S}}(p, q) = \int L_{\mathbf{S}}(p, q)dp$$

is schlicht on $\mathfrak{M}$ and maps it onto a subdomain of the sphere with maximum external area. Hence $\Psi_{\mathbf{S}}(p, q)$ is schlicht in a slightly larger domain $\mathfrak{M}'$. Under the hypothesis (6.5.28), the function

$$\Xi_{\mathbf{F}}(p, q) = \int \mathscr{L}_{\mathbf{F}}(p, q) dp$$

coincides with $\Psi_{\mathbf{S}}(p, q)$ in $\mathfrak{M}$ and therefore in $\mathfrak{M}'$. In particular, it follows that $\mathscr{L}_{\mathbf{F}}(p, q) \neq 0$ for $p \in \mathfrak{M}'$ and we conclude as before that $\mathfrak{M}$ must be simply-connected.

6.6. Bounds for the Operators T and $\tilde{T}$

We suppose that the hypotheses used in Section 6.5 are satisfied, namely: $\mathfrak{M}$ is properly and essentially embedded in $\mathfrak{R}$ and $\mathscr{L}_{\mathbf{F}}(p, q) = \mathscr{L}_{\mathbf{F}}(q, p)$ and $\mathbf{F}_{\mathfrak{M}}$ is minimal. We remark that if $\mathbf{F}_{\mathfrak{M}} = \mathbf{S}$, the assumption concerning essential imbedding may be dropped.

We recall that a differential df of $\mathbf{F}_{\mathfrak{M}}$ is defined to be zero in $\mathfrak{R} - \mathfrak{M}$, and therefore a differential of $\mathbf{F}_{\mathfrak{M}}$ is transformed by the operators into a differential of $\mathbf{E}_{\mathfrak{R}}$, and not necessarily into a differential of $\mathbf{F}_{\mathfrak{M}}$. We may, however, define a new differential which coincides with the transformed differential in $\mathfrak{M}$, and vanishes in $\mathfrak{R} - \mathfrak{M}$. This differential belongs therefore to $\mathbf{F}_{\mathfrak{M}}$. In this way we define new operators $t_f(q)$, $\tilde{t}_f(q)$ and $s_f(q)$, which have $\mathbf{F}_{\mathfrak{M}}$ as domain and range. Thus, for instance,

$$t_f(q) = \begin{cases} T_f(q), & q \in \mathfrak{M}, \\ 0, & q \in \mathfrak{R} - \mathfrak{M}. \end{cases}$$

In the following formulas we shall be concerned with operations in the domain $\mathfrak{M}$ only. Therefore norms and scalar products without subscripts designating the domain of integration are assumed to be taken over $\mathfrak{M}$.

Let the non-zero differentials of $\mathbf{F}_{\mathfrak{M}}$ be normalized by the condition that

$$N(df) = 1. \tag{6.6.1}$$

We shall prove the following theorem:

THEOREM 6.6.1. *There are real numbers λ and ϱ, each greater than unity, such that*

$$N(t_f) \leq \frac{1}{\lambda^2}, \tag{6.6.2}$$

$$N(\tilde{t}_f) \leq \frac{1}{\varrho^2}. \tag{6.6.3}$$

There is equality in (6.6.3) for at least one differential of $\mathbf{F}_{\mathfrak{M}}$ *satisfying* (6.6.1). *The same statement applies to* (6.6.2) *except in the case* (6.5.28). *Thus t and $\tilde{t}$ have bounds which are less than one.*

The proofs of (6.6.2) and (6.6.3) turn on the fact that, for $q \epsilon \mathfrak{M}$,

(6.6.4) $t_f(q) = - (l_{\mathbf{F}}(q, p), f'(p))_{\mathfrak{M}},$

(6.6.5) $\tilde{t}_f(q) = (\mathscr{L}_{\mathbf{F}}(q, \tilde{p}), (f'(p))^-)_{\mathfrak{M}},$

where $l_{\mathbf{F}}(p, q)$ and $\mathscr{L}_{\mathbf{F}}(q, \tilde{p})$ are regular in the closure of $\mathfrak{M}$. The regularity in $\mathfrak{M}$ of $\mathscr{L}_{\mathbf{F}}(q, \tilde{p})$ is a consequence of the proper embedding.

Let $\{df_\nu\}$ be a sequence of differentials of $\mathbf{F}_{\mathfrak{M}}$ satisfying the normalization condition (6.6.1), and suppose that df_ν converges pointwise to a differential df as ν becomes infinite, the convergence being uniform in any compact subdomain $\mathfrak{M}'$ lying in the interior of $\mathfrak{M}$. Then

$$N_{\mathfrak{M}'}(df) = \lim_{\nu \to \infty} N_{\mathfrak{M}'}(df_\nu) \leqq 1.$$

Letting $\mathfrak{M}'$ tend to $\mathfrak{M}$, we see that

(6.6.6) $N_{\mathfrak{M}}(df) = N(df) \leqq 1.$

Thus the limit differential also belongs to the class $\mathbf{F}_{\mathfrak{M}}$, with norm bounded by unity.

Given $\varepsilon > 0$, let $\mathfrak{M}'$ be chosen such that, uniformly in q,

(6.6.7) $\displaystyle\int_{\mathfrak{M}-\mathfrak{M}'} | \, l_{\mathbf{F}}(q, p) \, |^2 \, dA_z < \varepsilon^2.$

Writing

$$t_\nu(q) = t_{f_\nu}(q),$$

we have, for $q \epsilon \mathfrak{M}$,

(6.6.8)
$$| \, t_f(q) - t_\nu(q) \, | \leqq \int_{\mathfrak{M}'} | \, l_{\mathbf{F}}(q, p) \, | \; | \, f'(p) - f'_\nu(p) \, | \, dA_z$$
$$+ \int_{\mathfrak{M}-\mathfrak{M}'} | \, l_{\mathbf{F}}(q, p) \, | \; | \, f'(p) - f'_\nu(p) \, | \, dA_z,$$

By the uniform convergence there is a number $\nu_0 = \nu_0(\varepsilon)$ such that

(6.6.9) $\displaystyle\int_{\mathfrak{M}'} | \, l_{\mathbf{F}}(q, p) \, | \; | \, f'(p) - f'_\nu(p) \, | \, dA_z < \varepsilon$

if $\nu \geqq \nu_0$. By the Schwarz inequality and the triangle inequality

$$\int\limits_{\mathfrak{M}-\mathfrak{M}'} |\, l_{\mathbf{F}}(q, p)\,|\,|\, f'(p) - f'_\nu(p)\,|\, dA_z$$

(6.6.10)
$$\leqq \left\{ \int\limits_{\mathfrak{M}-\mathfrak{M}'} |\, l_{\mathbf{F}}(q, p)\,|^2 dA_z \right\}^{\frac{1}{2}} \left\{ \int\limits_{\mathfrak{M}-\mathfrak{M}'} |\, f'(p) - f'_\nu(p)\,|^2 dA_z \right\}^{\frac{1}{2}}$$

$$\leqq \left\{ \int\limits_{\mathfrak{M}-\mathfrak{M}'} |\, l_{\mathbf{F}}(q, p)\,|^2 dA_z \right\}^{\frac{1}{2}} \left\{ \left[\int\limits_{\mathfrak{M}-\mathfrak{M}'} |\, f'(p)\,|^2 dA_z \right]^{\frac{1}{2}} \right.$$

$$\left. + \left[\int\limits_{\mathfrak{M}-\mathfrak{M}'} |\, f'_\nu(p)\,|^2 dA_z \right]^{\frac{1}{2}} \right\} \leqq 2\varepsilon,$$

by (6.6.6) and (6.6.7). Hence by (6.6.8), (6.6.9) and (6.6.10) we have, for $q \,\epsilon\, \mathfrak{M}$,

(6.6.11) $$|\, t_f(q) - t_\nu(q)\,| < 3\varepsilon,$$

provided that $\nu \geqq \nu_0$. Thus $t_\nu(q)$ converges uniformly to $t_f(q)$ in the closure of $\mathfrak{M}$ and, in particular,

(6.6.12) $$N(t_f - t_\nu) \to 0, \quad N(t_\nu) \to N(t_f),$$

as ν becomes infinite. The same conclusions are valid for $\tilde{t}_f$.

We remark that the above reasoning depends in an essential way on the fact that $l_{\mathbf{F}}(q, p)$ and $\mathscr{L}_{\mathbf{F}}(q, \tilde{p})$ are regular throughout the closure of $\mathfrak{M}$. This is illustrated by the following instructive example. Let $\{d\varphi_\nu\}$ be a complete orthonormal system for the class $\mathbf{F}_{\mathfrak{M}}$, and write

$$I_\nu(q) = -\varphi'_\nu(q) = \int\limits_{\mathfrak{M}} L_{\mathbf{F}}(q, \tilde{p})\varphi'_\nu(p)dA_z.$$

We have

$$\sum_{\nu=1}^{\infty} |\, \varphi'_\nu(q)\,|^2 < \infty$$

for any q in the interior of $\mathfrak{M}$, so $\varphi'_\nu(q)$ tends to zero as ν tends to infinity and, in fact, uniformly in any compact subdomain lying in the interior of $\mathfrak{M}$. Therefore

$$I_\nu(q) \to I(q)$$

where $I(q) \equiv 0$, but

$$\lim_{\nu \to \infty} N(I_\nu) = 1 \neq N(I) = 0.$$

The reason for this behavior is centered in the fact that $L_{\mathbf{F}}(p, \tilde{p})$ becomes infinite as p approaches a boundary point of $\mathfrak{M}$.

Let d be the least upper bound of $N(t_f)$ for differentials f' of $\mathbf{F}_{\mathfrak{M}}$ satisfying the normalization condition (6.6.1), and let $\{f'_\nu\}$ be a sequence of differentials for which

$$N(t_\nu) \to d$$

as ν tends to infinity. A subsequence of the f'_ν which we may take to be the whole sequence, converges uniformly on any compact subdomain $\mathfrak{M}'$ of $\mathfrak{M}$ to a differential f' of $\mathbf{F}_{\mathfrak{M}}$ satisfying (6.6.6). By (6.6.12) we have

$$N(t_f) = d. \tag{6.6.13}$$

If d is positive, we conclude that $f'(p)$ is not identically zero and it then follows from (6.5.18) that

$$N(t_f) = d < N(df) \leqq 1. \tag{6.6.14}$$

For the differential f' satisfying (6.6.13) we must have $N(df) = 1$. Otherwise we could multiply df by a number $b > 1$ such that $N(bdf) = 1$ and we would have

$$N(t_{bf}) = b^2 d > d,$$

a contradiction. Taking $\lambda^2 = 1/d$, we obtain (6.6.2). The case $d = 0$ corresponds to (6.5.28). In this case, the inequality (6.6.2) holds for every choice of $\lambda > 1$, but there is never equality.

A precisely similar argument gives (6.6.3). In this case, we have $d > 0$ by (6.5.27).

For differentials which are not normalized by the condition (6.6.1), the inequalities become

$$N(t_f) \leqq \frac{N(df)}{\lambda^2}, \tag{6.6.2$'$}$$

$$N(\tilde{t}_f) \leqq \frac{N(df)}{\varrho^2}. \tag{6.6.3$'$}$$

6.7. SPECTRAL THEORY OF THE t-OPERATOR

Two types of convergence of sequences of differentials have been considered, namely, (a) pointwise convergence of a sequence df_ν, which is uniform in each compact interior subdomain of $\mathfrak{M}$, and (b) convergence of the elements df_ν in the metric of the Hilbert space $\mathbf{F}_{\mathfrak{M}}$. It is also useful to interpret the first type of convergence in terms of the concepts of Hilbert space. Let $d\varphi$ be an arbitrary element of this space. From the first type of convergence of the sequence df_ν to a limit differential df we immediately conclude that

$$\lim_{\nu\to\infty} (df_\nu, d\varphi) = (df, d\varphi).$$

This type of convergence of $\{df_\nu\}$ to an element df of the Hilbert space is called weak convergence. The significance of the formula (6.6.12) may then be expressed as follows: the operators t_f and $\bar{t}_f$ transform each weakly convergent sequence in the Hilbert space into a properly, or strongly, convergent one.

This property of operators was called by Hilbert "complete continuity" and a large part of his theory on forms of infinitely many variables was based on this concept. By the reasoning of the last section, one can show that a completely continuous operator is always bounded.

In the preceding section we proved that the operators t_f and $\bar{t}_f$, and therefore also s_f, are bounded. The general theory of Hermitian linear operators in Hilbert space guarantees the existence of a discrete spectrum of eigen-values for the operator $\bar{t}_f$. By similar reasoning, we would be able to derive the same result for the other two operators. However, in order to make our representation of the theory self-contained, we shall build up the spectral theory for all three operators from more elementary concepts, making use of the particularly simple properties of these operators. In this way we are also led to certain interesting realizations of the abstract Hilbert space.

We suppose that the assumptions made at the beginning of Section 6.6 are satisfied, and we investigate the eigen-values and eigen-differentials of the following three integral equations for differentials $df(q)$ of $\mathbf{F}_{\mathfrak{M}}$:

(6.7.1) $$df(q) = \lambda t_f(q),$$

(6.7.2) $$df(q) = \varrho \bar{t}_f(q),$$

(6.7.3) $$df(q) = \sigma s_f(q).$$

We begin with the eigen-value problem (6.7.1), and we observe that

(6.7.4) $$t_f(q) = (\mathscr{L}_{\mathbf{F}}(q, p), f'(p))_{\mathfrak{M}} = -(l_{\mathbf{F}}(q, p), f'(p))_{\mathfrak{M}},$$

(6.7.5) $$(t^2(q))_f = (\mathscr{L}_{\mathbf{F}}(q, p), t_f(p))_{\mathfrak{M}} = -(l_{\mathbf{F}}(q, p), t_f(p))_{\mathfrak{M}},$$

(6.7.6)
$$\begin{aligned}(t^2(q))_f &= \int_{\mathfrak{M}}\int_{\mathfrak{M}} l_{\mathbf{F}}(q, p_1)(l_{\mathbf{F}}(p_1, p_2))^{-} f'(p_2) dA_{z_1} dA_{z_2} \\ &= \int_{\mathfrak{M}} Q_{\mathbf{F}}(\bar{p}_2, q) f'(p_2) dA_{z_2}\end{aligned}$$

where by (5.3.20) and (5.4.1)

(6.7.7) $$Q_{\mathbf{F}}(\bar{p}, q) = \int_{\mathfrak{M}} l_{\mathbf{F}}(r, q)(l_{\mathbf{F}}(r, p))^{-} dA_\eta = -l_{\mathbf{F}}(\bar{p}, q) - \Gamma_{\mathbf{F}}(\bar{p}, q).$$

Here η is a local uniformizer at r. This shows that the iterated t-operator is Hermitian.

We have from (6.1.14) that

(6.7.8) $$(t_f, dg) = (t_g, df).$$

From (6.7.8) we observe that the t-operator is not self adjoint in the Hilbert space with the norm (df, dg). For just this reason we introduced in Section 6.1 another scalar product, namely

$$[df, dg] = \text{Re}\{(df, dg)\},$$

with respect to which the operator t is self-adjoint. We remark further that in the new metric based on $[df, dg]$, the norm of each differential is the same as before.

We shall now prove the existence of eigen-differentials and eigen-values of the integral equation (6.7.1) by considering extremum problems in the Hilbert space with the metric $[df, dg]$.

Suppose the first ν eigen-values $\lambda_1, \cdots, \lambda_\nu$ and an orthonormal set of eigen-differentials $d\varphi_1, \cdots, d\varphi_\nu$ are known. They satisfy

(6.7.9) $$d\varphi_k(q) = \lambda_k t_{\varphi_k}(q).$$

Consider now all differentials df of $\mathbf{F}_{\mathfrak{M}}$, measured by the scalar product $[df, dg]$. Since

$$[df + \lambda dh, df + \lambda dh] = [df, df] + 2\mathrm{Re}\{\bar{\lambda}(df, dh)\} + |\lambda|^2 [dh, dh],$$

we observe that these differentials form a Hilbert space if multiplication with real numbers only is permitted. For in this case the above identity takes the form

$$[df + \lambda dh, df + \lambda dh] = [df, df] + 2\lambda[df, dh] + \lambda^2 [dh, dh].$$

We are thus led to a "real" Hilbert space $F_{\mathfrak{M}}$ in which linear dependence is defined in the real sense. Each non-zero differential df of the Hilbert space $\mathbf{F}_{\mathfrak{M}}$ leads thus to two independent elements df and idf of $F_{\mathfrak{M}}$. In particular, setting $d\varphi_{-k} = i\,d\varphi_k$ and $\lambda_{-k} = -\lambda_k$, $k = 1, 2, \cdots, \nu$, we have an orthonormal set of 2ν eigen-differentials in $F_{\mathfrak{M}}$. We remark that a differential df which is orthogonal to both $d\varphi_k$ and $d\varphi_{-k}$ in $F_{\mathfrak{M}}$ is orthogonal to $d\varphi_k$ in $\mathbf{F}_{\mathfrak{M}}$, and conversely.

Among all differentials $df \in F_{\mathfrak{M}}$ satisfying the side conditions

$$N(df) = 1, \quad [df, d\varphi_k] = 0, \quad k = \pm 1, \pm 2, \cdots, \pm \nu, \tag{6.7.10}$$

there is at least one, say $df = d\varphi_{\nu+1}$, which maximizes

$$[df, t_f] = \mathrm{Re}\{(df, t_f)\}. \tag{6.7.11}$$

We remark that

$$\max\,[df, t_f] = (d\varphi_{\nu+1}, t_{\varphi_{\nu+1}}) \geqq 0.$$

For if df satisfies the conditions, so does the differential $e^{i\theta}df$, θ real, and we have

$$(e^{i\theta}df, t_{e^{i\theta}f}) = e^{2i\theta}(df, t_f). \tag{6.7.12}$$

Given any admissible df we can, by choice of θ, always make

$$[e^{i\theta}df, t_{e^{i\theta}f}] = \mathrm{Re}\{e^{2i\theta}(df, t_f)\} \geqq 0. \tag{6.7.13}$$

Moreover, if

$$(df, t_f) \neq 0,$$

we can choose θ such that there will be inequality in (6.7.13). Further, a maximizing differential $d\varphi_{\nu+1}$ must satisfy

$$[d\varphi_{\nu+1}, t_{\varphi_{\nu+1}}] = (d\varphi_{\nu+1}, t_{\varphi_{\nu+1}}),$$

for if $\operatorname{Im}\{(d\varphi_{\nu+1}, t_{\varphi_{\nu+1}})\} \neq 0$ we can choose θ such that, for $df = e^{i\theta} d\varphi_{\nu+1}$, the value $[df, t_f]$ exceeds $[d\varphi_{\nu+1}, t_{\varphi_{\nu+1}}]$. On the other hand, we observe from (6.6.2)′ that

$$|(df, t_f)| \leqq \{N(df)\}^{\frac{1}{2}} \{N(t_f)\}^{\frac{1}{2}} \leqq \frac{N(df)}{\lambda} = \frac{1}{\lambda}, \tag{6.7.14}$$

where $\lambda > 1$. Hence

$$0 \leqq \max\,[df, t_f] \leqq \frac{1}{\lambda}, \qquad \lambda > 1. \tag{6.7.15}$$

Writing

$$\max\,[df, t_f] = (d\varphi_{\nu+1}, t_{\varphi_{\nu+1}}) = \frac{1}{\lambda_{\nu+1}}, \tag{6.7.16}$$

we therefore have

$$1 < \lambda_{\nu+1} \leqq \infty. \tag{6.7.17}$$

Assume that $\lambda_{\nu+1} < \infty$. Let $d\varphi$ be a differential of $\mathbf{F}_{\mathfrak{M}}$ which is orthogonal to $d\varphi_1, d\varphi_2, \cdots, d\varphi_\nu$ and let $dg = d\varphi_{\nu+1} + \varepsilon d\varphi$. Then

$$[dg, t_g] = \frac{1}{\lambda_{\nu+1}} + 2\mathrm{Re}\{\varepsilon(d\varphi, t_{\varphi_{\nu+1}})\} + \mathrm{Re}\{\varepsilon^2(d\varphi, t_\varphi)\},$$

and it follows that

$$\begin{aligned}\frac{[dg, t_g]}{N(dg)} &= \frac{\dfrac{1}{\lambda_{\nu+1}} + 2\mathrm{Re}\{\varepsilon(d\varphi, t_{\varphi_{\nu+1}})\} + \mathrm{Re}\{\varepsilon^2(d\varphi, t_\varphi)\}}{1 + 2\mathrm{Re}\{\varepsilon(d\varphi, d\varphi_{\nu+1})\} + |\varepsilon|^2 N(d\varphi)} \\ &= \frac{1}{\lambda_{\nu+1}} + 2\mathrm{Re}\Big\{\varepsilon\Big[(d\varphi, t_{\varphi_{\nu+1}}) - \frac{1}{\lambda_{\nu+1}}(d\varphi, d\varphi_{\nu+1})\Big]\Big\} + O(|\varepsilon|^2) \\ &\leqq \frac{1}{\lambda_{\nu+1}}.\end{aligned}$$

Since ε is an arbitrary complex number, we have

$$(d\varphi, d\varphi_{\nu+1}) = \lambda_{\nu+1}(d\varphi, t_{\varphi_{\nu+1}}), \tag{6.7.18}$$

or

$$(d\varphi, d\varphi_{\nu+1} - \lambda_{\nu+1} t_{\varphi_{\nu+1}}) = 0. \tag{6.7.19}$$

From (6.7.19) we conclude that

$$d\varphi_{\nu+1} = \lambda_{\nu+1} t_{\varphi_{\nu+1}}. \tag{6.7.20}$$

In fact, the differential

$$df_{\nu+1} = d\varphi_{\nu+1} - \lambda_{\nu+1} t_{\varphi_{\nu+1}}$$

is orthogonal to all differentials $d\varphi$ which are orthogonal to $d\varphi_1, \cdots, d\varphi_\nu$. We can also show that $df_{\nu+1}$ is orthogonal to all $d\varphi_\varrho$, $\varrho = 1, 2, \cdots, \nu$. By (6.7.8) we have

$$(df_{\nu+1}, d\varphi_\varrho) = (d\varphi_{\nu+1}, d\varphi_\varrho) - \lambda_{\nu+1}(t_{\varphi_{\nu+1}}, d\varphi_\varrho)$$
$$= -\lambda_{\nu+1}(t_{\varphi_\varrho}, d\varphi_{\nu+1}).$$

Hence, by (6.7.9),

$$(df_{\nu+1}, d\varphi_\varrho) = -\lambda_{\nu+1}(\lambda_\varrho)^{-1}(d\varphi_\varrho, d\varphi_{\nu+1}) = 0.$$

Since the space $\mathbf{F}_{\mathfrak{M}}$ can be decomposed into the linear space of the $d\varphi_\varrho$, $\varrho = 1, 2, \cdots, \nu$, and its orthogonal complement, $df_{\nu+1}$ is orthogonal to every differential of $\mathbf{F}_{\mathfrak{M}}$ and hence is identically zero. Thus (6.7.20) is proved.

It is clear that $d\varphi_{-\nu-1} = i\, d\varphi_{\nu+1}$ minimizes (6.7.11) under the side conditions (6.7.10) and corresponds to the eigen-value $\lambda_{-\nu-1} = -\lambda_{\nu+1}$. Further

$$[d\varphi_{-\nu-1}, d\varphi_{\nu+1}] = \operatorname{Re}\{iN(d\varphi_{\nu+1})\} = 0;$$

that is, the differentials $d\varphi_k$, $k = \pm 1, \pm 2, \cdots, \pm(\nu + 1)$, form an orthonormal set in $F_{\mathfrak{M}}$. The differentials $d\varphi_k$, $k = 1, 2, \cdots, \nu + 1$, form an orthonormal set in $\mathbf{F}_{\mathfrak{M}}$.

Beginning with the lowest eigen-value λ_1, we can construct a sequence of eigen-values and eigen-differentials by repeating the above maximum problem. In this way, we obtain sequences $d\varphi_k$ and λ_k.

The k-th eigen-differential satisfies the equation

$$-\frac{1}{\lambda_k} d\varphi_k(q) = \int_{\mathfrak{M}} l_F(q, p)(\varphi_k'(p))^- dA_z.$$

Therefore the k-th Fourier coefficient of the differential $l_F(p, q)$ with respect to the orthonormal system $\{d\varphi_k\}$ is $-d\varphi_k/\lambda_k$. By formula (2.3.23) (Bessel's inequality)

$$\Sigma \frac{|\varphi_k'(q)|^2}{\lambda_k^2} \leq \int_{\mathfrak{M}} |l_F(p, q)|^2 dA_z. \tag{6.7.21}$$

Integrating (6.7.21) over $\mathfrak{M}$, we find that

$$\Sigma \frac{1}{\lambda_k^2} \leqq \int\limits_{\mathfrak{M}} \int\limits_{\mathfrak{M}} | l_F(p, q) |^2 \, dA_z \, dA_\zeta . \tag{6.7.22}$$

Hence, if there are infinitely many eigen-differentials, then λ_k tends to infinity with k. In particular, every finite eigen-value is of finite order.

Suppose now that there exists a differential df of class $\mathbf{F}_{\mathfrak{M}}$ with norm unity which is orthogonal to all the eigen-differentials. Then

$$[df, t_f] = 0. \tag{6.7.23}$$

In fact, if there are infinitely many eigen-differentials,

$$-\frac{1}{\lambda_k} \leqq [df, t_f] \leqq \frac{1}{\lambda_k}, \quad k = 1, 2, \cdots,$$

and (6.7.23) follows. If there are only finitely many eigen-differentials, (6.7.23) is still true; otherwise we could define a further eigen-value and eigen-differential by our maximum process. Since $e^{i\theta}f$, θ real, also has a norm equal to unity and is orthogonal to all eigen-differentials, we conclude from (6.7.12) and (6.7.23) that

$$(df, t_f) = 0. \tag{6.7.24}$$

Now

$$(d\varphi_\nu, t_f) = (df, t_{\varphi_\nu}) = \frac{1}{\lambda_\nu} (df, d\varphi_\nu) = 0 \tag{6.7.25}$$

by hypothesis, so t_f is orthogonal to all the eigen-differentials. If df_1 is any other non-zero differential which is orthogonal to all eigen-differentials we have

$$0 = (df_1 + df, t_{f_1+f}) = (df_1, t_{f_1}) + (df, t_{f_1}) + (df_1, t_f) + (df, t_f)$$
$$= (df, t_{f_1}) + (df_1, t_f)$$

by (6.7.24);

$$= 2(df_1, t_f)$$

by (6.7.8). Hence

$$(df_1, t_f) = 0. \tag{6.7.24$'$}$$

The orthonormal system $\{d\varphi_k\}$ of eigen-differentials can be made complete by adding an orthonormal set of differentials $df_1, df_2, \cdots$ which are orthogonal to all the eigen-differentials. By (6.7.24), (6.7.25) and (6.7.24)′ we see that t_f is orthogonal to all differentials of the complete orthonormal system, so

$$t_f(q) = 0. \tag{6.7.26}$$

Thus any differential df which is orthogonal to all eigen-differentials may be regarded as a solution of the integral equation (6.7.1) corresponding to the eigen-value $\lambda = \infty$. By admitting infinite eigen-values, we may therefore suppose that the set of eigen-differentials forms a complete orthonormal system.

In the case (6.5.28), the equation (6.7.26) holds for all df and any complete orthonormal system in $\mathbf{F}_{\mathfrak{M}}$ belongs to the eigen-value $\lambda = \infty$.

The kernel $l_{\mathbf{F}}(p, q)$ of the integral equation (6.7.1) may be expressed in terms of the complete orthonormal system $\{d\varphi_\nu\}$ of the eigen-differentials, and its k-th Fourier coefficient is $-d\varphi_k/\lambda_k$. Thus

$$l_{\mathbf{F}}(p, q) = -\sum_{\nu=1}^{\infty} \frac{1}{\lambda_\nu} d\varphi_\nu(p) d\varphi_\nu(q). \tag{6.7.27}$$

Substituting from (6.7.27) into (6.7.7), we obtain

$$Q_{\mathbf{F}}(\bar{p}, q) = Q_{\mathbf{F}}(q, \bar{p}) = \sum_{\nu=1}^{\infty} \frac{1}{\lambda_\nu^2} (d\varphi_\nu(p))^- d\varphi_\nu(q). \tag{6.7.28}$$

Iterating the equation (6.7.9), we have

$$d\varphi_\nu(q) = \lambda_\nu^2 (t^2(q))_{\varphi_\nu} = \lambda_\nu^2 \int_{\mathfrak{M}} Q_{\mathbf{F}}(q, \bar{p}) \varphi_\nu'(p) dA_z. \tag{6.7.29}$$

In the usual notation we may therefore write

$$Q_{\mathbf{F}}(q, \bar{p}) = l_{\mathbf{F}}^{(2)}(q, p) = -l_{\mathbf{F}}(q, \bar{p}) - \Gamma_{\mathbf{F}}(q, \bar{p}) \tag{6.7.30}$$

(second iterated kernel). From (4.9.6)

$$L_{\mathbf{F}}(p, \bar{q}) = -\sum_{\nu=1}^{\infty} d\varphi_\nu(p) (d\varphi_\nu(q))^-. \tag{6.7.31}$$

It is thus natural to define

(6.7.32) $$Q_{\mathbf{F}}^{(0)}(p, \tilde{q}) = -L_{\mathbf{F}}(p, \tilde{q}),\ Q_{\mathbf{F}}^{(2)}(p, \tilde{q}) = Q_{\mathbf{F}}(p, \tilde{q}),$$

and

(6.7.33) $$Q_{\mathbf{F}}^{(2\mu)}(p, \tilde{q}) = \int_{\mathfrak{M}} Q_{\mathbf{F}}^{(2\mu-2)}(p, \tilde{r}) Q_{\mathbf{F}}(r, \tilde{q}) dA_\eta, \quad \mu > 1.$$

Then

(6.7.34) $$Q_{\mathbf{F}}^{(2\mu)}(p, \tilde{q}) = \sum_{\nu=1}^{\infty} \frac{1}{\lambda_\nu^{2\mu}} d\varphi_\nu(p)(d\varphi_\nu(q))^-, \quad \mu = 0, 1, 2, \cdots.$$

The formula (6.7.34) gives the development of the iterated kernels in terms of the eigen-differentials. In the case $\mu = 0$ we obtain the development for the kernel $L_{\mathbf{F}}(p, \tilde{q})$. It is interesting to observe that in the general theory of integral equations the developments (6.7.34) for $\mu = 0$ and $\mu = 1$ do not in general converge. The Hilbert spaces of differentials with which we are concerned are spanned by orthonormal systems whose convergence properties are of an especially simple type.

6.8. Spectral Theory of the $\tilde{t}$-operator

We come now to the integral equation (6.7.2), and we have (see Section 6.6),

(6.8.1) $$\tilde{t}_f(q) = \int_{\mathfrak{M}} \mathscr{L}_{\mathbf{F}}(q, \tilde{p}) f'(p) dA_z,$$

(6.8.2) $$(\tilde{t}^2(q))_f = \int_{\mathfrak{M}} \mathscr{L}_{\mathbf{F}}(q, \tilde{p}) \tilde{t}_f(p) dA_z.$$

For differentials of class $\mathbf{F}_{\mathfrak{M}}$, $\tilde{t}_f(q) = \tilde{T}_f(q)$ but $(\tilde{t}^2(q))_f$ will not generally equal $(\tilde{T}^2(q))_f$. We have

(6.8.3) $$\begin{aligned}(\tilde{t}^2(q))_f &= \int_{\mathfrak{M}} \int_{\mathfrak{M}} \mathscr{L}_{\mathbf{F}}(q, \tilde{p}) \mathscr{L}_{\mathbf{F}}(p, \tilde{r}) f'(r) dA_z dA_\eta \\ &= \int_{\mathfrak{M}} P_{\mathbf{F}}(q, \tilde{r}) f'(r) dA_\eta,\end{aligned}$$

where

$$P_{\mathbf{F}}(q, \bar{p}) = \int_{\mathfrak{M}} \mathscr{L}_{\mathbf{F}}(r, \bar{p})(\mathscr{L}_{\mathbf{F}}(r, \tilde{q}))^{-} dA_{\eta}$$
$$= \int_{\mathfrak{R}} \mathscr{L}_{\mathbf{F}}(r, \bar{p})(\mathscr{L}_{\mathbf{F}}(r, \tilde{q}))^{-} dA_{\eta}$$
$$- \int_{\mathfrak{R}-\mathfrak{M}} \mathscr{L}_{\mathbf{F}}(r, \bar{p})(\mathscr{L}_{\mathbf{F}}(r, \tilde{q}))^{-} dA_{\eta}.$$

Now

$$\int_{\mathfrak{R}} \mathscr{L}_{\mathbf{F}}(r, \bar{p})(\mathscr{L}_{\mathbf{F}}(r, \tilde{q}))^{-} dA_{\eta} = -\mathscr{L}_{\mathbf{F}}(q, \bar{p}),$$

and

$$\int_{\mathfrak{R}-\mathfrak{M}} \mathscr{L}_{\mathbf{F}}(r, \bar{p})(\mathscr{L}_{\mathbf{F}}(r, \tilde{q}))^{-} dA_{\eta} = \Gamma_{\mathbf{F}}(q, \bar{p}).$$

Thus

$$P_{\mathbf{F}}(q, \bar{p}) = -\mathscr{L}_{\mathbf{F}}(q, \bar{p}) - \Gamma_{\mathbf{F}}(q, \bar{p}). \tag{6.8.4}$$

If we wanted to define an operator $\tilde{t}_f$ by applying the $\sim$-operation to t_f, we would have the choice of setting

$$\tilde{t}_f = (\mathscr{L}_{\mathbf{F}}(q, \bar{p}),\ (f'(p))^{-})_{\mathfrak{M}}$$

or

$$\tilde{t}_f = -(l_{\mathbf{F}}(q, \bar{p}),\ (f'(p))^{-})_{\mathfrak{M}}.$$

By our definitions of Section 6.6 we have committed ourselves to the first of these two possible definitions. This choice has the decisive advantage of providing us with a completely continuous operator $\tilde{t}_f$. This property does not hold for the other choice, since $l_{\mathbf{F}}(p, \tilde{q})$ is not regular for q and p in the closure of $\mathfrak{M}$.

As we have already remarked, in Section 6.1, $\tilde{T}$ is a self-adjoint operator in $\mathbf{F}_{\mathfrak{M}}$. The same clearly holds for $\tilde{t}_f$ in $\mathbf{F}_{\mathfrak{M}}$, hence we shall be able to construct a spectral theory for the $\tilde{t}_f$ operator in $\mathbf{F}_{\mathfrak{M}}$, where the scalar product is (df, dg).

Assume that the eigen-values $-\varrho_1, -\varrho_2, \cdots, -\varrho_\nu$ and eigen-differentials $d\psi_1, d\psi_2, \cdots, d\psi_\nu$ of the integral equation (6.7.2) have been defined, and consider all differentials df of $\mathbf{F}_{\mathfrak{M}}$ for which

$$(6.8.5)\qquad N(df) = 1, \quad (df, d\psi_k) = 0, \quad k = 1, 2, \cdots, \nu.$$

Among these differentials there is one, say $df = d\psi_{\nu+1}$, which maximizes the non-negative expression (compare (6.2.3)):

$$(6.8.6)\qquad -(df, \tilde{t}_f) = -\int_{\mathfrak{M}}\int_{\mathfrak{M}} \mathscr{L}_F(p, \bar{q})(\overline{f'(p)}) f'(q)\, dA_z dA_\zeta.$$

Writing

$$(6.8.7)\qquad \max\{-(df, \tilde{t}_f)\} = -(d\psi_{\nu+1}, \tilde{t}_{\psi_{\nu+1}}) = \frac{1}{\varrho_{\nu+1}},$$

we have

$$(6.8.8)\qquad 1 < \varrho_{\nu+1} \leqq \infty,$$

since

$$|(df, \tilde{t}_f)| \leqq \{N(df)\}^{\frac{1}{2}}\{N(\tilde{t}_f)\}^{\frac{1}{2}} \leqq \frac{1}{\varrho}$$

where $\varrho > 1$.

Suppose that $\varrho_{\nu+1} < \infty$. Let $d\varphi$ be a differential of $\mathbf{F}_{\mathfrak{M}}$ which is orthogonal to $d\psi_1, d\psi_2, \cdots, d\psi_\nu$ and write $dg = d\psi_{\nu+1} + \varepsilon d\varphi$. Then

$$-(dg, \tilde{t}_g) = \frac{1}{\varrho_{\nu+1}} - 2\mathrm{Re}\{\varepsilon(d\varphi, \tilde{t}_{\psi_{\nu+1}})\} - |\varepsilon|^2 (d\varphi, \tilde{t}_\varphi),$$

since

$$(6.8.9)\qquad (\tilde{t}_f, dg) = (df, \tilde{t}_g)$$

by (6.1.15). Hence

$$\frac{-(dg, \tilde{t}_g)}{N(dg)} = \frac{\dfrac{1}{\varrho_{\nu+1}} - 2\,\mathrm{Re}\{\varepsilon(d\varphi, \tilde{t}_{\psi_{\nu+1}})\} - |\varepsilon|^2(d\varphi, \tilde{t}_\varphi)}{1 + 2\,\mathrm{Re}\{\varepsilon(d\varphi, d\psi_{\nu+1})\} + |\varepsilon|^2 N(d\varphi)},$$

$$= \frac{1}{\varrho_{\nu+1}} - 2\,\mathrm{Re}\left\{\varepsilon\left[(d\varphi, \tilde{t}_{\psi_{\nu+1}}) + \frac{1}{\varrho_{\nu+1}}(d\varphi, d\psi_{\nu+1})\right]\right\} + O(|\varepsilon|^2)$$

$$\leqq \frac{1}{\varrho_{\nu+1}}.$$

Thus

$$(6.8.10)\qquad (d\varphi, d\psi_{\nu+1}) = -\varrho_{\nu+1}(d\varphi, \tilde{t}_{\psi_{\nu+1}}),$$

or

$$(d\varphi, d\psi_{\nu+1} + \varrho_{\nu+1} \bar{t}_{\psi_{\nu+1}}) = 0. \tag{6.8.11}$$

We conclude, as in Section 6.7, that

$$d\psi_{\nu+1}(q) = - \varrho_{\nu+1} \bar{t}_{\psi_{\nu+1}}(q). \tag{6.8.12}$$

Repeating the above maximum problem, we obtain a sequence of eigen-values $-\varrho_k$ and eigen-differentials $d\psi_k$. By arguments which are entirely similar to those used in Section 6.7 we see that

$$\Sigma \frac{1}{\varrho_k^2} < \infty.$$

In this case the set of eigen-differentials is complete. In fact, if df is orthogonal to all the eigen-differentials, then $\bar{t}_f = 0$ (compare (6.7.26)) and hence, by Section 6.5, df vanishes identically.

Since an eigen-differential $d\psi_k$ satisfies

$$d\psi_k = - \varrho_k \bar{T}_{\psi_k}$$

in $\mathfrak{M}$, we see that $d\psi_k$ can be extended to a differential of $\mathbf{F}_{\mathfrak{R}}$. Moreover,

$$\begin{aligned}(d\psi_\mu, d\psi_\nu)_{\mathfrak{R}} &= - \varrho_\mu \int_{\mathfrak{R}} \int_{\mathfrak{M}} \mathscr{L}_{\mathbf{F}}(q, \bar{p}) \psi'_\mu(p) (\psi'_\nu(q))^- dA_z dA_\zeta \\ &= \varrho_\mu \int_{\mathfrak{M}} \psi'_\mu(p) (\psi'_\nu(p))^- dA_z = \varrho_\mu (d\psi_\mu, d\psi_\nu)_{\mathfrak{M}}.\end{aligned} \tag{6.8.13}$$

We obtain in the $d\psi_\mu$ a system of differentials on $\mathfrak{R}$ which are orthogonal both with respect to integration over $\mathfrak{M}$ and over $\mathfrak{R}$. We call the $\{d\psi_\mu\}$ a doubly-orthogonal set with respect to the domains $\mathfrak{R}$ and $\mathfrak{M}$. These sets play an important role in the theory of continuation of a differential of $\mathbf{F}_{\mathfrak{M}}$ over $\mathfrak{R}$. We have:

THEOREM 6.8.1. *Let df be a differential of the class $\mathbf{F}_{\mathfrak{M}}$ and let*

$$f' = \sum_{\mu=1}^{\infty} a_\mu \psi'_\mu$$

be its Fourier development in terms of the eigen-differentials ψ'_μ. A *necessary and sufficient condition that $df \in \mathbf{F}_\mathfrak{R}$ is the convergence condition*

$$(6.8.14) \qquad \sum_{\mu=1}^{\infty} \varrho_\mu \,|\, a_\mu \,|^2 < \infty.$$

In fact, the differentials

$$d\psi_\mu^* = \frac{1}{\sqrt{\varrho_\mu}} \, d\psi_\mu$$

form an orthonormal system in $\mathbf{F}_\mathfrak{R}$ and (6.8.14) is the Bessel inequality for differentials of this class.

The kernel $\mathscr{L}_\mathbf{F}(q, \tilde{p})$ of the integral (6.8.1) may be expressed in terms of the complete orthonormal system $\{d\psi_\nu\}$ of the eigen-differentials; its k-th Fourier coefficient is $-(\psi_k'(p))^-/\varrho_k$. Thus

$$(6.8.15) \qquad \mathscr{L}_\mathbf{F}(q, \tilde{p}) = - \sum_{\nu=1}^{\infty} \frac{1}{\varrho_\nu} \, (\psi_\nu'(p))^- \, \psi_\nu'(q).$$

The iterated kernel $P_\mathbf{F}(q, \tilde{p})$ has the expansion

$$(6.8.16) \qquad P_\mathbf{F}(q, \tilde{p}) = \sum_{\nu=1}^{\infty} \frac{1}{\varrho_\nu^2} \, (\psi_\nu'(p))^- \, \psi_\nu'(q),$$

and from (4.9.6) we have also

$$(6.8.17) \qquad L_\mathbf{F}(q, \tilde{p}) = - \sum_{\nu=1}^{\infty} (\psi_\nu'(p))^- \, \psi_\nu'(q).$$

6.9. SPECTRAL THEORY OF THE s-OPERATOR

Finally, consider the integral equation (6.7.3) where

$$(6.9.1) \qquad s_f(q) = t_f(q) + \tilde{t}_f(q).$$

Obviously, s_f is self-adjoint in the Hilbert space $F_\mathfrak{M}$ with the metric $[df, dg]$. By (6.1.12) we observe that

$$(6.9.2) \quad (\lambda df, s_{\lambda f}) = (\lambda df, t_{\lambda f}) + (\lambda df, \tilde{t}_{\lambda f}) = \lambda^2(df, t_f) + |\,\lambda\,|^2 \, (df, \tilde{t}_f),$$

where λ is an arbitrary complex number. Since

$$(6.9.3) \qquad (df, \tilde{t}_f) \leq 0$$

and since λdf is always an element of $F_\mathfrak{M}$ if df is, we can always choose λ such that

$$[\lambda df, s_{\lambda f}] = \operatorname{Re}\{(\lambda df, s_{\lambda f})\} \leq 0.$$

We note the identity

$$(df, s_g) = (df, t_g) + (df, \bar{t}_g) = (dg, t_f) + ((dg, \bar{t}_f))^-,$$

that is,

$$[df, s_g] = [dg, s_f] = [s_f, dg]. \tag{6.9.4}$$

If there is a differential $df \in F_{\mathfrak{M}}$ for which $[df, s_f] > 0$, let

$$\max\,[df, s_f] = \frac{1}{\sigma_1}, \quad \sigma_1 > 0, \tag{6.9.5}$$

the maximum being taken with respect to all differentials in $\dot{F}_{\mathfrak{M}}$ satisfying $N(df) = 1$. Denote a maximizing differential by $d\chi_1$; if df is an arbitrary differential in $F_{\mathfrak{M}}$ and ε is an arbitrary real number, we have by the maximum property of $d\chi_1$,

$$[d\chi_1 + \varepsilon df, s_{\chi_1} + s_{\varepsilon f}] \leqq \frac{1}{\sigma_1} N(d\chi_1 + \varepsilon df).$$

Hence, by (6.9.4) and (6.9.5),

$$\frac{1}{\sigma_1} + 2[\varepsilon df, s_{\chi_1}] + [\varepsilon df, s_{\varepsilon f}] \leqq \frac{1}{\sigma_1} + \frac{2}{\sigma_1} \operatorname{Re}\{\varepsilon(d\chi_1, df)\} + \frac{\varepsilon^2}{\sigma_1} N(df),$$

that is,

$$\left[df, s_{\chi_1} - \frac{1}{\sigma_1} d\chi_1\right] = O(\varepsilon)$$

for arbitrary choice of $df \in F_{\mathfrak{M}}$ and ε. Thus, by the usual reasoning

$$\sigma_1 s_{\chi_1}(q) = d\chi_1(q).$$

We define now a sequence of eigen-differentials of (6.7.3) by the following recursive procedure. If $\chi_1, \cdots, \chi_\nu$ have been determined, belonging to the positive eigen-values $\sigma_1, \cdots, \sigma_\nu$, consider the class of all differentials in $F_{\mathfrak{M}}$ which satisfy

$$N(df) = 1, \quad [df, d\chi_\varrho] = 0, \quad \varrho = 1, \cdots, \nu. \tag{6.9.6}$$

If this class of differentials has elements df with $[df, s_f] > 0$, determine one differential $d\chi_{\nu+1}$ such that

$$(6.9.7) \qquad [d\chi_{\nu+1}, s_{\chi_{\nu+1}}] = \max [df, s_f] = \frac{1}{\sigma_{\nu+1}}.$$

From the maximum property of $d\chi_{\nu+1}$ we derive, as in the case of $d\chi_1$, that for any differential df satisfying (6.9.6):

$$\left[df, s_{\chi_{\nu+1}} - \frac{1}{\sigma_{\nu+1}} d\chi_{\nu+1}\right] = 0.$$

Moreover, in view of (6.9.4) and (6.9.6):

$$\left[d\chi_\varrho, s_{\chi_{\nu+1}} - \frac{1}{\sigma_{\nu+1}} d\chi_{\nu+1}\right] = [s_{\chi_\varrho}, d\chi_{\nu+1}] = \frac{1}{\sigma_\varrho}[d\chi_\varrho, d\chi_{\nu+1}] = 0,$$

for all $\varrho = 1, 2, \cdots, \nu$. Thus, the differential $s_{\chi_{\nu+1}} - \frac{1}{\sigma_{\nu+1}} d\chi_{\nu+1}$ has been proved to be orthogonal to all differentials of $F_{\mathfrak{M}}$ and is, therefore, identically zero. Thus

$$(6.9.8) \qquad d\chi_{\nu+1} = \sigma_{\nu+1} s_{\chi_{\nu+1}}.$$

We can therefore build an orthonormal system of eigen-differentials $d\chi_\nu$ by repeating the above extremum problem.

Suppose that $\sigma_{-1}, \sigma_{-2}, \cdots, \sigma_{-\nu}$ and $d\chi_{-1}, d\chi_{-2}, \cdots, d\chi_{-\nu}$ where $d\chi_{-k} = \sigma_{-k} s_{\chi_{-k}}$, $\sigma_{-k} < 0$, $k = 1, 2, \cdots, \nu$, have been defined, and consider differentials df of $F_{\mathfrak{M}}$ satisfying

$$N(df) = 1, \quad [df, d\chi_{-k}] = 0, \quad k = 1, 2, \cdots, \nu.$$

If $[df, s_f]$ can assume negative values under these conditions, let $d\chi_{-\nu-1}$ be a differential minimizing $[df, s_f]$,

$$[d\chi_{-\nu-1}, s_{\chi_{-\nu-1}}] = \frac{1}{\sigma_{-\nu-1}}, \quad \sigma_{-\nu-1} < 0.$$

As above we find that

$$(6.9.9) \qquad d\chi_{-\nu-1} = \sigma_{-\nu-1} s_{\chi_{-\nu-1}}.$$

We observe that the eigen-differentials $d\chi_h$, $h > 0$, and $d\chi_{-k}$, $k > 0$, are orthogonal to each other in the sense that $[d\chi_h, d\chi_{-k}] = 0$.

For

$$d\chi_h = \sigma_h s_{\chi_h}, \quad d\chi_{-k} = \sigma_{-k} s_{\chi_{-k}},$$

where $\sigma_h > 0$, $\sigma_{-k} < 0$. By (6.9.4),

$$[s_{\chi_h}, d\chi_{-k}] = [s_{\chi_{-k}}, d\chi_h];$$

that is,

$$\sigma_h[d\chi_h, d\chi_{-k}] = \sigma_{-k}[d\chi_h, d\chi_{-k}]. \tag{6.9.10}$$

Since $\sigma_h > 0$, $\sigma_{-k} < 0$, we conclude that $[d\chi_h, d\chi_{-k}] = 0$.

Now let the eigen-values be ordered according to increasing absolute values. In this order we denote the eigen-values by $\tau_1, \tau_2, \cdots$ and the corresponding eigen-differentials by $d\Phi_1, d\Phi_2, \cdots$. The $d\Phi_k$ form an orthonormal system in $F_{\mathfrak{M}}$, that is

$$[d\Phi_\mu, d\Phi_\nu] = \delta_{\mu\nu}. \tag{6.9.11}$$

Let

$$\Lambda_F(p, q, \bar{q}) = -l_F(q, p) + \mathscr{L}_F(p, \bar{q}).$$

Then

$$\text{Re}\,\{s_f(q)\} = \text{Re}\left\{\int_{\mathfrak{M}} \Lambda_F(p, q, \bar{q})(f'(p))^- dA_z\right\}.$$

The k-th Fourier coefficient of Λ_F is therefore equal to

$$\text{Re}\,\frac{\{\Phi'_k(q)\}}{\tau_k}.$$

We have by Bessel's inequality

$$\begin{aligned} \Sigma\left[\frac{\text{Re}\,\{\Phi'_k(q)\}}{\tau_k}\right]^2 &\leqq \int_{\mathfrak{M}} |\,\Lambda_F\,(p, q, \bar{q})\,|^2\, dA_z \\ &\leqq \left[\left(\int_{\mathfrak{M}} |\,l_F(q, p)\,|^2\, dA_z\right)^{\frac{1}{2}} + \left(\int_{\mathfrak{M}} |\,\mathscr{L}_F\,(p, \bar{q})\,|^2\, dA_z\right)^{\frac{1}{2}}\right]^2 \end{aligned} \tag{6.9.12}$$

by the triangle inequality;

$$\leqq 2\left[\int_{\mathfrak{M}} |\,l_F(q, p)\,|^2\, dA_z + \int_{\mathfrak{M}} |\,\mathscr{L}_F(p, \bar{q})\,|^2\, dA_z\right].$$

On the other hand, let

$$\Theta_{F}(p, q, \bar{q}) = i\,[l_{F}(q, p) + \mathscr{L}_{F}(p, \bar{q})].$$

Then

$$\operatorname{Im}\ \{s_f(q)\} = \operatorname{Re}\left\{\int_{\mathfrak{M}} \Theta(p, q, \bar{q})(f'(p))^{-} dA_z\right\}.$$

Hence the k-th Fourier coefficient of Θ_F is equal to

$$\frac{\operatorname{Im}\ \{\Phi'_k(q)\}}{\tau_k}$$

and Bessel's inequality gives

$$(6.9.13)\qquad \begin{aligned} \Sigma\left[\frac{\operatorname{Im}\ \{\Phi'_k(q)\}}{\tau_k}\right]^2 &\leqq \int_{\mathfrak{M}} |\ \Theta_F(p, q, \bar{q})\ |^2\, dA_z \\ &\leqq 2\left[\int_{\mathfrak{M}} |\ l_F(q, p)\ |^2\, dA_z + \int_{\mathfrak{M}} |\ \mathscr{L}_F(p, \bar{q})\ |^2\, dA_z\right]. \end{aligned}$$

By (6.9.12) and (6.9.13)

$$(6.9.14)\qquad \Sigma\frac{|\ \Phi'_k(q)\ |^2}{\tau_k^2} \leqq 4\left[\int_{\mathfrak{M}} |\ l_F(q, p)\ |^2\, dA_z + \int_{\mathfrak{M}} |\ \mathscr{L}_F(p, \bar{q})\ |^2\, dA_z\right].$$

Integrating (6.9.14) over $\mathfrak{M}$, we see that

$$(6.9.15)\qquad \begin{aligned} \Sigma\frac{1}{\tau_k^2} \leqq 4\Bigg[&\int_{\mathfrak{M}}\int_{\mathfrak{M}} |\ l_F(q, p)\ |^2\, dA_z dA_\zeta \\ &+ \int_{\mathfrak{M}}\int_{\mathfrak{M}} |\ \mathscr{L}_F(p, \bar{q})\ |^2\, dA_z dA_\zeta\Bigg]. \end{aligned}$$

Thus, if there are infinitely many eigen-differentials, then $|\ \tau_k\ |$ tends to infinity with k.

Suppose that there exists a differential df of class $F_{\mathfrak{M}}$ which is orthogonal to all the eigen-differentials, that is

$$[df, d\Phi_\nu] = 0, \quad \nu = 1, 2, \cdots.$$

Then (compare (6.7.23))

$$[df, s_f] = \operatorname{Re}\ \{(df, s_f)\} = 0.$$

We have

$$(6.9.16) \qquad [d\Phi_\nu, s_f] = [s_{\Phi_\nu}, df] = \frac{1}{\tau_\nu}[d\Phi_\nu, df] = 0,$$

by hypothesis. If df_1 is any other differential orthogonal to all eigen-differentials, we have

$$\begin{aligned} 0 &= [df_1 + df, s_{f_1+f}] = [df_1, s_{f_1}] + [df, s_{f_1}] + [df_1, s_f] + [df, s_f] \\ &= [df, s_{f_1}] + [df_1, s_f] = 2[df_1, s_f], \end{aligned}$$

that is,

$$(6.9.17) \qquad [df_1, s_f] = 0.$$

By adding orthonormal differentials $df_1, \cdots, df_k$ which are orthogonal to all the eigen-differentials, the orthonormal system of eigen-differentials can be made into a complete orthonormal system. By (6.9.16) and (6.9.17) we see that s_f is orthogonal to all differentials of this complete system, so

$$(6.9.18) \qquad s_f(q) \equiv 0.$$

Any differential df which is orthogonal to all eigen-differentials may therefore be regarded as an eigen-differential belonging to the eigen-value infinity. We admit infinite eigen-values, and we again denote the resulting complete orthonormal system of eigen-differentials by $\{d\Phi_\nu\}$.

We have

$$(6.9.19) \qquad -l_F(q, p) + \mathscr{L}_F(p, \bar{q}) = \sum_{\nu=1}^{\infty} \operatorname{Re}\left\{\frac{\Phi'_\nu(q)}{\tau_\nu}\right\}\Phi'_\nu(p),$$

$$(6.9.20) \qquad l_F(q, p) + \mathscr{L}_F(p, \bar{q}) = -i \sum_{\nu=1}^{\infty} \operatorname{Im}\left\{\frac{\Phi'_\nu(q)}{\tau_\nu}\right\}\Phi'_\nu(p).$$

Thus

$$(6.9.21) \qquad l_F(q, p) = -\frac{1}{2}\sum_{\nu=1}^{\infty}\frac{1}{\tau_\nu}\Phi'_\nu(p)\Phi'_\nu(q),$$

and

$$(6.9.22) \qquad \mathscr{L}_F(p, \bar{q}) = \frac{1}{2}\sum_{\nu=1}^{\infty}\frac{1}{\tau_\nu}\Phi'_\nu(p)(\Phi'_\nu(q))^-.$$

By the reproducing property

$$\int_{\mathfrak{M}} L_F(p, \bar{q})(\Phi_\nu'(p))^- dA_z = -(\Phi_\nu'(q))^-,$$

$$\int_{\mathfrak{M}} iL_F(p, \bar{q})(\Phi_\nu'(p))^- dA_z = -i(\Phi_\nu'(q))^-.$$

Hence

$$[L_F(p, \bar{q}), \Phi_\nu'] = -\operatorname{Re}\{\Phi_\nu'(q)\}, \tag{6.9.23}$$

$$[i L_F(p, \bar{q}), \Phi_\nu'] = -\operatorname{Im}\{\Phi_\nu'(q)\}. \tag{6.9.24}$$

Thus

$$L_F(p, \bar{q}) = -\sum_{\nu=1}^{\infty} \operatorname{Re}\{\Phi_\nu'(q)\}\, \Phi_\nu'(p), \tag{6.9.25}$$

$$-L_F(p, \bar{q}) = -i\sum_{\nu=1}^{\infty} \operatorname{Im}\{\Phi_\nu'(q)\}\, \Phi_\nu'(p). \tag{6.9.26}$$

Adding and subtracting (6.9.25) and (6.9.26), we see that

$$L_F(p, \bar{q}) = -\frac{1}{2}\sum_{\nu=1}^{\infty} \Phi_\nu'(p)(\Phi_\nu'(q))^-, \tag{6.9.27}$$

$$0 = \sum_{\nu=1}^{\infty} \Phi_\nu'(p)\Phi_\nu'(q). \tag{6.9.28}$$

We shall discuss later the eigen-differentials of the integral equation (6.7.3) which belong to the eigen-value $\sigma_\nu = -1$. In this case, we have

$$d\Phi_\nu(q) = -s_{\Phi_\nu}. \tag{6.9.29}$$

We showed in Section 6.2 that the S-transformation is norm-preserving that is

$$N_{\mathfrak{M}}(d\Phi_\nu) = N_{\mathfrak{M}}(S_{\Phi_\nu}) + N_{\mathfrak{R}-\mathfrak{M}}(S_{\Phi_\nu}). \tag{6.9.30}$$

Since $S_{\Phi_\nu} = s_{\Phi_\nu}$ in $\mathfrak{M}$, we derive from (6.9.30)

$$S_{\Phi_\nu} \equiv 0 \text{ in } \mathfrak{R} - \mathfrak{M}. \tag{6.9.31}$$

Thus, by our result at the end of Section 6.5,

(6.9.32) $$\Phi'_\nu(q) = i \sum_{\mu=1}^{G} b_\mu Z'_\mu(q), \quad b_\mu \text{ real.}$$

Conversely, every differential (6.9.32) is an eigen-differential of (6.7.3) with the eigen-value $\sigma_\nu = -1$.

We remark that the eigen-value $\sigma_\nu = 1$ is not possible, since this case would also imply (6.9.31).

6.10. Minimum-maximum Properties of the Eigen-differentials

The eigen-values of the equations (6.7.1)—(6.7.3) may be defined as minima maximorum in the usual way (see [3]). In all three cases, the eigen-values have been defined by maximizing or minimizing, under suitable side conditions, an expression $[df, n_f]$ where $n_f = t_f$, $\bar{t}_f$ or s_f. This was done explicitly in the first and third cases, but even the second case can be easily reduced to such a maximum problem in $F_{\mathfrak{M}}$.

We now consider the problem of maximizing $[f', n_f]$ for differentials f' of $F_{\mathfrak{M}}$ with $N(f') = 1$ which satisfy $k-1$ side conditions of the form

(6.10.1) $$[f', g'_\varrho] = 0, \quad \varrho = 1, 2, \cdots, k-1,$$

where $g'_1, \cdots, g'_{k-1}$ are arbitrarily chosen but fixed differentials of $F_{\mathfrak{M}}$. The maximum of $[f', n_f]$ depends on the choice of the differentials g'_ν, $\nu = 1, 2, \cdots, k-1$, and will be denoted by $1/K$, $K = K[g'_1, \cdots, g'_{k-1}]$.

We arrange the positive eigen-values of the operator n_f in increasing order and denote them by $\gamma_1, \gamma_2, \cdots$; similarly, the negative eigen-values will be arranged in decreasing order and denoted by γ_{-1}, $\gamma_{-2}, \cdots$. The eigen-differential corresponding to γ_ν will be denoted by χ'_ν.

Consider a differential in $F_{\mathfrak{M}}$ of the form

(6.10.2) $$f' = c_1\chi'_1 + \cdots + c_k\chi'_k.$$

The requirement $N(f') = 1$ leads to the condition

(6.10.2)' $$c_1^2 + \cdots + c_k^2 = 1.$$

We have in the k variables c_k enough parameters to satisfy the $k-1$

side conditions (6.10.1). The differential f' will, in general, be uniquely determined by (6.10.1) and (6.10.2)$'$. We have

$$n_f = \sum_{\nu=1}^{k} c_\nu n_{\chi_\nu} = \sum_{\nu=1}^{k} \frac{c_\nu}{\gamma_\nu} \chi'_\nu ,$$

and hence

$$(6.10.3) \qquad [f', n_f] = \sum_{\nu=1}^{k} \frac{c_\nu^2}{\gamma_\nu}.$$

In view of the condition (6.10.2)$'$, we obtain the estimate

$$(6.10.4) \qquad \frac{1}{K} \geqq [f', n_f] \geqq \frac{1}{\gamma_k}.$$

Let us choose now, in particular, $g'_1 = \chi'_1, \cdots, g'_{k-1} = \chi'_{k-1}$. Each differential f' with norm 1 which satisfies the side conditions (6.10.1) in this particular case has the form:

$$f' = \sum_{\nu=-1}^{-\infty} d_\nu \chi'_\nu + \sum_{\nu=k}^{\infty} d_\nu \chi'_\nu, \ \sum_{\nu=-\infty}^{\infty} d_\nu^2 = 1.$$

We easily compute

$$(6.10.5) \qquad [f', n_f] = \sum_{\nu=-1}^{-\infty} \frac{d_\nu^2}{\gamma_\nu} + \sum_{\nu=k}^{\infty} \frac{d_\nu^2}{\gamma_\nu} \leqq \frac{1}{\gamma_k}.$$

For $f' = \chi'_k$, we have equality in (6.10.5); thus, we have shown that $K(\chi'_1, \cdots, \chi'_{k-1})] = \gamma_k$. Combining this result with (6.10.4), we obtain:

THEOREM 6.10.1. *$1/\gamma_k$ is the minimum maximorum for the expression $[f', n_f]$ for differentials of $F_{\mathfrak{M}}$ with $N(df) = 1$ which satisfy any $k-1$ side conditions of the form (6.10.1). This result characterizes the k-th eigen-value γ_k of the problem considered.*

The same considerations apply to the negative eigen-values of the operator n_f. They may be characterized as maxima minimorum in the corresponding extremum problem.

Let us discuss these results in the case of the particular operators t, $\bar{t}$ and s. Since we are operating in the real Hilbert space $F_{\mathfrak{M}}$, we have to distinguish between the linearly independent eigen-differentials φ'_ν and $i\varphi'_\nu$ in the first problem, ψ'_ν and $i\psi'_\nu$ in the second. If φ'_ν belongs to the eigen-value λ_ν, $i\varphi'_\nu$ belongs to the eigen-value $-\lambda_\nu$. ψ'_ν and $i\psi'_\nu$ belong to the same negative eigen-value $-\varrho_\nu$. We remark finally that if χ'_ν is an eigen-differential of the third problem, $i\chi'_\nu$ will not, in general, be an eigen-differential.

In order to apply the above general theory to the case $n_f = t_f$, we put: $\chi'_\nu = \varphi'_\nu$, $\gamma_\nu = \lambda_\nu$ and $\chi'_{-\nu} = i\varphi'_\nu$, $\gamma_{-\nu} = -\lambda_\nu$. In the case $n_f = \bar{t}_f$, we define $\chi'_{-2\nu+1} = \psi'_\nu$, $\chi'_{-2\nu} = i\psi'_\nu$, and $\gamma_{-2\nu+1} = \gamma_{-2\nu} = -\varrho_\nu$. In the case $n_f = s_f$, the previous notation is preserved for the differentials, while $\gamma_\nu = \sigma_\nu$ for all ν.

We observe that

(6.10.6) $$[f', s_f] - [f', \bar{t}_f] = [f', t_f].$$

Hence, in view of the negative-definite character of $[f', \bar{t}_f]$:

(6.10.7) $$[f', s_f] \leqq [f', t_f].$$

Choose $g'_1 = \varphi'_1$, $g'_2 = i\varphi'_1, \cdots, g'_{2k-3} = \varphi'_{k-1}$, $g'_{2k-2} = i\varphi'_{k-1}$, where the φ'_ν are the eigen-differentials of the equation (6.7.1). Then clearly, for each normed differential which is orthogonal to all these differentials,

$$[f', t_f] \leqq \frac{1}{\lambda_k}.$$

On the other hand, the maximum value of the left side of (6.10.7) is not less than the minimum maximorum $1/\sigma_{2k-1}$. Thus we obtain the inequality

$$\frac{1}{\sigma_{2k-1}} \leqq \frac{1}{\lambda_k},$$

that is

(6.10.8) $$\lambda_k \leqq \sigma_{2k-1}, \quad k = 1, 2, \cdots.$$

Consider now the inequality

$$[f', s_f] = [f', \bar{t}_f] + [f', t_f] \geqq [f', \bar{t}_f] - |[f', t_f]|.$$

Choose $g'_1 = \varphi'_1$, $g'_2 = i\varphi'_1, \cdots, g'_{2j-1} = \varphi'_j$, $g'_{2j} = i\varphi'_j$,

$$g'_{2j+1} = \psi'_1,\ g'_{2j+2} = i\psi'_1, \cdots,\ g'_{2j+2k-3} = \psi'_{k-1},\ g'_{2(j+k-1)} = i\psi'_{k-1},$$

where the φ'_ν are the eigen-differentials of (6.7.1) and the ψ'_ν belong to (6.7.2). Then for each normed df which is orthogonal to all dg,

$$[f', \bar{t}_f] \geqq -\frac{1}{\varrho_k}, \quad |[f', t_f]| \leqq \frac{1}{\lambda_{j+1}};$$

therefore

$$[f', s_f] \geqq -\frac{1}{\varrho_k} - \frac{1}{\lambda_{j+1}}.$$

But

$$\min\,[f', s_f] \leqq \frac{1}{\sigma_{-2(j+k)+1}},$$

for all differentials $f' \in F_{\mathfrak{M}}$ which are orthogonal to $2(j+k-1)$ differentials g'_ν. Therefore

$$\left|\frac{1}{\sigma_{-2(j+k)+1}}\right| \leqq \frac{1}{\varrho_k} + \frac{1}{\lambda_{j+1}}, \quad j = 1, 2, \cdots, k = 1, 2, \cdots. \tag{6.10.9}$$

Thus, various inequalities among the eigen-values of the three operators could be derived from the above characterization of the k-th eigen-value.

6.11. The Hilbert Space with Dirichlet Metric

We shall now discuss the integral equation (6.7.3) from a different point of view. In the preceding sections we stressed the role of the differentials on a Riemann surface $\mathfrak{M}$ and considered eigen-value problems connected with them. It is equally possible to focus the attention on the class of all harmonic functions in a given domain and to develop an analogous theory. We shall show in this section that both approaches are essentially equivalent and we will translate the eigen-value problem for differentials on $\mathfrak{M}$ into an eigen-value problem for harmonic functions on $\mathfrak{M}$.

The ν-th eigen-differential satisfies the equation

$$\begin{aligned}\Phi'_\nu(q) &= -\tau_\nu \int_{\mathfrak{M}} [l_F(q, p)(\Phi'_\nu(p))^- - \mathscr{L}_F(q, \bar{p})\Phi'_\nu(p)]dA_z \\ &= -\tau_\nu \int_{\mathfrak{M}} [l_F(q, p)(\Phi'_\nu(p))^- + l_F(q, \bar{p})\Phi'_\nu(p)]dA_z - \tau_\nu\Phi'_\nu(q),\end{aligned} \tag{6.11.1}$$

by (4.11.7). Hence

$$\left(1 + \frac{1}{\tau_\nu}\right)\Phi'_\nu(q) = -\int_{\mathfrak{M}} [l_F(q, p)(\Phi'_\nu(p))^- + l_F(q, \bar{p})\Phi'_\nu(p)]dA_z. \tag{6.11.2}$$

It follows that

$$\left(1+\frac{1}{\tau_\nu}\right)P(\Phi'_\nu, K_\varrho) = -\int_{\mathfrak{M}} [P(l_F(q,p),K_\varrho)(\Phi'_\nu(q))^- \tag{6.11.3}$$
$$+ P(l_F(q,\bar{p}),K_\varrho)\Phi'_\nu(p)]\,dA_z,$$

where $P(l_F(q,p),K_\varrho)$ is the period of $l_F(q,p)$ around the cycle K_ϱ, p being held fixed, and similarly for $P(l_F(q,\bar{p}),K_\varrho)$. From (6.5.26) we have

$$P(l_F(q,\bar{p}),K_\varrho) = (P(l_F(q,p),K_\varrho))^-. \tag{6.11.4}$$

If $\tau_\nu \neq -1$, that is if

$$\Phi'_\nu(q) \neq i\sum_{\mu=1}^{G} b_\mu Z'_\mu(q), \qquad b_\mu \text{ real}, \tag{6.11.5}$$

we conclude from (6.11.3) that $P(\Phi'_\nu, K_\varrho)$ is real. Thus

$$\operatorname{Im}\{\Phi_\nu(p)\} = H_\nu(p), \tag{6.11.6}$$

where $H_\nu(p)$ is a single-valued harmonic function on $\mathfrak{M}$, if $\tau_\nu \neq -1$.

Let f'_1 and f'_2 be differentials of class $\mathbf{F}_{\mathfrak{M}}$ and write

$$f_k = u_k + iv_k, \quad k = 1, 2.$$

where u_k, v_k are real. We observe that

$$[f'_1, f'_2] = \operatorname{Re}\left\{\int_{\mathfrak{M}} \frac{\partial f_1}{\partial z}\left(\frac{\partial f_2}{\partial z}\right)^- dA_z\right\} = 4\operatorname{Re}\left\{\int_{\mathfrak{M}} \frac{\partial v_1}{\partial z}\left(\frac{\partial v_2}{\partial z}\right)^- dA_z\right\}$$
$$= \operatorname{Re}\left\{\int_{\mathfrak{M}} \left(\frac{\partial v_1}{\partial x} - i\frac{\partial v_1}{\partial y}\right)\left(\frac{\partial v_2}{\partial x} + i\frac{\partial v_2}{\partial y}\right)dA_z\right\},$$

where $z = x + iy$;

$$= \int_{\mathfrak{M}} \left(\frac{\partial v_1}{\partial x}\frac{\partial v_2}{\partial x} + \frac{\partial v_1}{\partial y}\frac{\partial v_2}{\partial y}\right) dA_z = D_{\mathfrak{M}}(v_1, v_2) = D(v_1, v_2).$$

That is,

$$[f'_1, f'_2] = D(v_1, v_2) \tag{6.11.7}$$

where $D(v_1, v_2)$ is the Dirichlet integral as in Section 2.8.

In analogy with (6.1.2) write

$$g(p,q) = G(p,q) - \mathcal{G}(p,q). \tag{6.11.8}$$

Here $G(p, q)$ is the Green's function of $\mathfrak{M}$, $\mathscr{G}(p, q)$ that of $\mathfrak{R}$. Now let us specialize by taking $\mathbf{F} = \mathbf{M}$. By (4.10.1) and (4.10.2) we have

$$l_{\mathbf{M}}(p, q) = -\frac{2}{\pi}\frac{\partial^2 g(p, q)}{\partial p \partial q}, \tag{6.11.9}$$

$$l_{\mathbf{M}}(p, \bar{q}) = \frac{2}{\pi}\frac{\partial^2 g(p, q)}{\partial p \partial \bar{q}}. \tag{6.11.10}$$

In the case $\mathbf{F} = \mathbf{M}$ there are precisely $G = 2h + m - 1$ eigen-differentials $\Phi'_1, \cdots, \Phi'_G$, belonging to the eigen-value $\tau = -1$. We have

$$2i\frac{\partial H_\nu(p)}{\partial p} = \Phi'_\nu(p), \quad -2i\frac{\partial H_\nu(p)}{\partial \bar{p}} = (\Phi'_\nu(p))^-, \tag{6.11.11}$$

and equation (6.11.2) with $\mathbf{F} = \mathbf{M}$ becomes

$$\begin{aligned}\left(1 + \frac{1}{\tau_\nu}\right)\frac{\partial H_\nu(q)}{\partial q} = &-\frac{2}{\pi}\int_{\mathfrak{M}} \frac{\partial^2 g(p, q)}{\partial p \partial q}\frac{\partial H_\nu(p)}{\partial \bar{p}}\, dA_z \\ &-\frac{2}{\pi}\int_{\mathfrak{M}} \frac{\partial^2 g(p, q)}{\partial \bar{p} \partial q}\frac{\partial H_\nu(p)}{\partial p}\, dA_z.\end{aligned} \tag{6.11.12}$$

Thus

$$\begin{aligned}\left(1 + \frac{1}{\tau_\nu}\right)\frac{\partial H_\nu(q)}{\partial q} &= -\frac{2}{\pi}\frac{\partial}{\partial q}\int_{\mathfrak{M}} \frac{\partial g(p, q)}{\partial p}\frac{\partial H_\nu(p)}{\partial \bar{p}}\, dA_z \\ &\quad -\frac{2}{\pi}\frac{\partial}{\partial q}\int_{\mathfrak{M}} \frac{\partial g(p, q)}{\partial \bar{p}}\frac{\partial H_\nu(p)}{\partial p}\, dA_z \\ &= -\frac{4}{\pi}\frac{\partial}{\partial q}\operatorname{Re}\left\{\int_{\mathfrak{M}} \frac{\partial g(p, q)}{\partial p}\frac{\partial H_\nu(p)}{\partial \bar{p}}\, dA_z\right\} \\ &= -\frac{1}{\pi}\frac{\partial}{\partial q} D(g(p, q),\ H_\nu(p)).\end{aligned} \tag{6.11.13}$$

Integrating both sides of (6.11.13) with respect to q and choosing the constant of integration suitably, we obtain, for $\tau_\nu \neq -1$,

$$H_\nu(q) = -\frac{1}{\pi\left(1 + \dfrac{1}{\tau_\nu}\right)} D(g(p, q),\ H_\nu(p)). \tag{6.11.14}$$

Thus $H_\nu(q)$ is an eigen-function of the integral equation

$$H_\nu(q) = \gamma_\nu D(g(p, q), H_\nu(p)). \tag{6.11.15}$$

We observe from (6.11.7) that

$$D(H_\mu, H_\nu) = [\Phi'_\mu, \Phi'_\nu] = \delta_{\mu\nu}, \tag{6.11.16}$$

so the H_μ form an orthonormal system.

The eigen-differentials of the integral equation (6.7.3) belonging to the eigen-value $\tau = -1$ are just the orthonormalized differentials $iZ'_1, \cdots, iZ'_G$. Since

$$[Z'_\mu, Z'_\nu] = \mathrm{Re}\, \Gamma_{\mu\nu}$$

by (4.3.10), we see from the Gram-Schmidt orthonormalization process that the eigen-differentials $\Phi'_1, \cdots, \Phi'_G$ belonging to the eigen-value $\tau = -1$ are given by the formula

$$\Phi'_\nu = d_{\nu-1}^{-\frac{1}{2}} d_\nu^{-\frac{1}{2}} \begin{vmatrix} \mathrm{Re}\,\Gamma_{11} \cdots \mathrm{Re}\,\Gamma_{1,\nu-1} & iZ'_1 \\ \vdots \qquad \vdots & \vdots \\ \mathrm{Re}\,\Gamma_{\nu 1} \cdots \mathrm{Re}\,\Gamma_{\nu,\nu-1} & iZ'_\nu \end{vmatrix}, \nu = 1, 2, \cdots, G, \tag{6.11.17}$$

where $d_0 = 1$ and

$$d_k = \begin{vmatrix} [Z'_1, Z'_1] \cdots [Z'_1, Z'_k] \\ \vdots \qquad \vdots \\ [Z'_k, Z'_1] \cdots [Z'_k, Z'_k] \end{vmatrix} = \begin{vmatrix} \mathrm{Re}\,\Gamma_{11} \cdots \mathrm{Re}\,\Gamma_{1k} \\ \vdots \qquad \vdots \\ \mathrm{Re}\,\Gamma_{k1} \cdots \mathrm{Re}\,\Gamma_{kk} \end{vmatrix}. \tag{6.11.18}$$

In particular, we derive from (6.11.17):

$$\begin{aligned} (\Phi'_\nu, iZ'_\varrho) &= i\, P(\Phi'_\nu, K_\varrho) \\ &= d_{\nu-1}^{-\frac{1}{2}} d_\nu^{-\frac{1}{2}} \begin{vmatrix} \mathrm{Re}\,\Gamma_{11} \cdots \mathrm{Re}\,\Gamma_{1,\nu-1} & \Gamma_{1\varrho} \\ \vdots \qquad \vdots & \vdots \\ \mathrm{Re}\,\Gamma_{\nu 1} \cdots \mathrm{Re}\,\Gamma_{\nu,\nu-1} & \Gamma_{\nu\varrho} \end{vmatrix}. \end{aligned} \tag{6.11.19}$$

The eigen-functions of (6.11.15) belonging to the eigen-value ∞ are

$$H_\nu = \mathrm{Im}\,\Phi_\nu = d_{\nu-1}^{-\frac{1}{2}} d_\nu^{-\frac{1}{2}} \begin{vmatrix} \mathrm{Re}\,\Gamma_{11} \cdots \mathrm{Re}\,\Gamma_{1,\nu-1} & \mathrm{Re}\, Z_1 \\ \vdots \qquad \vdots & \vdots \\ \mathrm{Re}\,\Gamma_{\nu 1} \cdots \mathrm{Re}\,\Gamma_{\nu,\nu-1} & \mathrm{Re}\, Z_\nu \end{vmatrix}, \nu = 1, 2, \cdots, G. \tag{6.11.20}$$

These eigen-functions are not single-valued. However, using (6.11.19)

we have

$$D(H_\nu, \operatorname{Re} Z_\varrho) = -P(H'_\nu, K_\varrho)$$

$$(6.11.21) \qquad = d_{\nu-1}^{-\frac{1}{2}} d_\nu^{-\frac{1}{2}} \begin{vmatrix} \operatorname{Re} \Gamma_{11} \cdots \operatorname{Re} \Gamma_{1,\nu-1} & \operatorname{Re} \Gamma_{1\varrho} \\ \vdots \qquad\qquad \vdots & \vdots \\ \operatorname{Re} \Gamma_{\nu 1} \cdots \operatorname{Re} \Gamma_{\nu,\nu-1} & \operatorname{Re} \Gamma_{\nu\varrho} \end{vmatrix}.$$

Thus

$$(6.11.22) \quad D(H_\nu, \operatorname{Re} Z_\varrho) = -P(H'_\nu, K_\varrho) = 0, \quad \varrho = 1, 2, \cdots, \nu-1.$$

The orthonormal system $\{H_\nu\}$ of the eigen-functions of (6.11.15) is complete. For suppose that H is a harmonic function on $\mathfrak{M}$ with $D(H) = 1$ such that

$$(6.11.23) \qquad D(H, H_\nu) = 0, \quad \nu = 1, 2, \cdots.$$

Then, writing

$$\Phi = R + iH,$$

we would have

$$(6.11.24) \qquad [\Phi', \Phi'_\nu] = D(H, H_\nu) = 0,$$

so the system $\{\Phi'_\nu\}$ would not be complete, a contradiction.

The vector field whose elements are the gradients of the harmonic functions H on $\mathfrak{M}$ with finite Dirichlet integrals forms a Hilbert space in the metric of the Dirichlet integral. We may consider the harmonic functions themselves as elements of a Hilbert space with the metric $D(H)$ if we normalize them by the condition that all functions vanish at a fixed point p_0 of $\mathfrak{M}$. Then the only element of the Hilbert space with zero norm is the zero element $H = 0$. Every harmonic function H with finite norm has a Fourier expansion

$$(6.11.25) \qquad H = \sum_{\nu=1}^{\infty} a_\nu (H_\nu(p) - H_\nu(p_0)),$$

where

$$(6.11.26) \qquad a_\nu = D(H, H_\nu).$$

In particular, by (6.11.14),

$$(6.11.27) \quad g(p, q) - g(p_0, q) = -\pi \sum_{\nu=1}^{\infty} \left(1 + \frac{1}{\tau_\nu}\right) H_\nu(q)(H_\nu(p) - H_\nu(p_0)).$$

Differentiating (6.11.27) with respect to p and q or p and $\tilde{q}$, we obtain (6.9.21) and (6.9.22), in the case $\mathbf{F} = \mathbf{M}$, using (6.9.27) and (6.9.28).

In the eigen-value theory for the harmonic functions on $\mathfrak{M}$, the Green's function played a distinguished role. It might be of interest to study an analogous theory where the Neumann's function takes the corresponding central role. By (4.2.22),

$$(6.11.28) \qquad -\frac{2}{\pi}\frac{\partial^2 N(p, q, q_0)}{\partial p \partial q} = -\frac{2}{\pi}\frac{\partial^2 G(p, q)}{\partial p \partial q} + \sum_{\mu, \nu=1}^{G} c_{\mu\nu} Z'_\mu(p) Z'_\nu(q)$$

where by (4.6.4)

$$(6.11.29) \qquad \| c_{\mu\nu} \| = \| \operatorname{Re} \Gamma_{\mu\nu} \|^{-1}.$$

By (4.11.1),

$$(6.11.30) \qquad L_{\mathbf{S}}(p, q) = -\frac{2}{\pi}\frac{\partial^2 G(p, q)}{\partial p \partial q} + \sum_{\mu, \nu=1}^{G} \gamma_{\mu\nu} Z'_\mu(p) Z'_\nu(q)$$

where

$$(6.11.31) \qquad \| \gamma_{\mu\nu} \| = \| \bar{\Gamma}_{\mu\nu} \|^{-1}.$$

Domains $\mathfrak{M}$ of genus zero, in particular multiply-connected domains of the plane, are characterized by the property that

$$(6.11.32) \qquad \Gamma_{\mu\nu} = \operatorname{Re} \Gamma_{\mu\nu}, \quad \mu, \nu = 1, 2, \cdots, G.$$

Therefore, in the case of domains of genus zero,

$$(6.11.33) \qquad L_{\mathbf{S}}(p, q) = -\frac{2}{\pi}\frac{\partial^2 N(p, q, q_0)}{\partial p \partial q}.$$

In other words, for domains of genus zero the Neumann's function is related to the class $\mathbf{S}$ of single-valued functions in the same way that the Green's function is related to the class $\mathbf{M}$. The symmetry of the class $\mathbf{S}$ ($L_{\mathbf{S}}(p, q) = L_{\mathbf{S}}(q, p)$) is a characteristic property of the domains of genus zero, and if $\mathfrak{R}$ is also of genus zero we may take $\mathbf{F} = \mathbf{S}$ in (6.11.2). Let the eigen-values τ_ν in the case of the class $\mathbf{S}$ be denoted by t_ν and let the corresponding eigen-functions be Ψ'_ν. Equation (6.11.2) becomes

$$(6.11.34) \qquad \left(1 + \frac{1}{t_\nu}\right)\Psi'_\nu(q) = -\int_{\mathfrak{M}} [l_{\mathbf{S}}(q, p)(\Psi'_\nu(p))^- + l_{\mathbf{S}}(q, \bar{p})\Psi'_\nu(p)]\, dA_z.$$

In the case of the class **S** we may drop the assumption that any boundary component of $\mathfrak{M}$ which is homologous to zero on $\mathfrak{R}$ is also homologous to zero on $\mathfrak{M}$. For this assumption was only used to show that if

$$f'(q) + S_f(q) = \sum_{\mu=1}^{G} a_\mu Z'_\mu(q),$$

then $a_\mu = 0$. This conclusion, however, is obvious in the case of class **S**. Let $\mathcal{N}(p, q, q_0)$ denote the Neumann's function of $\mathfrak{R}$ and write

$$n(p, q, q_0) = N(p, q, q_0) - \mathcal{N}(p, q, q_0). \tag{6.11.35}$$

Then

$$l_{\mathbf{S}}(p, q) = -\frac{2}{\pi}\frac{\partial^2 n(p, q, q_0)}{\partial p \partial q} \tag{6.11.36}$$

$$l_{\mathbf{S}}(p, \bar{q}) = -\frac{2}{\pi}\frac{\partial^2 n(p, q, q_0)}{\partial p \partial \bar{q}}. \tag{6.11.37}$$

We observe that these formulas differ from the analogous formulas (6.11.9) and (6.11.10) in that (6.11.37) has a minus sign where (6.11.10) has a plus sign. This difference in sign reflects the difference in the symmetry of the Green's and Neumann's functions as given by formulas (4.2.6), (4.2.8). This difference in symmetry has an important consequence in establishing the analogue of formula (6.11.14) for (6.11.35), namely that we must use the harmonic functions $R_\nu = \operatorname{Re} \Psi_\nu$ in place of $\operatorname{Im} \Psi_\nu$. Formula (6.11.34) then becomes

$$\begin{aligned}\left(1 + \frac{1}{t_\nu}\right)\frac{\partial R_\nu(q)}{\partial q} &= \frac{2}{\pi}\frac{\partial}{\partial q}\int_{\mathfrak{M}} \left[\frac{\partial n(p, q, q_0)}{\partial p}\frac{\partial R_\nu(p)}{\partial \bar{p}} + \frac{\partial n(p, q, q_0)}{\partial \bar{p}}\frac{\partial R_\nu(p)}{\partial p}\right] dA_z \\ &= \frac{4}{\pi}\frac{\partial}{\partial q}\operatorname{Re}\left\{\int_{\mathfrak{M}} \frac{\partial n(p, q, q_0)}{\partial p}\frac{\partial R_\nu(p)}{\partial \bar{p}} dA_z\right\} \\ &= \frac{1}{\pi}\frac{\partial}{\partial} q\, D(n(p,q,q_0),\ R_\nu(p)).\end{aligned} \tag{6.11.38}$$

Integrating both sides of this equation and using the fact that $n(p, q_0, q_0) \equiv 0$, we find

$$(6.11.39) \quad R_\nu(q) - R_\nu(q_0) = \frac{1}{\pi\left(1 + \dfrac{1}{t_\nu}\right)} D(n(p, q, q_0), R_\nu(p)).$$

Consider now the Hilbert space of all harmonic functions on $\mathfrak{M}$ which have a finite Dirichlet integral and which vanish at p_0. The function

$$n(p, q, q_0) - n(p_0, q, q_0)$$

belongs to this space and the functions $R_\nu(p) - R_\nu(p_0)$ form a complete orthonormal system in this space. We have therefore

$$(6.11.40) \quad \begin{aligned} &n(p, q, q_0) - n(p_0, q, q_0) \\ &= \pi \sum_{\nu=1}^{\infty} \left(1 + \frac{1}{t_\nu}\right)(R_\nu(q) - R_\nu(q_0))(R_\nu(p) - R_\nu(p_0)). \end{aligned}$$

We observe finally that we are operating within the class **S** and hence

$$(\Psi'_\nu, Z'_\varrho) = 0.$$

If $\mathfrak{R}$ is, moreover, simply-connected, we have

$$l_{\mathbf{S}}(p, q) = l_{\mathbf{M}}(p, q) + \sum_{\mu, \nu=1}^{G} c_{\mu\nu} Z'_\mu(p) Z'_\nu(q).$$

Thus, (6.11.34) can be written in the form

$$(6.11.34)' \quad \begin{aligned} &\left(1 + \frac{1}{t_\nu}\right)\Psi'_\nu(q) \\ &= -\int_{\mathfrak{M}} [l_{\mathbf{M}}(q, p)(\Psi'_\nu(p))^- + l_{\mathbf{M}}(q, \bar{p})\Psi'_\nu(p)]\, dA_z. \end{aligned}$$

This equation agrees with (6.11.2) if we put $\mathbf{F} = \mathbf{M}$, $\tau_\mu = t_\nu$, $\Phi'_\mu = \Psi'_\nu$. Thus, when $\mathfrak{R}$ is simply connected, the eigen-differentials $\{\Psi'_\nu\}$ form a subset of the eigen-differentials $\{\Phi'_\mu\}$. We observe that the eigen-value $t_\nu = -1$ does not occur, since the differentials Ψ'_ν must belong to the class **S**. Thus the $\{R_\nu(q)\}$ are the real parts of a subset of $\{\Phi_\mu(q)\}$. Hence the construction of the Neumann's function uses a subset of the analytic functions used in the construction of the Green's function.

We discussed in some detail the case of domains $\mathfrak{M}$ of genus zero imbedded in a domain $\mathfrak{R}$ which is simply-connected, since this case contains the important theory of conformal mapping of multiply-connected domains in the complex plane. In this case, $\mathfrak{M}$ is the multiply-connected domain and $\mathfrak{R}$ the Riemann sphere. This case has been studied by the preceding methods in [1].

6.12. Comparison with Classical Potential Theory

In this section, we want to relate our results to the classical methods of Poincaré-Fredholm used in the boundary-value problems for harmonic functions. We shall recognize that our eigen-functions $H_\nu(p)$ and the eigen-values τ_ν are closely related to corresponding eigen-functions and eigen-values of the integral equation theory of Fredholm. It will appear that the boundary values and the values of the normal derivative of H_ν on the boundary give rise to two sets of classical eigen-functions.

For a fixed q in the interior of $\mathfrak{M}$, we have by Green's formula

$$D(G(p, q), H_\nu(p)) = -\int_C G(p, q)\frac{\partial H_\nu(p)}{\partial n_p} ds_p, \tag{6.12.1}$$

where $(\partial H_\nu/\partial n_p) ds_p$ has the interpretation described at the beginning of Section 4.3. Since $G(p, q) = 0$ for q in $\mathfrak{M}$ and p on the boundary, we see that

$$D(G(p, q), H_\nu(p)) = 0. \tag{6.12.2}$$

The integral equation (6.11.14) therefore becomes

$$H_\nu(q) = \frac{1}{\pi\left(1+\dfrac{1}{\tau_\nu}\right)} D(\mathscr{G}(p, q), H_\nu(p))$$

$$= -\frac{1}{\pi\left(1+\dfrac{1}{\tau_\nu}\right)}\int_C \mathscr{G}(p, q)\frac{\partial H_\nu(p)}{\partial n_p} ds_p,$$

by Green's formula; that is,

$$(6.12.3)\qquad H_\nu(q) = -\frac{1}{\pi\left(1+\dfrac{1}{\tau_\nu}\right)}\int_C \mathscr{G}(p,q)\frac{\partial H_\nu(p)}{\partial n_p}\,ds_p.$$

Removing a small uniformizer circle at q, applying Green's formula, and then letting the radius of the circle tend to zero, we obtain

$$(6.12.4)\qquad \int_C \mathscr{G}(p,q)\frac{\partial H_\nu(p)}{\partial n_p}\,ds_p = \int_C H_\nu(p)\frac{\partial \mathscr{G}(p,q)}{\partial n_p}\,ds_p - 2\pi H_\nu(q).$$

Eliminating the integral on the left by means of formula (6.12.3), we see that

$$(6.12.5)\qquad H_\nu(q) = \frac{1}{\pi\left(1-\dfrac{1}{\tau_\nu}\right)}\int_C \frac{\partial \mathscr{G}(p,q)}{\partial n_p}H_\nu(p)\,ds_p.$$

We remark that the eigen-differentials $\Phi'_\nu(q)$ of the integral equation (6.11.1) are regular on the boundary C of $\mathfrak{M}$. This is a consequence of the fact that $l_{\mathbf{M}}(q,p)$ and $\mathscr{L}_{\mathbf{M}}(q,\tilde{p})$ are regular for q on the boundary and p any point in the closure of $\mathfrak{M}$. The regularity of $\mathscr{L}_{\mathbf{M}}(p,\tilde{q})$ follows from the assumption that $\mathfrak{M}$ is properly embedded in $\mathfrak{R}$. Hence we may let q tend to the boundary in (6.12.5). From known theorems of potential theory concerning the behavior of a double layer distribution on the boundary, we conclude that for $q \in C$,

$$H_\nu(q) = \frac{1}{\pi\left(1-\dfrac{1}{\tau_\nu}\right)}\int_C \frac{\partial \mathscr{G}(p,q)}{\partial n_p}H_\nu(p)\,ds_p + \frac{1}{1-\dfrac{1}{\tau_\nu}}H_\nu(q),$$

that is

$$(6.12.6)\qquad H_\nu(q) = -\frac{\tau_\nu}{\pi}\int_C \frac{\partial \mathscr{G}(p,q)}{\partial n_p}H_\nu(p)\,ds_p,\quad q\in C.$$

Here the integral must be interpreted in the principal-value sense. Similarly, writing

$$h_\nu(q) = \frac{\partial H_\nu(q)}{\partial n_q},\quad q\in C,$$

we conclude from (6.12.3) and the known behavior of the normal derivative of a simple distribution on the boundary that for $q \in C$,

$$h_\nu(q) = -\frac{1}{\pi\left(1+\dfrac{1}{\tau_\nu}\right)} \int_C \frac{\partial \mathscr{G}(p, q)}{\partial n_q} h_\nu(p) ds_p + \frac{1}{1+\dfrac{1}{\tau_\nu}} h_\nu(q),$$

that is

$$h_\nu(q) = -\frac{\tau_\nu}{\pi} \int_C \frac{\partial \mathscr{G}(p, q)}{\partial n_q} h_\nu(p) ds_p, \quad q \in C. \tag{6.12.7}$$

Equation (6.12.7) is not in invariant form, but can be made so by multiplying both sides by ds_q, for

$$h_\nu(q) ds_q = \frac{\partial H_\nu(q)}{\partial n_q} ds_q$$

is invariant.

We observe that the equations (6.12.6) and (6.12.7) are transposes of one another. Equations related in this way have the same eigen-values, and the eigen-functions of the two equations form a biorthogonal set, that is

$$\int_C H_\mu(q) h_\nu(q) ds_q = 0, \qquad \mu \neq \nu.$$

In our case, we know that $h_\nu(q) = \partial H_\nu(q)/\partial n_q$, and the biorthogonality follows at once from the fact that

$$\int_C H_\mu(q) h_\nu(q) ds_q = \int_C H_\mu(q) \frac{\partial H_\nu(q)}{\partial n_q} ds_q = -D(H_\mu, H_\nu) = 0$$

for $\mu \neq \nu$. In particular, we have shown that the eigen-functions of the equations (6.12.6) and (6.12.7) are closely related, the eigen-functions of one being the normal derivatives of the eigen-functions of the other. In other words we have shown that the first and second boundary-value problems of two-dimensional potential theory are equivalent. The equivalence of these problems, as a consequence of the Cauchy-Riemann equations applied to conjugate harmonic functions on the boundary, was pointed out in [1].

6.13. Relation between the Eigen-differentials of $\mathfrak{M}$ and $\mathfrak{R}-\mathfrak{M}$

We shall show that the eigen-differentials of (6.7.3), the equation for the domain $\mathfrak{M}$, automatically give rise to the eigen-differentials of the corresponding integral equation for the residual domain $\mathfrak{N}$. Thus we obtain simultaneously the spectral theory of the s-operator for a domain and for its complement with respect to the domain $\mathfrak{R}$ in which it is imbedded.

Let $\mathfrak{N} = \mathfrak{R} - \mathfrak{M}$, and assume that $\mathfrak{N}$ consists of a single component. Then, according to our definition, $\mathfrak{N}$ is a finite Riemann surface. However, we shall see below that the connectedness of $\mathfrak{N}$ is not essential. Let us follow our earlier convention that differentials are zero throughout any domain in which the definition is not explicitly stated. If Φ_ν' is an eigen-differential, we have

$$(6.13.1) \qquad S_{\Phi_\nu}(q) = \begin{cases} \dfrac{1}{\tau_\nu}\Phi_\nu'(q), & q \in \mathfrak{M}, \\ \theta_\nu'(q), & q \in \mathfrak{N}, \end{cases}$$

where $\Phi_\nu'(q) \equiv 0$ in $\mathfrak{N}$ and $\theta_\nu'(q) \equiv 0$ in $\mathfrak{M}$. By (6.3.8)

$$(6.13.2) \qquad (S^2(q))_{\Phi_\nu} = \frac{1}{\tau_\nu} S_{\Phi_\nu}(q) + S_{\theta_\nu}(q) = \begin{cases} \Phi_\nu'(q), & q \in \mathfrak{M}, \\ 0, & q \in \mathfrak{N}. \end{cases}$$

Hence

$$(6.13.3) \qquad S_{\theta_\nu}(q) = \begin{cases} \left(1 - \dfrac{1}{\tau_\nu^2}\right)\Phi_\nu'(q), & q \in \mathfrak{M}, \\ -\dfrac{1}{\tau_\nu}\theta_\nu'(q), & q \in \mathfrak{N}. \end{cases}$$

In particular, θ_ν' is an eigen-differential of the integral equation

$$(6.13.4) \qquad \theta_\nu'(q) = -\tau_\nu S_{\theta_\nu}(q).$$

By (6.2.8) and (6.5.16), since the class $\mathbf{F}_{\mathfrak{R}}$ is assumed symmetric,

$$\delta_{\mu\nu} = [\Phi_\mu', \Phi_\nu'] = [S_{\Phi_\mu}, S_{\Phi_\nu}] = [S_{\Phi_\mu}, S_{\Phi_\nu}]_{\mathfrak{M}} + [S_{\Phi_\mu}, S_{\Phi_\nu}]_{\mathfrak{N}}$$

$$= \frac{1}{\tau_\mu\tau_\nu}[\Phi_\mu', \Phi_\nu']_{\mathfrak{M}} + [\theta_\mu', \theta_\nu']_{\mathfrak{N}},$$

so

(6.13.5) $$[\theta'_\mu, \theta'_\nu]_\mathfrak{R} = [\theta'_\mu, \theta'_\nu] = \left(1 - \frac{1}{\tau_\mu \tau_\nu}\right)\delta_{\mu\nu}.$$

Thus

(6.13.5)′ $$[\theta'_\mu, \theta'_\mu] = \left(1 - \frac{1}{\tau_\mu^2}\right).$$

We observe from (6.13.5)′ that $|\tau_\nu| < 1$ is impossible. There are at most G eigen-differentials Φ'_μ belonging to the eigen-value $\tau = -1$. Let these eigen-differentials be $\Phi'_1, \cdots, \Phi'_G$, some of which may be zero. From (6.13.5)′ we observe that $\theta'_1(q) = \cdots = \theta'_G(q) = 0$. Writing

(6.13.6) $$\Theta'_\mu = \frac{\theta'_\mu}{\sqrt{1 - \dfrac{1}{\tau_\mu^2}}},$$

we see that $\Theta'_{G+1}, \Theta'_{G+2}, \cdots$ form an orthonormal system. Thus the eigen-differentials Φ'_μ of the integral equation

(6.13.7) $$\Phi'_\mu(q) = \tau_\mu S_{\Phi_\mu}(q)$$

automatically give rise to eigen-differentials of the integral equation (6.13.4).

Suppose that there is a differential Θ' in $\mathfrak{R}$ such that

(6.13.8) $$[\Theta', \Theta'_\nu] = 0, \quad \nu = G + 1, G + 2, \cdots.$$

Then

$$0 = [\Theta', \Theta'_\nu] = [S_\Theta, S_{\Theta_\nu}] = [S_\Theta, S_{\Theta_\nu}]_\mathfrak{M} + [S_\Theta, S_{\Theta_\nu}]_\mathfrak{R}$$

$$= \left(1 - \frac{1}{\tau_\nu^2}\right)^{\frac{1}{2}}[S_\Theta, \Phi'_\nu]_\mathfrak{M} - \frac{1}{\tau_\nu}[S_\Theta, \Theta'_\nu]_\mathfrak{R}$$

$$= \left(1 - \frac{1}{\tau_\nu^2}\right)^{\frac{1}{2}}[S_\Theta, \Phi'_\nu]_\mathfrak{M}$$

since

$$[S_\Theta, \Theta'_\nu]_\mathfrak{R} = [\Theta', S_{\Theta_\nu}]_\mathfrak{R} = -\frac{1}{\tau_\nu}[\Theta', \Theta'_\nu] = 0.$$

Thus, since $|\tau_\nu| > 1$ for $\nu = G+1, \cdots$, we have

$$[S_\Theta, \Phi'_\nu]_{\mathfrak{M}} = 0, \quad \nu = G+1, \cdots. \tag{6.13.9}$$

In other words, S_Θ is orthogonal to all the eigen-differentials Φ'_ν which correspond to eigen-values τ_ν, $|\tau_\nu| > 1$. Since the eigen-differentials Φ'_ν form a complete orthonormal system, we conclude that

$$S_\Theta(q) = \sum_{\nu=1}^{G} a_\nu \Phi'_\nu(q), \quad q \in \mathfrak{M}, \tag{6.13.10}$$

where a_ν is real, $\nu = 1, \cdots, G$. Therefore

$$S_\Theta(q) = \begin{cases} \sum_{\nu=1}^{G} a_\nu \Phi'_\nu(q), & q \in \mathfrak{M}, \\ R'(q), & q \in \mathfrak{N}, \end{cases} \tag{6.13.11}$$

where $R'(q)$ denotes a differential which vanishes identically in $\mathfrak{M}$. It follows that

$$(S^2(q))_\Theta = \begin{cases} S_R(q) - \sum_{\nu=1}^{G} a_\nu \Phi'_\nu(q), & q \in \mathfrak{M}, \\ S_R(q), \quad q \in \mathfrak{N}, \end{cases} \tag{6.13.12}$$

since, by Section 6.5, $S_{\Phi_\mu}(q) \equiv 0$ in $\mathfrak{N}$ and $S_{\Phi_\mu}(q) = -\Phi'_\mu(q)$ in $\mathfrak{M}$ for $\mu = 1, 2, \cdots, G$. On the other hand,

$$(S^2(q))_\Theta = \begin{cases} 0, & q \in \mathfrak{M}, \\ \Theta'(q), & q \in \mathfrak{N}. \end{cases} \tag{6.13.13}$$

Comparing (6.13.12) and (6.13.13), we see that

$$S_R(q) = \begin{cases} \sum_{\nu=1}^{G} a_\nu \Phi'_\nu(p), & q \in \mathfrak{M}, \\ \Theta'(q), \quad q \in \mathfrak{N}. \end{cases} \tag{6.13.14}$$

From (6.2.11) and (6.13.11),

$$N_{\mathfrak{R}}(S_\Theta) = \sum_{\nu=1}^{G} a_\nu^2 + N_{\mathfrak{N}}(R') = N_{\mathfrak{N}}(\Theta'); \tag{6.13.15}$$

from (6.2.11) and (6.13.14),

$$N_{\mathfrak{R}}(S_R) = \sum_{\nu=1}^{G} a_\nu^2 + N_{\mathfrak{N}}(\Theta') = N_{\mathfrak{N}}(R'). \tag{6.13.16}$$

Thus $a_1 = a_2 = \cdots = a_G = 0$ and we have

$$S_{\Theta}(q) = \begin{cases} 0, & q \in \mathfrak{M}, \\ R'(q), & q \in \mathfrak{N}, \end{cases} \tag{6.13.11)'$$

$$S_{R}(q) = \begin{cases} 0, & q \in \mathfrak{M}, \\ \Theta'(q), & q \in \mathfrak{N}. \end{cases} \tag{6.13.14)'$$

Hence

$$\Theta'(q) - R'(q) = - S_{\Theta - R}(q), \tag{6.13.17}$$

and

$$\Theta'(q) + R'(q) = S_{\Theta + R}(q). \tag{6.13.18}$$

If there is a non-trivial differential Θ' satisfying (6.13.8), we have shown that the equation (6.13.4) must have an eigen-value $\tau = 1$ or an eigen-value $\tau = -1$.

We have thus proved the theorem:

THEOREM 6.13.1. *Every eigen-differential Φ_ν' of* (6.7.3) *which belongs to an eigen-value $\tau_\nu \neq -1$ transforms by the S-operator into an eigen-differential for the corresponding integral equation in the complementary domain and the eigen-value $-\tau_\nu$. Conversely, every eigen-differential of the latter integral equation which belongs to an eigen-value $\neq \pm 1$ is obtained in this way.*

We want to show now that no eigen-differential in $\mathfrak{R}$ belongs to the eigen-value $+1$. We recall that a domain is properly imbedded in $\mathfrak{R}$ if each of its boundary points is an interior point of $\mathfrak{R}$. Since we have assumed that $\mathfrak{M}$ is properly imbedded, it is clear that $\mathfrak{N} = \mathfrak{R} - \mathfrak{M}$ is not properly imbedded and this means that at least one boundary component of $\mathfrak{N}$ coincides with a boundary component of $\mathfrak{R}$. Hence $\mathscr{L}_{\mathbf{F}}(q, \bar{p})$ is not regular in the closure of $\mathfrak{N}$. Let $L_{\mathbf{F}}^{*}(q, p)$ be the bilinear differential of $\mathfrak{N}$ which belongs to a symmetric class $\mathbf{F} = \mathbf{F}_{\mathfrak{N}}$ corresponding to the given symmetric class $\mathbf{F}_{\mathfrak{R}}$, and let Θ' be an eigen-differential belonging to the eigen-value $+1$; that is

$$\Theta'(q) = S_{\Theta}(q). \tag{6.13.18)'$$

Then

(6.13.19)

$$\Theta'(q) = S_{\Theta}(q) = \int_{\mathfrak{N}} \mathscr{L}_{\mathbf{F}}(q, p)(\Theta'(p))^{-} dA_z + \int_{\mathfrak{N}} \mathscr{L}_{\mathbf{F}}(q, \bar{p}) \Theta'(p) dA_z.$$

Let us write

$$l_{\mathbf{F}}^{*}(q, p) = L_{\mathbf{F}}^{*}(q, p) - \mathscr{L}_{\mathbf{F}}(q, p).$$

Since $L_{\mathbf{F}}^{*}(q, p)$ is orthogonal to the class $\mathbf{F}_{\mathfrak{N}}$ and since $L_{\mathbf{F}}^{*}(q, \tilde{p})$ reproduces under scalar multiplication we have in view of (6.13.19)

$$(6.13.20)\qquad \Theta'(q) = -\int_{\mathfrak{N}} l_{\mathbf{F}}^{*}(q, p)\,(\Theta'(p))^{-} dA_z + \int_{\mathfrak{N}} \mathscr{L}_{\mathbf{F}}(q, \tilde{p})\Theta'(p) dA_z$$

and

$$(6.13.20)'\qquad 2\Theta'(q) = -\int_{\mathfrak{N}} l_{\mathbf{F}}^{*}(q, p)(\Theta'(p))^{-} dA_z - \int_{\mathfrak{N}} l_{\mathbf{F}}^{*}(q, \tilde{p})\Theta'(p) dA_z.$$

It follows from (6.13.20) and (6.13.20)′ that $\Theta'(q)$ is regular in the closure of $\mathfrak{N}$. By (6.2.11) and (6.13.18)′ we have

$$N_{\mathfrak{R}}(S_{\Theta}) = N_{\mathfrak{M}}(S_{\Theta}) + N_{\mathfrak{N}}(S_{\Theta}) = N_{\mathfrak{M}}(S_{\Theta}) + N_{\mathfrak{N}}(\Theta') = N_{\mathfrak{N}}(\Theta').$$

Hence

$$(6.13.21)\qquad S_{\Theta}(q) = \begin{cases} 0, & q \in \mathfrak{M}, \\ \Theta', & q \in \mathfrak{N}. \end{cases}$$

We conclude in the usual fashion that $S_{\Theta}(q)$ has the boundary values $(\Theta'(q))^{-}$ on that portion of the boundary of $\mathfrak{N}$ which coincides with the boundary of $\mathfrak{M}$. By (6.13.18)′, $S_{\Theta}(q)$ also has the boundary values $\Theta'(q)$, so $\Theta'(q)$ is real on that part of the boundary of $\mathfrak{N}$ which coincides with the boundary of $\mathfrak{M}$. We understand, of course, that $\Theta'(q)$ on the boundary is expressed in terms of a boundary uniformizer. Now let q be a point on that part of the boundary of $\mathfrak{N}$ which coincides with the boundary of $\mathfrak{R}$. By (6.13.20)′,

$$\begin{aligned} 2\Theta'(q) &= -\int_{\mathfrak{N}} l_{\mathbf{F}}^{*}(q, p)(\Theta'(p))^{-} dA_z - \int_{\mathfrak{N}} l_{\mathbf{F}}^{*}(q, \tilde{p})\Theta'(p) dA_z \\ (6.13.22)\qquad &= -\int_{\mathfrak{N}} l_{\mathbf{F}}^{*}(\tilde{q}, p)(\Theta'(p))^{-} dA_z - \int_{\mathfrak{N}} l_{\mathbf{F}}^{*}(q, \tilde{p})\Theta'(p) dA_z \\ &= -\int_{\mathfrak{N}} (l_{\mathbf{F}}^{*}(q, \tilde{p}))^{-}(\Theta'(p))^{-} dA_z - \int_{\mathfrak{N}} l_{\mathbf{F}}^{*}(q, \tilde{p})\Theta'(p) dA_z. \end{aligned}$$

Thus $\Theta'(q)$, expressed in terms of boundary uniformizers, is real on the entire boundary of $\mathfrak{N}$ and is therefore a differential of $\mathfrak{N}$. If $h' \in \mathbf{F}_{\mathfrak{R}}$,

$$(S_{\Theta}, h')_{\mathfrak{R}} = (\Theta', h')_{\mathfrak{N}}.$$

But by (6.2.12)

$$(S_{\Theta}, h')_{\mathfrak{R}} = -(\Theta', h')_{\mathfrak{N}},$$

so

$$(\Theta', h')_{\mathfrak{N}} = 0, \tag{6.13.23}$$

for every h' of $\mathbf{F}_{\mathfrak{R}}$.

Now assume that $\mathfrak{N}$ is essentially embedded in $\mathfrak{R}$. We recall that a similar condition was imposed upon $\mathfrak{M}$. We can prove, however, that if $\mathfrak{M}$ is connected, and if $\mathfrak{M}$ and $\mathfrak{R}$ each have more than one boundary component, $\mathfrak{N}$ satisfies the assumption in any case. For, let b be a boundary cycle of $\mathfrak{N}$ which bounds on $\mathfrak{R}$. Since $\mathfrak{R}$ has more than one boundary component, b is not a boundary component of $\mathfrak{R}$. Hence it is a boundary component of $\mathfrak{M}$ which bounds on $\mathfrak{R}$ and also, by hypothesis, on $\mathfrak{M}$. But $\mathfrak{M}$ is connected and has more than one boundary component, so this is impossible. Hence no boundary cycle of $\mathfrak{N}$ bounds on $\mathfrak{R}$ in this case.

We can now apply the reasoning of Section 6.5, and we conclude that $\Theta'(q) \equiv 0$. Thus there are no eigen-differentials corresponding to the eigen-value $+1$. However, the above argument cannot be used to exclude the eigen-value -1. In fact, since $\mathscr{L}_{\mathbf{F}}(\bar{p}, q)$ is not regular in the closure of $\mathfrak{N}$, there may be infinitely many eigen-differentials belonging to the eigen-value -1. If we include these eigen-differentials, we obtain a complete orthonormal system of eigen-differentials for the class $\mathbf{F}_{\mathfrak{N}}$ which corresponds to $\mathbf{F}_{\mathfrak{R}}$.

In order to overcome the difficulty that infinitely many eigen-differentials may belong to the eigen-value -1, we introduce the sub-class $\mathbf{R}_{\mathfrak{N}}$ of $\mathbf{F}_{\mathfrak{N}}$ which is composed of those differentials Θ' of $\mathbf{F}_{\mathfrak{N}}$ which, expressed in terms of boundary uniformizers, are regular analytic and real on that portion of the boundary of $\mathfrak{N}$ which lies on the boundary of $\mathfrak{R}$ ($\mathfrak{R}$-boundary of $\mathfrak{N}$). Suppose that Φ' belongs to $\mathbf{F}_{\mathfrak{M}}$, $\Phi' \equiv 0$ in $\mathfrak{N}$; then for q on the $\mathfrak{R}$-boundary of $\mathfrak{N}$ we have

$$S_\Phi(q) = \int_{\mathfrak{M}} \mathscr{L}_{\mathrm{F}}(q, p)\,(\Phi'(p))^- dA_z + \int_{\mathfrak{M}} \mathscr{L}_{\mathrm{F}}(q, \tilde{p})\Phi'(p) dA_z$$

$$= \int_{\mathfrak{M}} \mathscr{L}_{\mathrm{F}}(\tilde{q}, p)\,(\Phi'(p))^- dA_z + \int_{\mathfrak{M}} \mathscr{L}_{\mathrm{F}}(q, \tilde{p})\Phi'(p) dA_z$$

$$= \int_{\mathfrak{M}} (\mathscr{L}_{\mathrm{F}}(q, \tilde{p}))^-(\Phi'(p))^- dA_z + \int_{\mathfrak{M}} \mathscr{L}_{\mathrm{F}}(q, \tilde{p})\Phi'(p) dA_z.$$

Thus S_Φ belongs to $\mathbf{R}_{\mathfrak{N}}$. In particular, the eigen-functions Θ'_μ defined by (6.13.6) belong to $\mathbf{R}_{\mathfrak{N}}$.

We remark that the differentials of $\mathbf{R}_{\mathfrak{N}}$ form a complete Hilbert space. For any differential Θ' of $\mathbf{R}_{\mathfrak{N}}$ is regular analytic in $\mathfrak{N} + \tilde{\mathfrak{N}}$ and on the boundary of $\mathfrak{R}$. Moreover,

$$N_{\mathfrak{N}+\tilde{\mathfrak{N}}}(\Theta') = 2N_{\mathfrak{N}}(\Theta'),$$

since Θ' takes values in $\tilde{\mathfrak{N}}$ which are conjugate to those taken in $\mathfrak{N}$. Hence points of the $\mathfrak{R}$-boundary of $\mathfrak{N}$ behave like interior points of the domain so far as convergence questions are concerned. A Cauchy sequence in $\mathbf{R}_{\mathfrak{N}}$ will therefore converge to a differential which is real on the $\mathfrak{R}$-boundary.

If the system of eigen-differentials Θ'_μ already constructed do not form a complete orthonormal system for $\mathbf{R}_{\mathfrak{N}}$, there is a differential Θ' of $\mathbf{R}_{\mathfrak{N}}$ which is orthogonal to all the Θ'_μ. But then we have shown that

$$\Theta'(q) = -S_\Theta(q), \tag{6.13.24}$$

where

$$S_\Theta(q) \equiv 0, \quad q \in \mathfrak{M}.$$

It follows in the usual way that

$$S_\Theta(q) = (\Theta'(q))^-$$

on the $\mathfrak{M}$-boundary of $\mathfrak{N}$. Here $\Theta'(q)$ is expressed in terms of a boundary uniformizer. But, by (6.13.24),

$$S_\Theta(q) = -\Theta'(q)$$

on the $\mathfrak{M}$-boundary, so $\Theta'(q)$ is imaginary on the $\mathfrak{M}$-boundary. Since Θ' belongs to $\mathbf{R}_{\mathfrak{N}}$, it is real and regular analytic on the $\mathfrak{R}$-

boundary of $\mathfrak{N}$, and it follows that

$$\{\Theta'(q)\}^2$$

is a quadratic differential of $\mathfrak{N}$. Let σ (a finite number) be the number of linearly independent quadratic differentials of $\mathfrak{N}$. Since the number of quadratic differentials which are squares of linear differentials does not exceed σ, we see that by adding a finite number of eigen-differentials belonging to the eigen-value -1, we obtain a complete orthonormal system for $\mathbf{R}_{\mathfrak{N}}$. Let this complete orthonormal system be denoted by $\{\Theta'_\nu\}$, where Θ'_ν is the eigen-differential of the integral equation

$$\Theta'(q) = \tau S_\Theta(q) \tag{6.13.25}$$

belonging to the eigen-value $\tau = -\tau_\nu$; that is

$$\Theta'_\nu(q) = -\tau_\nu S_{\Theta_\nu}(q).$$

The differential

$$\Lambda_{\mathbf{F}}(p, q, \tilde{q}) = -l^*_{\mathbf{F}}(q, p) - l^*_{\mathbf{F}}(p, \tilde{q})$$

clearly belongs to $\mathbf{R}_{\mathfrak{N}}$, and its ν-th Fourier coefficient is equal to

$$\operatorname{Re}\left\{\int_{\mathfrak{N}} \Lambda_{\mathbf{F}}(p, q, \tilde{q})(\Theta'_\nu(p))^- dA_z\right\} = \operatorname{Re}\{S_{\Theta_\nu}(q) + \Theta'_\nu(q)\}$$

$$= \left(1 - \frac{1}{\tau_\nu}\right) \operatorname{Re}\{\Theta'_\nu(q)\}.$$

Similarly, the differential

$$\lambda_{\mathbf{F}}(p, q, \tilde{q}) = i\,[l^*_{\mathbf{F}}(q, p) - l^*_{\mathbf{F}}(p, \tilde{q})]$$

belongs to $\mathbf{R}_{\mathfrak{N}}$, and its ν-th Fourier coefficient is

$$\operatorname{Re}\left\{\int_{\mathfrak{N}} \lambda_{\mathbf{F}}(p, q, \tilde{q})(\Theta'_\nu(p))^- dA_z\right\} = \operatorname{Im}\{S_{\Theta_\nu}(q) + \Theta'_\nu(q)\}$$

$$= \left(1 - \frac{1}{\tau_\nu}\right) \operatorname{Im}\{\Theta'_\nu(q)\}.$$

Thus

$$l^*_{\mathbf{F}}(q, p) + l^*_{\mathbf{F}}(p, \tilde{q}) = -\sum_{\nu=1}^{\infty}\left(1 - \frac{1}{\tau_\nu}\right)\operatorname{Re}\{\Theta'_\nu(q)\}\,\Theta'_\nu(p), \tag{6.13.26}$$

$$l^*_{\mathbf{F}}(q, p) - l^*_{\mathbf{F}}(p, \tilde{q}) = -i\sum_{\nu=1}^{\infty}\left(1 - \frac{1}{\tau_\nu}\right)\operatorname{Im}\{\Theta'_\nu(q)\}\,\Theta'_\nu(p). \tag{6.13.27}$$

Adding and subtracting these equations, we obtain

$$(6.13.28)\qquad l_{\mathbf{F}}^{*}(p, q) = -\frac{1}{2}\sum_{\nu=1}^{\infty}\left(1-\frac{1}{\tau_\nu}\right)\Theta_\nu'(p)\Theta_\nu'(q),$$

$$(6.13.29)\qquad l_{\mathbf{F}}^{*}(p, \bar{q}) = -\frac{1}{2}\sum_{\nu=1}^{\infty}\left(1-\frac{1}{\tau_\nu}\right)\Theta_\nu'(p)(\Theta_\nu'(q))^{-}.$$

We observe that the eigen-differentials Θ_ν' belonging to the troublesome eigen-value $\tau = -1$ do not enter into the sums (6.13.28) and (6.13.29). The remaining eigen-differentials have the form

$$(6.13.30)\qquad \Theta_\nu'(q) = \frac{1}{\sqrt{1-\dfrac{1}{\tau_\nu^2}}}S_{\Phi_\nu}(q), \quad q \in \mathfrak{N},$$

where the τ_ν are the eigen-values, the Φ_ν' the eigen-differentials, of the corresponding integral equation for $\mathfrak{M}$; that is

$$(6.13.31)\qquad \Phi_\nu'(q) = \tau_\nu S_{\Phi_\nu}(q), \quad q \in \mathfrak{M}.$$

The formulas (6.13.28) and (6.13.29) may therefore be written

$$(6.13.28)'\qquad l_{\mathbf{F}}^{*}(p, q) = -\frac{1}{2}\sum_{\nu=1}^{\infty}\frac{\tau_\nu}{\tau_\nu+1}S_{\Phi_\nu}(p)\,S_{\Phi_\nu}(q),$$

$$(6.13.29)'\qquad l_{\mathbf{F}}^{*}(p, \bar{q}) = -\frac{1}{2}\sum_{\nu=1}^{\infty}\frac{\tau_\nu}{\tau_\nu+1}S_{\Phi_\nu}(p)\,(S_{\Phi_\nu}(q))^{-}.$$

Therefore, by solving the integral equation

$$\Phi'(q) = \tau S_{\Phi}(q),$$

for $\mathfrak{M}$, we simultaneously determine the bilinear differentials for $\mathfrak{M}$ and for the complementary domain $\mathfrak{N}$.

6.14. Extension to Disconnected Surfaces

At the beginning of Section 6.13 we assumed that the complement $\mathfrak{N}$ of $\mathfrak{M}$ with respect to $\mathfrak{R}$ is a single domain. The question arises whether this assumption is essential, and it turns out that it is not. We shall now extend the main formulas and results of the preceding sections to the case in which the set $\mathfrak{M}$ imbedded in $\mathfrak{R}$ consists of several components $\mathfrak{M}_\mu$, each of which is a finite Riemann surface.

The necessary definitions of the kernel and of the difference kernel have been given in Section 5.2.

For the present case, we define

$$(6.14.1)\quad T_f(q) = (\mathscr{L}_{\mathbf{F}}(q, p), f'(p))_{\mathfrak{R}} = \sum_{\mu=1}^{k} (\mathscr{L}_{\mathbf{F}}(q, p), f'(p))_{\mathfrak{M}_\mu},$$

$$(6.14.2)\quad \tilde{T}_f(q) = (\mathscr{L}_{\mathbf{F}}(q, \tilde{p}), (f'(p))^-)_{\mathfrak{R}} = \sum_{\mu=1}^{k} (\mathscr{L}_{\mathbf{F}}(q, \tilde{p}), (f'(p))^-)_{\mathfrak{M}_\mu}.$$

Next we observe that

$$(6.14.3)\qquad T_f(q) = -(l_{\mathbf{F}}(q, p), f'(p))_{\mathfrak{R}}$$

since (even without assuming symmetry)

$$(L_{\mathbf{F}}(q, p), f'(p))_{\mathfrak{R}} = 0.$$

The following formulas are seen to be true:

$$(6.14.4)\qquad (T_{f_1}, \tilde{T}_{f_2})_{\mathfrak{R}} = 0,$$

$$(6.14.5)\quad (T_{f_1}, T_{f_2})_{\mathfrak{R}} = -\int_{\mathfrak{R}}\int_{\mathfrak{R}} l_{\mathbf{F}}(\tilde{p}_2, p_1)(f_1'(p_1))^- f_2'(p_2)\,dA_{z_1}dA_{z_2},$$

$$(6.14.6)\quad (\tilde{T}_{f_1}, \tilde{T}_{f_2})_{\mathfrak{R}} = -\int_{\mathfrak{R}}\int_{\mathfrak{R}} \mathscr{L}_{\mathbf{F}}(p_2, \tilde{p}_1)\, f_1'(p_1)\, (f_2'(p_2))^- dA_{z_1}\, dA_{z_2},$$

$$(6.14.7)\quad \begin{aligned} (S_{f_1}, S_{f_2})_{\mathfrak{R}} &= (f_2', f_1')_{\mathfrak{R}} - 2i \sum_{\mu,\nu=1}^{G_0} \operatorname{Im}\{\beta_{\mu\nu}\} (\mathscr{Z}'_\mu, f_1')_{\mathfrak{R}} ((\mathscr{Z}'_\nu, f_2')_{\mathfrak{R}})^- \\ &+ 2i \operatorname{Im}\left\{\int_{\mathfrak{R}}\int_{\mathfrak{R}} \mathscr{L}_{\mathbf{F}}(p_1, \tilde{p}_2)\, (f_1'(p_1))^- f_2'(p_2)\, dA_{z_1}\, dA_{z_2}\right\}, \end{aligned}$$

$$(6.14.8)\quad \begin{aligned} N_{\mathfrak{R}}(S_f) &= N_{\mathfrak{R}}(T_f) + N_{\mathfrak{R}}(\tilde{T}_f) \\ &= N_{\mathfrak{R}}(f') - 2i \sum_{\mu,\nu=1}^{G_0} \operatorname{Im}\{\beta_{\mu\nu}\} (\mathscr{Z}'_\mu, f')_{\mathfrak{R}} ((\mathscr{Z}'_\nu, f')_{\mathfrak{R}})^-. \end{aligned}$$

If $\mathscr{L}_{\mathbf{F}}(p, q) = \mathscr{L}_{\mathbf{F}}(q, p)$ (symmetry), then (6.14.8) becomes

$$(6.14.8)'\qquad N_{\mathfrak{R}}(S_f) = N_{\mathfrak{R}}(f').$$

Furthermore, when $\mathscr{L}_{\mathbf{F}}(p, q)$ is symmetric, we have the three formulas:

$$(6.14.9)\qquad (T_{f_1}, f_2') = (T_{f_2}, f_1'),$$

$$(6.14.10)\qquad (\tilde{T}_{f_1}, f_2') = (f_1', \tilde{T}_{f_2}),$$

$$(6.14.11) \qquad [S_{f_1}, f_2'] = [S_{f_2}, f_1'].$$

In general, we also have

$$(6.14.12) \qquad (T_f, h') = 0, \ (\tilde{T}_f, h') = -(f', h')_{\mathfrak{R}}$$

for any $h' \in \mathbf{F}_{\mathfrak{R}}$.

The formulas of Section 6.3 extend immediately to the more general case where $\mathfrak{M}$ is disconnected; that is,

$$(6.14.13) \qquad T^2 = I + \tilde{T},$$

$$(6.14.14) \qquad \tilde{T}^2 = -\tilde{T},$$

$$(6.14.15) \qquad T\tilde{T} = \tilde{T}T = 0,$$

$$(6.14.16) \qquad S^2 = I.$$

Let us now assume that $\mathfrak{M}$ is properly and essentially imbedded in $\mathfrak{R}$ and that $\mathscr{L}_{\mathbf{F}}(p, q) = \mathscr{L}_{\mathbf{F}}(q, p)$. Then all results proved in Section 6.5-Section 6.13 remain valid for the case in which $\mathfrak{M}$ is disconnected. In particular, we may drop the assumption that $\mathfrak{N} = \mathfrak{R} - \mathfrak{M}$ is connected; we must assume, however, that $\mathfrak{N}$ satisfies the condition that the imbedding is essential. The fact of the matter is that if $\mathfrak{N}$ satisfies this condition, the hypothesis "$\mathfrak{N}$ is connected" is not necessary.

6.15. Representation of Domain Functionals of $\mathfrak{M}$ in Terms of the Domain Functionals of $\mathfrak{R}$

For a given surface $\mathfrak{R}$ there are infinitely many imbedded surfaces $\mathfrak{M}$ each of which is characterized by its boundary C relative to $\mathfrak{R}$. We may regard the differentials of the surface $\mathfrak{M}$ as functionals of C or of $\mathfrak{M}$ and have then to establish a method for calculating them in terms of the differentials of the fixed surface $\mathfrak{R}$ which carries the surfaces $\mathfrak{M}$.

We suppose that $\mathfrak{M}$ is the union of a finite number of domains $\mathfrak{M}_\mu$ and that $\mathfrak{M}$ is properly and essentially imbedded in $\mathfrak{R}$. We shall consider only classes $\mathbf{F}$ of differentials on $\mathfrak{M}$ for which the corresponding class $\mathbf{F}_{\mathfrak{R}}$ is symmetric. We established in the preceding sections the fact that, under the above assumptions, all eigen-values of the equation (6.7.1) are greater than unity. This enables us to solve certain integral equations for the differentials on $\mathfrak{M}$ by means of Neumann-

Liouville series and to express the functionals of $\mathfrak{M}$ in terms of the functionals of the carrier surface $\mathfrak{R}$ and of integrals of the $\mathfrak{R}$-functionals over $\mathfrak{M}$ and the residual set $\mathfrak{R} - \mathfrak{M}$.

Let $Q_{\mathbf{F}}^{(2\mu)}(p, \bar{q})$ be defined as in Section 6.7. By the reproducing property of $L_{\mathbf{F}}(p, \bar{r})$ we have

$$(6.15.1) \qquad Q_{\mathbf{F}}^{(2)}(p, \bar{q}) = -\int_{\mathfrak{M}} Q_{\mathbf{F}}^{(2)}(r, \bar{q})\, L_{\mathbf{F}}(p, \bar{r})\, dA_{\eta}.$$

Using (6.7.30) we may write

$$(6.15.2) \quad I_{\mathbf{F}}(p, \bar{q}) = \mathscr{L}_{\mathbf{F}}(p, \bar{q}) - \Gamma_{\mathbf{F}}(p, \bar{q}) = L_{\mathbf{F}}(p, \bar{q}) + Q_{\mathbf{F}}^{(2)}(p, \bar{q})$$

and bring (6.15.1) into the form of an integral equation for $L_{\mathbf{F}}(p, \bar{q})$:

$$(6.15.3) \qquad I_{\mathbf{F}}(p, \bar{q}) = L_{\mathbf{F}}(p, \bar{q}) - \int_{\mathfrak{M}} Q_{\mathbf{F}}^{(2)}(r, \bar{q})\, L_{\mathbf{F}}(p, \bar{r})\, dA_{\eta}.$$

The term $I_{\mathbf{F}}(p, \bar{q})$ depends, by its definition, only on differentials on $\mathfrak{R}$ and upon the difference domain $\mathfrak{R} - \mathfrak{M}$, and will therefore be considered as known. We can invert (6.15.3) since the kernel $Q_{\mathbf{F}}^{(2)}(r, \bar{q})$ has all its eigen-values λ_ν^2 greater than unity. We define the reciprocal kernel of $Q_{\mathbf{F}}^{(2)}(r, \bar{q})$ by the Neumann-Liouville series

$$(6.15.4) \qquad Q_{\mathbf{F}}^{(-1)}(p, \bar{q}) = \sum_{\nu=0}^{\infty} Q_{\mathbf{F}}^{(2\nu)}(p, \bar{q})$$

which converges uniformly in each closed subdomain of $\mathfrak{M}$. Using the expression (6.7.34) of $Q_{\mathbf{F}}^{(2\mu)}$ in terms of its eigen-values and eigen-differentials, we obtain

$$(6.15.5) \qquad Q_{\mathbf{F}}^{(-1)}(p, \bar{q}) = \sum_{\nu=1}^{\infty} \left(1 - \frac{1}{\lambda_\nu^2}\right)^{-1} d\varphi_\nu(p)(d\varphi_\nu(q))^{-}.$$

By means of the reciprocal kernel $Q_{\mathbf{F}}^{(-1)}(p, \bar{q})$ we may now solve the integral equations (6.15.3) in the form:

$$(6.15.6) \qquad \begin{aligned} L_{\mathbf{F}}(p, \bar{q}) &= I_{\mathbf{F}}(p, \bar{q}) + \sum_{\nu=1}^{\infty} \int_{\mathfrak{M}} Q_{\mathbf{F}}^{(2\nu)}(r, \bar{q})\, I_{\mathbf{F}}(p, \bar{r})\, dA_{\eta} \\ &= \int_{\mathfrak{M}} Q_{\mathbf{F}}^{(-1)}(r, \bar{q})\, I_{\mathbf{F}}(p, \bar{r})\, dA_{\eta}. \end{aligned}$$

The difficulty with the above result comes from the fact that

$Q_F^{(2)}(p, \tilde{q})$ still contains the differential $L_F(p, \tilde{q})$ of $\mathfrak{M}$ and hence the series development (6.15.6) does not yield $L_F(p, \tilde{q})$ in terms of $\mathfrak{R}$-differentials alone. However, this obstacle can be overcome as follows. Because of the reproducing property of the kernel $L_F(p, \tilde{q})$, we have

$$Q_F^{(4)}(r, \tilde{q}) = -L_F(r, \tilde{q}) + 2I(r, \tilde{q}) + I_F^{(2)}(r, \tilde{q}) \tag{6.15,7}$$

where

$$I_F^{(2)}(r, \tilde{q}) = \int_{\mathfrak{M}} I_F(r, \tilde{p})\, I_F(p, \tilde{q})\, dA_z. \tag{6.15.8}$$

In general

$$Q_F^{(2\nu)}(r, \tilde{q}) = -L_F(r, \tilde{q}) + \sum_{\mu=1}^{\nu} \binom{\nu}{\mu} I_F^{(\mu)}(r, \tilde{q}) \tag{6.15.9}$$

with

$$I_F^{(\mu)}(r, \tilde{q}) = \int_{\mathfrak{M}} I_F^{(\mu-1)}(r, \tilde{p})\, I_F(p, \tilde{q})\, dA_z. \tag{6.15.10}$$

Thus, each term $Q_F^{(2\nu)}$ is composed of iterations of the known term $I_F(p, \tilde{q})$ and of the kernel $L_F(r, \tilde{q})$. We do not know this term but we do know its effect under scalar multiplication over $\mathfrak{M}$ on all differentials on $\mathfrak{R}$ or $\mathfrak{M}$. In particular, we can now calculate all terms in (6.15.6) and obtain:

$$L_F(p, \tilde{q}) = \sum_{\nu=0}^{\infty} \left\{ \sum_{\mu=0}^{\nu} \binom{\nu}{\mu} I_F^{(\mu+1)}(p, \tilde{q}) \right\}. \tag{6.15.11}$$

In (6.15.11) all right-side terms involve known differentials of $\mathfrak{R}$; we have thus obtained a representation of the desired type for the kernel $L_F(p, \tilde{q})$ on $\mathfrak{M}$. The basic term on the right-side has by (6.15.2), (6.7.31) and (6.7.34) the form

$$\begin{aligned} I_F(p, \tilde{q}) &= L_F(p, \tilde{q}) + Q_F^{(2)}(p, \tilde{q}) \\ &= -\sum_{\nu=1}^{\infty} \left(1 - \frac{1}{\lambda_\nu^2}\right) d\varphi_\nu(p) (d\varphi_\nu(q))^-. \end{aligned} \tag{6.15.12}$$

Its iterates have consequently the form

$$I_F^{(\mu)}(p, \tilde{q}) = (-)^\mu \sum_{\nu=1}^{\infty} \left(1 - \frac{1}{\lambda^2}\right)^\mu d\varphi_\nu(p) (d\varphi(q))^- \tag{6.15.12$'$}$$

and the identity may be checked formally by inserting (6.15.12)′ into (6.15.11) and then verifying the identity by rearranging the terms on the right-hand side. It is also instructive to express the bracketed terms occurring in (6.15.11) in terms of the eigen-values and eigen-differentials. We find:

$$\text{(6.15.13)} \quad \sum_{\mu=0}^{\nu} \binom{\nu}{\mu} I_{\mathbf{F}}^{(\mu+1)}(p, \tilde{q}) = \sum_{\varrho=1}^{\infty} \frac{1}{\lambda_\varrho^\nu}\left(1 - \frac{1}{\lambda_\varrho^2}\right) d\varphi_\varrho(p)(d\varphi_\varrho(q))^-.$$

This shows that the successive brackets converge to zero like $1/\lambda_\varrho^\nu$ and we obtain an estimate for the contribution of each bracket to the sum total of (6.15.11).

We summarize our main result in

THEOREM 6.15.1. *The kernel $L_{\mathbf{F}}(p, \tilde{q})$ of the domain $\mathfrak{M}$ can be developed into a series in terms of iterated integrals of the $\mathfrak{R}$-differential $I_{\mathbf{F}}(p, \tilde{q}) = \mathscr{L}_{\mathbf{F}}(p, \tilde{q}) - \Gamma_{\mathbf{F}}(p, \tilde{q})$, which is given by formula* (6.15.11).

It is clear that once $L_{\mathbf{F}}(p, \tilde{q})$ is constructed there is no difficulty in obtaining $L_{\mathbf{F}}(p, q)$. In fact, using the reproducing property of $L_{\mathbf{F}}(p, \tilde{q})$ under integration over $\mathfrak{M}$, we find by (4.11.6) and (4.11.7):

$$\begin{aligned} l_{\mathbf{F}}(p, q) = L_{\mathbf{F}}(p, q) - \mathscr{L}_{\mathbf{F}}(p, q) &= -(l_{\mathbf{F}}(r, q), L_{\mathbf{F}}(r, \tilde{p}))_{\mathfrak{M}} \\ &= (\mathscr{L}_{\mathbf{F}}(r, q), L_{\mathbf{F}}(r, \tilde{p}))_{\mathfrak{M}}. \end{aligned}$$

Thus, finally:

$$\text{(6.15.14)} \qquad L_{\mathbf{F}}(p, q) = \mathscr{L}_{\mathbf{F}}(p, q) + (\mathscr{L}_{\mathbf{F}}(r, q), L_{\mathbf{F}}(r, \tilde{p}))_{\mathfrak{M}}.$$

We have expressed $L_{\mathbf{F}}(p, q)$ in terms of the $\mathscr{L}_{\mathbf{F}}$-kernel on $\mathfrak{R}$ and an improper integral involving $L_{\mathbf{F}}(p, \tilde{q})$ over $\mathfrak{M}$. This integral is, of course, to be interpreted in the sense of (4.9.8).

From (6.15.11) we may derive immediately a series development for the differential $Z_\varrho'(q)$ on $\mathfrak{M}$ in terms of differentials on $\mathfrak{R}$. Let us define

$$\text{(6.15.15)} \qquad J_\varrho^{(\mu)}(q) = -\int_{\mathfrak{M}} I^{(\mu+1)}(q, \tilde{p})\, Z_\varrho'(p)\, dA_z;$$

(for the sake of simplicity, only the symmetric class **M** will be considered and we drop the index **M**). We have by (6.15.10)

$$J_\varrho^{(\mu)}(q) = -\int_{\mathfrak{M}} I^{(\mu)}(q, \bar{r})\left\{\int_{\mathfrak{M}} I(r, \bar{p}) Z'_\varrho(p)\, dA_z\right\} dA_\eta \tag{6.15.16}$$
$$= \int_{\mathfrak{M}} I^{(\mu)}(q, \bar{r})\, J_\varrho^{(0)}(r)\, dA_\eta.$$

The integral $J_\varrho^{(\mu)}(q)$ is the conjugate of the period of the differential $I^{(\mu+1)}(p, \bar{q})$ with respect to a cycle K_ϱ; we find by definition (6.15.2) and by (5.1.5), for the special case $\mu = 0$:

$$J_\varrho^{(0)}(q) = \mathcal{Y}'_\varrho(q) - \int_{\mathfrak{R}-\mathfrak{M}} \mathcal{L}(q, r)\mathcal{Y}'_\varrho(\bar{r})\, dA_\eta. \tag{6.15.17}$$

Thus $J_\varrho^{(0)}(q)$, and by (6.15.16) all other differentials $J_\varrho^{(\mu)}(q)$, can be expressed in terms of $\mathfrak{R}$-differentials.

Consider now the series (6.15.11) which converges uniformly in each closed sub-domain of $\mathfrak{M}$; determine on both sides of the equation the period with respect to a cycle K_ϱ. Since this determination can be done by integrating along an interior path in $\mathfrak{M}$, we obtain the identity

$$Z'_\varrho(q) = \sum_{\nu=0}^{\infty}\left\{\sum_{\mu=0}^{\nu}\binom{\nu}{\mu} J_\varrho^{(\mu)}(q)\right\} \tag{6.15.18}$$

where the right-hand sum converges uniformly in each closed sub-domain of $\mathfrak{M}$.

Let us determine next the period matrix of the $Z'_\varrho(p)$-differentials. We observe that by (6.15.10) and (6.15.15),

$$J_\varrho^{(\mu)}(q) = \int_{\mathfrak{M}} I(q, \bar{r})\, J_\varrho^{(\mu-1)}(r)\, dA_\eta, \quad \mu = 1, 2, \cdots. \tag{6.15.19}$$

Hence, we derive from (5.1.8), (6.15.16) and (6.15.19),

$$\int_{\mathfrak{M}} J_\varrho^{(\mu)}(r) Z'_\sigma(\bar{r})\, dA_\eta = -\int_{\mathfrak{M}} (J_\sigma^{(0)}(r))^- J_\varrho^{(\mu-1)}(r)\, dA_\eta, \tag{6.15.20}$$
$$\mu = 1, 2, \cdots,$$

and

$$\int_{\mathfrak{M}} J_\varrho^{(0)}(r) Z'_\sigma(\bar{r})\, dA_\eta = \Pi_{\varrho\sigma} - \int_{\mathfrak{R}-\mathfrak{M}} \mathcal{Y}'_\varrho(\bar{r})\mathcal{Y}'_\sigma(r)\, dA_\eta = H_{\varrho\sigma}. \tag{6.15.20$'$}$$

Consequently, using (4.3.10), we find the period relation:

$$(6.15.21)\quad \Gamma_{\varrho\sigma} = \sum_{\nu=0}^{\infty} \left\{ H_{\varrho\sigma} - \int_{\mathfrak{M}} (J_{\sigma}^{(0)}(r))^{-} \sum_{\mu=1}^{\nu} \binom{\nu}{\mu} J_{\varrho}^{(\mu-1)}(r)\, dA_{\eta} \right\}.$$

Thus, we have obtained series developments for the differentials Z'_{ϱ} and for their periods in terms of differentials on $\mathfrak{R}$ and of their periods.

The above formulas were obtained by using the integral equation involving T. Let us now try to carry through a similar program for the integral equation involving $\tilde{T}$.

By (6.8.15) and (6.8.17) we have

$$(6.15.22)\qquad l_{\mathbf{F}}(p, \tilde{q}) = - \sum_{\nu=1}^{\infty} \left(1 - \frac{1}{\varrho_{\nu}}\right) \psi'_{\nu}(p)\, (\psi'_{\nu}(q))^{-}.$$

This formula shows that the eigen-values of the integral equation

$$(6.15.23)\qquad \psi'_{\nu}(q) = - \varrho_{\nu}^{*} \int_{\mathfrak{M}} l_{\mathbf{F}}(q, \tilde{p})\, \psi'_{\nu}(p)\, dA_{z}$$

are

$$(6.15.24)\qquad \varrho_{\nu}^{*} = \frac{1}{1 - \dfrac{1}{\varrho_{\nu}}}, \quad \nu = 1, 2, \cdots.$$

Since there are no eigen-values $\varrho = \infty$, there are no eigen-values $\varrho^{*} = 1$. However, as ν tends to infinity, ϱ_{ν}^{*} tends to unity.

From the point of view of Hilbert space theory alone, let us try to calculate $L_{\mathbf{F}}(p, \tilde{q})$ from $\mathscr{L}_{\mathbf{F}}(p, \tilde{q})$. By (6.8.15),

$$(6.15.25)\qquad \mathscr{L}_{\mathbf{F}}^{(\sigma)}(p, \tilde{q}) = \sum_{\nu=1}^{\infty} \left(-\frac{1}{\varrho_{\nu}}\right)^{\sigma} \psi'_{\nu}(p) (\psi'_{\nu}(q))^{-},$$

where

$$(6.15.26)\qquad \mathscr{L}_{\mathbf{F}}^{(\sigma)}(p, \tilde{q}) = \int_{\mathfrak{M}} \mathscr{L}_{\mathbf{F}}^{(\sigma-1)}(r, \tilde{q}) (\mathscr{L}_{\mathbf{F}}(r, \tilde{p}))^{-} dA_{\eta}.$$

for $\sigma > 1$. We have

$$(6.15.27)\qquad \sum_{\sigma=1}^{N} A_{\sigma} \mathscr{L}_{\mathbf{F}}^{(\sigma)}(p, \tilde{q}) = \sum_{\nu=1}^{\infty} \left\{ \sum_{\sigma=1}^{N} A_{\sigma} \left(-\frac{1}{\varrho_{\nu}}\right)^{\sigma} \right\} \psi'_{\nu}(p) (\psi'_{\nu}(q))^{-}.$$

Let $p_N(x)$ be the polynomial with real coefficients A_σ:

$$(6.15.28)\qquad p_N(x) = \sum_{\sigma=1}^{N} A_\sigma x^\sigma.$$

In the interval $-1 \leqq x \leqq -\delta$, $\delta > 0$, it is possible to approximate the function $\frac{1}{x}$ uniformly by polynomials (theorem of Weierstrass). Hence it is possible to approximate the constant 1 in the interval $-1 \leqq x \leqq -\delta$ uniformly by polynomials of the form (6.15.28) which vanish at $x = 0$. Thus, given a positive integer ν_0, there is an N_0 such that for $N \geqq N_0$ it is possible to make the coefficients

$$\sum_{\sigma=1}^{N} A_\sigma \left(-\frac{1}{\varrho_\nu}\right)^\sigma, \quad 1 \leqq \nu \leqq \nu_0,$$

in the series on the right of (6.15.27) as close as we please to unity. For a fixed point q interior to $\mathfrak{M}$ we have

$$(6.15.29)\quad N_{\mathfrak{M}}\left(\sum_{\sigma=1}^{N} A_\sigma \mathscr{L}_{\mathbf{F}}^{(\sigma)}(p, \tilde{q})\right) = \sum_{\nu=1}^{\infty}\left\{\sum_{\sigma=1}^{N} A_\sigma \left(-\frac{1}{\varrho_\nu}\right)^\sigma\right\}^2 |\, \psi_\nu'(q)\,|^2,$$

where the right side can be made to tend to $-L_{\mathbf{F}}(q, \tilde{q})$ as N tends to infinity. Thus, we can represent $-L_{\mathbf{F}}(q, \tilde{q})$ in the form

$$(6.15.29)'\qquad -L_{\mathbf{F}}(q, \tilde{q}) = \lim_{N\to\infty} N_{\mathfrak{M}}\left(\sum_{\sigma=1}^{N} A_\sigma^{(N)} \mathscr{L}_{\mathbf{F}}^{(\sigma)}(p, \tilde{q})\right).$$

Using (6.15.26), we may bring this equation into the form

$$(6.15.29)''\quad -L_{\mathbf{F}}(q, \tilde{q}) = \lim_{N\to\infty}\left\{\sum_{\mu,\nu=1}^{N} A_\mu^{(N)} A_\nu^{(N)} \mathscr{L}_{\mathbf{F}}^{(\mu+\nu)}(q, \tilde{q})\right\}.$$

We have therefore proved

THEOREM 6.15.2. *The kernel $L_{\mathbf{F}}(q, \tilde{q})$ may be approximated arbitrarily closely by finite combinations of iterated kernels $\mathscr{L}_{\mathbf{F}}^{(\sigma)}(q, \tilde{q})$, uniformly in each closed subdomain of $\mathfrak{M}$.*

In these considerations we have used only the properties of the Hilbert space and we have not used all the information available concerning the imbedding of $\mathfrak{M}$ into $\mathfrak{R}$.

Let us now try to imitate the method used above in the case of the integral equation involving T. Since the eigen-values ϱ^* are not bounded away from 1, we cannot form the reciprocal kernel but we can write down its finite partial sums. Let

$$\tilde{Q}_{\mathbf{F}}^{(2)}(p, \tilde{q}) = \int_{\mathfrak{M}} l_{\mathbf{F}}(r, \tilde{q})(l_{\mathbf{F}}(r, \tilde{p}))^{-} dA_{\eta} \tag{6.15.30}$$
$$= -l_{\mathbf{F}}(p, \tilde{q}) - \int_{\mathfrak{R}-\mathfrak{M}} \mathscr{L}_{\mathbf{F}}(r, \tilde{q})(\mathscr{L}_{\mathbf{F}}(r, \tilde{p}))^{-} dA_{\eta},$$

by (5.3.7), and let

$$\tilde{I}_{\mathbf{F}}(p, \tilde{q}) = \mathscr{L}_{\mathbf{F}}(p, \tilde{q}) - \int_{\mathfrak{R}-\mathfrak{M}} \mathscr{L}_{\mathbf{F}}(r, \tilde{q})(\mathscr{L}_{\mathbf{F}}(r, \tilde{p}))^{-} dA_{\eta} \tag{6.15.31}$$
$$= L_{\mathbf{F}}(p, \tilde{q}) + \tilde{Q}_{\mathbf{F}}^{(2)}(p, \tilde{q}).$$

We have by the reproducing property of the kernel $L_{\mathbf{F}}(p, \tilde{q})$,

$$\tilde{I}_{\mathbf{F}}(p, \tilde{q}) = L_{\mathbf{F}}(p, \tilde{q}) - \int_{\mathfrak{M}} \tilde{Q}_{\mathbf{F}}^{(2)}(r, \tilde{q}) L_{\mathbf{F}}(p, \tilde{r}) dA_{\eta}. \tag{6.15.32}$$

We write

$$\tilde{Q}_{\mathbf{F}, N}^{(-1)}(p, \tilde{q}) = \sum_{\sigma=0}^{N} \tilde{Q}_{\mathbf{F}}^{(2\sigma)}(p, \tilde{q}), \tag{6.15.33}$$

where

$$\tilde{Q}_{\mathbf{F}}^{(2\sigma)}(p, \tilde{q}) = \int_{\mathfrak{M}} \tilde{Q}_{\mathbf{F}}^{(2\sigma-2)}(p, \tilde{r}) \tilde{Q}_{\mathbf{F}}(r, \tilde{q}) dA_{\eta}, \quad \sigma \geqq 2, \tag{6.15.34}$$

and

$$\tilde{Q}_{\mathbf{F}}^{(0)}(p, \tilde{q}) = -L_{\mathbf{F}}(p, \tilde{q}). \tag{6.15.35}$$

We have, by (6.15.22) and (6.15.30),

$$\tilde{Q}_{\mathbf{F}}^{(2\sigma)}(p, \tilde{q}) = \sum_{\nu=1}^{\infty} \left(1 - \frac{1}{\varrho_{\nu}}\right)^{2\sigma} \psi_{\nu}'(p)(\psi_{\nu}'(q))^{-}, \quad \sigma = 0, 1, \cdots; \tag{6.15.36}$$

hence by (6.15.33),

$$\tilde{Q}_{\mathbf{F}, N}^{(-1)}(p, \tilde{q}) = \sum_{\nu=1}^{\infty} \left[\frac{1 - \left(1 - \frac{1}{\varrho_{\nu}}\right)^{2N+2}}{1 - \left(1 - \frac{1}{\varrho_{\nu}}\right)^{2}} \right] \psi_{\nu}'(p)(\psi_{\nu}'(q))^{-}, \tag{6.15.37}$$

and

(6.15.38) $$\tilde{I}_{\mathbf{F}}(p,\tilde{q}) = -\sum_{\nu=1}^{\infty}\left\{1-\left(1-\frac{1}{\varrho_\nu}\right)^2\right\}\psi_\nu'(p)((\psi_\nu'(q))^-.$$

Hence

(6.15.39)
$$\int_{\mathfrak{M}} \tilde{Q}_{\mathbf{F},N}^{(-1)}(r,\tilde{q})\tilde{I}_{\mathbf{F}}(p,\tilde{r})dA_\eta = -\sum_{\nu=1}^{\infty}\left[1-\left(1-\frac{1}{\varrho_\nu}\right)^{2N+2}\right]\psi_\nu'(p)\,(\psi_\nu'(q))^-.$$

Thus, for fixed p, q in the interior of $\mathfrak{M}$, we have

(6.15.40) $$L_{\mathbf{F}}(p,\tilde{q}) = \lim_{N\to\infty}\int_{\mathfrak{M}} \tilde{Q}_{\mathbf{F},N}^{(-1)}(r,\tilde{q})\,\tilde{I}_{\mathbf{F}}(p,\tilde{r})dA_\eta.$$

6.16. The Combination Theorem

We conclude this chapter with some remarks concerning the case in which the domain $\mathfrak{M}$ is the intersection of two surfaces $\mathfrak{R}_1$ and $\mathfrak{R}_2$. For simplicity, we suppose that $\mathfrak{R}_1$, $\mathfrak{R}_2$ and $\mathfrak{M}$ are 2-cells. We further suppose that the boundary curves of $\mathfrak{R}_1$ and $\mathfrak{R}_2$ intersect in just two points, p_1 and p_2 say, and that $\mathfrak{M}$ is bounded by two arcs C_1 and C_2 joining p_1 and p_2 where C_1 lies in the boundary of $\mathfrak{R}_1$, C_2 in the boundary of $\mathfrak{R}_2$. Later we shall take $\mathfrak{R}_1$ and $\mathfrak{R}_2$ to be domains of the local uniformizers z_1 and z_2 respectively. Since $\mathfrak{R}_1$ and $\mathfrak{R}_2$ can be represented as domains of the z_1- and z_2-planes, their bilinear differentials $\mathscr{L}_1(p,\tilde{q})$ and $\mathscr{L}_2(p,\tilde{q})$ may be assumed known and we give a process by which the bilinear differential $\mathscr{L}(p,\tilde{q})$ of the union $\mathfrak{R}_1 \cup \mathfrak{R}_2$ can be determined. This process is a variant of the Schwarz alternating procedure, and yields a proof of the existence of analytic functions on an abstract Riemann surface. Although this proof is not simpler than the standard ones, we give it for the reason that it makes our approach self-contained; even the existence theorems of Chapter 2 can then be based on the methods developed here. Furthermore, the method which we now develop is a natural outgrowth of our previous investigations.

It should be remembered that the boundary uniformizers of $\mathfrak{M}$ at p_1 and p_2 are not admissible uniformizers for $\mathfrak{R}_1$ and $\mathfrak{R}_2$. The intersection angles of the arcs C_1, C_2 at p_1 and p_2 do not exceed π and we now make the assumption that each of these angles is greater than zero. Let z_1 be a uniformizer of $\mathfrak{R}_1$ at the point p_1 and let α be the angle at p_1 between C_1 and C_2 in the z_1-plane. Then $z = z_1^{\frac{\pi}{\alpha}}$ will be a boundary uniformizer of $\mathfrak{M}$ near p_1. A differential of $\mathfrak{R}_1$ will be multiplied by a factor $\alpha z^{\frac{\alpha}{\pi}-1}/\pi$ if expressed in terms of the boundary uniformizer of $\mathfrak{M}$. But since $0 < \alpha/\pi \leqq 1$, every differential of $\mathfrak{R}_1$ which is regular in the closure of $\mathfrak{R}_1$ will still lead to a differential on $\mathfrak{M}$ which is square integrable and which possesses a bounded integral function in $\mathfrak{M}$. This fact will permit us to carry out all the following considerations in spite of the fact that some differentials considered become infinite near p_1 or p_2.

Since $\mathfrak{M}$, $\mathfrak{R}_1$ and $\mathfrak{R}_2$ are simply-connected, there exists on each surface only the class **S**, so the **F**-subscript is superfluous since all classes are the same. In particular, **M** = **S** so all kernels are symmetric. Let $L(p, q)$ be the bilinear differential of $\mathfrak{M}$ and set

$$(6.16.1)\quad l_1(p, q) = L(p, q) - \mathscr{L}_1(p, q);\ l_2(p, q) = L(p, q) - \mathscr{L}_2(p, q).$$

Similarly, we define $l_1(p, \tilde{q})$ and $l_2(p, \tilde{q})$.

We begin by proving the formula

$$(6.16.2)\quad \int_{\mathfrak{M}} l_1(p, q)(l_2(p, r))^- dA_z = \int_{\mathfrak{M}} (l_1(p, \tilde{q}))^- l_2(p, \tilde{r}) dA_z,$$

valid for q and r in $\mathfrak{M}$. We have

$$\int_{\mathfrak{M}} l_1(p, q)(l_2(p, r))^- dA_z = \int_{\mathfrak{M}} [L(p, q) - \mathscr{L}_1(p, q)](l_2(p, r))^- dA_z$$
$$= -\int_{\mathfrak{M}} \mathscr{L}_1(p, q)(l_2(p, r))^- dA_z$$

by (4.10.8). Let $\mathfrak{k}$ correspond to the uniformizer circle $|\zeta| < a$ at q. Integrating by parts, we obtain

$$-\int_{\mathfrak{M}} \mathscr{L}_1(p, q)(l_2(p, r))^- dA_z = -\frac{1}{\pi i}\int_{C_1+C_2-\partial\mathfrak{k}} \frac{\partial \mathscr{G}_1(p, q)}{\partial q}(l_2(p, r))^- d\bar{z} + o(1)$$
$$= -\frac{1}{\pi i}\int_{C_2} \frac{\partial \mathscr{G}_1(p, q)}{\partial q}(l_2(p, r))^- d\bar{z}$$

since $\partial \mathscr{G}_1(p, q)/\partial q = 0$ on C_1 and the integral over $\partial \mathfrak{k}$ is $o(1)$ as $a \to 0$;

$$= -\frac{1}{\pi i} \int_{C_2} \frac{\partial \mathscr{G}_1(p, q)}{\partial q} (l_2(\bar{p}, r))^- dz$$

since $p = \bar{p}$ on the boundary;

$$= -\frac{1}{\pi i} \int_{C_2} \frac{\partial \mathscr{G}_1(p, q)}{\partial q} l_2(p, \bar{r}) dz$$

$$= -\frac{1}{\pi i} \int_{C_1 + C_2 - \partial \mathfrak{k}} \frac{\partial \mathscr{G}_1(p, q)}{\partial q} l_2(p, \bar{r}) dz - \frac{1}{\pi i} \int_{\partial \mathfrak{k}} \frac{\partial \mathscr{G}_1(p, q)}{\partial q} l_2(p, \bar{r}) dz.$$

As $a \to 0$, the second integral yields the value

$$-l_2(q, \bar{r}) = \int_{\mathfrak{M}} (L(p, \bar{q}))^- l_2(p, \bar{r}) dA_z \tag{6.16.3}$$

while the first becomes

$$-\int_{\mathfrak{M}} \mathscr{L}_1(q, \bar{p}) l_2(p, \bar{r}) dA_z \tag{6.16.4}$$

using (4.10.2). Combining these results, we obtain (6.16.2).

We introduce now the bilinear differential

$$Q(p, \bar{q}) = \int_{\mathfrak{M}} l_1(r, p)(l_2(r, q))^- dA_\eta \tag{6.16.5}$$

and the two operators

$$T_\psi^{(1)}(q) = (\mathscr{L}_1(q, p), \psi'(p))_{\mathfrak{M}}, \tag{6.16.6}$$

$$T_\psi^{(2)}(q) = (\mathscr{L}_2(q, p), \psi'(p))_{\mathfrak{M}}, \tag{6.16.7}$$

which act on differentials ψ' on $\mathfrak{M}$. We also define the operators

$$t_\psi^{(1)}(q) = -(l_1(q, p), \psi'(p))_{\mathfrak{M}}, \tag{6.16.6)'}$$

$$t_\psi^{(2)}(q) = -(l_2(q, p), \psi'(p))_{\mathfrak{M}}. \tag{6.16.7)'}$$

For q in $\mathfrak{M}$ these give the same result as the operators $T_\psi^{(1)}$ and $T_\psi^{(2)}$. Finally we define the iterated operator

$$t_\psi^{(12)}(q) = t^{(1)}_{t_\psi^{(2)}}(q). \tag{6.16.8}$$

By (6.16.5), this may be expressed in the form

(6.16.8)′ $$t_\psi^{(12)}(q) = (Q(q, \tilde{p}), (\psi'(p))^-)_{\mathfrak{M}}.$$

We now prove

THEOREM 6.16.1. *There exists a positive number* $\alpha < 1$ *such that for all differentials* $\psi'(p)$ *on* $\mathfrak{M}$ *we have the inequality*

(6.16.9) $$N_{\mathfrak{M}}(t_\psi^{(12)}) \leqq \alpha N_{\mathfrak{M}}(\psi').$$

In order to prove the theorem we start from the inequality

(6.16.10) $$N_{\mathfrak{M}}(t_\psi^{(12)}) \leqq N_{\mathfrak{M}}(t_\psi^{(2)}) \leqq N_{\mathfrak{M}}(\psi')$$

which is an immediate consequence of (6.2.9) and (6.2.11). If we can show that

$$N_{\mathfrak{M}}(t_\psi^{(12)}) < N_{\mathfrak{M}}(\psi')$$

unless $\psi'(q) \equiv 0$ in $\mathfrak{M}$, then the reasoning of Section 6.6 shows that there is a number α, $0 < \alpha < 1$, such that

$$N_{\mathfrak{M}}(t_\psi^{(12)}) \leqq \alpha$$

for all differentials ψ' satisfying

$$N_{\mathfrak{M}}(\psi') = 1.$$

The reasoning can be carried through because $Q(q, \tilde{p})$ is regular in the closure of $\mathfrak{M}$, except for singularities of known character at the points p_1 and p_2.

It remains to investigate the possibility that

(6.16.10)′ $$N_{\mathfrak{M}}(t_\psi^{(12)}) = N_{\mathfrak{M}}(\psi').$$

In this case we must also have

(6.16.11) $$N_{\mathfrak{M}}(t_\psi^{(2)}) = N_{\mathfrak{M}}(\psi').$$

But by (6.2.11)

$$N_{\mathfrak{R}_2}(T_\psi^{(2)}) + N_{\mathfrak{R}_2}(\tilde{T}_\psi^{(2)}) = N_{\mathfrak{M}}(\psi').$$

Hence (6.16.11) implies the two identities:

(6.16.12) $$\tilde{T}_\psi^{(2)} \equiv 0 \text{ in } \mathfrak{R}_2$$

and

(6.16.13) $$T_\psi^{(2)} \equiv 0 \text{ in } \mathfrak{R}_2 - \mathfrak{M}.$$

From these identities we derive the fact that a ifferential ψ' for which (6.16.10)′ holds is regular in the closure of $\mathfrak{M}$, except possibly at the two critical points p_1 and p_2. In fact, by (6.3.9), we have

$$(T^{(2)})^2 = I + \tilde{T}^{(2)}. \tag{6.16.14}$$

By (6.16.13),

$$(T^{(2)})^2 = (t^{(2)})^2 \text{ in } \mathfrak{M}$$

and by (6.16.12) and (6.16.14),

$$(t^{(2)})^2 = \psi'(q).$$

Since the kernel $Q(p, \tilde{q})$ of the $t^{(2)}$-transformation is regular in the closure of $\mathfrak{M}$ (except for the two possible singularities at p_1 and p_2) we have proved the asserted regularity of $\psi'(q)$.

By reasoning used in Section 6.5 we see that $t_\psi^{(2)}(q)$ has the boundary values $(\psi'(q))^-$ on C_1. To show that $t_\psi^{(2)}(q)$ has also the boundary value $(\psi'(q))^-$ on C_2, a modified type of reasoning is necessary. We have

$$\begin{aligned} t_\psi^{(2)}(q) &= -\int_{\mathfrak{M}} l_2(q, p)(\psi'(p))^- dA_z \\ &= -\frac{2}{\pi}\int_{\mathfrak{M}} \frac{\partial^2[\mathcal{G}_2(q, p) - G(q, p)]}{\partial p\, \partial q} (\psi'(p))^- dA_z \\ &= \frac{1}{\pi i}\int_{C_1} \frac{\partial \mathcal{G}_2(q, p)}{\partial q} (\psi'(p))^- d\bar{z}, \end{aligned} \tag{6.16.15}$$

since $\dfrac{\partial \mathcal{G}_2(q, p)}{\partial q}$ vanishes on C_2 and $\dfrac{\partial G(q, p)}{\partial q}$ vanishes on C_1 and on C_2. If q is an interior point of the arc C_2, we have

$$\begin{aligned} \frac{1}{\pi i}\int_{C_1} \frac{\partial \mathcal{G}_2(q, p)}{\partial q} (\psi'(p))^- d\bar{z} &= \frac{1}{\pi i}\int_{C_1} \frac{\partial \mathcal{G}_2(\tilde{q}, p)}{\partial \tilde{q}} (\psi'(p))^- d\bar{z} \\ &= \left(\frac{1}{\pi i}\int_{C_1} \frac{\partial \mathcal{G}_2(q, p)}{\partial q} \psi'(p)\, dz\right)^-. \end{aligned} \tag{6.16.16}$$

Let q^* be a point of $\mathfrak{M}$ which is near an interior point of C_2. Then

$$\frac{1}{\pi i}\int_{C_1}\frac{\partial\mathscr{G}_2(q^*,p)}{\partial q^*}\psi'(p)dz = \frac{1}{\pi i}\int_{C_1+C_2}\left[\frac{\partial\mathscr{G}_2(q^*,p)}{\partial q^*} - \frac{\partial G(q^*,p)}{\partial q^*}\right]\psi'(p)\,dz$$

$$(6.16.17)\qquad = -\int_{\mathfrak{M}} l_2(q^*,\tilde{p})\psi'(p)\,dA_z$$

$$= \psi'(q^*) + \tilde{T}^{(2)}_{\psi}(q^*) = \psi'(q^*)$$

by (6.16.12). Letting q^* tend to an interior point q of C_2, we see from (6.16.15) and (6.16.16) that $t^{(2)}_{\psi}(q)$ has the boundary values $(\psi'(q))^-$ on C_2. We conclude that $t^{(2)}_{\psi}(q) + \psi'(q)$ and $(t^{(2)}_{\psi}(q) - \psi'(q))/i$ are real on the boundary of $\mathfrak{M}$. Both expressions have finite integrals in $\mathfrak{M}$ with constant imaginary part on the boundary of $\mathfrak{M}$ (even at p_1 and p_2). Applying the maximum principle to these imaginary parts, we see that they are constant and this shows that the differentials corresponding to them vanish identically in $\mathfrak{M}$. Thus we have shown that (6.16.10)′ implies $\psi'(q) \equiv 0$ in $\mathfrak{M}$ and Theorem 6.16.1 is proved.

We observe that whenever a differential φ' can be expressed in the form

$$(6.16.18)\qquad \varphi'(q) = \int_{\mathfrak{M}} Q(q,\tilde{p})\psi'(p)dA_z,$$

then Theorem 6.16.1 implies

$$(6.16.19)\qquad N_{\mathfrak{M}}(\varphi') \leqq \alpha N_{\mathfrak{M}}(\psi').$$

Now let r be a fixed point in the interior of $\mathfrak{R}_2 - \mathfrak{M}$, and define

$$(6.16.20)\qquad \psi_0'(p) = \begin{cases}\mathscr{L}_2(p,r), & p\in\mathfrak{R}_2,\\ 0\quad, & p\in\mathfrak{R}_1-\mathfrak{M}.\end{cases}$$

For $n \geqq 1$, set

$$(6.16.21)\qquad \psi_{2n}'(p) = \begin{cases}\psi_{2n-1}'(p), & p\in\mathfrak{R}_1-\mathfrak{M},\\ -\int_{\mathfrak{R}_2}\mathscr{L}_2(p,\tilde{q})\psi_{2n-1}'(q)\,dA_\zeta + \mathscr{L}_2(p,r), & p\in\mathfrak{R}_2,\end{cases}$$

and

$$(6.16.22)\qquad \psi_{2n-1}'(p) = \begin{cases}-\int_{\mathfrak{R}_1}\mathscr{L}_1(p,\tilde{q})\psi_{2n-2}'(q)\,dA_\zeta, & p\in\mathfrak{R}_1,\\ \psi_{2n-2}'(p), & p\in\mathfrak{R}_2-\mathfrak{M}.\end{cases}$$

Observe that all $\psi'_\nu(p)$ have at r the same singularity as $\mathscr{L}_2(p, r)$.

Let p be a point of $\mathfrak{R}_2$. Then

$$\psi'_{2n}(p) = -\int\limits_{\mathfrak{R}_2-\mathfrak{M}} \mathscr{L}_2(p, \bar{q})\psi'_{2n-2}(q)\, dA_\zeta$$

$$- \int\limits_{\mathfrak{M}} \mathscr{L}_2(p, \bar{q})\psi'_{2n-1}(q)\, dA_\zeta + \mathscr{L}_2(p, r)$$

$$= -\int\limits_{\mathfrak{R}_2} \mathscr{L}_2(p, \bar{q})\psi'_{2n-2}(q)\, dA_\zeta + \int\limits_{\mathfrak{M}} \mathscr{L}_2(p, \bar{q})\psi'_{2n-2}(q)\, dA_\zeta$$

$$- \int\limits_{\mathfrak{M}} \mathscr{L}_2(p, \bar{q})\psi'_{2n-1}(q)\, dA_\zeta + \mathscr{L}_2(p, r)$$

$$= -\int\limits_{\mathfrak{R}_2} \mathscr{L}_2(p, \bar{q})[\psi'_{2n-2}(q) - \mathscr{L}_2(q, r)]\, dA_\zeta$$

$$+ \int\limits_{\mathfrak{M}} \mathscr{L}_2(p, \bar{q})\psi'_{2n-2}(q)\, dA_\zeta - \int\limits_{\mathfrak{M}} \mathscr{L}_2(p, \bar{q})\psi'_{2n-1}(q)\, dA_\zeta + \mathscr{L}_2(p, r),$$

by (4.10.8),

$$= \psi'_{2n-2}(p) - \mathscr{L}_2(p, r) + \int\limits_{\mathfrak{M}} \mathscr{L}_2(p, \bar{q})\psi'_{2n-2}(q)\, dA_\zeta$$

$$- \int\limits_{\mathfrak{M}} \mathscr{L}_2(p, \bar{q})\psi'_{2n-1}(q)\, dA_\zeta + \mathscr{L}_2(p, r)$$

by (4.10.8)′;

$$= \psi'_{2n-2}(p) - \int\limits_{\mathfrak{M}} \mathscr{L}_2(p, \bar{q})[\psi'_{2n-1}(q) - \psi'_{2n-2}(q)]\, dA_\zeta.$$

Thus, for $p \in \mathfrak{R}_2$,

$$\psi'_{2n}(p) - \psi'_{2n-2}(p) = -\int\limits_{\mathfrak{M}} \mathscr{L}_2(p, \bar{q})[\psi'_{2n-1}(q) - \psi'_{2n-2}(q)]\, dA_\zeta. \tag{6.16.23}$$

Now let p be a point of $\mathfrak{R}_1$. Then

$$\psi'_{2n-1}(p) = -\int_{\mathfrak{R}_1-\mathfrak{M}} \mathscr{L}_1(p, \tilde{q})\psi'_{2n-3}(q)\, dA_\zeta - \int_{\mathfrak{M}} \mathscr{L}_1(p, \tilde{q})\psi'_{2n-2}(q)\, dA_\zeta$$

$$= -\int_{\mathfrak{R}_1} \mathscr{L}_1(p, \tilde{q})\psi'_{2n-3}(q)\, dA_\zeta + \int_{\mathfrak{M}} \mathscr{L}_1(p, \tilde{q})\psi'_{2n-3}(q)\, dA_\zeta$$

$$- \int_{\mathfrak{M}} \mathscr{L}_1(p, \tilde{q})\psi'_{2n-2}(q)\, dA_\zeta$$

$$= \psi'_{2n-3}(p) - \int_{\mathfrak{M}} \mathscr{L}_1(p, \tilde{q})[\psi'_{2n-2}(q) - \psi'_{2n-3}(q)]\, dA_\zeta.$$

Thus, for $p \in \mathfrak{R}_1$,

$$\psi'_{2n-1}(p) - \psi'_{2n-3}(p) = -\int_{\mathfrak{M}} \mathscr{L}_1(p, \tilde{q})[\psi'_{2n-2}(q) - \psi'_{2n-3}(q)]\, dA_\zeta. \tag{6.16.24}$$

We observe that the formula (6.16.24) is valid in the case $n = 1$ if we define $\psi'_{-1}(p) \equiv 0$.

Write

$$d'_m(p) = \psi'_m(p) - \psi'_{m-1}(p), \quad m = 1, 2, \cdots. \tag{6.16.25}$$

For $p \in \mathfrak{M}$ we have

$$-\psi'_{2n-1}(p) + \psi'_{2n-2}(p) = \int_{\mathfrak{M}} L(p, \tilde{q})[\psi'_{2n-1}(q) - \psi'_{2n-2}(q)]\, dA_\zeta,$$

so (6.16.23) and (6.16.24) become, for $p \in \mathfrak{M}$,

$$\begin{cases} d'_{2n}(p) = \int_{\mathfrak{M}} l_2(p, \tilde{q}) d'_{2n-1}(q)\, dA_\zeta, \\ d'_{2n-1}(p) = \int_{\mathfrak{M}} l_1(p, \tilde{q}) d'_{2n-2}(q)\, dA_\zeta. \end{cases} \tag{6.16.26}$$

Hence, by (6.16.2) and (6.16.5),

$$\begin{cases} d'_{2n}(p) = \int_{\mathfrak{M}} (Q(q, \bar{p}))^- d'_{2n-2}(q)\, dA_\zeta, \\ d'_{2n-1}(p) = \int_{\mathfrak{M}} Q(p, \tilde{q}) d'_{2n-3}(q)\, dA_\zeta. \end{cases} \tag{6.16.27}$$

But then by (6.16.18) and (6.16.19),

$$N_{\mathfrak{M}}(d'_m) \leqq \alpha N_{\mathfrak{M}}(d'_{m-2}). \tag{6.16.28}$$

Therefore

$$\begin{cases} N_{\mathfrak{M}}(d'_{2m}) \leqq \alpha^{m-1} N_{\mathfrak{M}}(d'_2), \\ N_{\mathfrak{M}}(d'_{2m-1}) \leqq \alpha^{m-1} N_{\mathfrak{M}}(d'_1). \end{cases} \tag{6.16.29}$$

Thus

$$N_{\mathfrak{M}}(d'_m) \leqq A\alpha^{\frac{m}{2}}, \tag{6.16.30}$$

where A is a number independent of m. If $m < n$ we have

$$\{N_{\mathfrak{M}}(\psi'_m - \psi'_n)\}^{\frac{1}{2}} \leqq \sum_{\nu=m+1}^{n} \{N_{\mathfrak{M}}(d'_\nu)\}^{\frac{1}{2}} \leqq A^{\frac{1}{2}} \sum_{\nu=m+1}^{n} \alpha^{\frac{\nu}{4}} \leqq A^{\frac{1}{2}} \frac{\alpha^{\frac{m+1}{4}}}{1-\alpha^{\frac{1}{4}}},$$

so $N_{\mathfrak{M}}(\psi'_m - \psi'_n)$ tends to zero as m, n approach infinity. It follows from (6.16.23) that $\psi'_{2m}(p)$ converges to a limit $\Psi'_2(p)$ in $\mathfrak{R}_2$ and from (6.16.24) that $\psi'_{2m-1}(p)$ converges to a limit $\Psi'_1(p)$ in $\mathfrak{R}_1$. Since these sequences have a common limit $\Psi'(p)$ in $\mathfrak{M}$, we conclude that $\Psi'_1(p) \equiv \Psi'_2(p)$ in $\mathfrak{M}$, and $\Psi'_2(p)$ is the analytic continuation of $\Psi'_1(p)$ into $\mathfrak{R}_2 - \mathfrak{M}$. We have therefore established the existence of an analytic function $\Psi(p)$ over the union of $\mathfrak{R}_1$ and $\mathfrak{R}_2$. Since $\Psi(p)$ has a simple pole at the point r of $\mathfrak{R}_2 - \mathfrak{M}$, it cannot be constant over $\mathfrak{R}_1 \cup \mathfrak{R}_2$.

Let φ be any function regular in the union of $\mathfrak{R}_1$ and $\mathfrak{R}_2$ with

$$N(\varphi') < \infty. \tag{6.16.31}$$

Here the integration in $N(\varphi')$ is over the union of $\mathfrak{R}_1$ and $\mathfrak{R}_2$. Then by (4.10.8)

$$(\psi'_0, \varphi') = (\psi'_0, \varphi')_{\mathfrak{R}_2} = 0.$$

Further,

$$(\psi'_{2n}, \varphi') = (\psi'_{2n-1}, \varphi')_{\mathfrak{R}_1 - \mathfrak{M}} - \int_{\mathfrak{R}_2}\int_{\mathfrak{R}_2} \mathscr{L}_2(p, \bar{q})(\varphi'(p))^- \psi'_{2n-1}(q)\, dA_z\, dA_\zeta$$

$$+ \int_{\mathfrak{R}_2} \mathscr{L}_2(p, r)(\varphi'(p))^- dA_z$$

$$= (\psi'_{2n-1}, \varphi')_{\mathfrak{R}_1 - \mathfrak{M}} + (\psi'_{2n-1}, \varphi')_{\mathfrak{R}_2} = (\psi'_{2n-1}, \varphi'),$$

by (4.10.8) and (4.10.8)'. Similarly

$$(\psi'_{2n-1}, \varphi') = (\psi'_{2n-2}, \varphi').$$

Thus

$$(6.16.32) \qquad (\psi'_m, \varphi') - (\psi'_0, \varphi') = 0, \quad m = 1, 2, \cdots.$$

It follows that also for the limit function Ψ

$$(6.16.33) \qquad (\Psi', \varphi') = 0$$

for every φ' regular in the union of $\mathfrak{R}_1$ and $\mathfrak{R}_2$ which satisfies (6.16.31). Let $\mathscr{L}(p, r)$ be the bilinear differential of the union of $\mathfrak{R}_1$ and $\mathfrak{R}_2$. Then

$$(6.16.33)' \qquad (\mathscr{L}, \varphi') = 0.$$

Subtracting (6.16.33)' from (6.16.33), we obtain

$$(6.16.34) \qquad (\Psi' - \mathscr{L}, \varphi') = 0.$$

Taking, in particular, $\varphi' = \Psi' - \mathscr{L}$, we see that $\Psi' = \mathscr{L}$; that is,

$$(6.16.35) \qquad \Psi'(p) = \mathscr{L}(p, r).$$

We have thus proved the theorem:

THEOREM 6.16.2. *The recursive procedure* (6.16.20)—(6.16.22) *converges uniformly in each closed subdomain of the union* $\mathfrak{R}_1 \cup \mathfrak{R}_2$ *and leads to a representation of the* $\mathscr{L}$*-function of this union in terms of the* $\mathscr{L}$*-functions of the constituent domains* $\mathfrak{R}_1$ *and* $\mathfrak{R}_2$.

Since the union of $\mathfrak{R}_1$ and $\mathfrak{R}_2$ is simply-connected, the function belonging to the differential

$$\Psi' = \mathscr{L}(p, r)$$

maps the union onto the exterior of a circle, the point r going into infinity. We have therefore proved the "combination theorem" of H. A. Schwarz and C. Neumann: if two domains, each of which can be conformally mapped onto a circle, have a connected (and therefore a simply-connected) intersection, the union of the two domains can also be mapped onto a circle. Since the proof of the uniformization theorem essentially turns on the combination theorem (see [4]), we have established the uniformization theorem by the methods developed in this chapter.

REFERENCES.

1. S. BERGMAN and M. SCHIFFER, "Kernel functions and conformal mapping," *Compositio Math.*, 8 (1951), 205—249.

2. R. COURANT, *Dirichlet's Principle, conformal mapping, and minimal surfaces*, Interscience, New York, 1950.

3. R. COURANT and D. HILBERT, *Methoden der mathematischen Physik*, Vol. I, Springer, Berlin, 1931. (Reprint, Interscience, New York, 1943).

4. L. SCHLESINGER, *Automorphe Funktionen*, Göschens Lehrbücherei, Vol. 5, De Gruyter, Berlin, 1924.

7. Variations of Surfaces and of their Functionals

7.1. Boundary Variations

The theory of Riemann surfaces is mainly concerned with the study of the functionals on a given surface or on a given class of surfaces, and the dependence of the functionals on the surface itself in the sense of the functional calculus has never been investigated systematically. In this chapter we investigate the way in which the functionals change if the surface is altered.

So far as topological properties are concerned, every finite surface may be obtained from the sphere by performing, a finite number of times, one or more of the following three types of operations: (i) cutting out holes; (ii) attaching handles; (iii) attaching cross-caps. However, the conformal type of a Riemann surface may be altered by the topologically insignificant operation of "attaching a cell". To attach a cell, we first form a hole by removing a 2-cell from the surface, then we fill the hole with a new 2-cell which is attached by identifying its boundary with the boundary of the hole. We therefore add a fourth operation: (iv) interior deformation by cutting out a hole and attaching a cell. We shall compute the variation of the Green's function under these four operations. By differentiation we then obtain the variation of the bilinear differential $L_{\mathrm{M}}(p, q)$ and, by computing its periods around the basis cycles, we obtain the variation of the differentials of the first kind.

The operations (i), (ii), and (iii) will be performed in a special way, since their main function is to change the topology, and the resulting variations therefore will involve a minimal number of free parameters. The operation (iv), on the other hand, will supply the missing parameters and will provide the necessary generality for the variations. In particular, the operation (iv) may be performed in such a way that the conformal type of the surface is preserved. The fact that conformal type can be preserved has an important application in constructing variations of conformal maps of a given surface $\mathfrak{N}$ onto subdomains of a given surface $\mathfrak{R}$. In Chapter 5 we derived necessary and sufficient conditions for such a mapping; in the present chapter we give the variational procedure by means of which a given mapping from $\mathfrak{N}$ onto a subdomain $\mathfrak{M}$ of $\mathfrak{R}$ can be varied to produce a mapping from $\mathfrak{N}$ onto a neighboring subdomain $\mathfrak{M}^*$ of $\mathfrak{R}$. This provides a powerful tool in the investigation of extremal problems relating to one-one mappings of $\mathfrak{N}$ into $\mathfrak{R}$.

Every orientable finite Riemann surface with boundary is properly imbedded in its double, and may therefore be varied by shifting the position of its boundary on the double, the double being held fixed. This type of variation is called a "boundary variation", and is historically the oldest variation in conformal mapping. We therefore begin by discussing boundary variations, and we show that such variations arise automatically from the considerations of Chapters 5 and 6 by taking the domain $\mathfrak{M}$ to be such a large part of the domain $\mathfrak{R}$ in which it is imbedded that the residual domain $\mathfrak{R}-\mathfrak{M}$ is a thin strip.

In the case of plane domains bounded by analytic curves the variation of the Green's function can be expressed by Hadamard's classical formula [3]. Let $\mathfrak{M}$ be a domain of the plane bounded by analytic curves C_ν, $\nu = 1, 2, \cdots, n$, and let every point of the boundary be expressed by a parameter s which measures the lengths of the curves successively. A neighboring domain $\mathfrak{M}^*$ may be defined by displacing each boundary point of $\mathfrak{M}$ along the normal, thus defining new curves C_ν^*, $\nu = 1, 2, \cdots, n$, which bound $\mathfrak{M}^*$. Let $\delta n(s) = \varepsilon\nu(s)$ be a continuously differentiable function of s which determines the normal displacement of the boundary point $r = r(s)$ of the domain $\mathfrak{M}$. We suppose that $\delta n(s)$ is positive if the displacement is in the direction of the inner normal with respect to $\mathfrak{M}$. If $G(p, q)$ is the Green's function of $\mathfrak{M}$, $G^*(p, q)$ that of $\mathfrak{M}^*$, Hadamard's formula states that

$$(7.1.1)\quad G^*(p, q)=G(p, q)-\frac{\varepsilon}{2\pi}\int_C \frac{\partial G(r, p)}{\partial n}\frac{\partial G(r, q)}{\partial n}\nu(s)ds+o(\varepsilon).$$

Using the notation of the functional calculus, we may write (7.1.1) in the form

$$\delta G(p, q) = -\frac{1}{2\pi} \int_C \frac{\partial G(r, p)}{\partial n} \frac{\partial G(r, q)}{\partial n} \delta n ds. \tag{7.1.2}$$

If $\mathfrak{M}$ is an abstractly given Riemann surface with m boundary components C_ν, $\nu = 1, 2, \cdots, m$, we define a surface $\mathfrak{M}^*$ in the following manner. Let

$$G(p, q) + iH(p, q) \tag{7.1.3}$$

be the analytic completion of the Green's function $G(p, q)$ of $\mathfrak{M}$. If h is the genus of $\mathfrak{M}$, we have

$$\operatorname{Im} Z_{2h+\nu}(q) = \frac{1}{2\pi} \int_{C_\nu} \frac{\partial G(p, q)}{\partial n} ds = -\frac{1}{2\pi} \int_{C_\nu} dH(p, q) > 0. \tag{7.1.4}$$

For a fixed q_0 in the interior of $\mathfrak{M}$, the function

$$\zeta = \exp\left\{ -\frac{G(p, q_0) + iH(p, q_0)}{\operatorname{Im} Z_{2h+\nu}(q_0)} \right\} \tag{7.1.5}$$

is single-valued on C_ν and maps it onto the circumference $|\zeta| = 1$ in the plane of ζ, and it may therefore be regarded as a boundary uniformizer in the large. Each boundary component C_ν of $\mathfrak{M}$ is mapped onto a unit circumference in this manner, and we define the arc-length parameter s, $0 \leqq s \leqq 2\pi m$, in terms of these circumferences. A normal displacement $\delta n(s)$ in the planes of these circumferences may be defined as before and, as s runs over $|\zeta_\nu| = 1$, $\delta n(s)$ determines a neighboring curve γ_ν.

Let η be a small positive number, and add to $\mathfrak{M}$ the set of points corresponding to the m annular neighborhoods $1 < |\zeta_\nu| < 1 + \eta$. In this way we define a finite Riemann surface $\mathfrak{R}$ containing $\mathfrak{M}$ in its interior. It is clear that $\mathfrak{R}$ satisfies all the conditions imposed on the finite Riemann surfaces heretofore considered, and we call $\mathfrak{R}$ an enlargement of $\mathfrak{M}$. The curves γ_ν, $\nu = 1, 2, \cdots, m$, bound a subdomain $\mathfrak{M}^*$ of $\mathfrak{R}$ with Green's function $G^*(p, q)$, and Hadamard's formula is valid.

Since formula (7.1.2) has a form which is conformally invariant, we may express it in terms of any local boundary uniformizers $z = z(r)$, $z = x + iy$, and it becomes

$$(7.1.6) \qquad \delta G(p, q) = -\frac{1}{2\pi} \int_C \frac{\partial G(r, p)}{\partial y} \frac{\partial G(r, q)}{\partial y} \delta y dx.$$

The proof of formula (7.1.6) in the general case does not differ essentially from that for plane domains.

Since $G(p, q)$ does not change its value along C, we have

$$\frac{\partial G(r, p)}{\partial x} = 0, \quad \frac{\partial G(r, p)}{\partial z} = -\frac{\partial G(r, p)}{\partial \bar{z}} = -\frac{i}{2} \frac{\partial G(r, p)}{\partial y}.$$

Hence, we may put (7.1.6) into the equivalent form

$$(7.1.6)' \qquad \delta G(p, q) = -\frac{2}{\pi} \int_C \frac{\partial G(r, p)}{\partial z} \frac{\partial G(r, q)}{\partial \bar{z}} \delta y dx$$

or, invariantly,

$$(7.1.6)'' \qquad \delta G(p, q) = -\frac{2}{\pi} \int_C \frac{\partial G(r, p)}{\partial r} \frac{\partial G(r, q)}{\partial \bar{r}} \delta n ds.$$

Differentiating this formula with respect to p, q and $\tilde{q}$ we find, by (4.10.1) and (4.10.2),

$$(7.1.7) \qquad \delta L_{\mathbf{M}}(p, q) = -\int_C L_{\mathbf{M}}(r, p)\, L_{\mathbf{M}}(q, \tilde{r}) \delta n ds;$$

$$(7.1.8) \qquad \delta L_{\mathbf{M}}(p, \tilde{q}) = -\int_C L_{\mathbf{M}}(r, p)\, (L_{\mathbf{M}}(r, q))^{-} \delta n ds;$$

$$(7.1.8)' \qquad \delta L_{\mathbf{M}}(p, \tilde{q}) = -\int_C L_{\mathbf{M}}(\tilde{r}, p)\, L_{\mathbf{M}}(r, \tilde{q}) \delta n ds.$$

Taking in both sides of (7.1.7) the periods as q describes the cycle K_ν, we obtain by means of (4.10.15)

$$(7.1.9) \quad \delta Z'_\nu(p) = -\int_C L_{\mathbf{M}}(r, p) Z'_\nu(\tilde{r}) \delta n ds = -\int_C L_{\mathbf{M}}(\tilde{r}, p) Z'_\nu(r) \delta n ds.$$

If we take now the periods with respect to a cycle K_μ on both sides of (7.1.9), we derive by (4.3.10)

$$(7.1.10) \qquad \delta \Gamma_{\mu\nu} = \int_C Z'_\mu(\tilde{r}) Z'_\nu(r)\, \delta n ds.$$

From these variational formulas one can then readily derive

$$\delta L_F(q, p) = -\int_C L_F(r, p)\, L_F(q, \tilde{r})\, \delta n\, ds, \tag{7.1.11}$$

$$\delta L_F(\tilde{q}, p) = -\int_C L_F(r, p)\, (L_F(r, q))^{-}\, \delta n\, ds. \tag{7.1.12}$$

These formulas express the first variations of the functionals L_F, Z'_ν and $\Gamma_{\mu\nu}$ in their dependence on the domain. In the case of plane domains these formulas have been derived from Hadamard's formula in [6a].

7.2. Variation of Functionals as First Terms of Series Developments

Instead of deriving the formulas (7.1.7)—(7.1.12) from Hadamard's variation formula for the Green's function, we now proceed differently. Let $\mathfrak{M}$ be a subdomain of $\mathfrak{R}$ whose boundary is obtained from that of $\mathfrak{R}$ by an analytic deformation (defined below) depending on a parameter ε. For a fixed point q interior to $\mathfrak{M}$ we show that, under these circumstances,

$$l_F(p, q) = O(\varepsilon), \quad l_F(p, \tilde{q}) = O(\varepsilon),$$

uniformly for p in the closure of $\mathfrak{M}$. Formulas (7.1.7) and (7.1.8) for a positive δn then follow at once from formulas (5.3.27) and (5.3.27)′, and (7.1.9), (7.1.10) are obtained by computing periods from (7.1.7). By subtraction of the variation formulas for two domains embedded in $\mathfrak{R}$ we eliminate the condition that δn is positive. Finally, by applying the results of Section 6.15 we obtain formulas which enable us to compute the variations of the domain functionals to arbitrarily high orders, that is to say, to arbitrarily high powers of the parameter ε. Our method therefore yields more information than the one which is based on Hadamard's formula alone.

We suppose that $\mathfrak{R}$ is given and define a domain $\mathfrak{M}$ embedded in $\mathfrak{R}$ in the following manner. Let the ν-th boundary curve of $\mathfrak{R}$ be B_ν. The function

$$\zeta = \exp\left\{ -\frac{\mathscr{G}(p, q_0) + i\mathscr{H}(p, q_0)}{\operatorname{Im} \mathscr{Z}_{2h+\nu}(q_0)} \right\} \tag{7.2.1}$$

maps B_ν onto the circumference $|\zeta| = 1$. Moreover, it maps a boundary strip of $\mathfrak{R}$ near B_ν onto a ring $1 - 3\delta_\nu < |\zeta| < 1$, $\delta_\nu > 0$. Let $\lambda_\nu(\zeta)$ be regular in the closed ring $1 - 2\delta_\nu \leqq |\zeta| \leqq 1$ with

$$\text{Re}\left\{\frac{\lambda_\nu(\zeta)}{\zeta}\right\} \leqq -1 \tag{7.2.2}$$

there, and suppose that

$$\left|\frac{\lambda_\nu(\zeta_1) - \lambda_\nu(\zeta_2)}{\zeta_1 - \zeta_2}\right| \leqq M \tag{7.2.3}$$

for $1 - 2\delta_\nu \leqq |\zeta_1| \leqq 1$, $1 - 2\delta_\nu \leqq |\zeta_2| \leqq 1$. If ε is a sufficiently small positive number, the function

$$\zeta^* = \zeta + \varepsilon\lambda_\nu(\zeta) = \zeta\left[1 + \varepsilon\frac{\lambda_\nu(\zeta)}{\zeta}\right] \tag{7.2.4}$$

maps $|\zeta| = 1$ onto an analytic Jordan curve J_ν lying in the ring $1 - \delta_\nu < |\zeta| < 1$. In fact, for any two distinct points ζ_1, ζ_2 of the ring $1 - 2\delta_\nu \leqq |\zeta| \leqq 1$, we have by (7.2.3)

$$\zeta_1 - \zeta_2 + \varepsilon[\lambda_\nu(\zeta_1) - \lambda_\nu(\zeta_2)] = (\zeta_1 - \zeta_2)\left[1 + \varepsilon\frac{\lambda_\nu(\zeta_1) - \lambda_\nu(\zeta_2)}{\zeta_1 - \zeta_2}\right] \neq 0,$$

provided that $0 < \varepsilon < 1/M$. It follows that ζ^* is a schlicht function of ζ in $1 - 2\delta_\nu \leqq |\zeta| \leqq 1$ and, in particular, that the image of $|\zeta| = 1$ is a Jordan curve. The curve J_ν may be mapped back into $\mathfrak{R}$ by the inverse of the mapping (7.2.1), thus defining a curve C_ν in $\mathfrak{R}$. We suppose that functions $\lambda_\nu(\zeta)$, $\nu = 1, 2, \cdots, m$, satisfying (7.2.2) and (7.2.3) are given. Then there is an ε_0 such that for $0<\varepsilon<\varepsilon_0$ the curves C_ν defined in the above manner bound a subdomain $\mathfrak{M}$ of $\mathfrak{R}$.

Our first concern will be to obtain an estimate for the difference of the Green's functions of $\mathfrak{M}$ and $\mathfrak{R}$ which will be valid uniformly in the closure of $\mathfrak{M}$.

If ε_0 is sufficiently small, then for $0 \leqq \varepsilon \leqq \varepsilon_0$ the circumferences $|\zeta| = 1 - \delta_\nu$, $|\zeta| = 1 - 2\delta_\nu$ correspond by (7.2.1) to curves $C_\nu^{(1)}$, $C_\nu^{(2)}$ on $\mathfrak{R}$ which bound subdomains $\mathfrak{M}_1$, $\mathfrak{M}_2$ of $\mathfrak{R}$ satisfying

$$\mathfrak{M}_2 \subset \mathfrak{M}_1 \subset \mathfrak{M}.$$

Given a point p in the boundary strip $\mathfrak{R} - \mathfrak{M}_2$, let ζ be its image in one of the rings $1 - 2\delta_\nu \leqq |\zeta| \leqq 1$. The value ζ is transformed by (7.2.4) into a value ζ^* which will lie in $1 - 3\delta_\nu < |\zeta| < 1$ if ε_0 is small enough. The image of ζ^* in $\mathfrak{R}$ defined by (7.2.1) will be denoted by $p^*(p)$. It is clear that $p^*(p)$ lies in $\mathfrak{M}$ and that $p^*(p)$ is a boundary point of $\mathfrak{M}$ whenever p is a boundary point of $\mathfrak{R}$. Let $G(p, q)$ be the Green's function of $\mathfrak{M}$. Since the mapping defined by $p^*(p)$ is conformal, the function $G(p^*, q^*) = G(p^*(p), q^*(q))$ is a harmonic function of p^* or of q^* in $\mathfrak{R} - \mathfrak{M}_2$, with a singularity at $p^* = q^*$.

Let ε_0 be chosen so small that each curve J_ν lies in $1 - \frac{1}{4}\delta_\nu < |\zeta| < 1$, $\nu = 1, 2, \cdots, m$. Consider the difference

$$\mathscr{G}(p, q) - G(p, q)$$

where $p \in C^{(1)}$, $q \in \mathfrak{M}$. If $q \in C = \cup C_\nu$, then $G(p, q) = 0$. The remaining term $\mathscr{G}(p, q)$ may be estimated by a series development about the point $q \in C$. Suppose that q lies on the boundary component C_ν of $\mathfrak{M}$, in which case its image ζ by the mapping (7.2.1) lies on the curve J_ν. Let ζ_1 be the point on $|\zeta| = 1$ such that $\arg \zeta_1 = \arg \zeta$, and let q_1 be the point on B_ν which corresponds to ζ_1 by (7.2.1). Then

$$0 = \mathscr{G}(q_1, p) = \mathscr{G}(q, p) + 2\,\mathrm{Re}\left\{\frac{\partial \mathscr{G}(q, p)}{\partial \zeta}(\zeta_1 - \zeta)\right\} + \cdots,$$

so

$$\mathscr{G}(p, q) = O(\varepsilon), \quad p \in C^{(1)}, \quad q \in C. \tag{7.2.5}$$

Thus

$$\mathscr{G}(p, q) - G(p, q) = O(\varepsilon) \tag{7.2.6}$$

for $p \in C^{(1)}$, $q \in C$. Hence by the maximum principle the estimate (7.2.6) holds uniformly for $p \in \mathfrak{M}_1$, $q \in \mathfrak{M}$.

Next,

$$\begin{aligned} \mathscr{G}(p, q) - G(p^*, q^*) = {} & \mathscr{G}(p^*, q^*) - G(p^*, q^*) \\ & + \mathscr{G}(p, q) - \mathscr{G}(p^*, q^*). \end{aligned} \tag{7.2.7}$$

If $p \in C^{(1)}$, $q \in \mathfrak{R} - \mathfrak{M}_2$, then $p^* \in \mathfrak{M}_1$, $q^* \in \mathfrak{M}$. Hence, by (7.2.6),

$$\mathscr{G}(p^*, q^*) - G(p^*, q^*) = O(\varepsilon), \tag{7.2.8}$$

uniformly for $p \in C^{(1)}$, $q \in \mathfrak{R} - \mathfrak{M}_2$. The difference $\mathscr{G}(p, q) -$

$\mathscr{G}(p^*, q^*)$ is regular harmonic for p and q in $\mathfrak{R} - \mathfrak{M}_2$ since the logarithmic singularities cancel. Let $p \in C^{(1)}$. For $q \in C^{(2)}$ the difference $\mathscr{G}(p, q) - \mathscr{G}(p^*, q^*)$ is clearly $O(\varepsilon)$ and the same is true for $q \in B$. It therefore follows from the maximum principle that

$$\mathscr{G}(p, q) - \mathscr{G}(p^*, q^*) = O(\varepsilon) \tag{7.2.9}$$

for $p \in C^{(1)}$, $q \in \mathfrak{R} - \mathfrak{M}_2$. From (7.2.7), (7.2.8) and (7.2.9) we have

$$\mathscr{G}(p, q) - G(p^*, q^*) = O(\varepsilon), \ p \in C^{(1)}, \ q \in \mathfrak{R} - \mathfrak{M}_2.$$

Since $\mathscr{G}(p, q) - G(p^*, q^*)$ vanishes for $p \in B$, we conclude that

$$\mathscr{G}(p, q) - G(p^*, q^*) = O(\varepsilon) \tag{7.2.10}$$

uniformly for $p \in \mathfrak{R} - \mathfrak{M}_1$, $q \in \mathfrak{R} - \mathfrak{M}_2$, therefore uniformly for p, $q \in \mathfrak{R} - \mathfrak{M}_1$.

Since the difference (7.2.10) is zero for p or q on the boundary of $\mathfrak{R}$, it may be continued into some enlargement of $\mathfrak{R}$, say $\mathfrak{R}_1$, and the estimate (7.2.10) will be valid in $\mathfrak{R}_1 - \mathfrak{M}_1$. This remark shows that the partial derivatives of the difference (7.2.10) are also $O(\varepsilon)$; in particular,

$$-\frac{2}{\pi} \frac{\partial^2[\mathscr{G}(p, q) - G(p^*, q^*)]}{\partial p \partial q} = O(\varepsilon) \tag{7.2.11}$$

uniformly for p, $q \in \mathfrak{R} - \mathfrak{M}_1$. The constant implicit in the term $O(\varepsilon)$ occurring in (7.2.11) depends on the uniformizers used in forming the derivatives but is otherwise independent of p, q. Thus

$$\mathscr{L}_{\mathbf{M}}(p, q) - L_{\mathbf{M}}(p^*, q^*) \frac{dz^*}{dz} \frac{d\zeta^*}{d\zeta} = O(\varepsilon), \tag{7.2.12}$$

where z and ζ are uniformizers at p and q respectively. Since

$$\mathscr{L}_{\mathbf{M}}(p, q) - \mathscr{L}_{\mathbf{M}}(p^*, q^*) \frac{dz^*}{dz} \frac{d\zeta^*}{d\zeta} = O(\varepsilon),$$

we have

$$\mathscr{L}_{\mathbf{M}}(p^*, q^*) - L_{\mathbf{M}}(p^*, q^*) = O(\varepsilon);$$

that is,

$$l_{\mathbf{M}}(p, q) = O(\varepsilon) \tag{7.2.13}$$

uniformly for p, q in the closure of $\mathfrak{M}$.

Let q be in a fixed subdomain of $\mathfrak{M}$ and calculate the period of

$L_{\mathbf{M}}(p, q) - \mathscr{L}_{\mathbf{M}}(p, q)$ around the cycle K_ν. We have

$$P(L_{\mathbf{M}}(p, q) - \mathscr{L}_{\mathbf{M}}(p, q), K_\nu) = Z'_\nu(q) - \mathscr{Y}'_\nu(q). \tag{7.2.14}$$

Hence, by (7.2.13),

$$Z'_\nu(q) - \mathscr{Y}'_\nu(q) = O(\varepsilon), \tag{7.2.15}$$

uniformly for q in the closure of $\mathfrak{M}$. It follows from (7.2.15) that

$$\Gamma_{\mu\nu} - \Pi_{\mu\nu} = O(\varepsilon). \tag{7.2.16}$$

From (7.2.13), (7.2.15) and (7.2.16) we conclude that

$$l_{\mathbf{F}}(p, q) = O(\varepsilon), \tag{7.2.17}$$

uniformly for $p, q \in \mathfrak{M}$, where $\mathbf{F}$ denotes an arbitrary class of differentials.

If λ_ν, $\nu = 1, 2, \cdots$, are the eigen-values of the integral equation (6.7.1), we have from (6.7.27) and (7.2.13)

$$\sum_{\nu=1}^{\infty} \frac{1}{\lambda_\nu^2} = \int_{\mathfrak{M}} \int_{\mathfrak{M}} | l_{\mathbf{F}}(p, q) |^2 \, dA_z dA_\zeta = O(\varepsilon^2). \tag{7.2.18}$$

Thus λ_1, the smallest eigen-value, is of the order of magnitude $\dfrac{1}{\varepsilon}$.

If $\mathbf{F}$ is a symmetric class, we may apply the formulas of Section 6.15. By (6.15.6),

$$\begin{aligned} &L_{\mathbf{F}}(p, \tilde{q}) - \mathscr{L}_{\mathbf{F}}(p, \tilde{q}) \\ &= - \int_{\mathfrak{R}-\mathfrak{M}} \mathscr{L}_{\mathbf{F}}(r, p)(\mathscr{L}_{\mathbf{F}}(r, q))^- dA_\eta + \sum_{\nu=1}^{\infty} \int_{\mathfrak{M}} Q_{\mathbf{F}}^{(2\nu)}(r, \tilde{q}) I_{\mathbf{F}}(p, \tilde{r}) dA_\eta \end{aligned} \tag{7.2.19}$$

where

$$Q_{\mathbf{F}}^{(2\nu)}(r, \tilde{q}) = O(\varepsilon^{2\nu}). \tag{7.2.20}$$

Thus

$$\begin{aligned} L_{\mathbf{F}}(p, \tilde{q}) - \mathscr{L}_{\mathbf{F}}(p, \tilde{q}) &= - \int_{\mathfrak{R}-\mathfrak{M}} \mathscr{L}_{\mathbf{F}}(r, p)(\mathscr{L}_{\mathbf{F}}(r, q))^- dA_\eta \\ &\quad + \sum_{\nu=1}^{N} \int_{\mathfrak{M}} Q_{\mathbf{F}}^{(2\nu)}(r, \tilde{q}) I_{\mathbf{F}}(p, \tilde{r}) dA_\eta + O(\varepsilon^{2N+2}). \end{aligned} \tag{7.2.21}$$

This formula enables us to compute $L_{\mathbf{F}}(p, \tilde{q}) - \mathscr{L}_{\mathbf{F}}(p, \tilde{q})$ to any desired degree of accuracy.

In the case of the general class $\mathbf{F}$, which need not be symmetric, we go back to the formulas (5.3.27), (5.3.27)'. Using (7.2.17) we obtain:

$$(7.2.22) \quad \mathscr{L}_{\mathbf{F}}(\tilde{q}, p) - L_{\mathbf{F}}(\tilde{q}, p) = \int_{\Re - \mathfrak{M}} \mathscr{L}_{\mathbf{F}}(r, p)(\mathscr{L}_{\mathbf{F}}(r, q))^{-} dA_{\eta} + O(\varepsilon^2),$$

$$(7.2.23) \quad \mathscr{L}_{\mathbf{F}}(q, p) - L_{\mathbf{F}}(q, p) = \int_{\Re - \mathfrak{M}} \mathscr{L}_{\mathbf{F}}(r, p)(\mathscr{L}_{\mathbf{F}}(r, \tilde{q}))^{-} dA_{\eta} + O(\varepsilon^2).$$

Writing $\delta\mathscr{L}_{\mathbf{F}}(q, p)$ for the principal term of the difference

$$(7.2.24) \quad L_{\mathbf{F}}(q, p) - \mathscr{L}_{\mathbf{F}}(q, p),$$

we see that

$$(7.2.25) \quad \delta\mathscr{L}_{\mathbf{F}}(\tilde{q}, p) = - \int_{B} \mathscr{L}_{\mathbf{F}}(r, p)(\mathscr{L}_{\mathbf{F}}(r, q))^{-} \delta n ds,$$

$$(7.2.26) \quad \delta\mathscr{L}_{\mathbf{F}}(q, p) = - \int_{B} \mathscr{L}_{\mathbf{F}}(r, p)(\mathscr{L}_{\mathbf{F}}(r, \tilde{q}))^{-} \delta n ds.$$

Here δn expresses the normal displacement from the original boundary curves, and it is positive since the domain whose boundary results from the displacement lies inside the original domain.

We remark that the functions $\lambda_{\nu}(\zeta)$ which define the displacement may depend on ε provided that the conditions (7.2.2) and (7.2.3) are satisfied uniformly in ε.

Thus we have derived the Hadamard variational formulas (7.1.11) and (7.1.12) from the comparison formulas of Chapter 5. We want now to remove the restriction that only interior boundary shifts δn are admitted. We proceed as follows.

Let $\mathfrak{M}$ be a given domain, $\Re$ an enlargement of it. In the plane of ζ_{ν} we assume that $|\zeta_{\nu}| = 1$ corresponds to the ν-th boundary of $\Re$, while $|\zeta_{\nu}| = 1 - A_{\nu}\varepsilon$, A_{ν} a constant, corresponds to the ν-th boundary of $\mathfrak{M}$. As ζ_{ν} traverses $|\zeta_{\nu}| = 1$, let the boundary of the varied domain $\mathfrak{M}^*$ be defined by

$$(7.2.27) \quad \zeta_{\nu}^* = \zeta_{\nu} + \varepsilon\lambda_{\nu}(\zeta_{\nu}; \varepsilon),$$

where $\lambda_{\nu}(\zeta_{\nu}; \varepsilon)$ satisfies (7.2.2) and (7.2.3). Then

$$(7.2.28)\quad L_F(\tilde{q}, p) - \mathscr{L}_F(\tilde{q}, p) = -\int_B \mathscr{L}_F(r, p)(\mathscr{L}_F(r, q))^- \delta n_1\, ds,$$

$$(7.2.29)\quad L^*_F(\tilde{q}, p) - \mathscr{L}_F(\tilde{q}, p) = -\int_B \mathscr{L}_F(r, p)(\mathscr{L}_F(r, q))^- \delta n_2\, ds,$$

where δn_1 and δn_2, both positive, define the normal displacement of B into the boundaries of $\mathfrak{M}$, $\mathfrak{M}^*$ respectively. Subtracting (7.2.28) from (7.2.29), we obtain

$$(7.2.30)\quad L^*_F(\tilde{q}, p) - L_F(\tilde{q}, p) = -\int_B \mathscr{L}_F(r, p)(\mathscr{L}_F(r, q))^- \delta n\, ds,$$

where $\delta n = \delta n_2 - \delta n_1$ is the normal displacement of the boundary C of $\mathfrak{M}$ into the boundary C^* of $\mathfrak{M}^*$. With an error $O(\varepsilon^2)$ we may replace the boundary B of $\mathfrak{R}$ in the integral on the right of (7.2.30) by the boundary C of $\mathfrak{M}$, and then with a further error $O(\varepsilon^2)$ we may replace $\mathscr{L}_F$ by L_F. We thus obtain formula (7.1.7). Formula (7.1.8) is obtained in a similar way, and formulas (7.1.9), (7.1.10) are obtained by computing periods in the usual manner.

7.3. Variation by Cutting a Hole

Hadamard's formula gives the variation of the Green's function under a small displacement of the boundary of the domain. In this type of variation the original domain $\mathfrak{M}$ and the varied one $\mathfrak{M}^*$ are of the same topological type but not necessarily of the same conformal type. If, instead of displacing the m boundary curves of the domain $\mathfrak{M}$, we remove from $\mathfrak{M}$ a small hole, we obtain a domain $\mathfrak{M}^*$ with $m + 1$ boundary curves and therefore of different topological type. We now discuss a special variation of this kind.This type of variation will be applicable even to closed Riemann surfaces. In this section we shall assume that the surface $\mathfrak{M}$ is orientable and has a boundary, and therefore possesses a proper Green's function. In the following section we shall consider closed orientable surfaces.

Given the surface $\mathfrak{M}$ with Green's function $G(p, q)$, let q_1 be an interior point of $\mathfrak{M}$. If ε is a sufficiently small positive number, the level curve

$$(7.3.1)\qquad G(p, q_1) = \log \frac{1}{\varepsilon}$$

is a Jordan curve C_{m+1} which bounds a subdomain $\mathfrak{D}$ of $\mathfrak{M}$ containing q_1. We remove the domain $\mathfrak{D}$ from $\mathfrak{M}$ and obtain a surface $\mathfrak{M}^*$. Let $T(p, q)$ be the analytic function of p whose real part is the Green's function $G(p, q)$, and let ζ be a uniformizer at q_1 which is valid in a neighborhood of q_1 containing $\mathfrak{D}$. We assume that this uniformizer has been chosen once for all. We have

$$\exp\{-T(p, q_1)\} = a_1\zeta + a_2\zeta^2 + \cdots, \quad |a_1| \neq 0,$$

for p near q_1. Hence, if p is a point of C_{m+1},

$$|\zeta(p)| = \frac{\varepsilon}{|a_1|} + O(\varepsilon^2). \tag{7.3.2}$$

Thus, the image of C_{m+1} in the uniformizer plane is approximately a circle.

Let $G^*(p, q)$ be the Green's function of $\mathfrak{M}^*$, and let

$$h_\varepsilon(p, q, q_1) = G^*(p, q) - G(p, q) + \frac{G(p, q_1)G(q, q_1)}{\log \dfrac{1}{\varepsilon}}. \tag{7.3.3}$$

The function $h_\varepsilon(p, q, q_1)$ vanishes for p or q on the boundary C of $\mathfrak{M}$. Hold q fixed, and take p to be a point of C_{m+1}. We see that

$$\begin{aligned} h_\varepsilon(p, q, q_1) &= -G(p, q) + G(q_1, q) \\ &= -\operatorname{Re}\{T'(q_1, q)\zeta\} + O(\varepsilon^2), \end{aligned}$$

where the $'$ in $T'(p, q)$ denotes differentiation with respect to p, the differentiation being expressed in terms of ζ.

Let p and q be two arbitrary points near q_1 corresponding to the values ζ and ζ_0 of the uniformizing variable. Then

$$G(q, p) = \log\frac{1}{|\zeta_0 - \zeta|} + \text{regular terms}$$

and hence, by differentiation with respect to ζ_0,

$$2\frac{\partial G(q, p)}{\partial \zeta_0} = T'(q, p) = \frac{1}{\zeta - \zeta_0} + \text{regular terms}. \tag{7.3.3)'$$

In particular, we have

$$T'(q_1, p) = \frac{1}{\zeta} + \text{bounded terms}.$$

Thus, by (7.3.2), for $p \in C_{m+1}$,

$$h_\varepsilon(p, q, q_1) = -\operatorname{Re}\left\{T'(q_1, q)\frac{\varepsilon^2}{|a_1|^2\bar{\zeta}}\right\} + O(\varepsilon^2)$$
$$= -\operatorname{Re}\left\{\frac{\varepsilon^2}{|a_1|^2} T'(q_1, q)(T'(q_1, p))^-\right\} + O(\varepsilon^2).$$

By (7.3.3)′,

$$T'(q_1, p) = 2\frac{\partial G(q, p)}{\partial \zeta_0}\bigg|_{q=q_1}.$$

Hence $T'(q_1, p)$ vanishes when p is on the boundary C of $\mathfrak{M}$, and therefore

$$h_\varepsilon(p, q, q_1) + \operatorname{Re}\left\{\frac{\varepsilon^2}{|a_1|^2} T'(q_1, q))T'(q_1, p))^-\right\}$$

(7.3.4)
$$= \begin{cases} 0, & p \in C_\nu, \ \nu = 1, 2, \cdots, m, \\ O(\varepsilon^2), & p \in C_{m+1}. \end{cases}$$

We observe that

$$\omega_{m+1}(p) = \frac{G(p, q_1)}{\log \dfrac{1}{\varepsilon}} = \begin{cases} 0, & p \in C_\nu, \ \nu = 1, 2, \cdots, m, \\ 1, & p \in C_{m+1}, \end{cases} \tag{7.3.5}$$

so $\omega_{m+1}(p)$ is just the harmonic measure of C_{m+1} with respect to the point p of $\mathfrak{M}^*$. By the harmonic measure of a boundary component C_ν at a point p we mean the value at p of the everywhere regular single-valued harmonic function which is 1 on C_ν and 0 on the remaining boundary components. Since the left side of (7.3.4) is a harmonic function of p for fixed $q \in \mathfrak{M}$, we conclude that

$$h_\varepsilon(p, q, q_1) + \operatorname{Re}\left\{\frac{\varepsilon^2}{|a_1|^2} T'(q_1, q)(T'(q_1, p))^-\right\} = O(\varepsilon^2)\omega_{m+1}(p), \tag{7.3.6}$$

for $p \in \mathfrak{M}^*$. Here we have used the well-known and obvious fact that the absolute value of a harmonic function is sub-harmonic. It follows from (7.3.3), (7.3.5) and (7.3.6) that

(7.3.7)
$$G^*(p, q) = G(p, q) - \frac{G(p, q_1)G(q, q_1)}{\log \dfrac{1}{\varepsilon}}$$
$$- \operatorname{Re}\left\{\frac{\varepsilon^2}{|a_1|^2} T'(q_1, q)(T'(q_1, p))^-\right\} + o(\varepsilon^2)$$

in any fixed subdomain of $\mathfrak{M}$ whose closure does not contain a neighborhood of q_1.

Let $K_1, \cdots, K_{2h+m-1}$ be a homology basis for $\mathfrak{M}$. A basis for $\mathfrak{M}^*$ is then obtained by adding the cycle $K_{2h+m} = C_{m+1}$. As in Section 4.3, we define the integrals Z_ν, Z_ν^* of $\mathfrak{M}$, $\mathfrak{M}^*$ by the formulas

$$\text{(7.3.8)} \quad \operatorname{Im} Z_\nu(q) = \frac{1}{2\pi} \int_{K_\nu} \frac{\partial G(p, q)}{\partial n_p} ds_p, \quad \nu = 1, 2, \cdots, 2h + m - 1;$$

$$\text{(7.3.9)} \quad \operatorname{Im} Z_\nu^*(q) = \frac{1}{2\pi} \int_{K_\nu} \frac{\partial G^*(p, q)}{\partial n_p} ds_p, \quad \nu = 1, 2, \cdots, 2h + m.$$

Clearly

$$\text{(7.3.10)} \qquad \operatorname{Im} Z_{2h+m}^*(p) = \omega_{m+1}(p) = \frac{G(p, q_1)}{\log \dfrac{1}{\varepsilon}}.$$

For $\nu = 1, 2, \cdots, 2h + m - 1$,

$$\operatorname{Im} Z_\nu^*(q) - \operatorname{Im} Z_\nu(q) = \frac{1}{2\pi} \int_{K_\nu} \frac{\partial\{G^*(p, q) - G(p, q)\}}{\partial n_p} ds_p$$

$$= -\frac{G(q, q_1)}{\log \dfrac{1}{\varepsilon}} \operatorname{Im} Z_\nu(q_1) + O(\varepsilon^2),$$

by (7.3.7);

$$= -\operatorname{Im} Z_{2h+m}^*(q) \operatorname{Im} Z_\nu(q_1) + O(\varepsilon^2).$$

Computing periods in (7.3.7) around the ν-th cycle, we obtain

$$\text{(7.3.11)} \quad \begin{aligned} \operatorname{Im} Z_\nu^*(p) &= \operatorname{Im} Z_\nu(p) - \operatorname{Im} Z_{2h+m}^*(p) \operatorname{Im} Z_\nu(q_1) \\ &\quad - \operatorname{Im}\left\{\frac{\varepsilon^2}{|a_1|^2} Z_\nu'(q_1) (T'(q_1, p))^-\right\} + o(\varepsilon^2), \end{aligned}$$

in any fixed subdomain of $\mathfrak{M}$ whose closure does not contain a neighborhood of q_1

By the definition of complex differentiation,

$$\frac{\partial G(p, q)}{\partial p} = \frac{1}{2} T'(p, q).$$

The formulas (7.3.7) and (7.3.11) may therefore be written in the form

$$G^*(p, q) = G(p, q) - \log \frac{1}{\varepsilon} \operatorname{Im} Z^*_{2h+m}(p) \operatorname{Im} Z^*_{2h+m}(q)$$

$$(7.3.12) \qquad - 4 \operatorname{Re}\left\{ \frac{\varepsilon^2}{|a_1|^2} \frac{\partial G(q_1, q)}{\partial q_1} \frac{\partial G(q_1, p)}{\partial \tilde{q}_1} \right\} + o(\varepsilon^2),$$

$$\operatorname{Im} Z^*_\nu(p) = \operatorname{Im} Z_\nu(p) - \operatorname{Im} Z^*_{2h+m}(p) \operatorname{Im} Z_\nu(q_1)$$

$$(7.3.13) \qquad - 2 \operatorname{Im}\left\{ \frac{\varepsilon^2}{|a_1|^2} Z'_\nu(q_1) \frac{\partial G(q_1, p)}{\partial \tilde{q}_1} \right\} + o(\varepsilon^2).$$

We observe from (7.3.10) that the second term on the right in (7.3.12) has the correct order of magnitude $\left(\log \frac{1}{\varepsilon}\right)^{-1}$. Differentiating (7.3.12) with respect to p and q, (7.3.13) with respect to p, we find by (4.10.1) and (4.10.2) that

$$L^*_{\mathbf{M}}(p, q) = L_{\mathbf{M}}(p, q) - \frac{\log \frac{1}{\varepsilon}}{2\pi} Z^{*\prime}_{2h+m}(p) Z^{*\prime}_{2h+m}(q)$$

$$(7.3.14) \qquad - \frac{\pi \varepsilon^2}{|a_1|^2} \{L_{\mathbf{M}}(q, q_1) L_{\mathbf{M}}(p, \tilde{q}_1) + L_{\mathbf{M}}(q, \tilde{q}_1) L_{\mathbf{M}}(p, q_1)\} + o(\varepsilon^2),$$

$$Z^{*\prime}_\nu(p) = Z'_\nu(p) - Z^{*\prime}_{2h+m}(p) \operatorname{Im} Z_\nu(q_1)$$

$$(7.3.15) \qquad - \frac{\pi \varepsilon^2}{|a_1|^2} \{Z'_\nu(q_1) L_{\mathbf{M}}(p, \tilde{q}_1) + Z'_\nu(\tilde{q}_1) L_{\mathbf{M}}(p, q_1)\} + o(\varepsilon^2).$$

By (4.3.10)

$$\Gamma^*_{2h+m, \nu} = - P(Z^{*\prime}_{2h+m}, K_\nu) = - P\left(i \frac{T'(p, q_1)}{\log \frac{1}{\varepsilon}}, K_\nu\right)$$

$$(7.3.16) \qquad = \begin{cases} - \dfrac{2\pi}{\log \frac{1}{\varepsilon}} \operatorname{Im} Z_\nu(q_1), & \nu = 1, 2, \cdots, 2h + m - 1, \\[2ex] \dfrac{2\pi}{\log \frac{1}{\varepsilon}}, & \nu = 2h + m. \end{cases}$$

Formula (7.3.15) is also obtained immediately from (7.3.14) by computing the periods around K_ν, $1 \leqq \nu \leqq 2h+m-1$. Finally, taking periods in (7.3.15), we obtain the formula

$$\begin{aligned}\Gamma^*_{\mu\nu} &= \Gamma_{\mu\nu} + \frac{2\pi}{\log \frac{1}{\varepsilon}} \operatorname{Im} Z_\mu(q_1) \operatorname{Im} Z_\nu(q_1) \\ &\quad + \frac{2\pi\varepsilon^2}{|a_1|^2} \operatorname{Re} \{Z'_\mu(q_1) Z'_\nu(\bar{q}_1)\} + o(\varepsilon^2), \\ &\qquad \mu, \nu = 1, 2, \cdots, 2h+m-1.\end{aligned} \tag{7.3.17}$$

In particular, we observe that $\delta\Gamma_{\mu\nu}$ is real.

The reality of $\delta\Gamma_{\mu\nu}$ is, of course, an obvious consequence of our normalization of the period matrix.

Let

$$\begin{aligned}\Lambda^*_{\mu\nu} &= \Gamma^*_{\mu\nu} + \Gamma^*_{2h+m,\nu} \operatorname{Im} Z_\mu(q_1) \\ &= \Gamma^*_{\mu\nu} - \frac{2\pi}{\log \frac{1}{\varepsilon}} \operatorname{Im} Z_\mu(q_1) \operatorname{Im} Z_\nu(q_1), \\ &\qquad \mu, \nu = 1, 2, \cdots, 2h+m-1.\end{aligned} \tag{7.3.18}$$

For ε sufficiently small, $0 < \varepsilon \leqq \varepsilon_0$, the matrix $\| \Lambda^*_{\mu\nu} \|$ is clearly non-singular. Now let $\mathbf{F} = \mathbf{F}(i_1, i_2, \cdots, i_n)$ be the class of differentials f' on $\mathfrak{M}^*$ such that

$$P(f', K_{i_\nu}) = 0, \quad \nu = 1, 2, \cdots, n,$$

and suppose that the integer $2h+m$ is contained among the set $i_1, i_2, \cdots, i_n$. In other words, suppose that

$$P(f', K_{2h+m}) = 0, \quad f' \in \mathbf{F}. \tag{7.3.19}$$

Then, assuming that $i_n = 2h+m$, we have

$$\begin{aligned}L^*_{\mathbf{F}}(p, q) &= L^*_{\mathbf{M}}(p, q) + \frac{\log \frac{1}{\varepsilon}}{2\pi} Z^{*\prime}_{2h+m}(p) Z^{*\prime}_{2h+m}(q) \\ &\quad + \sum_{\mu,\nu=1}^{n-1} \alpha_{\mu\nu} [Z^{*\prime}_{i_\mu}(p) + \operatorname{Im} \{Z_{i_\mu}(q_1)\} Z^{*\prime}_{2h+m}(p)] \\ &\quad \cdot [Z^{*\prime}_{i_\nu}(q) + \operatorname{Im} \{Z_{i_\nu}(q_1)\} Z^{*\prime}_{2h+m}(q)],\end{aligned} \tag{7.3.20}$$

where

$$\|\alpha_{\mu\nu}\| = \|\Lambda^*_{i_\nu i_\mu}\|^{-1} = \|\bar{\Lambda}^*_{i_\mu i_\nu}\|^{-1}. \tag{7.3.21}$$

In fact, computing the period of $L^*_{\mathbf{F}}(p, q)$ around the cycle K_{2h+m}, we have by (7.3.16),

$$P(L^*_{\mathbf{F}}(p, q), K_{2h+m}) = Z^{*\prime}_{2h+m}(q) - Z^{*\prime}_{2h+m}(q)$$
$$- \sum_{\mu,\nu=1}^{n-1} \alpha_{\mu\nu}[\Gamma^*_{i_\mu, 2h+m} + \frac{2\pi}{\log \frac{1}{\varepsilon}} \operatorname{Im} Z_{i_\mu}(q_1)]$$
$$\cdot [Z^{*\prime}_{i_\nu}(q) + \operatorname{Im}\{Z_{i_\nu}(q_1)\} Z^{*\prime}_{2h+m}(q)] = 0.$$

Computing the period around the cycle K_{i_ϱ}, $1 \leqq \varrho < n$, we see that

$$P(L^*_{\mathbf{F}}(p, q), K_{i_\varrho}) = Z^{*\prime}_{i_\varrho}(q) + \operatorname{Im} Z_{i_\varrho}(q_1) Z^{*\prime}_{2h+m}(q)$$
$$- \sum_{\mu,\nu=1}^{n-1} \alpha_{\mu\nu} \Lambda^*_{i_\mu i_\varrho} [Z^{*\prime}_{i_\nu}(q) + \operatorname{Im}\{Z_{i_\nu}(q_1)\} Z^{*\prime}_{2h+m}(q)] = 0.$$

Now formulas (7.3.14), (7.3.15), (7.3.17), (7.3.20) and (7.3.21) show that

$$L^*_{\mathbf{F}}(p, q) = L_{\mathbf{F}}(p, q) + O(\varepsilon^2), \tag{7.3.22}$$

for any class $\mathbf{F}$ satisfying (7.3.19), while nothing better than

$$L^*_{\mathbf{F}}(p, q) = L_{\mathbf{F}}(p, q) + O\left(\log \frac{1}{\varepsilon}\right) \tag{7.3.23}$$

holds for any class $\mathbf{F}$ which does not satisfy (7.3.19). In Section 6.5 we were forced to make the assumption that any cycle C_ν of $\mathfrak{M}$ which bounds on $\mathfrak{R}$ must bound on $\mathfrak{M}$ (essential imbedding), and we remarked at the time that this assumption could be replaced by the condition that a cycle which is bounding on $\mathfrak{R}$ but not on $\mathfrak{M}$ be a cycle around which all functions of the class $\mathbf{F}$ on $\mathfrak{M}$ are single-valued. Here $\mathfrak{M}^*$ is imbedded in $\mathfrak{M}$ and the cycle $C_{m+1} = K_{2h+m}$ bounds on $\mathfrak{M}$ but not on $\mathfrak{M}^*$; the effect of the alternative hypothesis on the error term in the variation is striking. The difference in the order of magnitude of the error arises from the fact that, if (7.3.19) is not satisfied, then the eigen-values are not bounded away from unity.

On the basis of this observation we would expect the error term in the variation to be small for every class **F** provided that we form $\mathfrak{M}^*$ by removing a hole from a closed surface $\mathfrak{M}$. For in this case the boundary of the hole, which bounds on $\mathfrak{M}$, also bounds on $\mathfrak{M}^*$. In this case the estimate $L_\mathbf{M}(p, q) = L_\mathbf{M}^*(p, q) + O(\varepsilon^2)$ is in fact true, as we shall show in the next section.

We summarise the results of this section in the following two theorems:

THEOREM 7.3.1. *Let $\mathfrak{M}$ be an orientable surface with boundary whose Green's function is $G(p, q)$. If we remove from $\mathfrak{M}$ the subdomain* $G(p, q_1) \geqq \log \dfrac{1}{\varepsilon}$, *we obtain a surface $\mathfrak{M}^*$ whose Green's function $G^*(p, q)$ has the form*

$$G^*(p, q) = G(p, q) - \frac{G(p, q_1)G(q, q_1)}{\log \dfrac{1}{\varepsilon}}$$

(7.3.7)′
$$- \operatorname{Re}\left\{ 4\varepsilon^2 e^{2\gamma} \frac{\partial G(q, q_1)}{\partial q_1} \frac{\partial G(p, q_1)}{\partial \tilde{q}_1} \right\} + o(\varepsilon^2).$$

where $\gamma = \gamma(q_1)$ is defined by the series development

(7.3.24) $G(p, q_1) = \log \dfrac{1}{|\zeta|} + \gamma(q_1) + 0(|\zeta|).$

THEOREM 7.3.2. *Under the hypotheses of the preceding theorem, we have*

(7.3.22)″ $L_\mathbf{F}^*(p, q) = L_\mathbf{F}(p, q) + 0(\varepsilon^2)$

*for all classes **F** whose differentials have vanishing periods with respect to the cycle $G(p, q_1) = \log \dfrac{1}{\varepsilon}$. For any other class **F** we have*

(7.3.23)′ $L_\mathbf{F}^*(p, q) = L_\mathbf{F}(p, q) - \dfrac{\log \dfrac{1}{\varepsilon}}{2\pi} Z_{2h+m}^{*\prime}(p) \, Z_{2h+m}^{*\prime}(q) + 0(\varepsilon^2).$

7.4. VARIATION BY CUTTING A HOLE IN A CLOSED SURFACE

If $\mathfrak{M}$ is a closed surface, the method of Section 7.3 must be modified somewhat. However, we can readily compute the variation formulas for $L_{\mathbf{M}}(p, q)$, $Z'_\nu(p)$ and $\Gamma_{\mu\nu}$ when we cut a small hole from the closed surface $\mathfrak{M}$. Let $V(p, p_0; q, q_0)$ be the function defined by (4.2.23). For sufficiently small ε, $\varepsilon > 0$, the level curve

$$V(p, p_0; q_1, q_0) = \log \frac{1}{\varepsilon} \tag{7.4.1}$$

will be a Jordan curve $C_1 = C_1\,(p_0, q_0, q_1, \varepsilon)$ which bounds a domain $\mathfrak{D}$ of $\mathfrak{M}$ containing q_1 in its interior. Here p_0, q_0 are fixed points of $\mathfrak{M}$ distinct from q_1. Let p, q be two points lying in a neighborhood of q_1, and let the coordinates of p and q, expressed in terms of the same uniformizing variable, be ζ and η respectively. Then

$$\begin{aligned} &\exp\{\Omega_{q_0q}(p) - \Omega_{q_0q}(p_0)\} \\ &= a_1(\zeta - \eta) + a_2(\zeta - \eta)^2 + \cdots, \ |a_1| \neq 0, \end{aligned} \tag{7.4.2}$$

where $a_1 = a_1(q)$, $a_2 = a_2(q)$, etc. We suppose that ζ and η vanish at q_1, and by appropriate choice of the uniformizing variable we may suppose that $a_k(q_1) = 0$ for $k \geqq 2$. Taking logarithms of both sides of (7.4.2), we obtain

$$\begin{aligned} \Omega_{q_0q}(p) - \Omega_{q_0q}(p_0) &= \log(\zeta - \eta) + b_0 + b_1(\zeta - \eta) \\ &\quad + b_2(\zeta - \eta)^2 + \cdots, \end{aligned} \tag{7.4.3}$$

where $b_0 = \log a_1$, $b_1 = a_2/a_1$, $b_2 = (2a_1a_3 - a_2^2)/a_1^2$, etc. Differentiating the real part of (7.4.3) twice with respect to q and then taking $q = q_1$, $p \in C_1$, we obtain, since $b_\nu(q_1) = 0$ for $\nu > 0$,

$$2\frac{\partial V(q_1, q_0; p, p_0)}{\partial q_1} = \frac{1}{\zeta} - 2\frac{\partial \mathrm{Re}\, b_0}{\partial q_1} + O(\varepsilon), \tag{7.4.4}$$

$$2\frac{\partial^2 V(q_1, q_0; p, p_0)}{\partial {q_1}^2} = \frac{1}{\zeta^2} + O(1). \tag{7.4.5}$$

Taking $q = q_1$, $p \in C_1$, in (7.4.2) we find that

$$\zeta = \frac{1}{a_1}\exp\{\Omega_{q_0q_1}(p) - \Omega_{q_0q_1}(p_0)\}, \tag{7.4.6}$$

$$\zeta = \frac{\varepsilon^2}{|a_1|^2}\frac{1}{\bar{\zeta}}, \quad \zeta^2 = \frac{\varepsilon^4}{|a_1|^4}\frac{1}{\bar{\zeta}^2}. \tag{7.4.7}$$

Thus for $p \in C_1$ we have

$$\zeta = \frac{2\varepsilon^2}{|a_1|^2}\left\{\frac{\partial V(q_1, q_0; p, p_0)}{\partial \tilde{q}_1} + \frac{\partial \operatorname{Re} b_0}{\partial \tilde{q}_1} + O(\varepsilon)\right\}, \tag{7.4.8}$$

$$\zeta^2 = \frac{2\varepsilon^4}{|a_1|^4}\left\{\frac{\partial^2 V(q_1, q_0; p, p_0)}{\partial \tilde{q}_1^2} + O(1).\right\} \tag{7.4.9}$$

Let $\mathfrak{M}^*$ be the surface $\mathfrak{M} - \mathfrak{D}$, $G^*(p, q)$ the Green's function of $\mathfrak{M}^*$, and consider the function

$$\begin{aligned} h_\varepsilon(p, p_0; q, q_0; q_1) = {} & G^*(p, q) - G^*(p, q_0) \\ & - V(p, p_0; q, q_0) + V(q_1, p_0; q, q_0). \end{aligned} \tag{7.4.10}$$

As a function of p, h_ε is regular harmonic in $\mathfrak{M}^*$. For $p \in C_1$, q a fixed interior point of $\mathfrak{M}^*$, the first two terms on the right side of (7.4.10) vanish, and we have by Taylor's Theorem

$$\begin{aligned} h_\varepsilon &= - V(p, p_0; q, q_0) + V(q_1, p_0; q, q_0) \\ &= -2\operatorname{Re}\left\{\frac{\partial V(q_1, p_0; q, q_0)}{\partial q_1}\zeta + \frac{1}{2}\frac{\partial^2 V(q_1, p_0; q, q_0)}{\partial q_1^2}\zeta^2 + \cdots\right\} + O(\varepsilon^3) \\ &= -\frac{4\varepsilon^2}{|a_1|^2}\operatorname{Re}\left\{\frac{\partial V(q_1, p_0; q, q_0)}{\partial q_1}\left[\frac{\partial V(q_1, q_0; p, p_0)}{\partial \tilde{q}_1} + \frac{\partial \operatorname{Re} b_0}{\partial \tilde{q}_1}\right]\right\} \\ &\quad - \frac{2\varepsilon^4}{|a_1|^4}\operatorname{Re}\left\{\frac{\partial^2 V(q_1, p_0; q, q_0)}{\partial q_1^2}\frac{\partial^2 V(q_1, q_0; p, p_0)}{\partial \tilde{q}_1^2}\right\} + O(\varepsilon^3). \end{aligned}$$

Let

$$\begin{aligned} g^\varepsilon = h_\varepsilon &+ \frac{4\varepsilon^2}{|a_1|^2}\operatorname{Re}\left\{\frac{\partial V(q_1, p_0; q, q_0)}{\partial q_1}\left[\frac{\partial V(q_1, q_0; p, p_0)}{\partial \tilde{q}_1} + \frac{\partial \operatorname{Re} b_0}{\partial \tilde{q}_1}\right]\right\} \\ &+ \frac{2\varepsilon^4}{|a_1|^4}\operatorname{Re}\left\{\frac{\partial^2 V(q_1, p_0; q, q_0)}{\partial q_1^2}\frac{\partial^2 V(q_1, q_0; p, p_0)}{\partial \tilde{q}_1^2}\right\}. \end{aligned}$$

As a function of p, g_ε is regular harmonic throughout $\mathfrak{M}^*$ and it is $O(\varepsilon^3)$ on the boundary C_1 of $\mathfrak{M}^*$. From the maximum principle we conclude that $g_\varepsilon = O(\varepsilon^3)$ throughout $\mathfrak{M}^*$. In any compact subdomain of $\mathfrak{M}^*$ we therefore have

$$\begin{aligned} h_\varepsilon + \frac{4\varepsilon^2}{|a_1|^2}\operatorname{Re}\left\{\frac{\partial V(q_1, p_0; q, q_0)}{\partial q_1}\left[\frac{\partial V(q_1, q_0; p, p_0)}{\partial \tilde{q}_1} + \frac{\partial \operatorname{Re} b_0}{\partial \tilde{q}_1}\right]\right\} \\ = O(\varepsilon^3). \end{aligned} \tag{7.4.11}$$

Thus

$$G^*(p, q) = G^*(p, q_0) + V(p, p_0; q, q_0) - V(q_1, p_0; q, q_0)$$

$$(7.4.12) \quad -\frac{4\varepsilon^2}{|a_1|^2}\,\mathrm{Re}\left\{\frac{\partial V(q_1, p_0; q, q_0)}{\partial q_1}\left[\frac{\partial V(q_1, q_0; p, p_0)}{\partial \tilde{q}_1} + \frac{\partial\, \mathrm{Re}\, b_0}{\partial \tilde{q}_1}\right]\right\} + O(\varepsilon^3)$$

in any compact subdomain of $\mathfrak{M}^*$.

Differentiating (7.4.12) with respect to p and q, we have

$$L^*_{\mathbf{M}}(p, q) = L_{\mathbf{M}}(p, q) - \frac{\pi\varepsilon^2}{|a_1|^2}\left\{L_{\mathbf{M}}(q, q_1)L_{\mathbf{M}}(p, \tilde{q}_1) + L_{\mathbf{M}}(q, \tilde{q}_1)L_{\mathbf{M}}(p, q_1)\right\}$$

$$(7.4.13) \quad + o(\varepsilon^2).$$

This formula should be compared with the corresponding formula (7.3.14) which holds for a domain $\mathfrak{M}$ with boundary. On computing periods in (7.4.13) we find that

$$Z^{*\prime}_\mu(p) = Z'_\mu(p) - \frac{\pi\varepsilon^2}{|a_1|^2}\left\{Z'_\mu(q_1)L_{\mathbf{M}}(p, \tilde{q}_1) + Z'_\mu(\tilde{q}_1)L_{\mathbf{M}}(p, q_1)\right\}$$

$$(7.4.14) \quad + o(\varepsilon^2).$$

Finally, computing periods in (7.4.14), we obtain

$$(7.4.15) \quad \Gamma^*_{\mu\nu} = \Gamma_{\mu\nu} + \frac{2\pi\varepsilon^2}{|a_1|^2}\,\mathrm{Re}\{Z'_\mu(q_1)\, Z'_\nu(\tilde{q}_1)\} + o(\varepsilon^2),$$

$$(\mu, \nu = 1, 2, \cdots, 2h).$$

The variation formula for $L_{\mathbf{M}}(p, \tilde{q})$ is obtained on replacing q by $\tilde{q}$ in (7.4.13).

7.5. Attaching a Handle to a Closed Surface

We have constructed variations which arise when holes are cut out of surfaces. Every finite orientable surface with or without boundary can be obtained from the sphere by cutting holes and attaching handles, and it is therefore desirable to study the variations which arise by attaching handles. The handles attached will, like the holes, have a special form, but we shall later indicate a method whereby a surface can be varied without changing its topological type. By combining our variations it will then be possible to pass from the sphere to any other finite orientable Riemann surface of given conformal type.

We consider a closed surface $\mathfrak{M}$. Given two fixed points q_1, q_2 in $\mathfrak{M}$ and using an unessential parameter point p_0, we define two Jordan curves c_1 and c_2 surrounding q_1 and q_2 by

$$(7.5.1) \qquad V(p, p_0; q_1, q_2) = \log \frac{1}{\varepsilon}, \quad V(p, p_0; q_1, q_2) = \log \varepsilon,$$

respectively. Let $v(p, p_0; q_1, q_2)$ be an analytic function of p in $\mathfrak{M}$ whose real part is $V(p, p_0; q_1, q_2)$. We choose two arbitrary points $p_1^{(0)}$ and $p_2^{(0)}$ on c_1 and c_2 and have by (7.5.1)

$$(7.5.2) \qquad v(p_1^{(0)}, p_0; q_1, q_2) = v(p_2^{(0)}, p_0; q_1, q_2) + 2 \log \frac{1}{\varepsilon} - i\alpha$$

where α is real. Starting from $p_1^{(0)}$, $p_2^{(0)}$ we identify points p_1 on c_1 and p_2 on c_2 if they satisfy the equation

$$(7.5.3) \qquad v(p_1, p_0; q_1, q_2) = v(p_2, p_0; q_1, q_2) + 2 \log \frac{1}{\varepsilon} - i\alpha.$$

We can easily see that as p_1 moves around c_1 in the clockwise sense with respect to q_1, the identified point p_2 moves around c_2 in the counterclockwise sense with respect to q_2. The surface formed by removing the hole at q_1 bounded by c_1 and the hole at q_2 bounded by c_2 and then identifying points of c_1, c_2 according to the rule (7.5.3) will be denoted by $\mathfrak{M}^*$. It is clear that $\mathfrak{M}^*$ is formed from $\mathfrak{M}$ by attaching a handle. We denote by $\Omega^*_{qq_0}(p)$ and $V^*(p, p_0; q, q_0)$ the Abelian integral of the third kind and the Green's function of the surface $\mathfrak{M}^*$.

We state now a general theorem which is useful in all variational problems to be considered.

THEOREM 7.5.1. *Let $\mathfrak{M}$ and $\mathfrak{M}^*$ be two closed Riemann surfaces which have the domain $\mathfrak{M}_0$ in common. Suppose that there exists a set K of cycles K_ν in $\mathfrak{M}_0$ which is canonical with respect to $\mathfrak{M}$ and can be completed to a canonical set with respect to $\mathfrak{M}^*$. Then, for every quadruple p, p_0, q, q_0 in $\mathfrak{M}_0$, we have*

$$(7.5.4) \quad V^*(p, p_0; q, q_0) - V(p, p_0; q, q_0) = \operatorname{Re} \left\{ \frac{1}{2\pi i} \int_{\partial \mathfrak{M}_0} \Omega_{qq_0}(t) \, d\Omega^*_{pp_0}(t) \right\}.$$

Here V, V^* and Ω, Ω^* are the Green's functions and integrals of the third kind with respect to $\mathfrak{M}$ and $\mathfrak{M}^*$.

We prove this theorem by Riemann's classical method of contour integration. We select two fixed points q and q_0 in $\mathfrak{M}_0$ and connect them by a smooth curve $\gamma \in \mathfrak{M}_0$. Let $\mathfrak{J}$ be the domain obtained from $\mathfrak{M}_0$ by introducing cuts along K and γ. Obviously, $\Omega_{qq_0}(p)$ will be single-valued in $\mathfrak{J}$. Thus, we may apply the residue theorem as follows:

$$\frac{1}{2\pi i}\int_{\partial\mathfrak{J}} \Omega_{qq_0}(t)d\Omega^*_{pp_0}(t) = -\,\Omega_{qq_0}(p) + \Omega_{qq_0}(p_0). \tag{7.5.5}$$

On the other hand, $\partial\mathfrak{J} = \partial\mathfrak{M}_0 + K + \gamma$ so some of the integrations indicated in (7.5.5) can be carried out by introducing the known periods of the integrand. We observe first that the determinations of $\Omega_{qq_0}(t)$ on the two edges of γ differ by the amount $2\pi i$ so that the integration over both edges yields the contribution

$$-\,\Omega^*_{pp_0}(q) + \Omega^*_{pp_0}(q_0).$$

The integration over K can be carried out by the general method of Section 3.4. We obtain

$$\frac{1}{2\pi i}\int_K \Omega_{qq_0}(t)d\Omega^*_{pp_0}(t) = \sum_{\mu,\,\nu=1}^{2h} I(K_\mu, K_\nu)\cdot\frac{1}{2\pi i}\int_{K_\mu} d\Omega_{qq_0}\cdot\int_{K_\nu} d\Omega^*_{pp_0}.$$

We need not calculate this expression; it is sufficient to observe that Ω and Ω^* have single-valued real parts in $\mathfrak{M}_0$ by their definition. Hence their periods will be imaginary and the whole preceding term has the real part zero. Thus, using (4.2.23), we may put (7.5.5) into the form

$$\operatorname{Re}\left\{\frac{1}{2\pi i}\int_{\partial\mathfrak{M}_0} \Omega_{qq_0}(t)\; d\Omega^*_{pp_0}(t)\right\} = V^*(p, p_0; q, q_0) - V(p, p_0; q, q_0). \tag{7.5.6}$$

This proves the theorem.

Another elegant formula may be derived by the same considerations for the Abelian integral $\omega_{qq_0}(p)$ which is characterized by the property that its periods with respect to the cycles $K_{2\mu-1}$ are zero. This integral satisfies by (3.4.7)′ the symmetry law

$$W(p, p_0; q, q_0) = \omega_{qq_0}(p) - \omega_{qq_0}(p_0) = \omega_{pp_0}(q) - \omega_{pp_0}(q_0). \tag{7.5.7}$$

We find the following formula for W, analogous to (7.5.4):

$$W^*(p, p_0; q, q_0) - W(p, p_0; q, q_0)$$

(7.5.8)
$$= \frac{1}{2\pi i} \int_{\partial \mathfrak{M}_0} W(t, t_0; q, q_0) dW^*(t, t_0; p, p_0).$$

Let us apply Theorem 7.5.1 in order to calculate the Green's function V^* of the surface $\mathfrak{M}^*$ obtained by handle attachment. In this case, the domain $\mathfrak{M}_0$ consists of $\mathfrak{M}$ minus the interiors of the curves c_1 and c_2. Thus, we have:

(7.5.9)
$$V^*(p, p_0; q, q_0) - V(p, p_0; q, q_0) = \mathrm{Re}\left\{ \frac{1}{2\pi i} \int_{c_1 + c_2} \Omega_{qq_0}(t) d\Omega^*_{pp_0}(t) \right\}$$

where the integration along the curves c_1 and c_2 is to be understood in the positive sense with respect to $\mathfrak{M}_0$. In order to calculate the integrals we introduce the local uniformizers

(7.5.10)
$$\begin{cases} \zeta_1 = \exp\{-v(p, p_0; q_1, q_2)\}, & \text{for } p \text{ near } q_1, \\ \zeta_2 = \exp\{v(p, p_0; q_1, q_2)\}, & \text{for } p \text{ near } q_2. \end{cases}$$

Then the identification formula (7.5.3) establishes the following relation between the local uniformizers:

(7.5.11)
$$\zeta_1 = \frac{\eta \varepsilon^2}{\zeta_2},$$

where

(7512)
$$\eta = e^{i\alpha}, \quad |\eta| = 1.$$

Conversely, we may prescribe the parameter α in (7.5.3) or (7.5.12) arbitrarily and determine by this choice two particular points $p_1^{(0)}$ and $p_2^{(0)}$ on c_1 and c_2 which correspond under the identification.

In integrating in (7.5.9) over the curve c_1 we have to use the uniformizer ζ_1 in Ω but can use ζ_2 in Ω^* since Ω^* is an analytic function on the surface $\mathfrak{M}^*$. We may write, however,

(7.5.13)
$$\Omega_{qq_0}(t) = \Omega_{qq_0}(q_1) + \Omega'_{qq_0}(q_1)\zeta_1 + \frac{1}{2}\Omega''_{qq_0}(q_1)\zeta_1^2 + \cdots$$

and replace ζ_1 by ζ_2, using (7.5.11). We find

$$\begin{aligned}\frac{1}{2\pi i}\int_{c_1} \Omega_{qq_0}(t)\,d\Omega^*_{pp_0}(t) &= \Omega_{qq_0}(q_1)\cdot\frac{1}{2\pi i}\int_{c_1} d\Omega^*_{pp_0} \\ &+ \Omega'_{qq_0}(q_1)\cdot\frac{\varepsilon^2\eta}{2\pi i}\int_{c_2}\frac{1}{\zeta_2}\,d\Omega^*_{pp_0} \quad + \text{terms containing } \varepsilon^4 \text{ at least.}\end{aligned} \tag{7.5.14}$$

The integration over c_1 is to be carried out in the positive sense with respect to $\mathfrak{M}_0$, i.e. clockwise. Hence, the integrations over c_2 are to be taken in the counterclockwise sense. The integrations need not be carried out over the curve c_2 but may be taken over any homologous curve so long as neither p nor p_0 lies in the domain bounded by these two curves. This shows that the second righthand term in (7.5.14) is $O(\varepsilon^2)$ and the remainder term is $O(\varepsilon^4)$ so long as p and p_0 stay in a closed subdomain of $\mathfrak{M}_0$.

Let us consider, first, the term

$$\frac{1}{2\pi i}\int_{c_1} d\Omega^*_{pp_0} = -\frac{1}{2\pi}\int_{c_1}\frac{\partial V^*(t, t_0; p, p_0)}{\partial n_t}\,ds_t.$$

This expression is a real-valued harmonic function of p on $\mathfrak{M}^*$. If we approach c_1 from $\mathfrak{M}_0$ and cross this cycle, the value of the harmonic function will jump by unity. The function $V(p, p_0; q, q_0)$ is also harmonic but discontinuous in $\mathfrak{M}^*$ and jumps by the amount $2\log\varepsilon$. Let us assume that the original surface $\mathfrak{M}$ has h handles. Then, we may write

$$\operatorname{Im}\{Z^*_{2h+1}(p)\} = -\frac{1}{2\pi i}\int_{c_1} d\Omega^*_{pp_0} = \frac{V(p, p_0; q_1, q_2)}{2\log\dfrac{1}{\varepsilon}}. \tag{7.5.15}$$

We recognize that the integral (7.5.14) is $o(1)$ and that the analogous integral over c_2 converges to zero as $\varepsilon \to 0$. Thus, by (7.5.9), we have

$$V^*(p, p_0; q, q_0) - V(p, p_0; q, q_0) = o(1) \tag{7.5.16}$$

in each closed subdomain of $\mathfrak{M}_0$. If in (7.5.14) we replace the curve c_2 by a permissible homologous path of integration γ_2 in some fixed closed subdomain of $\mathfrak{M}_0$, we may replace $d\Omega^*$ in the integral by $d\Omega$ without affecting the result up to an order ε^2, since $\Omega^{*\prime}_{pp_0}(t) = 2\,\partial V^*(t, t_0; q, q_0)/\partial t$. Now, we may apply the residue theorem and we find that

$$
\text{(7.5.17)} \quad \frac{1}{2\pi i} \int_{c_1} \Omega_{qq_0}(t) d\Omega^*_{pp_0}(t) = - \Omega_{qq_0}(q_1) \, \mathrm{Im} \, \{Z^*_{2h+1}(p)\}
$$
$$
+ \, \varepsilon^2 \eta \, \Omega'_{qq_0}(q_1) \Omega'_{pp_0}(q_2) + O(\varepsilon^4).
$$

In a similar way, we may calculate the contribution of the integration over c_2. We observe that

$$
\frac{1}{2\pi i} \int_{c_1} d\Omega^*_{pp_0} = - \frac{1}{2\pi i} \int_{c_2} d\Omega^*_{pp_0}
$$

if both integrals are taken in the clockwise sense. Hence (7.5.9), (7.5.15) and (7.5.17) lead to the result

$$
\text{(7.5.18)} \quad V^*(p, p_0; q, q_0) = V(p, p_0; q, q_0) - \frac{V(q, q_0; q_1, q_2) V(p, p_0; q_1, q_2)}{2 \log \dfrac{1}{\varepsilon}}
$$
$$
+ \, \mathrm{Re} \, \{\varepsilon^2 \eta [\Omega'_{qq_0}(q_1) \Omega'_{pp_0}(q_2) + \Omega'_{qq_0}(q_2) \Omega'_{pp_0}(q_1)]\} + o(\varepsilon^2).
$$

Since $\Omega'_{qq_0}(q_1)$ can also be written as $2 \, \partial V(q_1, q_2; q, q_0)/\partial q_1$, we may express the variation of V in terms of V itself and of its derivatives:

$$
V^*(p, p_0; q, q_0) = V(p, p_0; q, q_0) - \frac{V(p, p_0; q_1, q_2) V(q, q_0; q_1, q_2)}{2 \log \dfrac{1}{\varepsilon}}
$$

$$
\text{(7.5.18)}' \quad + \, \mathrm{Re} \left\{ 4\varepsilon^2 \eta \left[\frac{\partial V(q_1, q_2; q, q_0)}{\partial q_1} \frac{\partial V(q_1, q_2; p, p_0)}{\partial q_2} \right. \right.
$$
$$
\left. \left. + \frac{\partial V(q_1, q_2; q, q_0)}{\partial q_2} \frac{\partial V(q_1, q_2; p, p_0)}{\partial q_1} \right] \right\} + o(\varepsilon^2).
$$

Differentiating (7.5.18)′ with respect to p and q and using definitions (4.2.25) and (4.10.1) we obtain the formula

$$
L^*_{\mathrm{M}}(p, q) = L_{\mathrm{M}}(p, q) - \frac{1}{\pi} \log \frac{1}{\varepsilon} Z^{*'}_{2h+1}(p) Z^{*'}_{2h+1}(q)
$$
$$
\text{(7.5.19)} \quad - \pi \varepsilon^2 \, \{\eta [L_{\mathrm{M}}(p, q_1) L_{\mathrm{M}}(q, q_2) + L_{\mathrm{M}}(p, q_2) L_{\mathrm{M}}(q, q_1)]
$$
$$
+ \, \bar{\eta} \, [L_{\mathrm{M}}(p, \tilde{q}_1) L_{\mathrm{M}}(q, \tilde{q}_2) + L_{\mathrm{M}}(p, \tilde{q}_2) L_{\mathrm{M}}(q, \tilde{q}_1)] \} + o(\varepsilon^2)
$$

where

$$(7.5.20)\qquad Z^*_{2h+1}(p) = \frac{i}{2\log\frac{1}{\varepsilon}}\left[\Omega_{q_1q_2}(p) - \Omega_{q_1q_2}(p_0)\right].$$

By formula (3.4.4)

$$(7.5.21)\qquad P(\Omega'_{q_1q_2}, K_\nu) = 2\pi i \operatorname{Im}\{Z_\nu(q_2) - Z_\nu(q_1)\},$$

so (7.5.20) gives:

$$(7.5.22)\qquad P(Z^{*\prime}_{2h+1}, K_\nu) = \frac{\pi}{\log\frac{1}{\varepsilon}}\operatorname{Im}\{Z_\nu(q_1) - Z_\nu(q_2)\},$$

that is

$$(7.5.23)\quad \Gamma^*_{2h+1,\nu} = -\frac{\pi}{\log\frac{1}{\varepsilon}}\operatorname{Im}\{Z_\nu(q_1) - Z_\nu(q_2)\},\quad \nu = 1, 2, \cdots, 2h.$$

It is interesting that the new periods $\Gamma^*_{2h+1,\nu}$ are independent of η, that is, of the particular way in which the curves c_1 and c_2 are identified.

We can also derive formulas for the change in the other integrals of the first kind by computing from (7.5.18) the periods of the analytic completions of the V-functions. The new period matrix $\Gamma^*_{\mu\nu}$ can then be obtained by considering the periods in the variational equations for the integrals of the first kind. We do not enter into these obvious calculations.

7.6. Attaching a Handle to a Surface with Boundary

We derived in the last section a formula for the change of the Green's function of a closed surface if a handle is attached to this surface. It is evident that this result enables us to derive an analogous formula for the case of a surface $\mathfrak{M}$ with boundary. In fact, we may imbed such a domain $\mathfrak{M}$ into the closed surface $\mathfrak{F} = \mathfrak{M} + \tilde{\mathfrak{M}}$ obtained by doubling $\mathfrak{M}$. If p and q are two points of $\mathfrak{M}$, the Green's function $G(p, q)$ has by (4.2.1) the form

$$(7.6.1)\qquad G(p, q) = \frac{1}{2} V(p, \tilde{p}; q, \tilde{q})$$

where V is the Green's function of the double $\mathfrak{F}$. Instead of considering the change of $\mathfrak{M}$ obtained by identifying points along two curves c_1 and c_2 in $\mathfrak{M}$, we may ask for the change of $\mathfrak{F}$ if we identify not only the points of these two curves but also the points of their images $\tilde{c}_1$ and $\tilde{c}_2$ in $\tilde{\mathfrak{M}}$ by the corresponding law. Such a variation will transform $\mathfrak{F}$ into a new closed surface $\mathfrak{F}^*$ and we will be able to compute the variation $V^* - V$ by the results of the last section.

We choose two points q_1, q_2 in $\mathfrak{M}$ and consider the function

$$U(p; q_1, q_2) = \frac{1}{2}[V(p, \tilde{p}; q_1, \tilde{q}_1) - V(p, \tilde{p}; q_2, \tilde{q}_2)] \tag{7.6.2}$$
$$= G(p, q_1) - G(p, q_2).$$

We consider the loci on $\mathfrak{F}$ where

$$U(p; q_1, q_2) = \log\frac{1}{\varepsilon}, \quad U(p; q_1, q_2) = \log \varepsilon. \tag{7.6.3}$$

They consist of the curves c_1 and c_2 in $\mathfrak{M}$ and, since $U(\tilde{p}; q_1, q_2) = -U(p; q_1, q_2)$, of the image curves $\tilde{c}_2$ and $\tilde{c}_1$ in $\tilde{\mathfrak{M}}$. Let $u(p; q_1, q_2)$ be that analytic function on $\mathfrak{F}$ which has $U(p; q_1, q_2)$ as real part. We identify points on c_1, c_2 and $\tilde{c}_1$, $\tilde{c}_2$ as in the last section by the relation

$$u(p_1; q_1, q_2) = u(p_2; q_1, q_2) + 2\log\frac{1}{\varepsilon} - i\alpha, \quad p_1 \in c_1, \quad p_2 \in c_2, \tag{7.6.3)'$$

where $\eta = e^{i\alpha}$,

$$u(\tilde{p}_1; q_1, q_2) = u(\tilde{p}_2; q_1, q_2) - 2\log\frac{1}{\varepsilon} + i\alpha, \quad \tilde{p}_1 \in \tilde{c}_1, \quad \tilde{p}_2 \in \tilde{c}_2. \tag{7.6.3)''$$

In the last section, we studied the effect of the attachment of one handle to a closed Riemann surface. If we had attached simultaneously several handles we should have been able to carry out the same considerations, integrating over several systems of identified curves instead of one. One can easily verify that, up to the order ε^2, the result of the attachment of several handles at the same time is obtained by adding the effects of each individual handle attachment. Thus, under the change (7.6.3)', (7.6.3)'' of $\mathfrak{F}$ into $\mathfrak{F}^*$, its Green's function will change according to the formula

$$
\begin{aligned}
V^*(p, p_0; q, q_0) &= V(p, p_0; q, q_0) \\
&- \frac{V(p, p_0; q_1, q_2)V(q, q_0; q_1, q_2)}{2\log\frac{1}{\varepsilon}} - \frac{V(p, p_0; \tilde{q}_1, \tilde{q}_2)V(q, q_0; \tilde{q}_1, \tilde{q}_2)}{2\log\frac{1}{\varepsilon}} \\
&+ \operatorname{Re}\left\{4\varepsilon^2\eta\left[\frac{\partial V(q_1, q_2; q, q_0)}{\partial q_1}\frac{\partial V(q_1, q_2; p, p_0)}{\partial q_2}\right.\right. \\
&\quad + \frac{\partial V(q_1, q_2; q, q_0)}{\partial q_2}\frac{\partial V(q_1, q_2; p, p_0)}{\partial q_1} \\
&\quad + \frac{\partial V(\tilde{q}_1, \tilde{q}_2; q, q_0)}{\partial \tilde{q}_1}\frac{\partial V(\tilde{q}_1, \tilde{q}_2; p, p_0)}{\partial \tilde{q}_2} \\
&\quad \left.\left. + \frac{\partial V(\tilde{q}_1, \tilde{q}_2; q, q_0)}{\partial \tilde{q}_2}\frac{\partial V(\tilde{q}_1, \tilde{q}_2; p, p_0)}{\partial \tilde{q}_1}\right]\right\} \\
&+ o(\varepsilon^2).
\end{aligned}
\tag{7.6.4}
$$

This lengthy formula reduces considerably if we put $p_0 = \tilde{p}$ and $q_0 = \tilde{q}$. We use the following simple identities:

$$V(q, \tilde{q}; \tilde{q}_1, \tilde{q}_2) = -V(q, \tilde{q}; q_1, q_2) \tag{7.6.5}$$

and

$$V(q, \tilde{q}; q_1, q_2) = G(q, q_1) - G(q, q_2) = U(q; q_1, q_2). \tag{7.6.6}$$

The first identity follows from the reality and antisymmetry of V, and the second identity becomes obvious if we observe that $V(q, \tilde{q}; q_1, q_2)$ is a harmonic function of q which vanishes on the boundary of $\mathfrak{M}$ and has simple logarithmic poles at q_1 and q_2. (7.6.6) may be considered as a generalization of (7.6.1). If we differentiate (7.6.6) with respect to q_1 and q_2, we obtain:

$$\frac{\partial V(q_1, q_2; q, \tilde{q})}{\partial q_1} = \frac{\partial G(q_1, q)}{\partial q_1}, \quad \frac{\partial V(q_1, q_2; q, \tilde{q})}{\partial q_2} = -\frac{\partial G(q_2, q)}{\partial q_2}. \tag{7.6.7}$$

Using all these relations, we obtain from (7.6.4)

$$
\begin{aligned}
G^*(p, q) &= G(p, q) - \frac{U(p; q_1, q_2)U(q; q_1, q_2)}{2\log\frac{1}{\varepsilon}} \\
&- \operatorname{Re}\left\{4\varepsilon^2\eta\left[\frac{\partial G(q_1, p)}{\partial q_1}\frac{\partial G(q_2, q)}{\partial q_2} + \frac{\partial G(q_2, p)}{\partial q_2}\frac{\partial G(q_1, q)}{\partial q_1}\right]\right\} + o(\varepsilon^2).
\end{aligned}
\tag{7.6.8}
$$

This formula gives the change of the Green's function of a domain with boundary if a handle attachments is performed by means of the identification (7.6.3)'.

Differentiating (7.6.8) with respect to p and q, we find that

$$L_M^*(p, q) = L_M(p, q) - \frac{1}{\pi} \log \frac{1}{\varepsilon} Z_{2h+1}^{*\prime}(p) Z_{2h+1}^{*\prime}(q)$$

(7.6.9)
$$- \pi \varepsilon^2 \{ \eta [L_M(p, q_1) L_M(q, q_2) + L_M(p, q_2) L_M(q, q_1)]$$
$$+ \bar{\eta} [L_M(p, \tilde{q}_1) L_M(q, \tilde{q}_2) + L_M(p, \tilde{q}_2) L_M(q, \tilde{q}_1)] \} + o(\varepsilon^2).$$

Here we write

(7.6.10)
$$Z_{2h+1}^*(p) = \frac{1}{2 \log \dfrac{1}{\varepsilon}} u(p; q_1, q_2)$$

which is indeed the new integral of the first kind added by the attachment of the new handle. Its real part is a single-valued harmonic function on $\mathfrak{M}^*$ but has a jump of amount 1 if we cross the cycle c_1, which equals c_2 after the identification.

We have by definition (7.6.2) and by (4.3.1)'

(7.6.11)
$$P(Z_{2h+1}^{*\prime}, K_\nu) = \frac{\pi}{\log \dfrac{1}{\varepsilon}} [\operatorname{Im} Z_\nu(q_1) - \operatorname{Im} Z_\nu(q_2)],$$
$$\nu = 1, 2, \cdots, 2h + m - 1.$$

That is,

(7.6.12)
$$\Gamma_{2h+1, \nu}^* = -P(Z_{2h+1}^{*\prime}, K_\nu) = -\frac{\pi}{\log \dfrac{1}{\varepsilon}} [\operatorname{Im} Z_\nu(q_1) - \operatorname{Im} Z_\nu(q_2)],$$
$$\nu = 1, 2, \cdots, 2h + m - 1.$$

Therefore, computing periods in (7.6.9), we have

$$Z_\nu^{*\prime}(q) = Z_\nu'(q) - Z_{2h+1}^*(q) \operatorname{Im} \{ Z_\nu(q_1) - Z_\nu(q_2) \}$$

(7.6.13)
$$- \pi \varepsilon^2 \{ \eta [Z_\nu'(q_1) L_M(q, q_2) + Z_\nu'(q_2) L_M(q, q_1)]$$
$$+ \bar{\eta} [Z_\nu'(\tilde{q}_1) L_M(q, \tilde{q}_2) + Z_\nu'(\tilde{q}_2) L_M(q, \tilde{q}_1)] \} + o(\varepsilon^2).$$

Finally, taking periods again in (7.6.13), we obtain

$$\Gamma^*_{\mu\nu} = \Gamma_{\mu\nu} + \frac{\pi}{\log\frac{1}{\varepsilon}} \operatorname{Im}\{Z_\mu(q_1) - Z_\mu(q_2)\} \cdot \operatorname{Im}\{Z_\nu(q_1) - Z_\nu(q_2)\}$$

$$\text{(7.6.14)} \qquad - 2\pi\varepsilon^2 \operatorname{Re}\{\eta\,[Z'_\mu(q_1)Z'_\nu(q_2) + Z'_\mu(q_2)Z'_\nu(q_1)]\} + o(\varepsilon^2).$$

Thus, we have obtained the variational formulas for the differentials of $\mathfrak{M}$.

7.7. Attaching a Cross-cap

For the sake of completeness, we now construct a variation which arises by attaching a cross-cap to the surface $\mathfrak{M}$. Means are then provided for passing, by infinitesimal steps, from the sphere to any finite Riemann surface, orientable or non-orientable. We recall that a non-orientable surface has a single-valued Green's function defined by (4.2.31).

Let $\mathfrak{M}$ be a surface (orientable or non-orientable) which has a single-valued Green's function $G(p, q)$. If $\mathfrak{M}$ is closed and non-orientable, it has no single-valued Green's function and we exclude this case, for the sake of simplicity. However, the treatment of the excluded case is the same, apart from the fact that care must be taken to specify the branches of the Green's function.

The method for the cross-cap attachment is closely related to the preceding method based on Theorem 7.5.1. However, instead of working with the analytic functions $\Omega_{pp_0}(t)$ we shall now work with the single-valued Green's function on the surface and instead of Cauchy's theorem we shall use Green's formula. We shall also make use of the Green's function in order to determine, in an invariant way, curves on $\mathfrak{M}$ to which the cross-cap is to be applied. For sufficiently small ε, $\varepsilon > 0$, the level curve

$$\text{(7.7.1)} \qquad G(p, q_1) = \log\frac{1}{\varepsilon}$$

is a Jordan curve c which bounds a subdomain of $\mathfrak{M}$ containing q_1. For p in a neighborhood of q_1, let

$$\text{(7.7.2)} \qquad \zeta = \exp\{-T(p, q_1)\}$$

where $T(p, q_1)$ is the analytic function whose real part is the Green's function. The curve c appears in the ζ-plane as the circle $|\zeta| = \varepsilon$.

We remove the interior of c from the surface $\mathfrak{M}$ and identify the point ζ of c with the point $-\varepsilon^2/\bar{\zeta}$, in this manner forming a new surface $\mathfrak{M}^*$ with a cross-cap. We observe that the identification is established by means of an anti-conformal (or indirect conformal) mapping. If ζ_1, ζ_2 are a pair of identified points on c, we have

$$\zeta_1 = -\frac{\varepsilon^2}{\bar{\zeta}_2}, \quad \zeta_2 = -\frac{\varepsilon^2}{\bar{\zeta}_1}. \tag{7.7.3}$$

The pair of identified points ζ_1, ζ_2 gives rise a single point of the surface $\mathfrak{M}^*$. In order to define a uniformizer at this point, let N_1 and N_2 be half-neighborhoods at ζ_1 and ζ_2 respectively which lie exterior to c and set

$$t = \begin{cases} \zeta, & \zeta \in N_1, \\ -\dfrac{\varepsilon^2}{\bar{\zeta}}, & \zeta \in N_2. \end{cases} \tag{7.7.4}$$

In the plane of t, the half-neighborhoods fit together to form a complete neighborhood, so t is the desired uniformizer.

Let $\partial/\partial n$ denote differentiation with respect to the normal which points into the exterior of the circle c. If r_1, r_2 are a pair of identified points on c with coordinates ζ_1, ζ_2 respectively, we have

$$\frac{\partial}{\partial n}(\zeta_2) ds_2 = -\frac{\partial}{\partial n}\left(\frac{-\varepsilon^2}{\bar{\zeta}_1}\right) ds_1 = -\frac{\partial}{\partial n}(\zeta_1) ds_1 \tag{7.7.5}$$

and

$$\frac{\partial G(r_1, q)}{\partial n_2} ds_2 = -\frac{\partial G(r_1, q)}{\partial n_1} ds_1. \tag{7.7.6}$$

Suppose first that $\mathfrak{M}$ is a non-orientable surface with boundary and let $G^*(p, q)$ be the single-valued Green's function of $\mathfrak{M}^*$ which is defined by means of the quadruple (Section 4.2). Then

$$G^*(r_2, p) = G^*(r_1, p), \quad \frac{\partial G^*(r_2, p)}{\partial n_2} ds_2 = -\frac{\partial G^*(r_1, p)}{\partial n_1} ds_1. \tag{7.7.7}$$

Hence, writing $c = c_1 + c_2$, where c_1 is the upper semi-circle, c_2 the lower semi-circle belonging to c, we have

$$\int_c \frac{\partial G^*(r, p)}{\partial n}\, ds = \int_{c_1} \frac{\partial G^*(r_1, p)}{\partial n_1}\, ds_1 + \int_{c_2} \frac{\partial G^*(r_2, p)}{\partial n_2}\, ds_2 = 0, \tag{7.7.8}$$

by (7.7.7). By Green's formula applied to the surface $\mathfrak{M}^*$ cut along the cycle c, we have

$$\begin{aligned} &G^*(p, q) - G(p, q) \\ &= \frac{1}{2\pi} \int_c \left[G^*(r_1, p) \frac{\partial G(r_1, q)}{\partial n_1} - G(r_1, q) \frac{\partial G^*(r_1, p)}{\partial n_1} \right] ds_1 \\ &= -\frac{1}{2\pi} \int_c \left[G^*(r_1, p) \frac{\partial G(r_1, q)}{\partial n_2} - G(r_1, q) \frac{\partial G^*(r_2, p)}{\partial n_2} \right] ds_2, \end{aligned} \tag{7.7.9}$$

by (7.7.6) and (7.7.7). We may develop $G(r, q)$ into a series about the point q_1, obtaining

$$G(r, q) = G(q_1, q) + 2 \operatorname{Re}\left\{ \frac{\partial G(q_1, q)}{\partial q_1} \zeta + \frac{1}{2} \frac{\partial^2 G(q_1, q)}{\partial q_1^2} \zeta^2 + \cdots \right\}. \tag{7.7.10}$$

We take $\zeta = \zeta_1 = -\varepsilon^2/\bar{\zeta}_2$ in (7.7.10) and substitute into (7.7.9). Dropping the subscript 2, we obtain

$$\begin{aligned} &G^*(p, q) - G(p, q) \\ &= \frac{1}{2\pi} \int_c \Bigg[G^*(r, p) \frac{\partial}{\partial n_r} \left(2 \operatorname{Re}\left\{ \frac{\partial G(q_1, q)}{\partial q_1} \frac{\varepsilon^2}{\bar{\zeta}} - \frac{1}{2} \frac{\partial^2 G(q_1, q)}{\partial q_1^2} \left(\frac{\varepsilon^4}{\bar{\zeta}^2} \right) + \cdots \right\} \right) \\ &\quad - \frac{\partial G^*(r, p)}{\partial n_r} \left(2 \operatorname{Re}\left\{ \frac{\partial G(q_1, q)}{\partial q_1} \frac{\varepsilon^2}{\bar{\zeta}} - \frac{1}{2} \frac{\partial^2 G(q_1, q)}{\partial q_1^2} \left(\frac{\varepsilon^4}{\bar{\zeta}^2} \right) + \cdots \right\} \right) \Bigg] ds_r \\ &\quad + G(q_1, q) \frac{1}{2\pi} \int_c \frac{\partial G^*(r, p)}{\partial n_r}\, ds_r. \end{aligned} \tag{7.7.11}$$

The second term on the right is zero by (7.7.8). To estimate the first term, let c' be the circle $|\zeta| = \alpha$, where α is a positive number independent of ε, $\alpha > \varepsilon$. By (7.7.11) and Green's formula

$$G^*(p, q) - G(p, q)$$

$$= \frac{1}{2\pi}\int_{c'}\left[G^*(r, p)\frac{\partial}{\partial n_r}\left(2\,\mathrm{Re}\left\{\frac{\partial G(q_1, q)}{\partial q_1}\frac{\varepsilon^2}{\zeta}\right\}\right)\right.$$

$$\left. - \frac{\partial G^*(r, p)}{\partial n_r}\left(2\,\mathrm{Re}\left\{\frac{\partial G(q_1, q)}{\partial q_1}\frac{\varepsilon^2}{\zeta}\right\}\right)\right]ds_r + O(\varepsilon^4)$$

$$= \frac{1}{2\pi}\int_{c'}\left[G(r, p)\frac{\partial}{\partial n_r}\left(2\,\mathrm{Re}\left\{\frac{\partial G(q_1, q)}{\partial q_1}\frac{\varepsilon^2}{\zeta}\right\}\right)\right.$$

$$\left. - \frac{\partial G(r, p)}{\partial n_r}\left(2\,\mathrm{Re}\left\{\frac{\partial G(q_1, q)}{\partial q_1}\frac{\varepsilon^2}{\zeta}\right\}\right)\right]ds_r + O(\varepsilon^4)$$

$$= \mathrm{Re}\left\{\frac{\partial G(q_1, q)}{\partial \bar{q}_1}\frac{2\varepsilon^2}{2\pi}\int_{c'}\left[G(r, p)\frac{\partial}{\partial n}\left(\frac{1}{\zeta}\right)\right.\right.$$

$$\left.\left. - \left(\frac{1}{\zeta}\right)\frac{\partial G(r, p)}{\partial n}\right]ds\right\} + O(\varepsilon^4)$$

$$= \mathrm{Re}\left\{\frac{\partial G(q_1, q)}{\partial \bar{q}_1}\frac{\partial}{\partial q_1}\frac{4\varepsilon^2}{2\pi}\int_{c'}\left[G(r, p)\frac{\partial}{\partial n}\log\frac{1}{|\zeta|}\right.\right.$$

$$\left.\left. - \log\left(\frac{1}{|\zeta|}\right)\frac{\partial G(r, p)}{\partial n}\right]ds\right\} + O(\varepsilon^4)$$

$$= -\,\mathrm{Re}\left\{4\varepsilon^2\frac{\partial G(q_1, p)}{\partial q_1}\frac{\partial G(q_1, q)}{\partial \bar{q}_1}\right\} + O(\varepsilon^4),$$

that is,

$$(7.7.12)\quad G^*(p, q) - G(p, q) = -\,\mathrm{Re}\left\{4\varepsilon^2\frac{\partial G(q_1, p)}{\partial q_1}\frac{\partial G(q_1, q)}{\partial \bar{q}_1}\right\} + O(\varepsilon^4).$$

In formula (7.7.12), we have obtained an asymptotic expression for the new Green's function $G^*(p, q)$ in terms of the original Green's function $G(p, q)$ and its derivatives. In all variational formulas the end result is of this form; but it is, in general obtained by first deriving an exact integro-differential equation between the new and the old Green's functions and by replacing G^* by G in some places, introducing a small error but making the result more easily applicable. We want now to replace the asymptotic formula (7.7.12) by an exact integral equation for $G^*(p, q)$.

To obtain such a formula, we go back to (7.7.9). By (7.7.8) we have

$$G^*(p, q) - G(p, q)$$

$$= \frac{1}{2\pi}\int_c \left[G^*(r, p)\frac{\partial}{\partial n_r}(G(r, q) - G(q_1, q)) - \frac{\partial G^*(r, p)}{\partial n_r}(G(r, q) - G(q_1, q))\right] ds_r, \tag{7.7.13}$$

$$= \frac{1}{2\pi}\int_c \left[G^*(r, p)\frac{\partial \Delta(r, q)}{\partial n_r} - \Delta(r, q)\frac{\partial G^*(r, p)}{\partial n_r}\right] ds_r$$

where

$$\Delta(r, q) = G(r, q) - G(q_1, q). \tag{7.7.14}$$

Let

$$\begin{aligned} T^*(r, p) &= G^*(r, p) + iH^*(r, p); \\ Q(r, q) &= T(r, q) - T(q_1, q) = \Delta(r, q) + iJ(r, q). \end{aligned} \tag{7.7.15}$$

Then

$$G^*(p, q) - G(p, q) = -\frac{1}{2\pi}\int_c [G^*(r, p)dJ(r, q) - \Delta(r, q)dH^*(r, p)]. \tag{7.7.16}$$

In this formula the integration around c is in the clockwise sense. By (7.7.8),

$$\int_c dH^*(r, p) = 0, \tag{7.7.17}$$

so, integrating by parts,

$$G^*(p, q) - G(p, q) = -\frac{1}{2\pi}\int_c [G^*(r, p)dJ(r, q) + H^*(r, p)\,d\Delta(r, q)]$$

$$= \mathrm{Re}\left\{\frac{i}{2\pi}\int_c [G^*(r, p)dQ(r, q) + iH^*(r, p)\,dQ(r, q)]\right\} \tag{7.7.18}$$

$$= -\mathrm{Re}\left\{\frac{1}{2\pi i}\int_c T^*(r, p)dQ(r, q)\right\} = \mathrm{Re}\left\{\frac{1}{2\pi i}\int_c Q(r, q)dT^*(r, p)\right\}.$$

That is,

$$G^*(p, q) - G(p, q) = \mathrm{Re}\left\{\frac{1}{2\pi i}\int_c Q(r, q)\,dT^*(r, p)\right\}. \tag{7.7.19}$$

Suppose, however, that $G^*(p, q)$ is the Green's function of $\mathfrak{M}^*$ defined by means of the double. In this case, $\mathfrak{M}$ may be either closed or with boundary. Instead of (7.7.7), we have

$$(7.7.7)' \quad G^*(r_2, p) = -G^*(r_1, p); \quad \frac{\partial G^*(r_2, p)}{\partial n_2} ds_2 = \frac{\partial G^*(r_1, p)}{\partial n_1} ds_1,$$

where r_1, r_2 are identified points of c. The function $G^*(p, q)$ is not single-valued on $\mathfrak{M}^*$, but it will be single-valued on $\mathfrak{M}^*$ cut along c. Writing

$$(7.7.20) \qquad I(p) = \frac{1}{2\pi} \int_c \frac{\partial G^*(r, p)}{dn_r} ds_r,$$

we do not have $I(p) \equiv 0$, but, instead,

$$(7.7.8)' \qquad I(p) = \frac{G(q_1, p)}{\log \frac{1}{\varepsilon}}.$$

In fact, if p is a point of c, the function $G^*(r, p)$ has a logarithmic pole not only at the point p but also at the point $\tilde{p}$ of c. Let c be oriented in the sense in which exterior area lies to the left, and let p', p'' be points on opposite edges of c, p' on the left edge, p'' on the right. Then $\tilde{p}'$, $\tilde{p}''$ lie respectively on the right and left edges of c. We have

$$(7.7.21) \quad I(p') = I(p'') + \frac{1}{2\pi} \int_k \frac{\partial G^*(r, p)}{\partial n_r} ds_r + \frac{1}{2\pi} \int_{\tilde{k}} \frac{\partial G^*(r, \tilde{p})}{\partial n_r} ds_r,$$

where k is a small circle surrounding the point $p = p' = p''$ and $\tilde{k}$ is a small circle surrounding the point $\tilde{p}$. In the integral over k the normal derivative is in the direction exterior to k while in the integral over $\tilde{k}$ it is in the opposite direction. Hence

$$(7.7.22) \qquad I(p') = I(p'') - 2.$$

Now let p approach the point r_1 of c from the exterior of c. Then the point $\tilde{p}$ approaches the identified point r_2 of c from the interior and we have

$$I(r_1) + I(r_2) = 0.$$

If, however, the value $I(r_2)$ is obtained by approaching r_2 from the exterior of c, then by (7.7.22)

$$(7.7.23) \qquad I(r_1) + I(r_2) = 2.$$

By (7.7.1) and (7.7.2)

$$(7.7.24)\quad G(q_1, r_2) = G(q_1, r_1); \quad \frac{\partial G(q_1, r_2)}{\partial n_2} ds_2 = \frac{\partial G(q_1, r_1)}{\partial n_1} ds_1.$$

Hence, writing

$$(7.7.25)\qquad u(p) = I(p) - \frac{G(q_1, p)}{\log \dfrac{1}{\varepsilon}},$$

we have

$$(7.7.26)\qquad u(r_1) + u(r_2) = 0, \quad \frac{\partial u(r_2)}{\partial n_2} ds_2 = \frac{\partial u(r_1)}{\partial n_1} ds_1.$$

Applying Green's formula to the surface $\mathfrak{M}^*$ cut along c and observing that both $u(p)$ and $G^*(p, q)$ vanish for p on the boundary of $\mathfrak{M}^*$, we have

$$(7.7.27)\quad u(q) = \frac{1}{2\pi} \int_c \left[G^*(r, q) \frac{\partial u(r)}{\partial n_r} - u(r) \frac{\partial G^*(r, q)}{\partial n_r} \right] ds_r = 0,$$

by (7.7.7)′ and (7.7.26). This is equivalent to (7.7.8)′.

By Green's formula

$$(7.7.28)\quad \begin{aligned} &G^*(p, q) - G(p, q) \\ &= \frac{1}{2\pi} \int_c \left[G^*(r_1, p) \frac{\partial G(r_1, q)}{\partial n_1} - G(r_1, q) \frac{\partial G^*(r_1, p)}{\partial n_1} \right] ds_1 \\ &= \frac{1}{2\pi} \int_c \left[G^*(r_2, p) \frac{\partial G(r_1, q)}{\partial n_2} - G(r_1, q) \frac{\partial G^*(r_2, p)}{\partial n_2} \right] ds_2, \end{aligned}$$

by (7.7.6) and (7.7.7)′. Substituting from (7.7.10) with $\zeta = \zeta_1 = -\varepsilon^2/\bar{\zeta}_2$, and dropping the subscript 2, we find by (7.7.8)′ that

$$\begin{aligned} &G^*(p, q) - G(p, q) \\ &= -\frac{1}{2\pi} \int_c \left[G^*(r, p) \frac{\partial}{\partial n_r} \left(2\,\mathrm{Re} \left\{ \frac{\partial G(q_1, q)}{\partial q_1} \frac{\varepsilon^2}{\bar{\zeta}} - \frac{1}{2} \frac{\partial^2 G(q_1, q)}{\partial q_1^2} \left(\frac{\varepsilon^4}{\bar{\zeta}^2} \right) + \cdots \right\} \right) \right. \\ &\quad \left. - \frac{\partial G^*(r, p)}{\partial n_r} \left(2\,\mathrm{Re} \left\{ \frac{\partial G(q_1, q)}{\partial q_1} \frac{\varepsilon^2}{\bar{\zeta}} - \frac{1}{2} \frac{\partial^2 G(q_1, q)}{\partial q_1^2} \left(\frac{\varepsilon^4}{\bar{\zeta}^2} \right) + \cdots \right\} \right) \right] ds_r \\ &\quad - \frac{G(q_1, p) G(q_1, q)}{\log \dfrac{1}{\varepsilon}}. \end{aligned}$$

If the sign of $G^*(p, q)$ is chosen properly, then by the same reasoning that led to (7.7.12) we obtain the formula

$$G^*(p, q) - G(p, q) \tag{7.7.29}$$

$$= -\frac{G(q_1, p)G(q_1, q)}{\log \frac{1}{\varepsilon}} + \operatorname{Re}\left\{ 4\varepsilon^2 \frac{\partial G(q_1, p)}{\partial q_1} \frac{\partial G(q_1, q)}{\partial \tilde{q}_1} \right\} + O(\varepsilon^4).$$

7.8. INTERIOR DEFORMATION BY ATTACHING A CELL. FIRST METHOD

To compensate for the special nature of the preceding variations, we now make a general interior deformation which results from "attaching a cell" to the surface $\mathfrak{M}$. In the remainder of this chapter we suppose, for simplicity, that $\mathfrak{M}$ is orientable. Let γ be an analytic Jordan curve in $\mathfrak{M}$. Although the hypothesis is not essential, we suppose for simplicity of notation that γ lies in the domain of a local uniformizer z. Let $r(z)$ be a function which is regular analytic in a complete neighborhood of γ. If ε is a sufficiently small positive number, the point defined by $z + \varepsilon r(z)$ will trace a neighboring Jordan curve γ_ε as z traces γ. Let $\mathfrak{M}'$ denote the subdomain of $\mathfrak{M}$ exterior to γ, $\mathfrak{M}''$ the subdomain of $\mathfrak{M}$ interior to γ_ε. In general, $\mathfrak{M}'$ and $\mathfrak{M}''$ will not be disjoint but will overlap. We attach the surface $\mathfrak{M}''$ to $\mathfrak{M}'$ by identifying the point z on the exterior edge of γ with the point $z + \varepsilon r(z)$ on the interior edge of γ_ε. The resulting surface $\mathfrak{M}^*$ is thus obtained by removing from $\mathfrak{M}$ the interior of γ and then filling up the hole so formed by attaching the piece of surface which consists of the interior of γ_ε.

This construction of the surface $\mathfrak{M}^*$ is expressed in terms of a particular uniformizer z. If we change to another uniformizer τ, we have

$$\tau(z + \varepsilon r(z)) = \tau + \varepsilon \varrho_\varepsilon(\tau), \quad \tau = \tau(z). \tag{7.8.1}$$

The arcs γ, γ_ε appear in the plane of τ as arcs $\bar{\gamma}$, $\bar{\gamma}_\varepsilon$, and the point τ of $\bar{\gamma}$ is identified with the point $\tau + \varepsilon \varrho_\varepsilon(\tau)$ of $\bar{\gamma}_\varepsilon$. We observe that

$$\varrho_\varepsilon(\tau) = \varrho_0(\tau) + O(\varepsilon) \tag{7.8.2}$$

where

$$\varrho_0(\tau) = r(z)\frac{d\tau}{dz}. \tag{7.8.3}$$

Thus, in the limit, as ε tends to zero, the function $r(z)$ transforms like a reciprocal differential.

Let us assume first that $\mathfrak{M}$ is a closed surface. In this case, $\mathfrak{M}^*$ will also be closed and both surfaces have the domain $\mathfrak{M}'$, exterior to γ, in common. If we again denote the integrals of the third kind for $\mathfrak{M}$ and $\mathfrak{M}^*$ by $\Omega_{qq_0}(p)$ and $\Omega^*_{qq_0}(p)$ and the corresponding Green's functions by $V(p, p_0; q, q_0)$ and $V^*(p, p_0; q, q_0)$ we may determine the variation of the Green's function under the deformation considered in Theorem 7.5.1. Let p, p_0, q, q_0 be a quadruple of points lying in a fixed closed subdomain of $\mathfrak{M}'$; then

$$V^*(p, p_0; q, q_0) - V(p, p_0; q, q_0) = \text{Re}\left\{\frac{1}{2\pi i}\int_\gamma \Omega_{qq_0}(t)\, d\Omega^*_{pp_0}(t)\right\}. \tag{7.8.4}$$

Consider next the interior $\mathfrak{M}''$ of γ_ε; here both functions $\Omega_{qq_0}(t)$ and $\Omega^*_{pp_0}(t)$ are regular analytic since their poles are, for ε small enough, outside $\mathfrak{M}''$. Hence, by Cauchy's theorem

$$0 = \frac{1}{2\pi i}\int_{\gamma_\varepsilon} \Omega_{qq_0}(t_1)\, d\Omega^*_{pp_0}(t_1). \tag{7.8.5}$$

Because of the identification between γ_ε and γ on $\mathfrak{M}^*$, we have for $t_1 \in \gamma_\varepsilon$ and $t \in \gamma$, $t_1 = t + \varepsilon r(t)$,

$$d\Omega^*_{pp_0}(t_1) = d\Omega^*_{pp_0}(t).$$

Replacing the variable of integration t_1 in (7.8.5) by t and subtracting (7.8.5) from (7.8.4), we obtain

$$\begin{aligned} &V^*(p, p_0; q, q_0) - V(p, p_0; q, q_0) \\ &= \text{Re}\left\{\frac{1}{2\pi i}\int_\gamma [\Omega_{qq_0}(t) - \Omega_{qq_0}(t + \varepsilon r(t))]\, d\Omega^*_{pp_0}(t)\right\}. \end{aligned} \tag{7.8.6}$$

Since all functions under the integral sign are analytic, we may replace the integration over γ by an integration over any fixed curve γ_0 in $\mathfrak{M}'$ which does not enclose any of the points p, p_0, q, q_0. On γ_0, we clearly have

$$\Omega_{qq_0}(t) - \Omega_{qq_0}(t + \varepsilon r(t)) = -\varepsilon r(t)\Omega'_{qq_0}(t) + O(\varepsilon^2).$$

Thus, $V^* - V$ will be $O(\varepsilon)$ for p, p_0, q, q_0 in the closed subdomain of $\mathfrak{M}'$. We have, moreover,

$$\Omega^{*\prime}_{pp_0}(t) = 2\frac{\partial}{\partial t}V^*(t, t_0; p, p_0); \tag{7.8.7}$$

hence, $\Omega^{*\prime} - \Omega'$ is $O(\varepsilon)$ in that subregion of $\mathfrak{M}'$, and we obtain from (7.8.6)

$$\begin{aligned} &V^*(p, p_0; q, q_0) - V(p, p_0; q, q_0) \\ &= \mathrm{Re}\left\{-\frac{\varepsilon}{2\pi i}\int_\gamma r(t)\Omega'_{qq_0}(t)\, d\Omega_{pp_0}(t)\right\} + O(\varepsilon^2). \end{aligned} \tag{7.8.8}$$

Using (7.8.7), we may bring this result into the final form:

$$\begin{aligned} &V^*(p, p_0; q, q_0) - V(p, p_0; q, q_0) \\ &= \mathrm{Re}\left\{-\frac{2\varepsilon}{\pi i}\int_\gamma r(t)\frac{\partial}{\partial t}V(t, t_0; p, p_0)\frac{\partial}{\partial t}V(t, t_0; q, q_0)dt\right\} + O(\varepsilon^2). \end{aligned} \tag{7.8.9}$$

This formula is our main result in the variation arising from cell attachment. The variational formulas for all other functions and differentials on $\mathfrak{M}$ are easily derived from it. Differentiating (7.8.9) with respect to p and q, we find that

$$\begin{aligned} L^*_{\mathbf{M}}(p, q) - L_{\mathbf{M}}(p, q) &= \frac{\varepsilon}{2i}\int_\gamma r(t)\, L_{\mathbf{M}}(t, p)\, L_{\mathbf{M}}(t, q)dt \\ &+ \left(\frac{\varepsilon}{2i}\int_\gamma r(t)L_{\mathbf{M}}(t, \bar{p})L_{\mathbf{M}}(t, \bar{q})dt\right)^- + O(\varepsilon^2). \end{aligned} \tag{7.8.10}$$

By computing periods in the usual way, we obtain the formulas:

$$\begin{aligned} Z^{*\prime}_\mu(q) - Z'_\mu(q) &= \frac{\varepsilon}{2i}\int_\gamma r(t)Z'_\mu(t)L_{\mathbf{M}}(t, q)dt \\ &+ \left(\frac{\varepsilon}{2i}\int_\gamma r(t)Z'_\mu(t)L_{\mathbf{M}}(t, \bar{q})dt\right)^- + O(\varepsilon^2) \end{aligned} \tag{7.8.11}$$

and

$$\begin{aligned} \Gamma^*_{\mu\nu} - \Gamma_{\mu\nu} &= -\frac{\varepsilon}{2i}\int_\gamma r(t)Z'_\mu(t)Z'_\nu(t)\,dt \\ &- \left(\frac{\varepsilon}{2i}\int_\gamma r(t)\, Z'_\mu(t)\, Z'_\nu(t)dt\right)^- + O(\varepsilon^2). \end{aligned} \tag{7.8.12}$$

We may avoid the integrals in the variational formulas if we choose $r(t)$ to be analytic inside γ except at the single point q_0 where it has a simple pole with residue a/π. In this case, all integrals may be evaluated by the residue theorem and yield:

$$L_{\mathbf{M}}^*(p, q) = L_{\mathbf{M}}(p, q) - \varepsilon[aL_{\mathbf{M}}(p, q_0)L_{\mathbf{M}}(q, q_0) + \bar{a}L_{\mathbf{M}}(p, \tilde{q}_0)L_{\mathbf{M}}(q, \tilde{q}_0)] + O(\varepsilon^2), \tag{7.8.13}$$

$$Z_\mu^{*\prime}(q) = Z'_\mu(q) - \varepsilon[aZ'_\mu(q_0)L_{\mathbf{M}}(q, q_0) + \bar{a}Z'_\mu(\tilde{q}_0)L_{\mathbf{M}}(q, \tilde{q}_0)] + O(\varepsilon^2), \tag{7.8.14}$$

$$\Gamma_{\mu\nu}^* = \Gamma_{\mu\nu} + 2\varepsilon \operatorname{Re}\{aZ'_\mu(q_0)Z'_\nu(q_0)\} + O(\varepsilon^2). \tag{7.8.15}$$

In these formulas, $r(z)$ is to be considered as a reciprocal differential. In particular, under a change of uniformizer the value a transforms in such a way that a/dz_0^2 is invariant.

Let us now consider the case of a domain $\mathfrak{M}$ with boundary. We may imbed $\mathfrak{M}$ in the closed surface $\mathfrak{F} = \mathfrak{M} + \tilde{\mathfrak{M}}$ and extend the deformation of $\mathfrak{M}$ by cell-attachment to a deformation of $\mathfrak{F}$ by attaching an analogous cell at the double $\tilde{\gamma}$ of the Jordan curve γ. We use there the function $(r(t))^-$ instead of $r(t)$ so that the deformation transforms $\mathfrak{M}$ and $\tilde{\mathfrak{M}}$ into two domains $\mathfrak{M}^*$ and $\tilde{\mathfrak{M}}^*$ which are related to each other by the $\sim$-operation. We easily calculate the variation of the Green's function G of $\mathfrak{M}$ under the variation considered. We use (7.6.1) and (7.6.6) and observe that the integrations over γ and over $\tilde{\gamma}$ yield the same contributions in the variational formula (7.8.9). We thus obtain:

$$G^*(p, q) = G(p, q) - \operatorname{Re}\left\{\frac{2\varepsilon}{\pi i}\int_\gamma r(t)\frac{\partial G(t, q)}{\partial t}\frac{\partial G(t, p)}{\partial t}\,dt\right\} + O(\varepsilon^2). \tag{7.8.16}$$

From this formula, we may again derive variational formulas for the Z'_μ and the $\Gamma_{\mu\nu}$. An easy calculation shows that we obtain the formulas (7.8.10)—(7.8.15), just as in the case of a closed surface.

We considered in Section 7.5 the function $W(p, p_0; q, q_0)$ which plays a central role in the theory of those Abelian integrals on a surface which satisfy the Riemann normalization, that is, which have vanishing periods with respect to the cycles $K_{2\mu-1}$. In (7.5.8) we gave a formula comparing W^* and W and now apply this result in the case of a closed surface $\mathfrak{M}$ which is subjected to a cell-attachment of the above form. An easy calculation leads to the result:

$$(7.8.17)\quad \begin{aligned} &W^*(p, p_0; q, q_0) - W(p, p_0; q, q_0) \\ &= -\frac{\varepsilon}{2\pi i}\int_\gamma r(t)W'(t, t_0; p, p_0)\, W'(t, t_0; q, q_0)dt + O(\varepsilon^2). \end{aligned}$$

where the prime on W indicates differentiation with respect to t.

If we use the particular choice of $r(t)$ which makes $r(t)$ regular analytic inside γ except for the pole at r with residue a, we obtain the particularly elegant formula:

$$(7.8.18)\quad \begin{aligned} W^*(p, p_0; q, q_0) &= W(p, p_0; q, q_0) \\ &\quad + \varepsilon a W'(r, r_0; p, p_0)W'(r, r_0; q, q_0) + O(\varepsilon^2) \end{aligned}$$

where r_0 is an arbitrary (unessential) point. This simple variational behavior makes the function $W(p, p_0; q, q_0)$ a useful tool in the study of closed Riemann surfaces and of the inter-relations between their various differentials and periods.

7.9. Interior Deformation by Attaching a Cell. Second Method

A somewhat different variation is obtained if, instead of taking γ to be a Jordan curve, we take it to be an analytic arc joining the points α, β of $\mathfrak{M}$. We assume that γ lies in the domain of a local uniformizer z, and we impose the condition that $r(z)$ is regular analytic in a complete neighborhood of γ with $r(\alpha) = r(\beta) = 0$. If ε is a sufficiently small positive number, the point defined by $z + \varepsilon r(z)$ will trace a neighboring arc γ_ε joining α with β. Let the arcs γ and γ_ε be oriented in the direction from α to β, and let $\mathfrak{M}'$ be the domain "lying over" $\mathfrak{M}$ which is bounded by the boundary of $\mathfrak{M}$ (if there is any), the left edge of γ and the right edge of γ_ε. In general, $\mathfrak{M}'$ will not be a subdomain of $\mathfrak{M}$ since two points of $\mathfrak{M}'$ may lie over a point of $\mathfrak{M}$ in the neighborhood of γ. By identifying the point z on the left edge of γ with the point $z + \varepsilon r(z)$ on the right edge of γ_ε, we obtain a surface $\mathfrak{M}^*$ which is topologically equivalent to $\mathfrak{M}$.

Let p', p'' be a pair of identified points where p' is an interior point of γ lying on its left edge, p'' a point on the right edge of γ_ε, and let N', N'' be half-neighborhoods at p', p'', N' lying on the left of γ, N'' on the right of γ_ε. Define

$$t = \begin{cases} z + \varepsilon r(z), & z \in N', \\ z, & z \in N''. \end{cases}$$

In the plane of t the half-neighborhoods fit together to form a complete neighborhood, so the variable t acts as a uniformizer at pairs of identified points which correspond to interior points of the arcs γ, γ_ε. A complete neighborhood of the point α on the surface $\mathfrak{M}^*$ can be mapped conformally onto the interior of a circle; this follows from the uniformization theorem. The fact that α maps into a point, and not into a proper continuum, follows readily (a proof is to be found in [5]). Therefore $\mathfrak{M}^*$ is a Riemann surface which is topologically equivalent to $\mathfrak{M}$.

Suppose first that $\mathfrak{M}$ is a closed surface. Using Theorem 7.5.1, we have the equation

$$V^*(p, p_0; q, q_0) - V(p, p_0; q, q_0) = \mathrm{Re}\left\{\frac{1}{2\pi i}\int_{\gamma_0} \Omega_{qq_0}(t)\, d\Omega^*_{pp_0}(t)\right\} \tag{7.9.1}$$

where p, p_0, q, q_0 are a quadruple of points lying in a closed subdomain of $\mathfrak{M}$ which does not contain the curves γ and γ_ε, and where γ_0 is a curve surrounding γ and γ_ε. Both $\Omega_{qq_0}(t)$ and $\Omega^*_{pp_0}(t)$ are regular analytic in that part of $\mathfrak{M}'$ which lies inside γ_0. Hence, we may deform the path of integration in (7.9.1) and take instead of γ_0 the boundary of $\mathfrak{M}'$, i.e. the curve system $\gamma - \gamma_\varepsilon$. By the identification in $\mathfrak{M}^*$, we have

$$d\Omega^*_{pp_0}(t_1) = d\Omega^*_{pp_0}(t), \quad t \in \gamma,\ t_1 = t + \varepsilon r(t). \tag{7.9.2}$$

Hence, (7.9.1) may be brought into the form

$$\begin{aligned} &V^*(p, p_0, q; q_0) - V(p, p_0; q, q_0) \\ &= \mathrm{Re}\left\{\frac{1}{2\pi i}\int_{\gamma} [\Omega_{qq_0}(t) - \Omega_{qq_0}(t + \varepsilon r(t))]\, d\Omega^*_{pp_0}(t)\right\}. \end{aligned} \tag{7.9.3}$$

This corresponds exactly to formula (7.8.6). Integration is along the left edge of the curve γ and we may deform the path of integration γ into another curve γ_1 which also connects α with β but otherwise lies to the left of γ. It can be shown [5] that, in the new integration $d\Omega^*_{pp_0}$ can be replaced by $d\Omega_{pp_0}$ with an error $o(1)$. Therefore, (7.9.3) has the form,

$$V^*(p, p_0; q, q_0) - V(p, p_0; q, q_0) = \operatorname{Re}\left\{-\frac{\varepsilon}{2\pi i}\int_\gamma r(t)\Omega'_{qq_0}(t)\,\Omega'_{pp_0}(t)\,dt\right\} + o(\varepsilon). \tag{7.9.4}$$

Thus, we arrive at a formula which is formally the same as that given in the previous section for a Jordan curve γ. All other results of the last section can now obtained be by reasoning entirely analogous to that of Section 7.8 and formulas (7.8.9)—(7.8.15) remain valid for the case in which γ is an arc.

7.10. THE VARIATION KERNEL

In order to construct certain further variations, it will be necessary to introduce a surface functional $n(p, q)$ which transforms like a reciprocal differential with respect to the point p, like a quadratic differential with respect to q, and which has a simple pole for p near q (see [7a]). Because of its fundamental role in the variational theory, we shall call this functional the variation kernel.

We first define the variation kernel for the four orientable surfaces whose algebraic genus G is either 0 or 1; namely the sphere, the cell, the ring domain, and the torus. The first three domains are of genus zero, and can therefore be mapped onto plane canonical domains, namely, the whole plane, the unit disc, and the circular ring.

There is a six-parameter group of mappings of the closed w-plane onto itself, given by all transformations of the form

$$w = \frac{aw_1 + b}{cw_1 + d}, \quad ad - bc \neq 0. \tag{7.10.1}$$

Let $w(p)$ give the conformal map of the abstract sphere onto the closed complex plane. We construct the variation kernel $n(p, q)$ in terms of the complex function $w(p)$ defined on the abstract sphere. Let q_0 be an arbitrary but fixed point of the abstract sphere, and define

$$n(p, q) = \frac{[w(p) - w(q_0)]^3}{[w(p) - w(q)]\,[w(q) - w(q_0)]^3}\,\frac{[w'(q)]^2}{w'(p)}. \tag{7.10.2}$$

We note that $n(p, q)$ so defined is a reciprocal differential in p and a quadratic differential in q. It also has a simple pole at $p = q$, where the residue is $+1$ if p and q are expressed in terms of the same local co-ordinate system.

The right side of (7.10.2) remains formally invariant under the transformations (7.10.1). Thus $n(p, q)$ is independent of the particular choice of the mapping function $w(p)$ and depends only on the parameter q_0. If we choose q_0 such that $w(q_0) = \infty$, the formula (7.10.2) takes the particularly simple form

$$n(p, q) = \frac{[w'(q)]^2}{w'(p)[w(p) - w(q)]}. \tag{7.10.2$'$}$$

Let us next consider the case of the cell. There exists a function $w(p)$ which maps the abstract cell onto the unit circle in the complex plane. Every function $w_1(p)$ of the form

$$w_1(p) = e^{i\lambda} \frac{w - w_0}{1 - \bar{w}_0 w}, \quad \lambda \text{ real}, \ |w_0| < 1, \tag{7.10.3}$$

also maps the cell onto the same domain. In order to fix the map uniquely, we choose an arbitrary fixed point q_0 on the abstract cell and require that $w(q_0) = 0$. If a function $w(q)$ is known which performs the map of the abstract cell onto the unit circle, but for which $w(q_0) = w_0 \neq 0$, then each linear transformation (7.10.3) will lead to another map function with the required normalization $w(q_0) = 0$. For map functions $w(q)$ normalized in this way let us define

$$n(p, q) = \frac{w(p)}{2[w(q)]^2} \frac{w(p) + w(q)}{w(p) - w(q)} \frac{[w'(q)]^2}{w'(p)}. \tag{7.10.4}$$

We observe that $n(p, q)$ is real if both p and q lie on the boundary of the disc and if boundary uniformizers are used as the coordinates of p and q. The distinguished role of the arbitrarily chosen point q_0 in the analytic character of the variation kernel is obvious.

Given an abstract ring domain, there exists a function $w(p)$ which maps the ring onto the circular ring domain $\mu < |w| < 1$, $\mu > 0$, where μ^{-1} is the modulus of the ring. There are infinitely many map functions of this type, but they are all interrelated by the transformation formulas $w_1 = e^{i\lambda} w$ (λ real) and $w_1 = \mu/w$. We proceed to the construction of the variation kernel as follows. Let $\zeta(w)$ denote the Weierstrass ζ-function corresponding to the periods $2\omega_1$, $2\omega_2$, where ω_1 is real and ω_2 pure imaginary. Let the ω's be related to the modulus μ^{-1} by the formula

$$\mu = e^{\pi i \frac{\omega_2}{\omega_1}}. \tag{7.10.5}$$

In this way ω_1 and ω_2 are determined up to a constant real factor. In the usual notation we set

$$\eta_1 = \zeta(\omega_1), \ \eta_2 = \zeta(w_2). \tag{7.10.6}$$

We define the variation kernel as follows:

$$n(p, q) = \frac{\omega_1}{\pi i}\left\{\zeta\left(\frac{\omega_1}{\pi i}\log\frac{w(p)}{w(q)}\right) - \frac{\eta_1}{\pi i}\log\frac{w(p)}{w(q)}\right\}\frac{w(p)}{[w(q)]^2}\frac{[w'(q)]^2}{w'(p)}. \tag{7.10.7}$$

Then $n(p, q)$ has a simple pole for $p = q$ and, if we use the same uniformizer for p and q in the neighborhood of the point q, then the residue is $+ 1$. If p and q lie on the same boundary component of the ring, $\log [w(p)/w(q)]$ is pure imaginary and hence the bracketed expression in (7.10.7) is real. If boundary uniformizers are used, then the whole expression is real as well. Suppose next that p and q lie on different boundary components and that $| w(p)/w(q) | = \mu$. Using (7.10.5) we can therefore write

$$\frac{\omega_1}{\pi i}\log\frac{w(p)}{w(q)} = \omega_2 + R(p, q),$$

where $R(p, q)$ is a real number depending on p and q. It should be observed that ω_2 is a pure imaginary, and that the bracketed term in (7.10.7) is not real. We use the theorem of elliptic function theory that

$$\zeta(z \pm \omega_2) = \zeta_2(z) \pm \eta_2, \tag{7.10.8}$$

(see [8], formula XII_5), where $\zeta_2(z)$ is real for real values of its argument. Hence the bracketed terms in (7.10.7) may be transformed into

$$\zeta(R + \omega_2) - \frac{\eta_1}{\omega_1}R - \frac{\eta_1\omega_2}{\omega_1} = \zeta_2(R) - \frac{\eta_1}{\omega_1}R + \frac{\eta_2\omega_1 - \eta_1\omega_2}{\omega_1}. \tag{7.10.9}$$

Using the Legendre relation

$$\eta_2\omega_1 - \eta_1\omega_2 = \frac{-\pi i}{2},$$

we find that the imaginary part of (7.10.9) is a constant, namely $-\pi/2\omega_1$. Thus, using boundary uniformizers chosen so that $w'/w = i$ on the boundary, we see that the imaginary part of $n(p, q)$ is constant if p and q lie on different boundary components.

In the case of the unit circle and the circular ring, $n(p, q)$ is essentially the complex Poisson kernel. For simplicity let $w(p) = z$ and let $u(\tau)$ be a real continuous function on the boundary C of the unit circle; then the analytic function $\Omega(z)$ in $|z| < 1$ whose real part assumes the values $u(\tau)$ on $|z| = 1$ is given by the formula

$$\Omega(z) = -\frac{1}{\pi i z}\int_C u(\tau)n(z, \tau)\tau d\tau + iK, \tag{7.10.10}$$

where K is an arbitrary real constant. In the case of the circular ring let $u(\tau)$ be a real-valued function defined on the boundary C of the ring which satisfies the condition

$$\int_C u(\tau)\frac{d\tau}{\tau} = 0. \tag{7.10.11}$$

Then the single-valued analytic function $\Omega(z)$ in the ring $\mu < |z| < 1$ whose real part assumes the given boundary values is given by the formula (Villat's formula)

$$\Omega(z) = -\frac{1}{\pi i z}\int_C u(\tau)\,n(z, \tau)\tau d\tau - \frac{1}{2\pi i}\int_{|\tau| = \mu} u(\tau)\,\frac{d\tau}{\tau} + iK, K \text{ real}. \tag{7.10.12}$$

Finally, we take up the variation kernel for the torus. Since the torus has genus 1, it cannot be mapped onto a subdomain of the sphere. However its universal covering surface, which is simply-connected, can be mapped conformally onto the complex plane punctured at infinity. Let $w(p)$ give the conformal map of the universal covering surface onto the w-plane. The single valued functions on the torus go over into elliptic functions of w whose periods we shall denote by $2\omega_1$ and $2\omega_2$. However, in the case of the general unsymmetrical torus, we can no longer assume that the period ratio is pure imaginary. Let p_0 and q_0 be arbitrary but fixed points of the torus, and define

$$(7.10.13) \qquad \begin{aligned} n(p, q) = [\zeta(w(p) - w(q)) - \zeta(w(p) - w(q_0)) \\ - \zeta(w(p_0) - w(q)) + \zeta(w(p_0) - w(q_0))] \cdot \frac{[w'(q)]^2}{w'(p)}. \end{aligned}$$

We note that the bracketed expression is single-valued on the torus. We also observe that $n(p, q)$ has a simple pole at $p = q$, with a residue $+ 1$ if the same uniformizer is used to express p and q. In addition, this function has fixed poles at $p = q_0$ and $q = p_0$.

In the general case in which the algebraic genus G exceeds 1, we have at least two linearly independent everywhere finite differentials $Z_1'(p)$, $Z_2'(p)$ (linear independence being understood in the complex sense). Let

$$(7.10.14) \qquad \Delta(p, q) = \begin{vmatrix} Z_1'(p) & Z_1'(q) \\ Z_2'(p) & Z_2'(q) \end{vmatrix}.$$

This expression is an everywhere finite bilinear differential of $\mathfrak{M}$ which vanishes when $p = q$.

Next, let

$$(7.10.15) \qquad Z(p) = \frac{dZ_1(p)}{dZ_2(p)}.$$

Since $\tilde{Z}(p) = Z(p)$, this is a function of $\mathfrak{M}$ in the sense of Section 2.2. Differentiating this function, we obtain

$$(7.10.16) \qquad dZ(p) = \frac{\Delta(p) dz^3}{[dZ_2(p)]^2}$$

where

$$(7.10.17) \qquad \Delta(p) = \begin{vmatrix} \dfrac{d^2 Z_1}{dz^2} & \dfrac{dZ_1}{dz} \\ \dfrac{d^2 Z_2}{dz^2} & \dfrac{dZ_2}{dz} \end{vmatrix}$$

and z is a uniformizer at p. Since $dZ(p)$, $[dZ_2(p)]^2$ are invariant, Δdz^3 is also invariant and is therefore an everywhere finite cubic differential of $\mathfrak{M}$ (differential of dimension 3).

For a surface of algebraic genus greater than 1 we define

$$n(p, q) = \pi L_{\mathbf{M}}(p, q)\frac{\Lambda(p, q)}{\Delta(p)}. \tag{7.10.18}$$

If $\mathfrak{M}$ has a boundary, we observe that

$$\frac{\tilde{n}(p, q)d\tau^2}{dz} = \left(\frac{n(\tilde{p}, \tilde{q})\, d\tilde{\tau}^2}{d\tilde{z}}\right)^- = \frac{n(p, q)d\tau^2}{dz}, \tag{7.10.19}$$

where z and τ are uniformizers at p and q, respectively. In this sense $n(p, q)d\tau^2/dz$ is a differential of $\mathfrak{M}$, reciprocal with respect to p, quadratic with respect to q. Let p and q be close together, and let t be a uniformizer in a neighborhood containing both p and q. Regarding p as variable, we have

$$\frac{n(p, q)dt_0^2}{dt} = \frac{1}{t}\frac{dt_0^2}{dt} + \text{regular terms}, \quad t = t(p), \quad t_0 = t(q) = 0. \tag{7.10.20}$$

The principal part is invariant if p and q are represented by the same uniformizer.

In addition to the singularity at $p = q$, there are poles at the points where $\Delta(p)$ vanishes. Let p_k be such a point. If p_k is a simple zero of $\Delta(p)$, we have

$$n(p, q) = \frac{\pi L_{\mathbf{M}}(p_k, q)\, \Lambda(p_k, q)}{\Delta'(p_k)}\frac{1}{z} + \cdots, \tag{7.10.21}$$

where $z = z(p)$ and $z(p_k) = 0$. The important point is that the coefficient of $1/z$ is a quadratic differential in its dependence on q. Let

$$Q_k(p) = \pi\, L_{\mathbf{M}}(p_k, p)\, \Lambda(p_k, p). \tag{7.10.22}$$

Since

$$\Lambda(p_k, p_k) = 0; \quad \left.\frac{\partial \Lambda(p_k, p)}{\partial p}\right|_{p=p_k} = -\Delta(p_k) = 0, \tag{7.10.23}$$

we see that Q_k is everywhere finite. More generally, we observe that (symbolically)

$$\left.\left(\frac{\partial}{\partial p} + \frac{\partial}{\partial q}\right)^n \frac{\partial}{\partial p}\Lambda(p, q)\right|_{q\to p} = \frac{d^n}{dp^n}\Delta(p). \tag{7.10.24}$$

Assume now that $\Delta(p)$ has a zero of order λ at the point p_k of $\mathfrak{M}$, and let z be a uniformizer at p_k, $z(p_k) = 0$. Then

$$n(p, q) = \frac{\pi L_M(p, q)}{\Delta(p)} \Delta(p, q) = \frac{1}{z^\lambda} \cdot \sum_{\nu=0}^{\infty} A_\nu(q) z^\nu.$$

If q does not lie near p_k, we have by the residue theorem

$$A_\nu(q) = \frac{1}{2\pi i} \int_c n(p, q) z^{\lambda-\nu-1} dz$$

where c is a small circle around the point p_k. This shows that $A_\nu(q)$ is a quadratic differential in q, just as $n(p, q)$ is, in its dependence on its second variable. We have to determine the behavior of $A_\nu(q)$ at the point p_k. Suppose that q lies in the neighborhood of p_k, and let $\tau = z(q)$. If K is a contour lying in the domain of the uniformizer z which encloses in its interior both p_k and q, but no other pole of $n(p, q)$, we have for $0 \leqq \nu \leqq \lambda - 1$,

$$\text{(7.10.25)} \qquad A_\nu(q) = \tau^{\lambda-\nu-1} + \frac{1}{2\pi i} \int_K n(p, q) z^{\lambda-\nu-1} dz.$$

Letting q tend to p_k, we see that $A_\nu(q)$ remains finite and, since $z(p_k) = 0$,

$$\text{(7.10.26)} \qquad A_\nu(p_k) = \frac{1}{2\pi i} \int_K n(p, p_k) z^{\lambda-\nu-1} dz, \quad 0 \leqq \nu \leqq \lambda - 1.$$

Thus at a point p_k of $\mathfrak{M}$ where $\Delta(p)$ has a zero of order λ, $\lambda \geqq 1$, we have

$$\text{(7.10.27)} \qquad n(p, q) = \frac{A_0(q)}{z^\lambda} + \frac{A_1(q)}{z^{\lambda-1}} + \cdots + \text{regular terms}$$

where $A_0(q)$, $A_1(q)$, $\cdots$, $A_{\lambda-1}(q)$ are everywhere finite quadratic differentials on the surface $\mathfrak{M}$.

The sum of the multiplicities of the zeros of $\Delta(p)$ on the double $\mathfrak{F}$ of $\mathfrak{M}$ is given by formula (3.5.1)′:

$$\text{(7.10.28)} \qquad \text{ord}\,(\Delta dz^3) = 6(G - R^0) = 2(6h + 3m - 6), \; m > 0.$$

Here G is the algebraic genus. In the case $m = 0$, the double consists of two components, on each of which

$$\text{ord}\,(\Delta\, dz^3) = 6h - 6.$$

If $G > 1$, then by (3.7.1) and (3.7.2),

(7.10.29) $$\sigma = 6h + 3m - 6$$

where σ denotes the dimension of the space of all classes of conformally equivalent finite Riemann surfaces. Thus

(7.10.30) $$\text{ord}\,(\Delta\, dz^3) = 2\sigma, \quad G > 1,$$

or, by (3.8.5)

(7.10.31) $$\text{ord}\,(\Delta\, dz^3) = 2 \dim\,(dZ^2), \quad G > 1.$$

The number of coefficients in the principal parts of $n(p, q)$ at the zeros of $\Delta(p)$ is equal to ord $(\Delta\, dz^3)$, and these coefficients are quadratic differentials. By formula (7.10.31) only half of these quadratic differentials can be linearly independent in the complex sense.

If we consider $n(p, q)$ as a function of p, it has one pole at $p = q$ and the remaining poles are at points p_k, where the p_k are independent of q. The latter poles will be called p-poles for simplicity. As a function of q, $n(p, q)$ has a pole at $q = p$ and, in the case where $G = 0$ or 1, it may have an additional pole at a point q_0, q_0 independent of p. In (7.10.2)′, q_0 is the point at ∞, in (7.10.4) it is the point at the origin, while in (7.10.13) it is the arbitrary point q_0. This pole of $n(p, q)$, if it exists, will be called a q-pole. It should be remarked that the coefficient of $1/z - \tau_0$ at the p-pole in (7.10.13) is an everywhere finite quadratic differential in q. Hence in all cases the coefficients occurring in the principal parts at the p-poles are everywhere finite quadratic differentials.

7.11. Identities Satisfied by the Variation Kernel

We now establish some useful identities involving integrals of $n(p, q)$. Let γ be the Jordan curve considered in Section 7.8 and let $r(p)$ be a local reciprocal differential which is regular analytic in a complete neighborhood of γ. Let $Q_1, \cdots, Q_\sigma$ be a basis for the everywhere finite quadratic differentials of $\mathfrak{M}$, where σ is the number of real moduli of $\mathfrak{M}$:

(7.11.1) $$\sigma = \varrho + 6h + 3c + 3m - 6.$$

We impose upon $r(p)$ the orthogonality conditions

$$\int_\gamma r(p)\, Q_\nu(p) dz = 0, \quad \nu = 1, 2, \cdots, \sigma. \tag{7.11.2}$$

Let us consider first the case $m > 0$. We then have a Green's function $G(p, q)$ and we denote by $T(p, q)$ the analytic function of p whose real part is this Green's function. By (4.2.1), we have

$$T'(p, q) = \frac{\partial T(p, q)}{\partial p} = 2\frac{\partial G(p, q)}{\partial p} = \Omega'_{q\tilde{q}}(p). \tag{7.11.3}$$

Let the Jordan curve $\tilde{\gamma}$ on $\tilde{\mathfrak{M}}$ consist of the conjugate points of γ. We define the conjugate $\tilde{r}(p)$ of the reciprocal differential $r(p)$. by the formula (compare (2.2.3) and (2.2.3)′):

$$\tilde{r}(\tilde{p}) d\tilde{z}^{-1} = (r(p) dz^{-1})^{-}. \tag{7.11.4}$$

Let

$$h(p) = \frac{1}{2\pi i}\int_\gamma r(p_1)\, n(p, p_1) dz_1 - \frac{1}{2\pi i}\int_{\tilde{\gamma}} \tilde{r}(\tilde{p}_1) n(p, \tilde{p}_1) d\tilde{z}_1. \tag{7.11.5}$$

By (7.11.2) and (7.10.27) we see that $h(p)$ is regular analytic at the p-poles of $n(p, q)$. It is, therefore, a reciprocal differential which is regular analytic in $\mathfrak{M} - \gamma$. If p lies on the boundary of $\mathfrak{M}$ and if we use boundary uniformizers $h(p)$ is real.

We state now the following theorem:

THEOREM 7.11.1. *If $r(p)$ is a reciprocal differential near γ which satisfies the conditions* (7.11.2) *and $h(p)$ is the reciprocal differential* (7.11.5), *we have the identity*

$$\begin{aligned} &\operatorname{Re}\left\{\frac{1}{2\pi i}\int_\gamma r(p_1) T'(p_1, p) T'(p_1, q) dz_1\right\} \\ &\quad = \operatorname{Re}\{h(p) T'(p, q) + h(q) T'(q, p)\}. \end{aligned} \tag{7.11.6}$$

This identity has many applications in the variational calculus of Riemann surfaces and serves to transform and simplify formulas involving reciprocal differentials and derivatives of Green's functions.

In order to prove the identity (7.11.6), we denote the left-hand side of (7.11.6) by $U_l(p, q)$ and the right-hand side by $U_r(p, q)$. Both functions are harmonic in both variables, except on the curve γ. One readily verifies that $U_r(p, q)$ is also regular harmonic for $p = q$

since the singularities of its elements cancel for $p = q$. Let us choose a fixed point q in the interior of $\mathfrak{M} - \gamma$ and study the dependence of these functions upon p. Since $G(p_1, p)$ is identically zero in p_1 for p on the boundary of $\mathfrak{M}$, we have:

$$T'(p_1, p) = 0 \qquad \text{for } p \text{ on the boundary of } \mathfrak{M}.$$

Hence $U_\iota(p, q)$ will vanish on the boundary of $\mathfrak{M}$. But the same will also be true for the function $U_r(p, q)$. In fact, using boundary uniformizers we have

$$h(p) = \text{real}, \; T'(p, q) = \text{imaginary}, \quad \text{for } p \text{ on the boundary of } \mathfrak{M}.$$

Since, moreover, $T'(q, p)$ also vanishes on the boundary, we find $U_r(p, q) = 0$ if p lies on the boundary of $\mathfrak{M}$.

It remains to investigate the discontinuity behaviour of U_ι and U_r across the Jordan curve γ. Since $r(p)$ is analytic in the neighborhood of γ and since the singularities of $T'(p_1, p)$ and $n(p, p_1)$ on the curve have the same expression $1/(z(p) - z(p_1))$ in the uniformizer on γ, we derive easily from Cauchy's theorem that the expressions

$$\frac{1}{2\pi i} \int_\gamma r(p_1) T'(p_1, p) T'(p_1, q) dz_1 - [r(p) T'(p, q)] \delta$$

and

$$h(p)\, T'(p, q) - [r(p)\, T'(p, q)]\, \delta,$$

where $\delta = 1$ for p inside γ and $\delta = 0$ for p outside, are regular analytic in a complete neighborhood of γ. This proves that the harmonic function $U_r(p, q) - U_\iota(p, q)$ is regular harmonic in $\mathfrak{M}$ and, since this function vanishes on the boundary of $\mathfrak{M}$, it is identically zero. Thus, Theorem 7.11.1 is proved.

Let us proceed to the case $m = 0$. Now the Green's function is replaced by the function $V(p, p_0; q, q_0)$ defined in (4.2.23). We have:

$$2 \frac{\partial}{\partial p} V(p, p_0; q, q_0) = \Omega'_{qq_0}(p). \tag{7.11.7}$$

We choose a fixed point $q_0 \in \mathfrak{M}$ which lies outside a neighborhood of the Jordan curve γ. We define the reciprocal differential

$$h(p) = \frac{1}{2\pi i} \int_\gamma r(p_1) n(p, p_1) dz_1 \tag{7.11.8}$$

and again subject the (local) reciprocal differential $r(p)$ to the orthogonality conditions (7.11.2). We then have the theorem:

THEOREM 7.11.2. *Let $\mathfrak{M}$ be a closed surface and $r(p)$ a reciprocal differential which is defined near the Jordan curve γ and which satisfies* (7.11.2). *If the reciprocal differential $h(p)$ is defined by* (7.11.8), *we have the identity:*

$$\operatorname{Re}\left\{\frac{1}{2\pi i}\int_{\gamma} r(p_1)\Omega'_{pq_0}(p_1)\Omega'_{qq_0}(p_1)dz_1\right\}$$

$$= \Phi(q_0) + \operatorname{Re}\left\{\Omega'_{qq_0}(p)h(p) + \Omega'_{pq_0}(q)h(q) + 2h(q_0)\frac{\partial}{\partial q_0}\operatorname{Re}\left\{\Omega_{qq_0}(p) + \Omega_{pq_0}(q)\right\}\right\}. \tag{7.11.9}$$

Here $\Phi(q_0)$ is a real number which depends on $h(p)$ and the choice of q_0 but is independent of p and q.

In order to prove this identity let us observe that both sides of (7.11.9) represent harmonic functions of p and q in $\mathfrak{M}-\gamma$. The right-hand side is easily seen to be regular harmonic even if p approaches q or if either p or q converge to q_0.

We verify as before that the difference of the two sides of (7.11.9) is regular harmonic across the Jordan curve γ. Hence, the two sides can differ only by a constant, i.e. a number which cannot depend on p or q. This proves the theorem.

The constant $\Phi(q_0)$ is obtained in the easiest way by the following remark. In view of (3.4.7), we have

$$\Omega'_{qq_0}(p) = 2\frac{\partial}{\partial p}\operatorname{Re}\left\{\Omega_{pp_0}(q) - \Omega_{pp_0}(q_0)\right\} \tag{7.11.10}$$

and, letting $q \to q_0$, we find

$$\lim_{q\to q_0}\Omega'_{qq_0}(p) \equiv 0. \tag{7.11.11}$$

This remark shows that the left side of (7.11.9) vanishes for $q = q_0$ or $p = q_0$. Thus $-\Phi(q_0)$ is the limit of the real part on the right as $p \to q_0$ or $q \to q_0$.

From the identities (7.11.6) and (7.11.9) we may derive further identities. We take on both sides the normal derivatives of the harmonic functions and integrate them around the basis cycle K_μ.

Comparison of the results leads to the new identities. We make use of the formula

$$\int_{K_\mu} \frac{\partial}{\partial n_p} T'(q, p) ds_p = -2\pi i \, Z'_\mu(q) \tag{7.11.12}$$

and correspondingly, for the case of (7.11.9),

$$\int_{K_\mu} \frac{\partial}{\partial n_p} \Omega'_{pp_0}(q) ds_p = -2\pi i \, Z'_\mu(q), \tag{7.11.13}$$

and derive from (7.11.6) the identity

$$\operatorname{Im}\left\{ \frac{1}{2\pi i} \int_\gamma r(p_1) Z'_\mu(p_1) T'(p_1, q) dz_1 \right\} = \operatorname{Im}\{h(q) Z'_\mu(q)\} \tag{7.11.14}$$

for the case of a surface with boundary.

We can make an analogous calculation in (7.11.9). We observe that

$$\operatorname{Re}\left\{ \int_{K_\mu} \frac{\partial}{\partial n_p} \Omega_{qq_0}(p) ds_p \right\} = 2\pi \operatorname{Im}\left\{ \int_q^{q_0} dZ_\mu \right\}, \tag{7.11.15}$$

since the real part of $\Omega_{qq_0}(p)$ is single-valued. Further by (3.4.7)

$$\begin{aligned} \operatorname{Re}\left\{ \int_{K_\mu} \frac{\partial}{\partial n_p} \Omega_{pq_0}(q) ds_p - \int_{K_\mu} \frac{\partial}{\partial n_p} \Omega_{pq_0}(q_1) ds_p \right\} \\ = \operatorname{Re}\left\{ \int_{K_\mu} \frac{\partial}{\partial n_p} \Omega_{qq_1}(p) ds_p \right\}. \end{aligned} \tag{7.11.15$'$}$$

From (7.11.15) and (7.11.15)′ we deduce easily that

$$2 \frac{\partial}{\partial q_0} \operatorname{Re}\left\{ \int_{K_\mu} \frac{\partial}{\partial n_p} [\Omega_{qq_0}(p) + \Omega_{pq_0}(q)] ds_p \right\} = \alpha(q_0) \tag{7.11.16}$$

is a complex number which is independent of q. Hence, we derive from (7.11.9) the equation

$$\operatorname{Im}\left\{ \frac{1}{2\pi i} \int_\gamma r(p_1) Z'_\mu(p_1) \Omega'_{qq_0}(p_1) dz_1 \right\} = \operatorname{Im}\{Z'_\mu(q) h(q) + \beta(q_0)\}. \tag{7.11.17}$$

The constant $\beta(q_0)$ can be determined by means of (7.11.11). We see that the left side of (7.11.17) vanishes for $q = q_0$. Hence $\beta(q_0) = -Z'_\mu(q_0)h(q_0)$, and we obtain the final identity

$$\operatorname{Im}\left\{\frac{1}{2\pi i}\int_\gamma r(p_1)Z'_\mu(p_1)\Omega'_{qq_0}(p_1)dz_1\right\} \tag{7.11.18}$$
$$= \operatorname{Im}\{Z'_\mu(q)h(q) - Z'_\mu(q_0)h(q_0)\}.$$

Formulas (7.11.14) and (7.11.18) will be used in the sequel. Their significance is obvious; the left side of each formula is a harmonic function of q and the right side indicates the simple form of its analytic completion.

From these integral identities we can derive formulas connecting the variation kernel $n(p, q)$ with the complex completion of the Green's function $G(p, q)$ and with the differentials of the first kind $Z'_\mu(p)$. These formulas are given for completeness, but may be omitted by the reader if he is so inclined.

The formula (7.11.14) may be written in the form

$$\operatorname{Im}\left\{\frac{1}{2\pi i}\int_\gamma r(p_1)Z'_\mu(p_1)T'(p_1, q)dz_1\right\} \tag{7.11.19}$$
$$= \operatorname{Im}\left\{Z'_\mu(q)\frac{1}{2\pi i}\int_\gamma r(p_1)n(q, p_1)dz_1 - Z'_\mu(\tilde{q})\frac{1}{2\pi i}\int_\gamma r(p_1)\,n(\tilde{q}, p_1)\,dz_1\right\}.$$

Here we have used the fact that

$$\left(Z'_\mu(q)\frac{1}{2\pi i}\int_{\tilde{\gamma}} \bar{r}(\tilde{p}_1)n(q, \tilde{p}_1)d\tilde{z}_1\right)^{-} = -Z'_\mu(\tilde{q})\frac{1}{2\pi i}\int_\gamma r(p_1)n(\tilde{q}, p_1)dz_1, \tag{7.11.20}$$

by (7.10.19). If $r(p)$ satisfies the conditions (7.11.2), so does $ar(p)$ where a is any complex factor, and it follows from (7.11.14) that

$$\frac{1}{2\pi i}\int_\gamma r(p_1)Z'_\mu(p_1)T'(p_1, q)dz_1 \tag{7.11.21}$$
$$= Z'_\mu(q)\frac{1}{2\pi i}\int_\gamma r(p_1)n(q, p_1)dz_1 - Z'_\mu(\tilde{q})\frac{1}{2\pi i}\int_\gamma r(p_1)n(\tilde{q}, p_1)dz_1.$$

Formula (7.11.21) is valid provided that $r(p)$ is regular analytic in a neighborhood of γ and that (7.11.2) is satisfied.

In the same way, we derive from (7.11.18)

$$\frac{1}{2\pi i}\int_\gamma r(p_1)Z'_\mu(p_1)\Omega'_{qq_0}(p_1)dz_1 \tag{7.11.22}$$
$$= Z'_\mu(q)\frac{1}{2\pi i}\int_\gamma r(p_1)n(q, p_1)dz_1 - Z'_\mu(q_0)\frac{1}{2\pi i}\int_\gamma r(p_1)n(q_0, p_1)dz_1.$$

It is now easy to derive from the integral identities (7.11.21) and (7.11.22) new equations connecting the functionals themselves. For this purpose we choose a function $r(z)$ of the uniformizer z which is regular analytic inside the whole Jordan curve γ except for N points p_ϱ, where it has simple poles with residues α_ϱ. We are not quite free in our selection of the function $r(z)$. We must satisfy the conditions (7.11.2) which have now, by virtue of the residue theorem, the form

$$\sum_{\varrho=1}^{N}\alpha_\varrho Q_\nu(p_\varrho) = 0, \qquad \nu = 1, 2, \cdots, \sigma. \tag{7.11.23}$$

On the other hand every function $r(z)$ satisfying (7.11.23) will be a permissible choice.

We introduce $r(z)$ into (7.11.21) and by means of the residue theorem we obtain

$$\sum_{\varrho=1}^{N}\alpha_\varrho\,[Z'_\mu(p_\varrho)T'(p_\varrho, q) - Z'_\mu(q)n(q, p_\varrho) + Z'_\mu(\tilde{q})n(\tilde{q}, p_\varrho)] = 0. \tag{7.11.24}$$

Similarly, we obtain from (7.11.22)

$$\sum_{\varrho=1}^{N}\alpha_\varrho\,[Z'_\mu(p_\varrho)\Omega'_{qq_0}(p_\varrho) - Z'_\mu(q)n(q, p_\varrho) + Z'_\mu(q_0)n(q_0, p_\varrho)] = 0. \tag{7.11.25}$$

Since we can choose the poles p_ϱ and their residues α_ϱ arbitrarily except for the linear conditions (7.11.23), we derive from (7.11.24)

$$Z'_\mu(p)T'(p, q) = Z'_\mu(q)n(q, p) - Z'_\mu(\tilde{q})n(\tilde{q}, p) + \sum_{\nu=1}^{\sigma} a_{\mu\nu}(q)Q_\nu(p). \tag{7.11.26}$$

To evaluate the coefficients $a_{\mu\nu}(q)$, we first take q on the boundary of $\mathfrak{M}$ and we obtain

$$(7.11.27)\qquad \sum_{\nu=1}^{\sigma} a_{\mu\nu}(q)Q_\nu(p) = 0, \qquad \mu = 1, 2, \cdots.$$

Since this equation holds for arbitrary $p \in \mathfrak{M}$, we conclude from the linear independence of the $Q_\nu(p)$ that all $a_{\mu\nu}(q)$ vanish on the boundary of $\mathfrak{M}$. As q approaches a q-pole of $n(p, q)$, the sum

$$\sum_{\nu=1}^{\sigma} a_{\mu\nu}(q)Q_\nu(p)$$

must tend to infinity. Let the q-poles of $n(q, p)$ which lie in the interior of $\mathfrak{M}$ be q_ϱ, $\varrho = 1, 2, \cdots, M$, $M \leqq \sigma - k$. For simplicity, assume that all these q-poles are simple; if this assumption is not satisfied, the calculation is similar. For q near q_ϱ, let ζ be a uniformizer at q_ϱ, $\zeta(q) = \zeta$, $\zeta(q_\varrho) = 0$.

By (7.10.21) we have

$$(7.11.28)\qquad n(q, p) = \frac{P_\varrho(p)}{\zeta} + \text{regular terms}$$

where $P_\varrho(p)$ is a quadratic differential. We readily conclude that

$$(7.11.29)\qquad \sum_{\nu=1}^{\sigma} a_{\mu\nu}(q)Q_\nu(p) = -\sum_{\varrho=1}^{M} Z'_\mu(q_\varrho)T'(q_\varrho, q)P_\varrho(p).$$

Thus, formula (7.11.26) becomes finally

$$(7.11.30)\qquad \begin{aligned} Z'_\mu(p)T'(p, q) &= Z'_\mu(q)n(q, p) - Z'_\mu(\tilde{q})n(\tilde{q}, p) \\ &\quad - \sum_{\varrho=1}^{M} Z'_\mu(q_\varrho)T'(q_\varrho, q)P_\varrho(p). \end{aligned}$$

If not all q-poles of $n(p, q)$ are simple, higher derivatives of $T(q_\varrho, q)$ with respect to q_ϱ will occur in the sum on the right side of (7.11.30).

From (7.11.25), similar considerations lead to

$$(7.11.31)\quad Z'_\mu(p)\Omega'_{qq_0}(p) = Z'_\mu(q)n(q, p) - Z'_\mu(q_0)n(q_0, p) + \sum_{\nu=1}^{\sigma} b_{\mu\nu}(q)Q_\nu(p).$$

We observe that $b_{\mu\nu}(q)$ vanishes for $q = q_0$; for q near a q-pole of $n(p, q)$, the last sum must again counteract the increase of the first right side term. Assuming that all q-poles of $n(p, q)$ are simple, we readily verify that

$$Z'_\mu(p)\Omega'_{qq_0}(p) = Z'_\mu(q)n(q, p) - Z'_\mu(q_0)n(q_0, p) - \sum_{\varrho=1}^{M} Z'_\mu(q_\varrho)\Omega'_{qq_0}(q_\varrho)P_\varrho(p). \tag{7.11.32}$$

The identities (7.11.30) and (7.11.32) between the various types of differentials on a surface $\mathfrak{M}$ are due to the fact that the product of a differential and a reciprocal differential is a function on the surface and can, therefore, be expressed in terms of the fundamental functions on $\mathfrak{M}$.

7.12. Conditions for Conformal Equivalence under a Deformation

Let γ be the Jordan curve described in Section 7.8 and let z be a uniformizer which is valid in a complete neighborhood of γ. Let $r_0(z)$ be a function of z which is regular analytic in this neighborhood of γ and which satisfies the orthogonality conditions (7.11.2). We then show (under certain mild restrictions) that we can find a function $\varrho_\varepsilon(z)$ such that, if

$$r(z) = r_0(z) + \varepsilon\varrho_\varepsilon(z), \tag{7.12.1}$$

then the surface $\mathfrak{M}^*$ formed from $\mathfrak{M}$ by attaching a cell in the manner of Section 7.8 is conformally equivalent to $\mathfrak{M}$. Moreover,

$$|\varrho_\varepsilon(z)| \leqq M, \tag{7.12.2}$$

where M is independent of ε, $0 \leqq \varepsilon \leqq \varepsilon_0$.

We show first that an everywhere finite differential $Z'(p)$ of $\mathfrak{M}$ may always be found which has only simple zeros. Moreover, if $\mathfrak{M}$ has a boundary, all zeros of $Z'(p)$ lie interior to $\mathfrak{M}$.

If $\mathfrak{M}$ has a boundary and if $2h < \mu \leqq G = 2h + m - 1$, $\operatorname{Im} Z_\mu$ is single-valued on $\mathfrak{M}$ and constant on each boundary component. It follows readily from the maximum principle that $Z'_\mu(p)$ is non-vanishing on the boundary of $\mathfrak{M}$.

Let $Z'_\mu(p)$ be a basis differential of $\mathfrak{M}$, where $2h < \mu \leqq 2h + m - 1$ if $\mathfrak{M}$ has a boundary, and let $Z'_\mu(p)$ have a zero of order λ, $\lambda > 1$, at the point p_k. Suppose that there is another basis differential $Z'_\nu(p)$ which does not vanish at p_k and consider the combination

$$Z'_\mu(p) + \eta Z'_\nu(p) = Z'(p), \tag{7.12.3}$$

where $\eta > 0$. If z is a uniformizer at p_k, $z(p_k) = 0$, we have

$$Z'(p) = \eta(a_0 + a_1 z + \cdots + a_\lambda z^{\lambda-1} + \cdots) + b_\lambda z^\lambda + \cdots,$$

where $a_0 \neq 0$, $b_\lambda \neq 0$. For sufficiently small η, it is clear that $Z'(p)$ has λ simple zeros in a neighborhood of p_k. Moreover, if $\mathfrak{M}$ has a boundary, then by choosing η still smaller if necessary, all zeros of $Z'(p)$ will be in the interior of $\mathfrak{M}$. Hence an everywhere finite differential $Z'(p)$ of $\mathfrak{M}$ having the desired properties exists provided that there is no point of $\mathfrak{M}$ where all basis differentials vanish simultaneously.

Suppose that p_0 is a point of $\mathfrak{M}$ where all basis differentials vanish. Then every differential has a zero at p_0. Let $t_{p_0}(p)$ be the elementary integral of the second kind on the double $\mathfrak{F}$ of $\mathfrak{M}$ which is normalized by the condition that its periods around one of the two dual sets of cycles vanish. From (3.4.3)′ we conclude that $t_{p_0}(p)$ is single-valued on $\mathfrak{F}$, and hence $\mathfrak{M}$ is either the sphere or a simply-connected domain in which cases all differentials of the first kind vanish identically. We have therefore proved that, if there are any non-trivial differentials of $\mathfrak{M}$, there is at least one, say

$$Z'(p) = \sum_{\mu=1}^{G} c_\mu Z'_\mu(p), \tag{7.12.4}$$

which has only simple zeros and which does not vanish on the boundary of $\mathfrak{M}$ (if $\mathfrak{M}$ has a boundary). Our proof shows that the numbers c_μ, $\mu = 1, 2, \cdots, G$, occurring in (7.12.4) may be taken real (even in the case where $\mathfrak{M}$ is closed and $G = h$). We therefore suppose that the numbers c_μ in (7.12.4) are real and, once chosen, we shall assume that they are fixed.

Let $Z'(p)$ be the differential (7.12.4) for $\mathfrak{M}$. Then

$$Z^{*\prime}(p) = \sum_{\mu=1}^{G} c_\mu Z^{*\prime}_\mu(p), \tag{7.12.5}$$

where $Z^{*\prime}_\mu(p)$ is given by (7.8.11), is a corresponding differential for $\mathfrak{M}^*$ which, if ε is small enough, has only simple zeros interior to $\mathfrak{M}$.

Let $p_1, p_2, \cdots, p_N$ be the zeros of Z' in $\mathfrak{M}$, and $p_1^*, p_2^*, \cdots, p_N^*$ the zeros of $Z^{*\prime}$ in $\mathfrak{M}^*$. From (3.6.3) we have

$$N = G - R^0. \tag{7.12.6}$$

We formulate now the following theorem:

THEOREM 7.12.1. *Let $\mathfrak{M}$ be a Riemann surface, with boundary, of algebraic genus $G > 1$ and let $Z'(p)$ be an everywhere finite differential on $\mathfrak{M}$ which has only simple zeros in $\mathfrak{M}$ and no zero on the boundary of $\mathfrak{M}$. Let $\mathfrak{M}^*$ be obtained from $\mathfrak{M}$ by a variation and let $Z^{*\prime}(p)$ be the differential on $\mathfrak{M}^*$ obtained from $Z'(p)$ by this variation. If p_ν, $\nu = 1, 2, \cdots, N$, and p_ν^*, $\nu = 1, 2, \cdots, N$, are the zeros of $Z'(p)$ and $Z^{*\prime}(p)$, respectively, and if $Z'(p)$ is expressed in the form (7.12.4) we have the following necessary and sufficient conditions for the conformal equivalence of $\mathfrak{M}$ and $\mathfrak{M}^*$:*

(a) $$\operatorname{Im} Z(p_\nu) = \operatorname{Im} Z^*(p_\nu^*), \quad \nu = 1, 2, \cdots, N;$$

(b) $$\operatorname{Re}\left\{\int_{p_1}^{p_\nu} dZ\right\} = \operatorname{Re}\left\{\int_{p_1^*}^{p_\nu^*} dZ\right\}, \quad \nu = 2, 3, \cdots, N;$$

(c) $$\operatorname{Re}\{P(dZ, K_\mu)\} = \operatorname{Re}\{P(dZ^*, K_\mu)\}, \quad \mu = 1, 2, \cdots, G.$$

We remark in connection with (*b*) and (*c*) that $\mathfrak{M}^*$ should be regarded as lying over $\mathfrak{M}$. Points of $\mathfrak{M}^*$ and $\mathfrak{M}$ may then be denoted by the same symbol, the point of $\mathfrak{M}^*$ over the point p of $\mathfrak{M}$ also being denoted by p. This convention enables us to use the same symbol K_μ in the left and right sides of (*c*). So far as (*b*) is concerned, we suppose that the path of integration on the right lies over the path on the left except for small neighborhoods of p_1 and p_ν.

In order to prove the theorem we define the following relation between points $p \in \mathfrak{M}$ and $p^* \in \mathfrak{M}^*$:

$$\int_{p_1}^{p} dZ = \int_{p_1^*}^{p^*} dZ^*. \tag{7.12.8}$$

If the conditions (a), (b) and (c) are fulfilled this correspondence can be extended in a unique way over $\mathfrak{M}$ so as to give an everywhere conformal mapping of $\mathfrak{M}$ onto $\mathfrak{M}^*$. Thus, the conditions are clearly sufficient for equivalence. They are also necessary, since the harmonic functions $\operatorname{Im} Z_\mu(p)$, $\mu = 1, 2, \cdots, G$, are uniquely determined modulo 1 and since for small enough variation there is no room for the discrete increments which $\operatorname{Im} Z_\mu^*(p)$ might possibly have with respect to $\operatorname{Im} Z_\mu(p)$. Thus, conditions (a), (b) and (c) express just the conformal invariance of the differential $Z'(p)$ and are, therefore, necessary.

If $\mathfrak{M}$ is closed we have a very similar situation. However in this case $\operatorname{Im} Z_\mu(p)$ is not uniquely determined. We define therefore

$$\operatorname{Im} Z_\mu(q) = \frac{1}{2\pi} \int_{K_\mu} \frac{\partial V(p, p_0; q, q_0)}{\partial n_p} \, ds_p$$

with

$$V(p, p_0; q, q_0) = \operatorname{Re} \{\Omega_{qq_0}(p) - \Omega_{qq_0}(p_0)\},$$

that is, we require $\operatorname{Im} Z_\mu(q_0) \equiv 0 \pmod 1$. We may choose q_0 to be one of the zeros of the differential $Z'(q)$ on $\mathfrak{M}$ and of $Z^{*\prime}(q)$ on $\mathfrak{M}^*$. Thus, one equation in condition (a) can be satisfied by arbitrary normalization. The remaining set of conditions will again lead to a necessary and sufficient condition for conformal equivalence.

If $\mathfrak{M}$ has a boundary, there are $G - 1$ conditions (a), $G - 2$ conditions (b), and G conditions (c), giving $3G - 3 = 6h + 3m - 6$ conditions in all. If $\mathfrak{M}$ is closed and if one of the conditions (a) is eliminated as above, there are $G - 3$ conditions (a), $G - 3$ conditions (b) and G conditions (c), giving $3G - 6 = 6h - 6$ conditions in all. In both cases the number of conditions to be satisfied therefore agrees with (3.7.1).

If $G = 1$, there is a single basis differential Z_1' and it nowhere vanishes. In this case, the condition (c) is necessary and sufficient for conformal equivalence.

7.13. Construction of the Variation which Preserves Conformal Type

In the foregoing section we have formulated a set of necessary and sufficient conditions for the conformal equivalence of the surface $\mathfrak{M}$ and the surface $\mathfrak{M}^*$ which arises from it by the interior deformation. We come now to the proof that a function $\varrho_\varepsilon(z)$ satisfying the conditions stated at the beginning of Section 7.12. actually exists.

In order to give this existence proof it will not be sufficient to work with the asymptotic formulas derived in Section 7.8; we shall have to use exact integral equations connecting the differentials of the surfaces considered. We go back to the equation (7.8.6) for the variation of the Green's function of a closed Riemann surface.

We assume that $\mathfrak{M}$ is a surface with boundary and apply (7.8.6) to the function V of the closed surface $\mathfrak{M} + \tilde{\mathfrak{M}}$. Putting $p_0 = \tilde{p}$ and $q_0 = \tilde{q}$, we obtain from (7.8.6) by use of (7.6.1),

$$(7.13.1) \quad G^*(p, q) - G(p, q) = -\operatorname{Re}\left\{\frac{1}{2\pi i}\int_{\gamma} Q(z, q)dT^*(z, p)\right\}$$

where

$$T(p, q) = \Omega_{q\tilde{q}}(p), \; T^*(p, q) = \Omega^*_{q\tilde{q}}(p)$$

are analytic functions of p whose real parts are $G(p, q)$ and $G^*(p, q)$ respectively, and where

$$(7.13.1)' \qquad Q(z, q) = T(z + \varepsilon r(z), q) - T(z, q).$$

In obtaining (7.13.1), we have used (7.8.6) for cell attachments along γ and $\tilde{\gamma}$, and these cell attachments give equal contributions. This is an exact relation between G^* and G and we now derive from it corresponding relations between Z^*_μ and Z_μ. Keeping q (and therefore q^*) fixed, we compute the integrals of the normal derivatives of both sides of (7.13.1) around one of the basis cycles K_μ. Using the formulas

$$(7.13.2) \quad \frac{1}{2\pi}\int_{K_\mu}\frac{\partial G^*(p,q)}{\partial n_p}ds_p = \operatorname{Im} Z^*_\mu(q); \quad \frac{1}{2\pi}\int_{K_\mu}\frac{\partial G(p, q)}{\partial n_p}ds_p = \operatorname{Im} Z_\mu(q),$$

$$(7.13.2)' \quad \frac{1}{2\pi}\int_{K_\mu}\frac{\partial T^{*\prime}(q, p)}{\partial n_p}ds_p = -iZ^{*\prime}_\mu(q), \quad T^{*\prime}(q, p) = \frac{\partial T^*(q, p)}{\partial q},$$

we find that

$$(7.13.3) \quad \operatorname{Im} Z^*_\mu(q) = \operatorname{Im} Z_\mu(q) - \operatorname{Im}\left\{\frac{1}{2\pi i}\int_{\gamma} Q(z, q)\, dZ^*_\mu(z)\right\}.$$

Since the coefficients c_μ in (7.12.4) and (7.12.5) are real, we obtain (on multiplying both sides of (7.13.3) by c_μ and summing from $\mu = 1$ to $\mu = G$)

$$(7.13.4) \quad \operatorname{Im} Z^*(q) = \operatorname{Im} Z(q) - \operatorname{Im}\left\{\frac{1}{2\pi i}\int_{\gamma} Q(z, q)dZ^*(z)\right\}.$$

The conditions (a), (b) and (c), are equivalent to the existence of a conformal map

$$(7.13.5) \qquad M: \; p^* \to p$$

from $\mathfrak{M}^*$ onto $\mathfrak{M}$. If the map (7.13.5) exists, we have (modulo the numbers $c_1, c_2, \cdots, c_G$)

$$(7.13.6) \qquad \operatorname{Im} Z(p) = \operatorname{Im} Z^*(p^*).$$

Replacing q by p^* and then $\operatorname{Im} Z^*(p^*)$ by $\operatorname{Im} Z(p)$ in (7.13.4), we have

$$(7.13.7) \quad \operatorname{Im} Z(p) = \operatorname{Im} Z(p^*) - \operatorname{Im}\left\{\frac{1}{2\pi i}\int_\gamma Q(z, p^*)\, dZ^*(z)\right\}.$$

We assume that γ does not contain any zero point p_ν of $Z'(p)$. Taking $p = p_\nu$ in (7.13.7), we obtain

$$(7.13.8) \quad \operatorname{Im} Z(p_\nu) = \operatorname{Im} Z(p_\nu^*) - \operatorname{Im}\left\{\frac{1}{2\pi i}\int_\gamma Q(z, p_\nu^*) dZ^*(z)\right\}.$$

The equations (7.13.8) for $\nu = 1, 2, \cdots, N$ are seen to be equivalent to the conditions (a).

We observe that

$$(7.13.9) \qquad \frac{\partial}{\partial n_q} \operatorname{Im} Z(q) ds_q = d(\operatorname{Re} Z(q)).$$

Hence, integrating the normal derivatives of both sides of (7.13.4) from p_1 to p_ν, we obtain

$$(7.13.10) \quad \operatorname{Re}\left\{\int_{p_1}^{p_\nu} dZ^*\right\} = \operatorname{Re}\left\{\int_{p_1}^{p_\nu} dZ\right\} - \operatorname{Im}\left\{\frac{1}{2\pi i}\int_\gamma I_\nu(z) dZ^*(z)\right\},$$

where

$$(7.13.11) \qquad I_\nu(z) = \int_{p_1}^{p_\nu} \frac{\partial Q(z, p)}{\partial n_p}\, ds_p.$$

We use again the fact that under a map (7.13.5)

$$(7.13.12) \qquad \int_{p_1^*}^{p^*} dZ^*(p^*) = \int_{p_1}^{p} dZ(p).$$

Hence, we may put (7.13.10) into the form

$$\text{(7.13.13)}\quad \operatorname{Re}\left\{\int_{p_1}^{p_\nu} dZ^*\right\} = \operatorname{Re}\left\{\int_{p_1^*}^{p_\nu^*} dZ^*\right\} - \operatorname{Im}\left\{\frac{1}{2\pi i}\int_\gamma I_\nu(z)\, dZ^*(z)\right\}.$$

The equations (7.13.13) for $\nu = 2, 3, \cdots, N$ express the conditions (b) of the last section and are equivalent to them.

Finally, integrating the normal derivatives of both sides of (7.13.4) around a basis cycle K_μ and writing

$$\text{(7.13.14)}\qquad \Gamma_\mu = P(dZ, K_\mu), \quad \Gamma_\mu^* = P(dZ^*, K_\mu),$$

we obtain

$$\text{(7.13.15)}\qquad \operatorname{Re}\Gamma_\mu^* = \operatorname{Re}\Gamma_\mu - \operatorname{Im}\left\{\frac{1}{2\pi i}\int_\gamma J_\mu(z)\, dZ^*(z)\right\}$$

where

$$\text{(7.13.16)}\qquad J_\mu(z) = \int_{K_\mu} \frac{\partial Q(z, p)}{\partial n_p}\, ds_p.$$

The conditions (c) are therefore equivalent to the equations

$$\text{(7.13.17)}\quad \operatorname{Im}\left\{\frac{1}{2\pi i}\int_\gamma J_\mu(z)\, dZ^*(z)\right\} = 0, \quad \mu = 1, 2, \cdots, G.$$

Thus, the surfaces $\mathfrak{M}$ and $\mathfrak{M}^*$ will be conformally equivalent if the following conditions are fulfilled:

$$\text{(7.13.18)}\quad \operatorname{Im} Z(p_\nu) = \operatorname{Im} Z(p_\nu^*) - \operatorname{Im}\left\{\frac{1}{2\pi i}\int_\gamma Q(z, p_\nu^*)\, dZ^*(z)\right\}, \quad \nu = 1, 2, \cdots, N;$$

$$\text{(7.13.19)}\quad \operatorname{Re}\left\{\int_{p_1}^{p_\nu} dZ^*\right\} = \operatorname{Re}\left\{\int_{p_1^*}^{p_\nu^*} dZ^*\right\} - \operatorname{Im}\left\{\frac{1}{2\pi i}\int_\gamma I_\nu(z)\, dZ^*(z)\right\}, \quad \nu = 2, 3, \cdots N;$$

$$\text{(7.13.20)}\qquad \operatorname{Im}\left\{\frac{1}{2\pi i}\int_\gamma J_\mu(z)\, dZ^*(z)\right\} = 0, \qquad \mu = 1, 2, \cdots, G.$$

Let $r_0(z)$ be a function, assumed given, which is regular analytic in a complete neighborhood of γ, and which satisfies the orthogonality conditions (7.11.2). We set

$$r(z) = r_0(z) + \varepsilon\varrho_\varepsilon(z) \tag{7.13.21}$$

and we seek the conditions imposed on $\varrho_\varepsilon(z)$ by (7.13.18), (7.13.19) and (7.13.20).

By (7.13.1)′ and (7.13.21),

$$\begin{aligned} Q(z, p^*) &= T(z + \varepsilon\, r(z), p^*) - T(z, p^*) \\ &= \varepsilon r_0(z) T'(z, p^*) + O(\varepsilon^2). \end{aligned} \tag{7.13.22}$$

Substituting from (7.13.22) into (7.13.7), we find by (7.8 14) that

$$\begin{aligned} \operatorname{Im} Z(p) &= \operatorname{Im} Z(p^*) \\ &- \operatorname{Im}\left\{\frac{\varepsilon}{2\pi i}\int_\gamma r_0(z) T'(z, p^*) Z'(z)\, dz\right\} + o(\varepsilon). \end{aligned} \tag{7.13.23}$$

Using the identity (7.11.14) and assuming that $r_0(z)$ transforms like a reciprocal differential, we may write (7.13.23) in the form

$$\operatorname{Im} Z(p) = \operatorname{Im} Z(p^*) - \varepsilon \operatorname{Im}\{Z'(p^*)h(p^*)\} + o(\varepsilon). \tag{7.13.24}$$

Since the coefficient of ε in (7.13.24) is the imaginary part of an analytic function, we have

$$Z(p) = Z(p^*) - \varepsilon Z'(p^*)h(p^*) + C + o(\varepsilon), \tag{7.13.25}$$

where C is a real constant. We now evaluate C.

Suppose first that $G = 2h + m - 1$, $m > 1$. Then, by (7.12.6), $N \geqq 1$. Differentiating both sides of (7.13.25) with respect to p^* and then choosing p^* such that the image p in the mapping (7.13.5) is a point p_ν, we obtain, since $Z'(p_\nu) = 0$,

$$0 = Z'(p_\nu^*)[1 - \varepsilon h'(p_\nu^*)] - \varepsilon Z''(p_\nu^*)h(p_\nu^*) + o(\varepsilon). \tag{7.13.26}$$

Thus

$$Z'(p_\nu^*) = O(\varepsilon). \tag{7.13.27}$$

Let the local coordinates of p_ν and p_ν^*, expressed in terms of the same uniformizer, be z_ν and z_ν^* respectively. Since $Z'(p)$ has a simple zero at the point p_ν, we conclude from (7.13.27) that

$$z_\nu = z_\nu^* + O(\varepsilon). \tag{7.13.28}$$

We develop $Z(p)$ around the point p_ν into a power series of the local uniformizer z. Since $Z'(p_\nu) = 0$, we have by (7.13.28)

$$Z(p_\nu) = Z(p_\nu^*) + O(\varepsilon^2). \tag{7.13.29}$$

On the other hand, for $p^* = p_\nu^*$, formula (7.13.25) gives

$$Z(p_\nu) = Z(p_\nu^*) + C + o(\varepsilon). \tag{7.13.30}$$

Comparing (7.13.29) and (7.13.30), we conclude that $C = o(\varepsilon)$ and formula (7.13.25) may be written

$$Z(p) = Z(p^*) - \varepsilon Z'(p^*)h(p^*) + o(\varepsilon). \tag{7.13.31}$$

Therefore, if the local coordinates of p and p^* are expressed in terms of the same uniformizer, we have the formula

$$z = z^* - \varepsilon h(p^*) + o(\varepsilon). \tag{7.13.32}$$

In the case $G = 1$ (doubly-connected domain) the constant C occurring in (7.13.25) is not necessarily $o(\varepsilon)$. For in this case there is a one-parameter group of conformal mappings of $\mathfrak{M}$ onto itself. Each element of the group transforms a point p of $\mathfrak{M}$ into a point p_1, where $\operatorname{Im} Z(p) = \operatorname{Im} Z(p_1)$. However, we are still at liberty to normalize the mapping (7.13.25) by requiring that a given point p_0^* goes into a point p_0, where

$$\operatorname{Im} Z(p_0) = \operatorname{Im} Z(p_0^*); \quad \operatorname{Re} Z(p_0) = \operatorname{Re} Z(p_0^*) + K.$$

For this normalization we have

$$C = K + \varepsilon \operatorname{Re} \{Z'(p_0^*)h(p_0^*)\} + o(\varepsilon).$$

Choosing, in particular,

$$K = -\varepsilon \operatorname{Re} \{Z'(p_0^*)\, h(p_0^*)\},$$

we see that $C = o(\varepsilon)$.

In the case $G = 0$ (simply-connected domain) the problem of conformal equivalence is vacuous. However, if $\mathfrak{M}$ is represented as a domain of the z-sphere and if the mapping $z^* \to z$ is suitably normalized, it may be shown that formula (7.13.32) is valid. Since a proof of this case is to be found in [5], we omit details here.

We summarize the results obtained so far.

THEOREM 7.13.1. *Let $\mathfrak{M}$ be an orientable surface with boundary. If $\mathfrak{M}$ can be transformed into a conformally equivalent surface $\mathfrak{M}^*$ by means of a cell attached along a Jordan curve γ using a reciprocal differential $r(p)$ defined in a neighborhood of γ, then the conformal* *mapping can be realized in terms of a local uniformizer of* $\mathfrak{M}$ *by means of the correspondence*

$$z^* = z + \varepsilon h(z) + o(\varepsilon).$$

where h is defined by (7.11.5). *This relation holds uniformly in each closed subdomain of* $\mathfrak{M}$ *which does not contain the curve* γ.

In deriving formula (7.13.32), in which z and z^* are local coordinates of p and p^* expressed in terms of the same uniformizer, we have assumed that $r(z)$ is of the form (7.13.21), where $\varrho_\varepsilon(z)$ is bounded independently of ε, and we have made use of formula (7.13.7) which in turn depends on (7.13.6). Moreover, in (7.13.26), we have implicitly assumed that $p_\nu^* \to p_\nu$ under the mapping (7.13.5). Hence, essentially, the proof of (7.13.32) is based on the validity of conditions (7.13.18)—(7.13.20). We have also made certain normalizing assumptions in the cases $G = 0$ and $G = 1$.

Suppose now that the conditions (7.13.18)—(7.12.20) are fulfilled and that $r(z)$ is of the form (7.13.21). We are then justified in assuming that there is a mapping which, expressed in local coordinates, has the form (7.13.32). The relation (7.13.32) was derived from (7.13.18) by comparing terms of order ε on both sides of the equation. We shall now return to the same equation but work with a higher order of precision in ε and will find conditions on $\varrho_\varepsilon(z)$ in order that $\mathfrak{M}$ and $\mathfrak{M}^*$ be conformally equivalent.

By (7.13.1)′ and (7.13.32),

$$\begin{aligned} Q(z, p_\nu^*) &= \varepsilon r(z) T'(z, p_\nu) \\ &\quad + \varepsilon^2 \left\{ r(z) \left[\frac{\partial T'(z, p_\nu)}{\partial p_\nu} h(p_\nu) + \frac{\partial T'(z, p_\nu)}{\partial \bar{p}_\nu} h(\bar{p}_\nu) \right] \right. \\ &\qquad \left. + \frac{1}{2} [r(z)]^2 T''(z, p_\nu) \right\} + o(\varepsilon^2), \end{aligned} \tag{7.13.33}$$

and, by (7.13.31),

$$\operatorname{Im} Z(p_\nu^*) = \operatorname{Im} Z(p_\nu) + \frac{\varepsilon^2}{2} \operatorname{Im} \{ Z''(p_\nu) [h(p_\nu)]^2 \} + o(\varepsilon^2), \tag{7.13.34}$$

$$Z^*(p) = Z(p) - \varepsilon Z'(p) h(p) + o(\varepsilon). \tag{7.13.35}$$

Substituting these relations in (7.13.18), we find that

$$\operatorname{Im}\left\{\frac{\varepsilon}{2\pi i}\int_\gamma r(z)T'(z, p_\nu)Z'(z)dz\right.$$
$$-\varepsilon^2\left[\frac{1}{2}Z''(p_\nu)[h(p_\nu)]^2+\frac{1}{2\pi i}\int_\gamma r(z)T'(z, p_\nu)d(Z'(z)h(z))\right.$$
$$-\frac{1}{2\pi i}\int_\gamma r(z)\left(\frac{\partial T'(z, p_\nu)}{\partial p_\nu}h(p_\nu)+\frac{\partial T'(z,p_\nu)}{\partial \bar{p}_\nu}h(\bar{p}_\nu)\right)Z'(z)dz \tag{7.13.36}$$
$$\left.\left.-\frac{1}{2\pi i}\int_\gamma \frac{1}{2}[(r(z)]^2T''(z, p_\nu)Z'(z)dz\right]\right\}+o(\varepsilon^2)=0.$$

Dividing this equation by ε and letting ε tend to 0, we obtain

$$\operatorname{Im}\left\{\frac{1}{2\pi i}\int_\gamma r_0(z)T'(z, p_\nu)Z'(z)dz\right\}=0. \tag{7.13.37}$$

Since $Z'(p_\nu)=0$, the expression $iT'(q, p_\nu)Z'(q)$ is an everywhere finite quadratic differential on $\mathfrak{M}$ and (7.13.37) is a consequence of (7.11.2). Thus no new condition is imposed on $r_0(z)$ by the ε-term. On the other hand, consideration of the higher order terms will lead to conditions on $\varrho_\varepsilon(z)$ and we shall show that these conditions can be satisfied for any $r_0(z)$ which satisfies (7.11.2).

However, before entering into these consideration, we want to investigate in an analogous way the conditions imposed upon $r_0(z)$ by the equations (7.13.19) and (7.13.20). We have by (7.13.1)′ and (7.13.11)

$$I_\nu(z)=\varepsilon r(z)\int_{p_1}^{p_\nu}\frac{\partial T'(z, p)}{\partial n_p}ds_p+\frac{\varepsilon^2}{2}(r(z))^2\int_{p_1}^{p_\nu}\frac{\partial T''(z, p)}{\partial n_p}ds_p+\cdots$$
$$=-\varepsilon r(z)2\pi i W'_{p_1p_\nu}(z)-\frac{\varepsilon^2}{2}(r(z))^2\,2\pi i W''_{p_1p_\nu}(z)+\cdots, \tag{7.13.38}$$

where

$$\int_{p_1}^{p_\nu}\frac{\partial T'(q, p)}{\partial n_p}ds_p=2\frac{d}{dq}\int_{p_1}^{p_\nu}\frac{\partial G(q, p)}{\partial n_p}ds_p=2\frac{d}{dq}\operatorname{Im}2\pi W_{p_1p_\nu}(q)$$
$$=-2\pi i W'_{p_1p_\nu}(q). \tag{7.13.39}$$

The expression $W_{p_1 p_\nu}(q)$ is a differential with simple poles at p_1 and p_ν. Moreover

$$\mathrm{Re}\left\{\int_{p_1^*}^{p_\nu^*} dZ^* - \int_{p_1}^{p_\nu} dZ^*\right\} = -\frac{\varepsilon^2}{2}\mathrm{Re}\{Z''(p_1)(h(p_1))^2 - Z''(p_\nu)(h(p_\nu))^2\} \tag{7.13.40}$$
$$+ o(\varepsilon^2).$$

Substituting from (7.13.38) and (7.13.40) into (7.13.19), we have

$$\mathrm{Im}\left\{\varepsilon\int_\gamma r(z)W'_{p_1 p_\nu}(z)Z'(z)dz - \frac{i\varepsilon^2}{2}[Z''(p_1)(h(p_1))^2 - Z''(p_\nu)(h(p_\nu))^2]\right.$$
$$\left. - \varepsilon^2\int_\gamma r(z)W'_{p_1 p_\nu}(z)d(Z'(z)h(z)) + \frac{\varepsilon^2}{2}\int_\gamma (r(z))^2 W''_{p_1 p_\nu}(z)Z'(z)dz\right\} \tag{7.13.41}$$
$$+ o(\varepsilon^2) = 0.$$

Dividing by ε and letting ε tend to zero, we obtain

$$\mathrm{Im}\left\{\int_\gamma r_0(z)W'_{p_1 p_\nu}(z)Z'(z)dz\right\} = 0. \tag{7.13.42}$$

Since $Z'(p_1) = 0$, $Z'(p_\nu) = 0$, the quadratic differential $W_{p_1 p_\nu}(q)Z'(q)$ is everywhere finite and (7.13.42) is a consequence of (7.11.2).

Finally

$$J_\mu(z) = -\varepsilon r(z)\, 2\pi i\, Z'_\mu(z) - \frac{\varepsilon^2}{2}(r(z))^2\, 2\pi i\, Z''_\mu(z) + o(\varepsilon^2). \tag{7.13.43}$$

Substituting into (7.13.20) we obtain

$$\mathrm{Im}\left\{\varepsilon\int_\gamma r(z)\, Z'_\mu(z)\, Z'(z)dz - \varepsilon^2\int_\gamma r(z)\, Z'_\mu(z)\, d(Z'(z)\, h(z))\right. \tag{7.13.44}$$
$$\left. + \frac{\varepsilon^2}{2}\int_\gamma (r(z))^2\, Z''_\mu(z)\, Z'(z)dz\right\} + o(\varepsilon^2) = 0.$$

On dividing by ε and letting ε tend to zero:

$$\mathrm{Im}\left\{\int_\gamma r_0(z)\, Z'_\mu(z)\, Z'(z)dz\right\} = 0, \tag{7.13.45}$$

and this again is a consequence of (7.11.2).

The conditions (7.13.18)—(7.13.20) may therefore be written in the form

$$\text{(7.13.46)} \quad \operatorname{Im}\left\{\int_\gamma \varrho_\varepsilon(z) iT'(z, p_\nu)\, Z'(z)dz + F_{\nu\varepsilon}(r_0 + \varepsilon\varrho_\varepsilon)\right\} = 0,$$

$$\text{(7.13.47)} \quad \operatorname{Im}\left\{\int_\gamma \varrho_\varepsilon(z)\, W'_{p_1 p_\nu}(z) Z'(z)dz + G_{\nu\varepsilon}(r_0 + \varepsilon\varrho_\varepsilon)\right\} = 0,$$

$$\text{(7.13.48)} \quad \operatorname{Im}\left\{\int_\gamma \varrho_\varepsilon(z)\, Z'_\mu(z)\, Z'(z)dz + H_{\mu\varepsilon}(r_0 + r\varrho_\varepsilon)\right\} = 0,$$

where $F_{\nu\varepsilon}$, $G_{\nu\varepsilon}$ and $H_{\mu\varepsilon}$ are functionals of $r_0 + \varepsilon\varrho_\varepsilon = r$. For $\varepsilon = 0$ we have by (7.13.36), (7.13.41) and (7.13.44),

$$F_{\nu 0}(r_0) = \pi Z''(p_\nu)(h(p_\nu))^2 - i\int_\gamma r_0(z)\, T'(z, p_\nu)\, d(Z'(z)h(z))$$

$$\text{(7.13.49)} \quad + \int_\gamma r_0(z) i\left[\frac{\partial T'(z, p_\nu)}{\partial p_\nu} h(p_\nu) + \frac{\partial T'(z, \bar{p}_\nu)}{\partial \bar{p}_\nu} h(\bar{p}_\nu)\right] Z'(z)dz$$

$$+ \int_\gamma \frac{1}{2}(r_0(z))^2 iT''(z, p_\nu) Z'(z)dz,$$

$$\text{(7.13.50)} \quad \begin{aligned} G_{\nu 0}(r_0) &= \frac{i}{2}[Z''(p_\nu)(h(p_\nu))^2 - Z''(p_1)(h(p_1))^2] \\ &- \int_\gamma r_0(z) W'_{p_1 p_\nu}(z) d(Z'(z)h(z)) + \frac{1}{2}\int_\gamma (r_0(z))^2 W''_{p_1 p_\nu}(z)\, Z'(z)dz, \end{aligned}$$

(7.13.51)

$$H_{\mu 0}(r_0) = -\int_\gamma r_0(z)\, Z'_\mu(z) d(Z'(z)h(z)) + \frac{1}{2}\int_\gamma (r_0(z))^2\, Z''_\mu(z)\, Z'(z)dz.$$

If $G > 1$, the number $\sigma = 3G - 3 = 6h + 3m - 6$ is the number of real moduli of the surface $\mathfrak{M}$ (which has a boundary, by hypothesis). If $G = 1$, the number σ of real moduli is 1. The everywhere finite quadratic differentials

$$(7.13.52)\qquad \begin{cases} iT'(q, p_\nu)Z'(q), & \nu = 1, 2, \cdots, G-1, \\ W'_{p_1 p_\nu}(q)Z'(q), & \nu = 2, 3, \cdots, G-1, \\ Z'_\mu(q)Z'(q), & \mu = 1, 2, \ldots, G, \end{cases}$$

form a basis for the quadratic differentials of $\mathfrak{M}$. For they are σ in number, and each is real on the boundary of $\mathfrak{M}$ when expressed in terms of a boundary uniformizer. The real linear independence of the differentials (7.13.52) is a consequence of the real independence of the linear differentials

$$(7.13.53)\qquad \begin{cases} \text{(a)}\ iT'(q, p_\nu) & \nu = 1, 2, \cdots, G-1, \\ \text{(b)}\ W'_{p_1 p_\nu}(q), & \nu = 2, 3, \cdots, G-1, \\ \text{(c)}\ Z'_\mu(q), & \mu = 1, 2, \cdots, G. \end{cases}$$

The independence of these differentials follows from the fact that the residue of $iT'(q, p_\nu)$ at p_ν is $-i$, while the residues of $W'_{p_1 p_\nu}(q)$ at p_1, p_ν are ± 1. If a linear combination of the differentials (7.13.53) with real coefficients vanishes identically, the residues vanish and therefore no differential (a) or (b) can have a coefficient different from zero. Since the Z'_μ are independent, all coefficients vanish. The basis (7.13.52) will be denoted by $Q_1, Q_2, \cdots, Q_\sigma$.

The term $\varrho_\varepsilon(q)$ behaves near γ like a reciprocal differential. We will try to represent it, therefore, as a linear combination of reciprocal differentials $n(p, q)$. We assume without loss of generality that the Jordan curve γ does not enclose any p-pole of $n(p, q)$. We then select σ points q_ν inside γ for which the determinant

$$| \operatorname{Im} Q_i(q_\nu) |_{i, \nu=1, 2, \ldots, \sigma} \neq 0.$$

There are always such points q_ν, for otherwise we would have an identity

$$(7.13.54)\qquad \operatorname{Im}\{ \sum_{i=1}^{\sigma} \lambda_i Q_i(q)\} \equiv 0, \quad \lambda_i \text{ real, for all } q \in \mathfrak{M},$$

which is impossible because of the linear independence of the Q_i. We then set

$$(7.13.55)\qquad \varrho_\varepsilon(p) = \sum_{\nu=1}^{\sigma} a_\nu(\varepsilon) n(p, q_\nu), \quad a_\nu(\varepsilon) \text{ real,}$$

and try to choose the functions $a_\nu(\varepsilon)$ in such a way that the σ equations

(7.13.46)—(7.13.48) are fulfilled. Using the residue theorem, we may bring these equations into the form:

$$\sum_{\nu=1}^{\sigma} a_\nu(\varepsilon) \operatorname{Im} Q_i(q_\nu) - \varepsilon \Phi_i(a_1, \cdots, a_\sigma; \varepsilon) = F_i(r_0). \tag{7.13.56}$$

Here the functions $-F_i(r_0)$ are the set $F_{\nu 0}(r_0)$, $G_{\nu 0}(r_0)$, $H_{\mu 0}(r_0)$ considered above. We can calculate from (7.13.56) the values of $a_\nu(0)$.

We observe next that the functions $\Phi_i(a_1, \cdots, a_\sigma; \varepsilon)$ are continuously differentiable functions of all their arguments. This depends ultimately on the existence of continuous derivatives of arbitrary order of the Green's function $G^*(p, q)$ with respect to its variables and with respect to the parameter ε. Indeed one sees from (7.13.1) that one can develop $G^*(p, q) - G(p, q)$ into powers of ε up to an arbitrary order ε^α with an error term $O(\varepsilon^{\alpha+1})$. We obtain in this way more precise variational formulas; but we have confined ourselves to the first order term for simplicity.

We now apply the following theorem on implicit functions [1, page 9]:

Consider the system of equations

$$f_\nu(x_1, x_2, \cdots, x_\sigma; t) = 0, \quad \nu = 1, 2, \cdots, \sigma. \tag{7.13.57}$$

Let a set $x_1^0, \cdots, x_\sigma^0$ be given such that

$$f_\nu(x_1^0, x_2^0, \cdots, x_\sigma^0); 0) = 0, \quad \nu = 1, 2, \cdots, \sigma.$$

Let all $\sigma(\sigma + 1)$ functions $\partial f_\nu / \partial x_\varrho$, $\partial f_\nu / \partial t$ be continuous in their $\sigma + 1$ variables and let the determinant $| \partial f_\nu / \partial x_\varrho |$ be different from zero at the point $(x_1^0, \cdots, x_\sigma^0, 0)$. There exists a uniquely determined set of continuous functions $g_\varrho(t)$ defined for sufficiently small values of t such that

$$g_\varrho(0) = x_\varrho^0, \quad \varrho = 1, 2, \cdots \sigma.$$

and that

$$f_\nu(g_1(t), g_2(t), \cdots, g_\sigma(t); t) \equiv 0$$

is satisfied identically in t. Moreover, the first derivatives $dg_\nu(t)/dt$ exist and are continuous in the t-interval considered.

Applying this result to our special problem we recognize that we may determine the reciprocal differential $r_0(p)$ arbitrarily except for the conditions (7.11.2). We can then always find a function $\varrho_\varepsilon(p)$ which is continuously differentiable in ε and is a reciprocal differential of p such that all conditions (7.13.18)—(7.13.20) are fulfilled identically for sufficiently small ε.

We have assumed in this section that $\mathfrak{M}$ has a boundary; the same reasoning can be applied in the case of a closed surface. In this case the role of the boundary is taken over by a point q_0 of the surface with respect to which we normalize the Z_μ by the condition $\operatorname{Im} Z_\mu(q_0) \equiv 0 \pmod 1$. It is convenient in this case to require of $r_0(p)$ in addition to (7.11.2) also the condition:

$$h(q_0) = \frac{1}{2\pi i} \int_\gamma r_0(p_1) n(q_0, p_1) dz_1 = 0. \tag{7.13.58}$$

This simplifies several formulas and by (7.13.32) also leads to the result that q_0 and q_0^* have the same coordinate in any local uniformizer.

We summarize the results proved in this section in the form of a theorem.

THEOREM 7.13.2. *Let γ be a Jordan curve lying in the domain of a local uniformizer z on the surface $\mathfrak{M}$. Let $r_0(z)$ be a function which is regular analytic in a complete neighborhood of γ and which satisfies the orthogonality conditions*

$$\int_\gamma r_0(z)\, Q_\nu(z) dz = 0, \quad \nu = 1, 2, \cdots, \sigma, \tag{7.13.59}$$

where $Q_1, Q_2, \cdots, Q_\sigma$ are a basis for the everywhere finite quadratic differentials of $\mathfrak{M}$. In its dependence on the uniformizer, $r_0(z)$ transforms like a reciprocal differential. Let m denote the number of boundary components of $\mathfrak{M}$ and write

$$h(p) = \begin{cases} \dfrac{1}{2\pi i} \displaystyle\int_\gamma r_0(p_1) n(p, p_1) dz_1 - \dfrac{1}{2\pi i} \displaystyle\int_{\tilde\gamma} \tilde r_0(\tilde p_1) n(p, \tilde p_1) d\tilde z_1, & m > 0, \\ \dfrac{1}{2\pi i} \displaystyle\int_\gamma r_0(p_1) n(p, p_1) dz_1, & m = 0. \end{cases} \tag{7.13.60}$$

If $\mathfrak{M}$ is closed, we impose on $r_0(p)$ the further condition

$$h(q_0) = 0 \tag{7.13.61}$$

where q_0 is the point for which $\operatorname{Im} Z_\mu(q_0) \equiv 0$ *(mod 1),* $\mu = 1, 2, \cdots, 2G,\ G = h$.

Under these assumptions there exists, for all sufficiently small ε, *a function* $\varrho_\varepsilon(z)$ *which is bounded independently of* ε *such that the attachment of a cell to* $\mathfrak{M}$ *along* γ *by means of the reciprocal differential*

$$r(z) = r_0(z) + \varepsilon\varrho_\varepsilon(z) \tag{7.13.62}$$

in the manner described in Section 7.8, *leads to a domain* $\mathfrak{M}^*$ *which is conformally equivalent to* $\mathfrak{M}$. *That is, there exists a one-one conformal mapping in which the point* p^* *of* $\mathfrak{M}^*$ *goes into the point* p *of* $\mathfrak{M}$.

If the local coordinates of p^* and p are expressed in terms of the same uniformizer z of $\mathfrak{M}$, $z^* = z(p^*)$, $z = z(p)$, we have by Theorem 7.13.1 the relation

$$z^* = z + \varepsilon h(p) + o(\varepsilon). \tag{7.13.62}$$

Theorem 7.13.2 justifies the statement, made at the end of Section 3.8, that the quadratic differentials are connected with the moduli of the Riemann surface $\mathfrak{M}$.

7.14. Variational Formulas for Conformal Mapping

In Chapter 5 we developed a theory of the mappings of a given surface $\mathfrak{N}$ into another surface $\mathfrak{R}$. We obtained necessary and sufficient conditions for such mappings, given locally by certain power series. These conditions are expressed in terms of the coefficients of the power series considered; but they soon become so complicated that it is difficult to answer even relatively simple questions by means of them. In the case that $\mathfrak{N}$ is the unit circle and $\mathfrak{R}$ the complex plane the above conditions should, for example, solve the coefficient problem for schlicht functions. It has been impossible, however, to deduce from them bounds even for the third coefficient of the power series for a schlicht function.

It is, therefore, useful to apply variational methods and we shall characterize maps which extremalize a given functional of the map. Clearly, the main tool in such a variational approach is a sufficiently general formula for the construction of nearby mapping functions. We proceed now to establish such formulas by generalizing the methods of Section 7.3.

Let $\mathfrak{N}$ be a given surface which is mapped into a subdomain $\mathfrak{M}$ of $\mathfrak{R}$. The surface $\mathfrak{R}$ may have a boundary or may be closed. If $\mathfrak{N}$

is closed, $\mathfrak{R}$ is necessarily closed and is itself the image of $\mathfrak{N}$, that is $\mathfrak{M} = \mathfrak{R}$. An interesting problem arises if, for example, $\mathfrak{N}$ is a multiply-connected domain and $\mathfrak{R}$ is the sphere. This leads to the theory of schlicht functions in multiply-connected domains and will be treated in the next chapter.

We shall perform on $\mathfrak{N}$ an interior deformation of the type described in Section 7.13. The given mapping of $\mathfrak{N}$ onto $\mathfrak{M}$ then defines an interior deformation, by cell attachment, of $\mathfrak{M}$ and therefore of $\mathfrak{R}$. If we have enough parameters at our disposal, we can find an interior deformation which preserves the conformal type of $\mathfrak{N}$ and $\mathfrak{R}$ simultaneously. The mapping from the domain $\mathfrak{N}$ into the varied subdomain $\mathfrak{M}^* \subset \mathfrak{R}$ serves as an infinitesimally near comparison map for the original map of $\mathfrak{N}$ onto $\mathfrak{M}$.

Let N be a mapping from $\mathfrak{N}$ onto an arbitrary subdomain $\mathfrak{M}$ of the given surface $\mathfrak{R}$. Let $Q_1(p), Q_2(p), \cdots, Q_{\sigma_1}(p)$ and $\mathcal{Q}_1(q), \mathcal{Q}_2(q), \cdots, \mathcal{Q}_{\sigma_2}(q)$ be real bases for the everywhere finite quadratic differentials of $\mathfrak{N}$ and $\mathfrak{R}$ respectively; σ_1 is the number of real moduli of $\mathfrak{N}$ and σ_2 the number of real moduli of $\mathfrak{R}$. We suppose that the point p of $\mathfrak{N}$ corresponds to the point q of $\mathfrak{R}$ under the mapping N. We consider the linear combination

$$\sum_{\nu=1}^{\sigma_1} c_\nu Q_\nu(p) + \left(\frac{d\zeta}{dz}\right)^2 \sum_{\nu=1}^{\sigma_2} C_\nu \mathcal{Q}_\nu(q), \tag{7.14.1}$$

where z is a uniformizer at the point p of $\mathfrak{N}$, ζ a uniformizer at the point q of $\mathfrak{R}$. If there exist $\sigma_1 + \sigma_2$ real numbers $c_1, \cdots c_{\sigma_1}, C_1, \cdots, C_{\sigma_2}$, not all zero, such that the linear combination vanishes identically, we say that the $\sigma_1 + \sigma_2$ differentials are real linearly dependent with respect to N.

Suppose first that the differentials are independent with respect to N. Let γ be a Jordan curve in the domain of a local uniformizer z on the surface $\mathfrak{N}$, and let $r_0(p)$ be a reciprocal differential which is regular analytic in a neighborhood of γ. Let q be the image point of p in the mapping N, ζ a uniformizer at q, and write

$$r_0(p) = R_0(q)\frac{dz}{d\zeta}. \tag{7.14.2}$$

The Jordan curve γ represented in the plane of ζ will be denoted by Γ. We suppose that

$$(7.14.3)\qquad \int_\gamma r_0(p)Q_\nu(p)dz = 0, \quad \nu = 1, 2, \cdots, \sigma_1,$$

$$(7.14.4)\qquad \int_\gamma r_0(p)\,\mathcal{Q}_\nu(q)\left(\frac{d\zeta}{dz}\right)^2 dz = \int_\Gamma R_0(q)\,\mathcal{Q}_\nu(q)d\zeta = 0, \quad \nu = 1, 2, \cdots, \sigma_2.$$

Let m_1, m_2 be the numbers of boundaries, and let $n(p_1, p_2)$, $N(q_1, q_2)$ be the n-functionals of $\mathfrak{N}$, $\mathfrak{R}$ respectivley. We write

$$(7.14.5)\qquad h(p) = \begin{cases} \dfrac{1}{2\pi i}\displaystyle\int_\gamma r_0(p_1)n(p, p_1)dz_1 - \dfrac{1}{2\pi i}\displaystyle\int_{\tilde{\gamma}} \tilde{r}_0(\tilde{p}_1)n(p, \tilde{p}_1)d\tilde{z}_1, & m_1 > 0, \\ \dfrac{1}{2\pi i}\displaystyle\int_\gamma r_0(p_1)n(p, p_1)dz_1, & m_1 = 0, \end{cases}$$

and

$$(7.14.6)\qquad H(q) = \begin{cases} \dfrac{1}{2\pi i}\displaystyle\int_\Gamma R_0(q_1)N(q, q_1)d\zeta_1 - \dfrac{1}{2\pi i}\displaystyle\int_{\tilde{\Gamma}} \tilde{R}_0(\tilde{q}_1)N(q, \tilde{q}_1)d\tilde{\zeta}_1, & m_2 > 0, \\ \dfrac{1}{2\pi i}\displaystyle\int_\Gamma R_0(q)N(q, q_1)d\zeta_1, & m_2 = 0. \end{cases}$$

If either $\mathfrak{N}$ or $\mathfrak{R}$ is closed, we suppose that the corresponding condition (7.13.58) is satisfied. In other words, we impose upon $r_0(p)$ the requirements of Theorem 7.13.2 with respect to both surfaces $\mathfrak{N}$ and $\mathfrak{R}$.

To the σ_1 conditions (7.13.46)—(7.13.48) for the function ϱ_ε imposed by the surface $\mathfrak{N}$ we add the σ_2 corresponding conditions for ϱ_ε imposed by $\mathfrak{R}$, giving $\sigma_1 + \sigma_2$ conditions in all for the function ϱ_ε. Since the quadratic differentials involved are independent in the real sense, the reasoning of Section 7.13 shows that for all sufficiently small ε there exists a function $\varrho_\varepsilon(z)$ which is bounded independently of ε and which has the following properties. Writing

$$(7.14.7)\qquad r(z) = r_0(z) + \varepsilon\varrho_\varepsilon(z),$$

we identify the point z of γ with the point $z + \varepsilon r(z)$ of the neighboring curve γ_ε in the manner described in Section 7.8. Since z acts as a uniformizer for $\mathfrak{R}$ as well as for $\mathfrak{N}$, we simultaneously form in this way surfaces $\mathfrak{N}^*$ and $\mathfrak{R}^*$ from $\mathfrak{N}$ and $\mathfrak{R}$ respectively. The function $\varrho_\varepsilon(z)$ has the property that not only $\mathfrak{N}$ and $\mathfrak{N}^*$, but also $\mathfrak{R}$ and $\mathfrak{R}^*$ are of the same conformal type.

Let the mapping from $\mathfrak{N}^*$ onto $\mathfrak{N}$, in which the point p^* goes into the point p, be denoted by N_ε, and let the mapping from $\mathfrak{R}^*$ onto $\mathfrak{R}$, in which q^* goes into q, be denoted by R_ε. In the mapping N from $\mathfrak{N}$ into $\mathfrak{R}$, the point p with local coordinate z, say, goes into the point q with local coordinate ζ. By inversion of (7.13.32) we have

$$z^* = z + \varepsilon h(p^*) + o(\varepsilon) = z + \varepsilon h(p) + o(\varepsilon) \tag{7.14.8}$$

and, by the analogue of (7.13.32) for $\mathfrak{R}$,

$$\zeta = \zeta^* - \varepsilon H(q^*) + o(\varepsilon) = \zeta^* - \varepsilon H(q) + o(\varepsilon). \tag{7.14.9}$$

The composite mapping $N_\varepsilon^{-1} N R_\varepsilon$ is a one-one conformal mapping from $\mathfrak{N}$ onto a subdomain $\mathfrak{M}^\Delta$ of $\mathfrak{R}$ which sends the point p of $\mathfrak{N}$ into a point q of $\mathfrak{R}$ and, expressed in terms of local uniformizers, it is of the form

$$\zeta^\Delta = \zeta + \varepsilon h(p) \frac{d\zeta}{dz} - \varepsilon H(q) + o(\varepsilon). \tag{7.14.10}$$

Here $h(p)$ is expressed in terms of the uniformizer z while $H(q)$ is expressed in terms of ζ.

Thus, given a mapping N from $\mathfrak{N}$ onto a subdomain $\mathfrak{M}$ of $\mathfrak{R}$, it is possible to construct a varied mapping (7.14.10) in which $\mathfrak{N}$ is carried onto a subdomain $\mathfrak{M}^\Delta$ of $\mathfrak{R}$ where $\mathfrak{M}^\Delta$ is an ε-variation of $\mathfrak{M}$. The possibility of varying conformal maps in this fashion enables us to apply the calculus of variations to extremal problems in conformal mapping.

We have to make sure, however, that the mapping (7.14.10) does not reduce to the identity $\zeta^\Delta = \zeta$ or even to a mapping $o(\varepsilon)$, that is $\zeta^\Delta = \zeta + o(\varepsilon)$. We must, therefore, investigate the possibility that for each permissible choice of $r_0(p)$ we have identically

$$\begin{aligned} &\frac{1}{2\pi i}\int_\gamma r_0(p_1)n(p, p_1)dz_1 - \frac{1}{2\pi i}\int_{\tilde{\gamma}} \tilde{r}_0(\tilde{p}_1)n(p, \tilde{p}_1)d\tilde{z}_1 \\ &= \frac{dz}{d\zeta}\left[\frac{1}{2\pi i}\int_\Gamma R_0(q_1)N(q, q_1)d\zeta_1 - \frac{1}{2\pi i}\int_{\tilde{\Gamma}} \tilde{R}_0(\tilde{q}_1)N(q, \tilde{q}_1)d\tilde{\zeta}_1\right]. \end{aligned} \tag{7.14.11}$$

For the sake of simplicity we restrict ourselves to the case that $m_1 > 0$ and $m_2 > 0$; the same reasoning would also be valid in all other cases. We may assume without loss of generality that $\mathfrak{N}$ coincides with $\mathfrak{M}$, i.e. is already imbedded in $\mathfrak{R}$. Thus, we may put $z = \zeta$, $p = q$, $\gamma = \Gamma$, $r_0(q) = R_0(q)$. Hence (7.14.11) has the simpler form:

$$\frac{1}{2\pi i}\int_{\gamma} r_0(q_1)[N(q, q_1) - n(q, q_1)]d\zeta_1 \tag{7.14.11$'$}$$
$$= \frac{1}{2\pi i}\int_{\tilde{\gamma}} \tilde{r}_0(\tilde{q}_1)[N(q, \tilde{q}_1) - n(q, \tilde{q}_1)]d\tilde{\zeta}_1.$$

This equation is assumed to hold for arbitrary choice of $r_0(q)$ so long as (7.14.3) and (7.14.4) are fulfilled. One derives easily that this condition can be fulfilled only if

$$N(q, q_1) - n(q, q_1) = \sum_{\nu=1}^{\sigma_1} \alpha_\nu(q)Q_\nu(q_1) + \sum_{\nu=1}^{\sigma_2} \beta_\nu(q)\mathfrak{Q}_\nu(q_1)$$

and

$$N(q, \tilde{q}_1) - n(q, \tilde{q}_1) = \sum_{\nu=1}^{\sigma_1} \gamma_\nu(q)Q_\nu(\tilde{q}_1) + \sum_{\nu=1}^{\sigma_2} \delta_\nu(q)\mathfrak{Q}_\nu(\tilde{q}_1)$$

identically in q_1 for arbitrary choice of q. However, the last equation is clearly impossible in the case where q is a boundary point of $\mathfrak{M}$ which is neither a q-pole of $n(q, q_1)$ nor a boundary point of $\mathfrak{R}$. In this case, $n(q, \tilde{q}_1)$ would become infinite as $\tilde{q}_1$ approaches q whereas $N(q, \tilde{q}_1)$ remains finite since q is not a boundary point of $\mathfrak{R}$.

Thus, we have proved:

THEOREM 7.14.1. *If a domain $\mathfrak{N}$ can be mapped into a subdomain $\mathfrak{M}$ of $\mathfrak{R}$ and if there is no real linear dependence between the basic quadratic differentials of $\mathfrak{M}$ and of $\mathfrak{R}$, then there exists an infinity of mappings of $\mathfrak{N}$ into $\mathfrak{R}$ of the form* (7.14.10).

We have assumed, however, that the $\sigma_1 + \sigma_2$ quadratic differentials arising from the surfaces $\mathfrak{N}$ and $\mathfrak{R}$ are linearly independent in the real sense. Suppose now that this hypothesis is not fulfilled. Then there exist real numbers $c_1, \cdots, c_{\sigma_1}, C_1, \cdots, C_{\sigma_2}$, not all zero, such that

$$\sum_{\nu=1}^{\sigma_1} c_\nu Q_\nu(p) + \left(\frac{d\zeta}{dz}\right)^2 \sum_{\nu=1}^{\sigma_2} C_\nu \mathfrak{Q}_\nu(q) \equiv 0. \tag{7.14.1$'$}$$

It is clear, because of the linear independence of the $\mathcal{Q}_\nu$, that not all coefficients c_ν are zero. Similarly, not all C_ν are zero. Writing

(7.14.12) $$Q(p) = \sum_{\nu=1}^{\sigma_1} c_\nu Q_\nu(p), \quad \mathcal{Q}(q) = -\sum_{\nu=1}^{\sigma_2} C_\nu \mathcal{Q}_\nu(q),$$

we have the equation

(7.14.13) $$\mathcal{Q}_\zeta(q)\left(\frac{d\zeta}{dz}\right)^2 = Q_z(p)$$

where the subscript denotes the uniformizer in terms of which the corresponding differential is expressed. This equation shows that the surface $\mathfrak{R}$ is mapped onto a subdomain $\mathfrak{M}$ of $\mathfrak{R}$ whose boundary consists of finitely many analytic arcs on each of which we have

(7.14.14) $$\mathcal{Q}_\zeta(q)d\zeta^2 > 0 \text{ or } \mathcal{Q}_\zeta(q)d\zeta^2 < 0.$$

A proof of this statement depends only on the local behavior near the boundary and is therefore the same as the argument given in [5], Chapter VI, where $\mathcal{Q}$, Q are quadratic differentials of special type. For this reason a proof is omitted here. We may also interpret equation (7.14.13) as stating that $\mathcal{Q}_\zeta(q)d\zeta^2$ is a quadratic differential of $\mathfrak{M}$ as well as of $\mathfrak{R}$, in which case (7.14.13), written in the form

(7.14.13)′ $$\mathcal{Q}_\zeta(q)d\zeta^2 = Q_z(p)dz^2,$$

expresses the invariance of the quadratic differential under a change of uniformizer.

In the case that the domain $\mathfrak{M} \subset \mathfrak{R}$ possesses a quadratic differential which is finite everywhere in $\mathfrak{R}$ and is also a quadratic differential for $\mathfrak{R}$, it might well happen that $\mathfrak{M}$ cannot be varied within $\mathfrak{R}$ under preservation of conformal type. In this case we say the domain $\mathfrak{M}$ is rigidly imbedded in $\mathfrak{R}$.

Consider the following example of a rigid imbedding. Let $\mathfrak{R}$ be a triply-connected domain in the complex plane bounded by three curves B_ν, $\nu = 1, 2, 3$; we suppose that B_1 encloses the two other curves B_2 and B_3. If we connect B_2 with B_3 by a continuum in $\mathfrak{R}$ we obtain a doubly-connected subdomain $\mathfrak{M}$ of $\mathfrak{R}$. This domain can be mapped upon a circular ring in the w-plane, $1 < |w| < \mu(\mathfrak{M})$. $\mu(\mathfrak{M})$ is called the modulus of the doubly-connected domain $\mathfrak{M}$. It is now easy to show that there exists a subdomain $\mathfrak{M} \subset \mathfrak{R}$ obtained from $\mathfrak{R}$ by removing an arc between B_2 and B_3 such that its modulus be the largest possible among all doubly-connected domains imbedded in $\mathfrak{R}$ by the same method. The domain $\mathfrak{M}$ is rigidly imbedded in $\mathfrak{R}$ in the sense that it cannot be mapped into any neighboring domain in $\mathfrak{R}$ of the same conformal type; indeed any other domain in $\mathfrak{R}$ which lies near enough to $\mathfrak{M}$ has a lesser modulus and cannot be equivalent to $\mathfrak{M}$. The only possible conformal mapping in $\mathfrak{M}$ is a mapping of $\mathfrak{M}$ into itself, which always exists in the case of a doubly-connected domain. It is clear from our general theory that $\mathfrak{M}$ and $\mathfrak{R}$ must have a quadratic differential in common. This can easily be shown by characterizing the extremum property of $\mathfrak{M}$ by variational methods. We shall now do this, providing at the same time an illustration for the use of our general formulas.

Let $w = f(z)$ be the schlicht function in $\mathfrak{M}$ which maps $\mathfrak{M}$ upon the circular ring in the w-plane. Clearly, the function

$$Z_1(z) = i\,\frac{\log f(z)}{\log \mu} \tag{7.14.14}$$

represents the only Abelian integral of the first kind in $\mathfrak{M}$; indeed, its imaginary part is single-valued in $\mathfrak{M}$ and has the boundary values 0 and 1, respectively. The period Γ_{11} of $Z_1(z)$ is obviously

$$\Gamma_{11} = -\frac{2\pi}{\log \mu}. \tag{7.14.14$'$}$$

Thus, instead of maximizing the modulus of $\mathfrak{M}$ we may try to maximize the period Γ_{11} of $\mathfrak{M}$.

We draw a Jordan curve γ in $\mathfrak{M}$ and perform a cell attachment along γ by means of an arbitrary function $r_0(z)$ which need only be orthogonal to all finite quadratic differentials of $\mathfrak{R}$. By (7.8.12) we will obtain a new domain $\mathfrak{M}^* \subset \mathfrak{R}$ with

$$\Gamma_{11}^* = \Gamma_{11} - 2\,\mathrm{Re}\left\{\frac{\varepsilon}{2i}\int_\gamma r_0(t)(Z_1'(t))^2 dt\right\} + O(\varepsilon^2). \tag{7.14.15}$$

The extremum requirement $\Gamma_{11}^* \leqq \Gamma_{11}$ and the arbitrary value of ε lead necessarily to the conclusion

$$\int_\gamma r_0(t)(Z_1'(t))^2 dt = 0 \tag{7.14.16}$$

if $r_0(t)$ is orthogonal to all finite quadratic differentials of $\mathfrak{R}$. But this implies by the usual reasoning of the calculus of variations that $(Z_1'(z))^2$ is itself a finite quadratic differential of $\mathfrak{R}$.

If the closure of $\mathfrak{M}$ does not fill $\mathfrak{R}$, a simpler type of variation is obtained by choosing the curve Γ in the interior of $\mathfrak{R}$ and in the exterior of $\mathfrak{M}$. In this case we assume that the reciprocal differential satisfies only the requirements of Theorem 7.13.2 with respect to $\mathfrak{R}$; then the mapping NR_ε is obviously a one-one conformal mapping from $\mathfrak{N}$ onto a subdomain $\mathfrak{M}^\Delta$ of $\mathfrak{R}$ which has the form

$$\zeta^\Delta = \zeta - \varepsilon H(q) + o(\varepsilon). \tag{7.14.17}$$

7.15. Variations of Boundary Type

The variations constructed in Section 7.14 are of "interior type"; that is to say, they depend only on the character of the mapping as defined over the interior of $\mathfrak{M}$ and not on the behavior of the mapping at the boundary. The advantage of this type of variation in extremal problems is apparent, since it is not clear a priori that an extremal mapping will be well-behaved on the boundary. However, variations of "boundary type", in which it is presupposed that the mapping is regular, or at least fairly smooth, on the boundaries, are useful in obtaining further information concerning an extremal mapping once it has been established by a variation of interior type that its behavior on the boundary is sufficiently regular. Variations of boundary type are in general much easier to construct than variations of interior type.

We indicate here a simple variation of boundary type for a function $f(p)$ defined over a Riemann surface $\mathfrak{M}$ with boundary, and we assume for simplicity that f is regular analytic up to and including the boundary of $\mathfrak{M}$. The variation of boundary type can be obtained from the variation of interior type by taking the arc γ along the boundary of $\mathfrak{M}$. Then the term $\varepsilon H(q)$ in (7.14.10) can be dropped since the regularity in the interior of $\mathfrak{M}$ is no longer disturbed by the presence of a singular term. However, since γ lies on the boundary of $\mathfrak{M}$, the conjugate arc $\tilde{\gamma}$ coincides with it and the two integrals in $h(p)$ therefore cancel, causing $h(p)$ to vanish identically. For this reason we choose, in place of $h(p)$, the reciprocal differential

(7.15.1) $$\chi(p) = \frac{1}{2\pi i}\int_\gamma r_0(p_1)n(p, p_1)dz_1.$$

Let γ lie in the domain of a boundary uniformizer z, and define

(7.15.2) $$r_0(z) = -2i\nu(z)$$

where $\nu(z)$ is regular analytic in a complete neighborhood of γ, real on γ, and zero at the end points of γ. We assume that $\nu(z)$ is a reciprocal differential in its dependence on the uniformizer. Hence, if z_1 is an arbitrary uniformizer valid in a neighborhood of γ, we have

(7.15.2)′ $$\nu_1(z_1) = \nu(z)\frac{dz_1}{dz}.$$

Thus $\nu(z)$ will be real on γ if and only if the uniformizer in terms of which it is expressed is a boundary uniformizer.

If p is a boundary point of $\mathfrak{M}$ which is not on γ, we see from the definition of r_0 and from (7.10.19) that $\chi(p)/dz$ is real. Now let p approach an interior point of the arc γ. In the neighborhood of this point we deform the path of integration in (7.15.1) into a semi-circular arc lying in the exterior of $\mathfrak{M}$ on the double $\mathfrak{F}$. Letting the radius of this circle approach zero, we have

(7.15.3) $$\chi(p) = \frac{1}{2\pi i}\int_\gamma r_0(p_1)n(p, p_1)dz_1 + i\nu(z),$$

where the integral is to be interpreted as a Cauchy principle value. Here we assume that the integration along γ is in the sense in which interior points of $\mathfrak{M}$ lie to the left. Thus, at points p on the boundary of $\mathfrak{M}$ which do not lie on γ, $\chi(p)/dz$ is real, while at points of γ it is equal to a real quantity plus an imaginary term $i\nu(z)/dz$.

Now suppose that r_0 satisfies the orthogonality conditions

(7.15.4) $$\int_\gamma r_0(p)Q_\nu(p)dz = 0, \quad \nu = 1, 2, \cdots, \sigma.$$

Then $\chi(p)$ is regular analytic throughout the interior of $\mathfrak{M}$ by (7.10.27). Hence, setting

(7.15.5) $$f^\Delta(p) = f(p) + \varepsilon f'(p)\chi(p),$$

we see that f^{Δ} is regular analytic and single-valued on $\mathfrak{M}$ if f is. With the notation

$$\delta z = i\varepsilon\nu(z);\ \delta w = \delta w(p) = f'(p)\delta z;\ \delta f(p) = f^{\Delta}(p) - f(p), \tag{7.15.6}$$

formula (7.15.5) becomes

$$\delta f(p) = -\frac{f'(p)}{\pi i}\int_{\gamma} n(p, p_1)\frac{\delta w(p_1)dw(p_1)}{w'(p_1)^2}, \tag{7.15.7}$$

where $w = f(p)$, and $\delta w(p)$ is the normal shift of the boundary in the plane of w. When $\varepsilon\nu$ is positive, the shift is in the direction of the inner normal.

The arc γ in formula (7.15.7) may be taken to be the whole boundary curve on which it lies, and in this case the restriction concerning the vanishing of $\nu(z)$ may be dropped.

Formula (7.15.7) is a generalization of Julia's well-known variational formula for the unit circle (see [4]). It may be applied when the boundary in the w-plane is piece-wise analytic or composed partly of piece-wise analytic slits provided that $\nu(p)$ and a suitable number of its derivatives vanish at the end points of analytic arcs or slits.

REFERENCES

1. C. CARATHÉODORY, *Variationsrechnung*, Teubner, Leipzig, 1935. (Reprint, Edwards, Ann Arbor, 1945).
2. P. R. GARABEDIAN and M. SCHIFFER, "Identities in the theory of conformal mapping," *Trans. Amer. Math. Soc.* 65 (1949), 187—238.
3. J. HADAMARD, *Mémoire sur le problème d'analyse relatif à l'équilibre des plaques élastiques encastrées*, Mémoires présentés par divers savants à l'Académie des Sciences, 33 (1908).
4. G. JULIA, „Sur une équation aux derivées fonctionelles liée à la représentation conforme," *Annales Sci. de l'Ecole Normale Sup.* (3), 39 (1922), 1—28.
5. A. C. SCHAEFFER and D. C. SPENCER, *Coefficient regions for schlicht functions*, Colloquium Publications, Vol. 35, Amer. Math. Soc., New York, 1950.
6. M. SCHIFFER, (a) "Hadamard's formula and variation of domain-functions," *Amer. Jour. of Math.*, 68 (1946), 417—448. (b) "The kernel function of an orthonormal system," *Duke Math. Jour.*, 13 (1946), 529—540.
7. M. SCHIFFER and D. C. SPENCER, (a) "The coefficient problem for multiply-connected domains," *Annals of Math.*, 52 (1950), 362—402. (b) "A variational calculus for Riemann surfaces," *Ann. Acad. Sci. Fenn. A. I*, 93 (1951).
8. J. TANNERY and J. MOLK, *Eléments de la théorie des fonctions elliptiques*, Tableau des formules, Vol. IV, Gauthiers-Villars, Paris, 1902.

8. Applications of the Variational Method

8.1. Identities for Functionals

We shall show in this section how the variational formulas may be applied in order to investigate the relations among the various functions, differentials and their periods on a closed Riemann surface $\mathfrak{M}$. It will be sufficient to use for this purpose a particularly simple type of variation of $\mathfrak{M}$.

We choose an arbitrary point $t \in \mathfrak{M}$ and introduce a local uniformizer z which vanishes at t. We describe in the z-plane a circle of radius ϱ around the origin, lying in the image of the uniformizer neighborhood of t. Let γ be the curve of $\mathfrak{M}$ which corresponds to the circumference $|z| = \varrho$; we perform a cell attachment along γ, of the type described in Section 7.8, and use $r(z) = e^{2i\varphi}/z$, $\varepsilon = \varrho^2$ for the deformation. Since the point $z = \varrho e^{i\alpha}$ on the circumference is shifted into the point

$$(8.1.1) \qquad z^* = z + e^{2i\varphi}\varrho^2 \frac{1}{z} = \varrho e^{i\varphi}[e^{i(\alpha-\varphi)} + e^{i(\varphi-\alpha)}] = 2\varrho e^{i\varphi} \cos(\alpha-\varphi),$$

we may describe the deformation of the surface $\mathfrak{M}$ as follows. We draw the diameter of the circle $|z| = \varrho$ which has the direction of the complex vector $e^{i\varphi}$ and identify points on the circumference of the circle which lie on the same normal to this diameter. This procedure leads to a new Riemann surface $\mathfrak{M}^*$ and we may use the coordinates p, q for the points of $\mathfrak{M}^*$ as well as for the points of $\mathfrak{M}$ so long as we stay outside of γ.

We can now express the various functions and differentials of $\mathfrak{M}^*$ in terms of functions and differentials of $\mathfrak{M}$ by means of the variational formulas of Section 7.8. In order to obtain suitable formulas for our particular purpose it is convenient to study the behavior of those Abelian integrals on $\mathfrak{M}$ which have their periods normalized with respect to the cycles $K_{2\mu-1}$. We consider, in particular, the integral of the third kind $W(p, p_0; q, q_0)$ which has zero periods with respect to the odd cycles and which is symmetric in its two pairs of variables. Let $\omega_{qq_0}(p)$ be the Abelian integral of the third kind which has zero periods with respect to the $K_{2\mu-1}$ and which was discussed in Section 3.3. Then

$$(8.1.2) \qquad W(p, p_0; q, q_0) = \omega_{qq_0}(p) - \omega_{qq_0}(p_0).$$

We denote corresponding quantities on $\mathfrak{M}^*$ by the same letter with an asterisk. In this notation, we obtain easily the following formula (see (7.8.18)):

$$(8.1.3) \quad W^*(p, p_0; q, q_0) = W(p, p_0; q, q_0) + e^{2i\varphi}\varrho^2\omega'_{pp_0}(t)\omega'_{qq_0}(t) + o(\varrho^2).$$

Now let p describe a cycle $K_{2\mu}$ and compare the corresponding periods on both sides. Using the period relations (3.4.4)′, we obtain

$$(8.1.4) \quad w^*_\mu(q) - w^*_\mu(q_0) = w_\mu(q) - w_\mu(q_0) + e^{2i\varphi}\varrho^2 w'_\mu(t)\omega'_{qq_0}(t) + o(\varrho^2),$$

where $w_\mu(q)$ is the μ-th Abelian integral of the first kind defined in Section 3.3. Thus we have obtained a variational formula for these integrals too.

We denote the period matrix of the differentials $w'_\mu(q)$ with respect to the cycles $K_{2\nu}$ by $\|\gamma_{\mu\nu}\|$. If q in the identity (8.1.4) describes a cycle $K_{2\nu}$, we derive the period variational formula:

$$(8.1.5) \qquad \gamma^*_{\mu\nu} = \gamma_{\mu\nu} + e^{2i\varphi}\varrho^2 \cdot 2\pi i w'_\mu(t) w'_\nu(t) + o(\varrho^2).$$

Finally, we can derive from (8.1.3) a particularly elegant variational formula for the bilinear differential

$$(8.1.6) \qquad \lambda(p, q) = \frac{\partial^2 W(p, p_0; q, q_0)}{\partial p\, \partial q} = \frac{\partial}{\partial q}\omega'_{qq_0}(p)$$

which is symmetric in p and q and has a double pole for $p = q$. We obtain from (8.1.3) by differentiation with respect to p and q

$$(8.1.7) \qquad \lambda^*(p, q) = \lambda(p, q) + e^{2i\varphi}\varrho^2\lambda(t, p)\lambda(t, q) + o(\varrho^2).$$

The formulas (8.1.3)—(8.1.7) stand in close analogy to the formulas of Section 7.8. The advantage of the Riemann normalization of the differentials lies in the fact that the differentials themselves occur in the variational formulas and not the real part of some expression constructed from them.

We now apply these formulas in order to show that the periods $\gamma_{\mu\nu}$ of the differentials of the first kind depend in a differentiable way upon the moduli of the surface $\mathfrak{M}$ considered in the last chapter.

We introduced in Theorem 7.12.1 a set of $6h-6$ real parameters, which we now denote by μ_j, which depend on the zeros and the periods of a differential $Z'(p)$ on $\mathfrak{M}$ and which characterize the conformal type of $\mathfrak{M}$ in a unique way. We calculated in Section 7.13 the variation of these numbers under a general cell attachment and in the case of our particular variation, we have

$$\mu_j^* = \mu_j + \operatorname{Re}\{e^{2i\varphi}\varrho^2 Q_j(t)\} + o(\varrho^2), \tag{8.1.8}$$

where $Q_j(t)$, $j = 1, 2, \cdots, 6h-6$, form a linearly independent set of quadratic differentials, in the real sense.

We prove first:

THEOREM 8.1.1. *Every quadratic differential $w'_\mu(t)w'_\nu(t)$ can be expressed in terms of the real basis $Q_j(t)$ in the form*

$$w'_\mu(t)w'_\nu(t) = \sum_{j=1}^{6h-6} A_{\mu\nu,j} Q_j(t) \tag{8.1.9}$$

with real coefficients $A_{\mu\nu,j}$.

In order to prove this theorem let us assume conversely that, for a given choice of μ and ν, the $6h-5$ quadratic differentials $w'_\mu(t)w'_\nu(t)$, $Q_j(t)$ are linearly independent in the real sense. We can then determine $12h-11$ points t_α in $\mathfrak{M}$ and $12h-11$ real numbers c_α such that

$$\sum_{\alpha=1}^{12h-11} c_\alpha Q_j(t_\alpha) = 0, \qquad j = 1, 2, \cdots, 6h-6,$$
$$\sum_{\alpha=1}^{12h-11} c_\alpha w'_\mu(t_\alpha) w'_\nu(t_\alpha) \neq 0. \tag{8.1.10}$$

We then construct a variation of $\mathfrak{M}$ which is composed of $12h-11$ cell attachments at the points t_α with functions $r_\alpha(z) = c_\alpha e^{2i\varphi}/z$, $\varepsilon = \varrho^2$ on each circle around the corresponding point t_α. Using the methods of Section 7.13 we may correct this variation by an additional term $o(\varrho^2)$ in such a way that $\mathfrak{M}$ goes over into a conformally equivalent surface $\mathfrak{M}^*$; for we can arrange the correction so that the moduli μ_j do not change under the variation. On the other hand, this same variation will lead to a change in $\gamma_{\mu\nu}$ by virtue of (8.1.5) and (8.1.10). This is obviously a contradiction since the period matrix $\gamma_{\mu\nu}$ is a conformal invariant. Thus, the theorem has been proved.

In a similar way, we prove the formula

$$(8.1.9)' \qquad i w'_\mu(t) w'_\nu(t) = \sum_{j=1}^{6h-6} B_{\mu\nu,j} Q_j(t), \quad B_{\mu\nu,j} \text{ real.}$$

From (8.1.5), (8.1.8), (8.1.9) and (8.1.9)′ we derive finally:

$$(8.1.10) \qquad \begin{aligned} \operatorname{Re}\{\gamma^*_{\mu\nu} - \gamma_{\mu\nu}\} &= \sum_{j=1}^{6h-6} 2\pi B_{\mu\nu,j}(\mu^*_j - \mu_j) + o(\varrho^2), \\ \operatorname{Im}\{\gamma^*_{\mu\nu} - \gamma_{\mu\nu}\} &= \sum_{j=1}^{6h-6} 2\pi A_{\mu\nu,j}(\mu^*_j - \mu_j) + o(\varrho^2). \end{aligned}$$

These relations, which prove that the $\gamma_{\mu\nu}$ possess partial derivatives with respect to the μ_j, can be written in the form

$$(8.1.11) \qquad \frac{\partial \operatorname{Re} \gamma_{\mu\nu}}{\partial \mu_j} = 2\pi B_{\mu\nu,j}, \qquad \frac{\partial \operatorname{Im} \gamma_{\mu\nu}}{\partial \mu_j} = 2\pi A_{\mu\nu,j}.$$

Thus, we have proved

THEOREM 8.1.2. *The period matrix $\gamma_{\mu\nu}$ depends differentiably upon the $6h - 6$ real moduli μ_j.*

In a similar way it is possible to show that the differentials on $\mathfrak{M}$ possess derivatives with respect to the moduli μ_j. The dependence of the differentials of a Riemann surface upon its moduli has been studied intensively since the time of B. Riemann and L. Fuchs. The principal tool has been the theory of the theta-functions and a considerable number of relations and differential equations have been derived. The theory of the dependence of the differentials on the moduli has, however, always suffered from the great complexity of the formulas arising. The variational formulas (8.1.3), (8.1.4), (8.1.5) and (8.1.7) contain all these relations in principle and it is easy to derive them from the variational formulas. We shall illustrate the method by one particular application.

Let $w'_1(p)$ be the first differential of the first kind with Riemann normalization and let

$$(8.1.12) \qquad \Phi(p, q) = \frac{\lambda(p, q)}{w'_1(p) w'_1(q)}.$$

This expression depends on p and q; we may use the values of $w_1(p)$ as local uniformizers and consider $\Phi(p, q)$ as an analytic function of w_1:

$$\Phi(p, q) = F(w, \omega; \mu_j), \quad w = w_1(p), \; \omega = w_1(q); \tag{8.1.13}$$

Φ also depends, of course, on the moduli of $\mathfrak{M}$. In order to study the nature of the function F, we perform the special deformation of $\mathfrak{M}$ into $\mathfrak{M}^*$ considered above. Using formulas (8.1.3)—(8.1.7), we find

$$\begin{aligned}\Phi^*(p, q) &= F(w^*, \omega^*; \mu_j^*) = \Phi(p, q) \\ &+ e^{2i\varphi}\varrho^2[\Phi(t, p)\Phi(t, q) - \Phi(p, q)(\Phi(t, p) + \Phi(t, q))]w_1'(t)^2 + o(\varrho^2).\end{aligned} \tag{8.1.14}$$

In order to obtain this simple result we had to make the normalization that at the fixed point $q_0 \in \mathfrak{M}$ and at the image point $q_0 \in \mathfrak{M}^*$ the values of $w_1(q)$ should be the same, for the determination of this integral considered in the choice of uniformizer. By (8.1.4) and our normalization, we may use as local uniformizers belonging to the points $p, q \in \mathfrak{M}^*$ the quantities

$$\begin{aligned} w^* &= w + e^{2i\varphi}\varrho^2 w_1'(t)\,\omega_{pq_0}'(t) + o(\varrho^2), \\ \omega^* &= \omega + e^{2i\varphi}\varrho^2 w_1'(t)\,\omega_{qq_0}'(t) + o(\varrho^2). \end{aligned} \tag{8.1.15}$$

Inserting these into $F(w^*, \omega^*; \mu_j^*)$ and developing into a Taylor's series, we find

$$\begin{aligned}\Phi(p, q) &+ e^{2i\varphi}\varrho^2[\Phi(t, p)\Phi(t, q) - \Phi(p, q)(\Phi(t, p) + \Phi(t, q))]w_1'(t)^2 \\ &= \Phi(p, q) + e^{2i\varphi}\varrho^2\left[\frac{\partial F}{\partial w}\omega_{pq_0}'(t) + \frac{\partial F}{\partial \omega}\omega_{qq_0}'(t)\right]w_1'(t) \\ &\quad + \sum_{j=1}^{6h-6}\frac{\partial F}{\partial \mu_j}\,\mathrm{Re}\,\{e^{2i\varphi}\varrho^2 Q_j(t)\} + o(\varrho^2).\end{aligned} \tag{8.1.16}$$

Cancelling the finite terms, dividing by ϱ^2 and passing to the limit $\varrho = 0$, we obtain

$$\begin{aligned}\Phi(t, p)\Phi(t, q) &- \Phi(p, q)\{\Phi(t, p) + \Phi(t, q)\} \\ &= \frac{\partial F(w, \omega; \mu_j)}{\partial w}\frac{\omega_{pq_0}'(t)}{w_1'(t)} + \frac{\partial F(w, \omega; \mu_j)}{\partial \omega}\frac{\omega_{qq_0}'(t)}{w_1'(t)} \\ &+ \sum_{j=1}^{6h-6}\frac{\partial F(w, \omega; \mu_j)}{\partial \mu_j}e^{-2i\varphi}\frac{1}{(w_1'(t))^2}\,\mathrm{Re}\,\{e^{2i\varphi}Q_j(t)\}.\end{aligned} \tag{8.1.17}$$

Since this identity must hold for arbitrary choice of φ, we find the identity

$$\sum_{j=1}^{6h-6}\frac{\partial F(w, \omega; \mu_j)}{\partial \mu_j}(Q_j(t))^- = 0. \tag{8.1.18}$$

If we let t tend toward q_0, all terms except the first two on the right side of (8.1.17) remain finite. Hence, their singularities must cancel, which leads to the condition:

$$\frac{\partial F(w, \omega; \mu_j)}{\partial w} + \frac{\partial F(w, \omega; \mu_j)}{\partial \omega} = 0. \tag{8.1.19}$$

This partial differential equation shows that F is a function of $w - \omega$. We have proved

$$\Phi(p, q) = F(w_1(p) - w_1(q); \mu_j) \tag{8.1.20}$$

and the identity

$$\Phi(t, p)\Phi(t, q) - \Phi(p, q)(\Phi(t, p) + \Phi(t, q)) = F'(w - \omega; \mu_j)\frac{\omega'_{pq_0}(t) - \omega'_{qq_0}(t)}{w'_1(t)} + \sum_{j=1}^{6h-6} \frac{1}{2} \frac{\partial F(w - \omega; \mu_j)}{\partial \mu_j} \frac{Q_j(t)}{(w'_1(t))^2}. \tag{8.1.21}$$

In order to understand this identity better, we simplify the notation as follows. We put $q = q_0$ and assume moreover, without loss of generality, that $w_1(q_0) = 0$. We let $w_1(t) = z$ and write (8.1.21) in the form

$$\begin{aligned} &F(z - w; \mu_j)F(z; \mu_j) - F(w; \mu_j)[F(z; \mu_j) + F(z - w; \mu_j)] \\ &= F'(w; \mu_j)\int_0^w F(\tau - z; \mu_j)d\tau + \sum_{j=1}^{6h-6} \frac{1}{2}\frac{\partial F(w; \mu_j)}{\partial \mu_j}\frac{Q_j(t)}{(w'_1(t))^2}. \end{aligned} \tag{8.1.22}$$

The function $F(w; \mu_j)$ is an even function of w which has at the origin (corresponding to $q_0 \in \mathfrak{M}$) a double pole and the series development

$$F(w; \mu_j) = -\frac{1}{w^2} + a + bw^2 + \cdots \tag{8.1.23}$$

where the coefficients $a, b, \cdots$ depend on the moduli μ_j. Let w tend to 0 on both sides of (8.1.22) and compute the limit equation by means of (8.1.23). We obtain

$$F(z; \mu_j)^2 - 2aF(z; \mu_j) + \frac{1}{6}F''(z; \mu_j) = \sum_{j=1}^{6h-6} \frac{1}{2}\frac{\partial a}{\partial \mu_j}\frac{Q_j(t)}{(w'_1(t))^2}. \tag{8.1.24}$$

This equation may be considered as a second order differential equation for the function $F(z; \mu_j)$, with coefficients depending in a simple way on the quadratic differentials of $\mathfrak{M}$.

Let us differentiate (8.1.22) with respect to z; in this way we can eliminate the integral on the right-hand side of the identity and obtain

$$\frac{d}{dz}[F(z-w;\mu_j)F(z;\mu_j)-F(w;\mu_j)(F(z;\mu_j)+F(z-w;\mu_j))]$$

(8.1.25)
$$=F'(w;\mu_j)[F(z;\mu_j)-F(z-w;\mu_j)]+\sum_{j=1}^{6h-6}\frac{1}{2}\frac{\partial F(w;\mu_j)}{\partial\mu_j}\frac{d}{dz}\left(\frac{Q_j(t)}{w_1'(t)^2}\right).$$

In order to clarify the meaning of the identities obtained it is useful to specialize them to the case of a surface $\mathfrak{M}$ of genus 1. In this case, our formulas will reduce to well-known relations between elliptic functions. We uniformize $\mathfrak{M}$ by mapping its universal covering surface into the u-plane; each copy of $\mathfrak{M}$ will appear as a parallelogram in this plane. Let ω_1 and ω_2 be the periods of the parallelogram net; then there exists one integral of the first kind on $\mathfrak{M}$ which, because of the Riemann normalization, has the form

$$w_1=\frac{1}{\omega_1}u. \tag{8.1.26}$$

The integral of the third kind will be

$$\omega_{pp_0}(t)=\log\frac{\sigma[\omega_1(z-w_0)]}{\sigma[\omega_1(z-w)]}-\omega_1\eta_1(w-w_0)z, \tag{8.1.27}$$

where $\sigma(z)$ is the Weierstrass entire function and represents a multiplicative integral on $\mathfrak{M}$. By differentiation with respect to z and w, we obtain by (8.1.6) the expression $\lambda(p,t)$. We derive easily:

$$\lambda(p,t)=F(z-w;\omega)=-\omega_1^2\left[\wp(\omega_1(z-w))+\frac{\eta_1}{\omega_1}\right], \tag{8.1.28}$$

where $\omega=\omega_2/\omega_1$ is the only modulus of the surface $\mathfrak{M}$. ω is a complex modulus whose real and imaginary parts may serve as the real moduli μ_1 and μ_2. Equation (8.1.28) states that F depends analytically on $\mu_1+i\mu_2$ and hence $a=a(\omega)$. We easily find

$$a(\omega)=-\omega_1\eta_1,\quad Q_1(t)=2\pi i(w_1'(t))^2,\quad Q_2(t)=2\pi(w_1'(t))^2, \tag{8.1.29}$$

and derive from (8.1.24)

$$2\pi i\frac{1}{\omega_1^4}\frac{\partial}{\partial\omega}(\eta_1\omega_1)=\left[\frac{1}{6}\wp''(z)-\wp(z)^2\right]+\frac{\eta_1^2}{\omega_1^2}. \tag{8.1.30}$$

This relation contains on the one hand the differential equation of the function $\wp(z)$,

$$2\wp''(z) = 12\wp(z)^2 - g_2 \tag{8.1.31}$$

and on the other hand the dependence of the expression $\eta_1\omega_1$ on the modulus ω,

$$2\pi i \frac{\partial}{\partial\omega}(\eta_1\omega_1) = \eta_1^2\omega_1^2 - \frac{1}{12}\omega_1^4 g_2. \tag{8.1.32}$$

This is a well-known differential equation for the periods of the elliptic functions in their dependence on the period-ratio ω.

We might develop a systematic theory of the relations between the various differentials on $\mathfrak{M}$ by the method briefly outlined above. This theory will again lead to rather heavy formulas. On the other hand, we possess in the variational equations for W, w, $\gamma_{\mu\nu}$ and $\lambda(p, q)$ a simple parametrization of these relations and these fundamental equations are the root of all the other relationships which may be established.

8.2. The Coefficient Problem for Schlicht Functions

A classical problem in elementary function theory is the following. Consider the class of all functions $f(z)$ which are univalent or schlicht in the unit circle and have a series development of the normal form

$$f(z) = z + \sum_{\nu=2}^{\infty} a_\nu z^\nu. \tag{8.2.1}$$

The problem is to determine bounds for the values $|a_\nu|$; it is known that $|a_2| \leqq 2$, $|a_3| \leqq 3$ and that these estimates are the best possible. It is conjectured that $|a_n| \leqq n$ holds for all integers n and this estimate would be the best possible since the function

$$\frac{z}{(1-z)^2} = z + \sum_{\nu=2}^{\infty} \nu z^\nu \tag{8.2.1$'$}$$

is schlicht and for this function the sign of equality holds in the above estimate.

Another problem has been considered in this connection also. We plot the sets $a_2, a_3, \cdots, a_{n+1}$ for all schlicht functions (8.2.1) as points in a real $2n$-dimensional space and obtain a certain subspace, called the coefficient space V_n of the schlicht functions. One may investigate the structure of V_n and, in particular, its boundary B_n. The points of B_n may be considered as defining schlicht functions $f(z)$ with certain extremum properties and these functions may be studied by variational methods. It is quite possible that the solution of the coefficient problem in the strict sense, that is the establishing of exact bounds for the $|a_n|$, may be facilitated by the treatment of the more general problem of V_n which is, in addition, of interest in itself.

We shall consider instead a more general problem which can be treated by the same methods as the problems described above and which leads by specialization to several results in complex analysis. We consider a Riemann surface $\mathfrak{N}$ with boundary and assume that it can be mapped conformally into a subdomain $\mathfrak{M}$ of another given surface $\mathfrak{R}$. Let $\pi_0 \in \mathfrak{N}$ correspond to the point $p_0 \in \mathfrak{R}$; we introduce local uniformizers at π_0 and at p_0 such that $z(\pi_0) = 0$, $z(\pi) = z$ and $w(p_0) = 0$, $w(p) = w$. The mapping $p = p(\pi)$ can then be expressed in terms of the uniformizers in the form

$$w = f(z) = b_1 z + b_2 z^2 + \cdots + b_n z^n + \cdots, \tag{8.2.2}$$

and we may consider $f(z)$ as a function on $\mathfrak{N}$ which is schlicht relative to $\mathfrak{R}$. The problem arises of determining bounds for the coefficients b_ν and, more generally, of studying the coefficient body V_n arising from all possible n-tuples $b_1, \cdots, b_n$ in the $2n$-dimensional Euclidean space. The body V_n need not be connected and can, in fact, consist of disjoint pieces. But we can easily prove that V_n is bounded and closed except for the case in which $\mathfrak{R}$ is the Riemann sphere of complex numbers.

In fact, if $\mathfrak{R}$ is not the sphere then there exists a uniformizer W on $\mathfrak{R}$ which maps the universal covering surface schlicht on the interior of a fixed circle or upon the W-plane punctured at the point at infinity. We may assume without loss of generality that W has at p_0 the series development

$$W(w) = w + B_2 w^2 + B_3 w^3 + \cdots; \tag{8.2.3}$$

putting $w = f(z)$, we obtain an analytic function of z,

$$F(z) = W[f(z)] = b_1 z + (b_2 + B_2 b_1^2) z^2 + \cdots \tag{8.2.3$'$}$$

which is regular analytic in a fixed circle $|z| < r$ and schlicht in the strict sense. It follows easily from the theory of schlicht functions in a fixed circle that the $F(z)$ form a normal family, that is that in each infinite set $F_\nu(z)$, $\nu = 1, 2, \cdots$, we can find a subset which converges uniformly in each circle $|z| \leq r_0 < r$ to a schlicht function $F(z)$. We conclude that the functions $f(z)$ themselves form a normal family, too. This proves that the coefficient body V_n is bounded and closed, as asserted.

In order to study the structure of V_n we take an admissible function (8.2.2) which can be extended over $\mathfrak{N}$ in order to lead to an imbedding of $\mathfrak{N}$ in $\mathfrak{R}$. We try to construct infinitesimally near map functions $f^\Delta(z)$ by means of the variational method of Section 7.14. Using (7.14.10), we see that we can find a function

$$f^\Delta(z) = f(z) + \varepsilon h(z) f'(z) - \varepsilon H(f(z)) + o(\varepsilon) \tag{8.2.4}$$

in an ε-neighborhood of $f(z)$ which can also be extended to a schlicht mapping of $\mathfrak{N}$ into $\mathfrak{R}$. We have to make one assumption, namely that the imbedding $\mathfrak{N} \to \mathfrak{M} \subset \mathfrak{R}$ does not lead to a subdomain $\mathfrak{M}$ of $\mathfrak{R}$ which is bounded by curves in $\mathfrak{R}$ satisfying the differential equation

$$\mathcal{Q}(p)dp^2 \lessgtr 0, \quad \mathcal{Q}(p) = \text{finite quadratic differential on } \mathfrak{R}. \tag{8.2.5}$$

If $f(z)$ leads to a mapping of $\mathfrak{N}$ into $\mathfrak{R}$ of this particular type, we cannot vary $f(z)$ freely and it is, in this sense, an extremum function.

In the general case, however, we may choose

$$h(z) = \frac{1}{2\pi i}\int_\gamma r_0(\pi_1)n(z, \pi_1)dz_1 - \frac{1}{2\pi i}\int_{\tilde{\gamma}} \tilde{r}_0(\tilde{\pi}_1)n(z, \tilde{\pi}_1)d\tilde{z}_1 \tag{8.2.6}$$

and, taking $p_1 = f(\pi_1)$,

$$\begin{aligned} H[f(z)] &= \frac{1}{2\pi i}\int_\gamma r_0(\pi_1)\left(\frac{dp_1}{d\pi_1}\right)^2 N(f(z), p_1)dz_1 \\ &\quad - \frac{1}{2\pi i}\int_{\tilde{\gamma}} \tilde{r}_0(\tilde{\pi}_1)\overline{\left(\left(\frac{dp_1}{d\pi_1}\right)^2\right)} N(f(z), \tilde{p}_1)d\tilde{z}_1 \end{aligned} \tag{8.2.6$'$}$$

with an arbitrary Jordan curve γ in $\mathfrak{N}$ and a reciprocal differential $r_0(\pi)$ which has only to satisfy the conditions

$$\int_\gamma r_0(\pi) Q_\nu(\pi) dz = 0, \qquad \nu = 1, 2, \cdots, \sigma_1, \tag{8.2.7}$$

and

$$\int_\gamma r_0(\pi) \left(\frac{dp}{d\pi}\right)^2 \mathfrak{Q}_\nu(p) dz = 0, \qquad \nu = 1, 2, \cdots, \sigma_2, \tag{8.2.8}$$

where the $Q_\nu(\pi)$ for a real basis for the finite quadratic differentials of $\mathfrak{N}$ and the $\mathfrak{Q}_\nu(\pi)$ an analogous basis for $\mathfrak{R}$. By taking $H(f(z))$ in the form (8.2.6)′ we have assumed that $\mathfrak{R}$ possesses a boundary; we do this only for the sake of definiteness but the reasoning will hold in every case.

The variational formula (8.2.4) creates for each schlicht function (8.2.2) a whole neighborhood of schlicht functions. If we want them to have the same normalization as $f(z)$ we must require

$$h(0) f'(0) = H(0) + o(1), \tag{8.2.9}$$

which leads to the following condition on $r_0(\pi)$:

$$\begin{aligned} &\frac{1}{2\pi i} \int_\gamma r_0(\pi) \left[n(0, \pi) f'(0) - \left(\frac{dp}{d\pi}\right)^2 N(0, p) \right] dz \\ &= \frac{1}{2\pi i} \int_\gamma \tilde{r}_0(\tilde{\pi}) \left[n(0, \tilde{\pi}) f'(0) - \overline{\left(\left(\frac{dp}{d\pi}\right)^2\right)} N(0, \tilde{p}) \right] d\tilde{z}. \end{aligned} \tag{8.2.9'}$$

Let us now compute the coefficient b_ν^Δ of $f^\Delta(z)$. For this purpose, we develop the functions $N(f(z), p)$, $N(f(z), \tilde{p})$ at $z = 0$ and the functions $n(z, \pi)$, $n(z, \tilde{\pi})$ around the same point. We shall have

$$n(z, \pi) = \sum_{\varrho=0}^{\infty} a_\varrho(\pi) z^\varrho, \quad n(z, \tilde{\pi}) = \sum_{\varrho=0}^{\infty} a_\varrho(\tilde{\pi}) z^\varrho; \tag{8.2.10}$$

$$N(f(z), p) = \sum_{\varrho=0}^{\infty} A_\varrho(p) z^\varrho, \quad N(f(z), \tilde{p}) = \sum_{\varrho=0}^{\infty} A_\varrho(\tilde{p}) z^\varrho. \tag{8.2.11}$$

The coefficients $a_\varrho(\pi)$ and $A_\varrho(p)$ will play a central role in the theory of the coefficients of functions in $\mathfrak{N}$ which are schlicht relative to $\mathfrak{R}$. Let us consider them in greater detail. The coefficients $a_\varrho(\pi)$ are to be understood as given with the domain $\mathfrak{N}$ while the $A_\varrho(p)$ depend not only upon the domain $\mathfrak{R}$ but also upon the coefficients of the function $f(z)$ in question. It can be easily seen that $A_\varrho(p)$ depends on the first ϱ coefficients $b_1, \cdots, b_\varrho$ of $f(z)$. Since $n(z, \pi)$ depends on π like a quadratic differential, the coefficients $a_\varrho(\pi)$ will be quadratic differentials on $\mathfrak{N}$. They will not be finite, however, on the whole surface $\mathfrak{N}$, but will have poles at the point π_0. In fact, we have by (8.2.10),

$$(8.2.10)' \qquad a_\varrho(\pi) = \frac{1}{\varrho!}\left[\frac{d^\varrho}{dz^\varrho}n(z, \pi)\right]_{z=0}$$

and since $n(0, \pi)$ has a simple pole at π_0, we find that $a_\varrho(\pi)$ has a pole of order $\varrho + 1$ at π_0. The coefficients $a_\varrho(\tilde{\pi})$, on the other hand, are everywhere finite quadratic differentials on the double of $\mathfrak{N}$. They permit us to define on $\mathfrak{N}$ the everywhere regular quadratic differential

$$(8.2.12) \qquad \tilde{a}_\varrho(\pi) = (a_\varrho(\tilde{\pi}))^-$$

which, when expressed in terms of boundary uniformizers, satisfies

$$(8.2.13) \qquad a_\varrho(\pi) = ((\tilde{a}_\varrho(\pi))')^-$$

on the boundary of $\mathfrak{N}$.

We can make an analogous statement with respect to the quadratic differentials on $\mathfrak{R}$, namely $A_\varrho(p)$ and

$$(8.2.14) \qquad \tilde{A}_\varrho(p) = (A_\varrho(\tilde{p}))^-.$$

$A_\varrho(p)$ has a pole of order $\varrho + 1$ at p_0 while $\tilde{A}_\varrho(p)$ is regular everywhere in $\mathfrak{R}$, and on the boundary of $\mathfrak{R}$ we have, using boundary uniformizers,

$$(8.2.15) \qquad A_\varrho(p) = (\tilde{A}(p))^-.$$

From (8.2.4)—(8.2.6) we can now calculate the coefficients b_ν^Δ of $f^\Delta(z)$. We easily find:

$$(8.2.16) \quad b_\nu^\Delta = b_\nu + \varepsilon \cdot \frac{1}{2\pi i}\int_\gamma r_0(\pi)U_\nu(\pi)dz - \varepsilon \cdot \frac{1}{2\pi i}\int_{\tilde{\gamma}} \tilde{r}_0(\tilde{\pi})(V_\nu(\pi))^- d\tilde{z} + o(\varepsilon),$$

with

$$(8.2.17) \qquad U_\nu(\pi) = \sum_{\varrho=1}^{\nu+1} \varrho b_\varrho a_{\nu+1-\varrho}(\pi) - A_\nu(p)\left(\frac{dp}{d\pi}\right)^2$$

and

$$(8.2.18) \qquad V_\nu(\pi) = \sum_{\varrho=1}^{\nu+1} \varrho(b_\varrho)^- \tilde{a}_{\nu+1-\varrho}(\pi) - \tilde{A}_\nu(p)\left(\frac{dp}{d\pi}\right)^2.$$

We shall call a point $(b_1, \cdots, b_n)$ an interior point of the coefficient body V_n if we can choose $r_0(\pi)$ in such a way that each point of a sufficiently small sphere around this point in Euclidean $2n$-space is attained by a variation (8.2.16) of the coefficients. We shall call the point a boundary point of V_n if we cannot fill a whole sphere around it by these variations. We may put (8.2.16) into the form

$$\text{Re}\,\{b_\nu^\Delta - b_\nu\} = \varepsilon\,\text{Re}\left\{\frac{1}{2\pi i}\int_\gamma r_0(\pi)[U_\nu(\pi) + V_\nu(\pi)]dz\right\} + o(\varepsilon),$$

(8.2.19)

$$\text{Im}\,\{b_\nu^\Delta - b_\nu\} = \varepsilon\,\text{Re}\left\{\frac{1}{2\pi i}\int_{\tilde\gamma} r_0(\pi)\cdot\frac{1}{i}[U_\nu(\pi) - V_\nu(\pi)]dz\right\} + o(\varepsilon),$$

and formulate the condition (8.2.9)′ on $r_0(\pi)$ as follows:

$$\frac{1}{2\pi i}\int_\gamma r_0(\pi)U_0(\pi)dz = \frac{1}{2\pi i}\int_{\tilde\gamma} \tilde r_0(\tilde\pi)(V_0(\pi))^- d\tilde z. \tag{8.2.20}$$

Since $r_0(\pi)$ is quite arbitrary except for the conditions (8.2.7), (8.2.8) and (8.2.20), it is obvious that we can attain every point in the neighborhood of the initial point $(b_1, b_2, \cdots, b_n)$ so long as there is no linear dependence, with real coefficients, between the $2n+2+\sigma_1+\sigma_2$ quantities

$$U_\nu(\pi) + V_\nu(\pi),\quad \frac{1}{i}(U_\nu(\pi) - V_\nu(\pi)),\quad \nu = 0, 1, \cdots, n;$$

(8.2.21)

$$Q_\nu(\pi),\ \nu = 1, \cdots, \sigma_1;\quad \mathfrak{Q}_\mu(p)\left(\frac{dp}{d\pi}\right)^2,\quad \mu = 1, \cdots, \sigma_2.$$

Thus, a boundary point of V_n can be characterized by the existence of $n + 1$ complex numbers $\lambda_0, \lambda_1 \cdots, \lambda_n$ such that

$$\sum_{\nu=0}^{n} [\lambda_\nu U_\nu(\pi) + \bar\lambda_\nu V_\nu(\pi)] = Q(\pi) + \mathfrak{Q}(p)\left(\frac{dp}{d\pi}\right)^2, \tag{8.2.22}$$

where $Q(\pi)$ and $\mathfrak{Q}(p)$ are two finite quadratic differentials on $\mathfrak{N}$ and $\mathfrak{R}$, respectively.

We are now able to characterize the domains $\mathfrak{M} \subset \mathfrak{R}$ which belong to the extremum functions $f(z)$ in $\mathfrak{N}$ whose first n coefficients lead to a boundary point on V_n. Let π be a boundary point of $\mathfrak{N}$; by virtue of (8.2.13) we may then write (8.2.22) in the form:

$$(8.2.23) \qquad \left(\sum_{\nu=0}^{n} [\lambda_\nu A_\nu(p) + \bar{\lambda}_\nu \tilde{A}_\nu(p)] + \mathcal{Q}(p)\right) \left(\frac{dp}{d\pi}\right)^2 = \text{real}$$

if π lies on the boundary of $\mathfrak{N}$. But the expression

$$(8.2.24) \qquad \mathscr{W}(p) = \mathcal{Q}(p) + \sum_{\nu=0}^{n} [\lambda_\nu A_\nu(p) + \bar{\lambda}_\nu \tilde{A}_\nu(p)]$$

is a quadratic differential on $\mathfrak{R}$ with a pole of order $n+1$ at most at the point p_0. Thus, the boundary of the image domain $\mathfrak{M}$ may be characterized by the simple differential equation on $\mathfrak{R}$:

$$(8.2.25) \qquad \mathscr{W}(p) \left(\frac{dp}{d\pi}\right)^2 \lesseqgtr 0.$$

We conclude, in particular, that the extremum functions $f(z)$ give domains $\mathfrak{M} \subset \mathfrak{R}$ with boundary curves which are composed of analytic arcs with respect to uniformizers on $\mathfrak{R}$.

We may characterize the extremum mappings in the following form. We define the quadratic differential on $\mathfrak{N}$

$$(8.2.26)\ Y(\pi) = \sum_{\nu=0}^{n} \left[\lambda_\nu \sum_{\varrho=1}^{\nu+1} \varrho b_\varrho a_{\nu+1-\varrho}(\pi) + \bar{\lambda}_\nu \sum_{\varrho=1}^{\nu+1} \varrho \bar{b}_\varrho \tilde{a}_{\nu+1-\varrho}(\pi)\right] - Q(\pi).$$

This differential has a pole of order $n+1$ at most at π_0 and is regular on the rest of $\mathfrak{N}$. The condition (8.2.22) may then be expressed in the form:

$$(8\ 2.27) \qquad Y(\pi) d\pi^2 = \mathscr{W}(p) dp^2.$$

Thus we have

THEOREM 8.2.1. *The mappings of a domain $\mathfrak{N}$ into a domain $\mathfrak{R}$ which lead to boundary points of the coefficient body V_n are characterized by the differential equation*

$$Y(\pi) d\pi^2 = \mathscr{W}(p) dp^2$$

where $Y(\pi)$ and $\mathscr{W}(p)$ are quadratic differentials of $\mathfrak{N}$ and $\mathfrak{R}$ respectively which are regular on their surfaces except for a pole of order $n+1$ at most at the corresponding points π_0 and p_0.

This statement covers also the case that $\mathfrak{N}$ is rigidly imbedded in $\mathfrak{R}$. In this case, we must consider the point $(b_1, \cdots, b_n)$ of the V_n as a boundary point; on the other hand, we showed in Section 7.14 that the necessary condition for a rigid imbedding of $\mathfrak{N}$ in $\mathfrak{R}$ is the differential equation

$$Q(\pi)d\pi^2 = \mathfrak{Q}(p)dp^2 \tag{8.2.28}$$

where $Q(\pi)$ and $\mathfrak{Q}(p)$ are two finite quadratic differentials on $\mathfrak{N}$ and $\mathfrak{R}$, respectively.

We state next

THEOREM 8.2.2. *If $\mathfrak{M} \subset \mathfrak{R}$ is the image of $\mathfrak{N}$ by an extremum mapping then there are no exterior points of $\mathfrak{M}$ on $\mathfrak{R}$, that is, every extremum function maps $\mathfrak{N}$ into a slit domain on $\mathfrak{R}$.*

We know already from Theorem 8.2.1 that the boundary slits of $\mathfrak{M}$ are all composed of analytic arcs with respect to uniformizers on $\mathfrak{R}$.

In order to prove Theorem 8.2.2 let us assume conversely that there exists a point $p_e \in \mathfrak{R}$ exterior to the image $\mathfrak{M}$ of $\mathfrak{N}$. We draw a Jordan curve Γ in the neighborhood of p_e which also lies outside of $\mathfrak{M}$ and deform $\mathfrak{R}$ by a cell attachment along Γ by means of a reciprocal differential $R_0(p)$ defined in a neighborhood of Γ. We put on $R_0(p)$ the restrictions

$$\int_\Gamma R_0(p)\mathfrak{Q}_\nu(p)dw = 0, \quad \nu = 1, 2, \cdots, \sigma_2, \tag{8.2.29}$$

where the $\mathfrak{Q}_\nu(p)$ are a basis for all finite quadratic differentials on $\mathfrak{R}$. Then the deformation of $\mathfrak{R}$ leads to a conformally equivalent surface $\mathfrak{R}^\Delta$. We may map $\mathfrak{R}^\Delta$ back into $\mathfrak{R}$ by means of a correspondence

$$\begin{aligned} w = w_0 &- \frac{\varepsilon}{2\pi i}\int_\Gamma R_0(p_1)N(w_0, p_1)dw_1 \\ &+ \frac{\varepsilon}{2\pi i}\int_{\tilde{\Gamma}} \tilde{R}_0(\tilde{p}_1)N(w_0, \tilde{p}_1)d\tilde{w}_1 + o(\varepsilon), \end{aligned} \tag{8.2.30}$$

which is valid for every choice of the uniformizer w. The function $f(\pi)$ maps $\mathfrak{N}$ into $\mathfrak{M}$; this domain is not affected by the cell attachment at a curve outside of it, but it will change into a domain $\mathfrak{M}^\Delta$ under the corrective mapping (8.2.30). Thus, instead of the schlicht function $f(\pi)$ mapping $\mathfrak{N}$ into $\mathfrak{M}$ we will have the nearby schlicht function

$$f^{\Delta}(\pi) = f(\pi) - \frac{\varepsilon}{2\pi i}\int\limits_{\Gamma} R_0(p_1)N(f(\pi), p_1)\,dw_1 + \frac{\varepsilon}{2\pi i}\int\limits_{\tilde{\Gamma}} \tilde{R}_0(\tilde{p}_1)N(f(\pi), \tilde{p}_1)\,d\tilde{w}_1 + o(\varepsilon) \tag{8.2.31}$$

which maps $\mathfrak{R}$ into $\mathfrak{M}^{\Delta}$. This formula corresponds to (8.2.4) but the term $h(z)$ is omitted. We may carry out the same calculations as before and find for the ν-th coefficient of $f^{\Delta}(\pi)$ the formula:

$$b_\nu^{\Delta} = b_\nu - \frac{\varepsilon}{2\pi i}\int\limits_{\Gamma} R_0(p)A_\nu(p)\,dw + \frac{\varepsilon}{2\pi i}\int\limits_{\tilde{\Gamma}} \tilde{R}_0(\tilde{p})(\hat{A}_\nu(p))^{-}\,d\tilde{w} + o(\varepsilon). \tag{8.2.32}$$

Reasoning just as before we conclude that $(b_1, b_2, \cdots, b_n)$ will be an interior point of V_n except in the case that there exist $n+1$ complex numbers $\lambda_0, \lambda_1, \cdots, \lambda_n$ such that

$$\sum_{\nu=0}^{n} [\lambda_\nu A_\nu(p) + \bar{\lambda}_\nu \tilde{A}_\nu(p)] = \mathcal{Q}(p), \tag{8.2.33}$$

where $\mathcal{Q}(p)$ is a finite quadratic differential on $\mathfrak{R}$. But such an equation is impossible since $A_\nu(p)$ has at p_0 a pole of exactly order $\nu+1$ and the poles in this equation cannot cancel each other. Thus, our assumption of an exterior point p_e of the extremum domain $\mathfrak{M}$ leads us to a contradiction. If $\mathfrak{M}$ belongs to an extremum function $f(\pi)$ there can be no point p_e exterior to it on $\mathfrak{R}$ and the Theorem 8.2.2 is proved.

Let us return now to the linear relation (8.2.22) and interpret it in geometric terms. We write briefly $\delta b_\nu = b_\nu^{\Delta} - b_\nu - o(\varepsilon)$ and may then derive from (8.2.22) the equation

$$\operatorname{Re}\left\{\sum_{\nu=1}^{n} \lambda_\nu \delta b_\nu\right\} = 0 \tag{8.2.34}$$

for all variations considered. This means that by our variations we can cover only a $(2n-1)$-dimensional part of V_n, namely a surface element which is orthogonal to the vector $(\operatorname{Re}\lambda_1, -\operatorname{Im}\lambda_1, \operatorname{Re}\lambda_2, \cdots, -\operatorname{Im}\lambda_n)$. The reason for this fact lies in the special nature of our variation. Since we are at a boundary point of V_n, the image domain $\mathfrak{M}$ on $\mathfrak{R}$ is by Theorem 8.2.2 a slit domain. All variations by cell attachment used so far will transform a slit on $\mathfrak{R}$ into a slit since they are regular everywhere on $\mathfrak{R}$ except in the neighborhood of the curve Γ. It is clear, however, that there are deformations of $\mathfrak{M}$ which do not change conformal type but destroy the slit nature of the domain. Since we know already that the boundary of an extremum domain consists of analytic arcs, we may use the Julia variation of the boundary of $\mathfrak{M}$ which was described in Section 7.15.

Let γ be an analytic arc on the boundary of $\mathfrak{N}$ and let Γ be its image on $\mathfrak{R}$; Γ will, therefore, be an analytic arc on the boundary of $\mathfrak{M}$. Let ζ be a boundary uniformizer on γ and ω a boundary uniformizer on Γ; let $\delta\nu(\zeta)$ be a real-valued analytic function of ζ on γ and let $\delta\omega = i\omega'(\zeta)\ \delta\nu$. We shift every point of Γ by an amount $\delta\omega$; this will lead to the boundary Γ^{Δ} of a new domain $\mathfrak{M}^{\Delta}$ which is conformally equivalent to $\mathfrak{M}$ if the following condition is fulfilled:

$$\int_{\gamma} \delta\nu(\zeta)\, Q(\zeta)d\zeta = 0 \tag{8.2.35}$$

for all finite quadratic differentials on $\mathfrak{N}$. By (7.15.7), the function $f^{\Delta}(\pi)$ which maps $\mathfrak{N}$ on $\mathfrak{M}^{\Delta}$ has the form

$$f^{\Delta}(z) = f(z) - f'(z)\frac{1}{\pi}\int_{\gamma} n(z,\zeta)\delta\nu(\zeta)\, d\zeta + o(\delta\nu). \tag{8.2.36}$$

Since $\mathfrak{M}$ is a slit domain we are forced to choose $\delta\nu \geqq 0$, that is, we can shift boundary points of $\mathfrak{M}$ only in the direction of the interior normal.

We again compute the coefficients b_ν^{Δ} of the varied function. Using (8.2.10), we find

$$b_\nu^{\Delta} = b_\nu - \frac{1}{\pi}\int_{\gamma} \delta\nu(\zeta)\left(\sum_{\varrho=1}^{\nu+1} \varrho b_\varrho a_{\nu+1-\varrho}(\zeta)\right) d\zeta + o(\delta\nu). \tag{8.2.37}$$

If we wish to keep the normalization (8.2.2) for $f^{\Delta}(z)$ also we must restrict $\delta\nu(\zeta)$ further by the condition

$$\int_{\gamma} \delta\nu(\zeta)\, a_0(\zeta)d\zeta = 0. \tag{8.2.38}$$

Then, using the notation (8.2.26), we have in view of (8.2.35) and (8.2.38):

$$\text{Re}\left\{\sum_{\nu=1}^{n} \lambda_\nu \delta b_\nu\right\} = -\frac{1}{2\pi}\int_\gamma Y(\zeta)\,\delta\nu(\zeta)\,d\zeta. \tag{8.2.39}$$

We may also consider variations composed of an interior cell attachment along a Jordan curve γ_1 in $\mathfrak{N}$ and a Julia variation along a boundary arc γ of $\mathfrak{N}$. This type of variation is very convenient since the conditions (8.2.35) and (8.2.38) may be difficult to satisfy with $\delta\nu > 0$, as necessary under a pure boundary variation. In a variation of the mixed type, however, we may choose $\delta\nu > 0$ quite arbitrarily on the boundary of $\mathfrak{N}$ and then adapt the interior cell attachment in such a way that the moduli are preserved and that the point $\pi_0 \in \mathfrak{N}$ is still mapped into $p_0 \in \mathfrak{R}$. The possibility of such adjustment can be shown by the same considerations that were used in Section 7.13. There are only two cases in which this adjustment is impossible, namely the case that $\mathfrak{M}$ is rigidly imbedded in $\mathfrak{R}$ and the case that some expression $\lambda_0 U_0(\pi) + \bar{\lambda}_0 V_0(\pi)$ depends linearly on the finite quadratic differentials of $\mathfrak{N}$ and $\mathfrak{R}$. In this case, the boundary of $\mathfrak{M}$ satisfies the differential equation

$$\mathscr{W}_1(p)\left(\frac{dp}{dt}\right)^2 \gtrless 0 \tag{8.2.5$'$}$$

where $\mathscr{W}_1(p)$ is a quadratic differential on $\mathfrak{R}$ which has a simple pole at most at the point p_0 and where t is a real parameter. There is, indeed, the possibility that the domain $\mathfrak{M}$ is rigidly imbedded in $\mathfrak{R}$, if we require additionally that the point $p_0 \in \mathfrak{R}$ be kept fixed.

We assume that we may vary $\mathfrak{M}$ even with preservation of p_0. In this case, we may choose $\delta\nu(\zeta) > 0$ arbitrarily on γ. On the other hand, we know that an interior cell attachment changes the term $\text{Re}\left\{\sum_{\nu=1}^{n} \lambda_\nu \delta b_\nu\right\}$ only in infinitesimals of higher order. Thus, the total effect of a mixed variation with arbitrary $\delta\nu > 0$ which preserves the moduli and keeps $\delta b_0 = 0$ will still be given by (8.2.39).

We see that we may indeed vary the value of $\text{Re}\left\{\sum_{\nu=1}^{n} \lambda_\nu \delta b_\nu\right\}$. In general, we shall even be able to increase or decrease this number by appropriate Julia variations. This will be impossible only in the case that the quadratic differential $Y(\zeta)$ does not change its sign on the entire boundary of $\mathfrak{N}$.

Let $F(\text{Re } b_1, \text{Im } b_1, \cdots, \text{Re } b_n, \text{Im } b_n)$ be a continuously differentiable real valued function in some open set of the Euclidean $2n$-space which contains V_n. Then F must attain its maximum in V_n at some point $(b_1, \cdots, b_n)$ and we must clearly have

$$\sum_{\nu=1}^{n} \frac{\partial F}{\partial \text{ Re } b_\nu} \text{Re}\{\delta b_\nu\} + \sum_{\nu=1}^{n} \frac{\partial F}{\partial \text{ Im } b_\nu} \{\text{Im } \delta b_\nu\} \leq 0 \tag{8.2.40}$$

for all permissible variations within the family of schlicht functions on $\mathfrak{N}$ relative to $\mathfrak{R}$. Thus, the extremum function $f(z)$ which leads to the maximum of F must satisfy a differential equation (8.2.27) in which the differentials do not change their sign on the boundary of $\mathfrak{N}$.

Thus we have proved:

THEOREM 8.2.3. *Let $f(\pi)$ be a mapping function of $\mathfrak{N}$ into $\mathfrak{R}$ whose first n coefficients $b_1, b_2, \cdots, b_n$ maximize a continuously differentiable function defined in an open set containing V_n. Then $p = f(\pi)$ will satisfy a differential equation*

$$Y(\pi)d\pi^2 = \mathscr{W}(p)dp^2$$

where $Y(\pi)$ and $\mathscr{W}(p)$ are quadratic differentials of $\mathfrak{N}$ and $\mathfrak{R}$ which have poles of order $n+1$ at most at the points π_0 and p_0 and which do not change their signs on the boundary of $\mathfrak{N}$ and on the boundary of $\mathfrak{M}$, respectively.

An extremum mapping $p = f(\pi)$ may also be considered as a realization of the given surface $\mathfrak{R}$ in the surface $\mathfrak{N}$. In fact, if we identify those boundary points of $\mathfrak{N}$ which correspond to the same point of a boundary slit of $\mathfrak{M}$ we obtain another replica of $\mathfrak{R}$. The process of identification can easily be carried out in $\mathfrak{N}$ by means of the quadratic differential $Y(\pi)$. In fact, let π_1 and π_1' be a pair of boundary points of $\mathfrak{N}$ which have already been identified and suppose that both correspond to the boundary point p_1 of $\mathfrak{M}$. If we run from p_1 along an arc of the boundary till we come to a point p_2, we will have one image arc on the boundary of $\mathfrak{N}$ running from π_1 to π_2 and another arc running from π_1' to π_2'. But because of the differential equation we will have

$$(8.2.41)\qquad \int_{\pi_1}^{\pi_2} \sqrt{|Y(\pi)|}\,d\pi = \int_{\pi_1'}^{\pi_2'} \sqrt{|Y(\pi)|}\,d\pi = \int_{p_1}^{p_2} \sqrt{|\mathscr{W}(p)|}\,dp.$$

Thus, we have to identify points on the boundary of $\mathfrak{N}$ which give equal arc length on the boundary in terms of the metric based on the quadratic differential $Y(\pi)$.

Since $\mathfrak{R}$ may be a surface of high genus and $\mathfrak{N}$ can be chosen even simply-connected and planar, the extremum problems considered often lead to very useful realizations of a complicated surface on a simple domain by boundary identification. We shall give examples of such realizations in the next section where we shall consider particular applications of the general theory developed here.

8.3. Imbedding a Circle in a Given Surface

In order to illustrate our general result, we make the following particular application. We assume that the surface $\mathfrak{N}$ is a disc and suppose that it is realized over the unit circle of the z-plane. We ask for the coefficients b_1 of all functions which map the unit circle into a given surface $\mathfrak{R}$ such that its center corresponds to a given point $p_0 \in \mathfrak{R}$ at which a fixed uniformizer $w(p)$ is prescribed.

First it is clear that $\mathfrak{N}$ can be mapped into $\mathfrak{R}$; for let $|w| < \varrho$ be a neighborhood of p_0 in the uniformizer plane. Then $\mathfrak{N}$ can be mapped into $\mathfrak{R}$ by the simple correspondence

$$(8.3.1)\qquad w = \varrho z.$$

If $f(z)$ maps $\mathfrak{N}$ into $\mathfrak{R}$ then every function $f(az)$ with $|a| < 1$ will lead to another imbedding of $\mathfrak{N}$ in $\mathfrak{R}$. In fact, the correspondence $z' = az$ maps $\mathfrak{N}$ into a subdomain $\mathfrak{N}' \subset \mathfrak{N}$ and the mapping by f transforms $\mathfrak{N}'$ into a subdomain of the image $\mathfrak{M} \subset \mathfrak{R}$ of $\mathfrak{N}$.

If b_1 is the coefficient of a function $f(z)$ which imbeds $\mathfrak{N}$ in $\mathfrak{R}$ then every number ab_1 with $|a| < 1$ will be an admissible coefficient, too. This shows clearly that the coefficient region V_1 is, in our special case, a circle around the origin in the complex plane and there remains only the question of determining its radius.

In order to solve this problem it is sufficient to ask for the mapping function $f(z)$ whose first coefficient is positive and has the largest possible value. In other words, we have to pose the extremum problem of maximizing Re b_1. We can now apply the general theory of the preceding section, but here we have additional information about the functionals of $\mathfrak{R}$. In a simply-connected domain no finite quadratic differentials exist. By (7.10.4) the variation kernel of the unit circle has the form:

$$n(z, \zeta) = \frac{z}{2\zeta^2} \frac{z+\zeta}{z-\zeta}, \qquad n(z, \bar{\zeta}) = -\frac{z}{2\bar{\zeta}^2} \frac{1+z\bar{\zeta}}{1-z\bar{\zeta}}. \tag{8.3.2}$$

Hence, we find from the definitions (8.2.10) and (8.2.12):

$$a_0(\zeta) = 0, \ \tilde{a}_0(\zeta) = 0; \ a_1(\zeta) = -\frac{1}{2\zeta^2}, \ \tilde{a}_1(\zeta) = -\frac{1}{2\zeta^2}. \tag{8.3.3}$$

If we want to exhibit more clearly the dependence of the problem on $f(z)$ it will be convenient to use instead of (8.2.11) the series developments:

$$N(w, p) = \sum_{\varrho=0}^{\infty} \alpha_\varrho(p) w^\varrho, \quad N(w, \tilde{p}) = \sum_{\varrho=0}^{\infty} \alpha_\varrho(\tilde{p}) w^\varrho \tag{8.3.4}$$

which represent the variation kernel of $\mathfrak{R}$ in a neighborhood of p_0 in terms of the uniformizer w. The $\alpha_\varrho(p)$ depend only on the choice of the uniformizer w while the coefficients $A_\varrho(p)$ in (8.2.11) depend also on the unknown extremum function $f(z)$. We insert $w = f(z)$ into (8.3.4) and compare the resulting series developments in z with the series (8.2.11). We obtain:

$$\begin{aligned} \alpha_0(p) &= A_0(p), & \tilde{\alpha}_0(p) &= (\tilde{\alpha}_0(\tilde{p}))^- = \tilde{A}_0(p), \\ b_1\alpha_1(p) &= A_1(p), & b_1\tilde{\alpha}_1(p) &= (b_1\alpha_1(\tilde{p}))^- = \tilde{A}_1(p). \end{aligned} \tag{8.3.5}$$

By Section 8.2, the function $p = f(z)$ with maximum value of Re b_1 maps the domain $\mathfrak{N}$ into $\mathfrak{R}$ such that the following differential equation holds:

$$[b_1(\alpha_1(p) + \tilde{\alpha}_1(p)) + \lambda_0\alpha_0(p) + \bar{\lambda}_0\tilde{\alpha}_0(p) + \mathcal{Q}(p)]dp^2 = -\frac{b_1}{z^2}dz^2. \tag{8.3.6}$$

Here, $\mathcal{Q}(p)$ denotes again a finite quadratic differential on $\mathfrak{R}$. The image domain $\mathfrak{M}$ of $\mathfrak{N}$ in $\mathfrak{R}$ will be a slit domain according to Theorem 8.2.2. We verify in our special case the assertion of Theorem 8.2.3 that the quadratic differential $Y(z) = 1/z^2$ does not change its sign on the boundary of $\mathfrak{N}$.

We can readily determine the parameter λ_0 which occurs in (8.3.6) by comparing the singularities on both sides for $z = 0$ and $w = 0$. By (7.10.29) and (8.2.10)' we can easily show that

(8.3.7) $\alpha_0(p) = -\dfrac{1}{w} + \text{regular terms}, \quad \alpha_1(p) = -\dfrac{1}{w^2} + \text{regular terms}$

near $w = 0$. Inserting $p = f(z)$ into (8.3.6) and (8.3.7) and developing in powers of z, we obtain by comparison of coefficients the condition

(8.3.8) $$\lambda_0 = -2\frac{b_2}{b_1}.$$

We have shown that the extremum function $f(z)$ maps $\mathfrak{N}$ into a slit domain on $\mathfrak{R}$ in such a way that the differential equation

(8.3.9) $$\mathscr{W}(p)dp^2 = -\left(\frac{dz}{z}\right)^2$$

holds, where $\mathscr{W}(p)$ is a quadratic differential on $\mathfrak{R}$ which has a double pole at the point p_0.

The slits which bound the image domain $\mathfrak{M} \subset \mathfrak{R}$ of $\mathfrak{N}$ cannot end or begin at points interior to $\mathfrak{R}$. For if a slit had a free end tip at a point $w \in \mathfrak{R}$ the derivative dp/dz would have to vanish at the corresponding boundary point of $\mathfrak{N}$; but this is impossible in view of (8.3.9). We see also that no zero point of the quadratic differential $\mathscr{W}(p)$ can lie inside the image domain $\mathfrak{M}$ since $1/z^2$ does not vanish and since dp/dz is regular in the whole unit circle. Hence these zeros must lie on the boundary of $\mathfrak{M}$. Thus, we see that the boundary of $\mathfrak{M}$ consists of analytic arcs whose endpoints lie either on the boundary of $\mathfrak{R}$ or at the zero points of the quadratic differential $\mathscr{W}(p)$ of $\mathfrak{R}$. This system of arcs may be considered as a set of analytic cross-cuts in $\mathfrak{R}$ which transform the surface into a disc $\mathfrak{M}$.

The conformal mapping of $\mathfrak{M}$ onto the unit circle is given by the function

(8.3.10) $$z(p) = \exp\left\{ i\int_{p_1}^{p} \sqrt{\mathscr{W}(p)}\, dp \right\}$$

where p_1 is an arbitrary boundary point of $\mathfrak{M}$. The point p_0 corresponds under this map to $z(p_0) = 0$. Now let p' be an arbitrary point on a boundary slit of $\mathfrak{M}$; corresponding to the two edges of the slit there will exist two points z_1' and z_2' on the periphery of the unit circle which map into p'. If p' moves along the slit into another point p'' its images will describe arcs on the circumference $|z| = 1$ with angles

$$\text{(8.3.11)} \qquad \begin{aligned} \arg \frac{z_1''}{z_1'} &= \operatorname{Re}\left\{ \int_{p'}^{p''} \sqrt{\mathscr{W}(p)}\,dp \right\} \\ \arg \frac{z_2''}{z_2'} &= -\operatorname{Re}\left\{ \int_{p'}^{p''} \sqrt{\mathscr{W}(p)}\,dp \right\}. \end{aligned}$$

In fact, the points z_1' and z_2' move around the circumference in opposite senses.

We derive from this result two facts. First we see that the cut system in $\mathfrak{R}$ is such that $\sqrt{W(p)}$ has opposite signs on the two edges of every slit. Secondly, we have shown that the mapping (8.3.10) maps $\mathfrak{M}$ onto the unit circle in such a way that the two edges of any arc on a boundary slit are mapped into two arcs of the unit circumference in the z-plane which have the same angle but opposite orientations.

The function $z(p)$ maps the boundary of $\mathfrak{R}$, which also belongs to the boundary of $\mathfrak{M}$, into arcs of the circumference $|z| = 1$. The slit boundary of $\mathfrak{M}$ goes, as we have seen, into circular arcs with two edges of the same slit going into a pair of arcs with equal length and opposite orientation such that all angular distances between images of corresponding points on the two edges of the slit are the same. Therefore, the image of $\mathfrak{R}$ in $\mathfrak{R}$ is obtained by identifying the corresponding arcs. We obtain in this way an important theorem on the uniformization of an arbitrary finite Riemann surface $\mathfrak{R}$, which was discovered by F. Klein in the case of algebraic surfaces.

THEOREM 8.3.1. *Every finite Riemann surface $\mathfrak{R}$ can be mapped onto the interior of the unit circle with appropriate identification of certain boundary arcs. This identification is such that corresponding subarcs have equal length.*

We are able to characterize the canonical mapping of $\mathfrak{R}$ onto the unit circle with boundary identification in the following way. Consider in $\mathfrak{M}$ the regular harmonic function

$$H(p) = \operatorname{Im}\left\{ \int_{p_1}^{p} \sqrt{\mathscr{W}(p)}\, dp \right\}. \tag{8.3.12}$$

It is easily seen that $H(p)$ becomes infinite at the point p_0 with the development

$$H(p) = \log \frac{1}{|w|} + \text{harmonic terms} \tag{8.3.13}$$

and that $H(p)$ vanishes on the boundary of $\mathfrak{M}$. In other words, $H(p)$ is the Green's function of the domain $\mathfrak{M}$, with the logarithmic pole at p_0; this is also obvious from (8.3.10). It can also be seen from the character of the function $\int_{p_1}^{p} \sqrt{\mathscr{W}(p)}\, dp$ on $\mathfrak{M}$ that if we approach a point p' on a boundary slit of $\mathfrak{M}$ from opposite sides we arrive with the determinations $+ H(p')$ and $- H(p')$. Thus, coming from the left edge of the slit with a determination $H(p)$ we may continue H analytically beyond the slit by giving it the determination $- H(p)$ where $H(p)$ is the value in $\mathfrak{M}$ of this harmonic function on the right side of the slit. Thus, except for a change of sign $H(p)$ can be continued over the original surface $\mathfrak{R}$.

There exists a well-defined two sheeted covering of $\mathfrak{R}$ on which $H(p)$ is single-valued and harmonic. This surface $\hat{\mathfrak{R}}$ will be constructed as follows. We consider the universal covering surface $\mathfrak{A}$ of $\mathfrak{R}$ and the fundamental group of transformations of $\mathfrak{A}$ into itself. We introduce a basis of transformations $T_1, \cdots, T_\tau$ such that every transformation of the group can be written in the form $T_{\alpha_1}^{n_1} T_{\alpha_2}^{n_2} \cdots T_{\alpha_k}^{n_k}$ with positive or negative integers n_i and such that this representation is unique. We identify all points of $\mathfrak{A}$ which are obtained by a transformation with Σn_i even; we call the point $\hat{p}$ associated with p if it is obtained from p by a transformation of the group with Σn_i odd. It is clear that we obtain by this process of identification on the universal covering surface a two-sheeted covering of $\mathfrak{R}$ and it is also easily seen that $H(p)$ is single-valued on $\hat{\mathfrak{R}}$. It has a logarithmic pole of the type (8.3.13) at the point $p_0 \in \hat{\mathfrak{R}}$ and a logarithmic pole of the opposite sign at the associated point $\hat{p}_0 \in \hat{\mathfrak{R}}$. Moreover, $H(p)$ is harmonic elsewhere on $\hat{\mathfrak{R}}$ and vanishes on the boundary of $\hat{\mathfrak{R}}$ if there is any. These properties of $H(p)$ permit us to identify it with certain Green's functions of $\hat{\mathfrak{R}}$.

In fact, suppose at first that $\mathfrak{R}$ has no boundary. In this case, $\hat{\mathfrak{R}}$ will have no boundary either but it will have a Green's function $V(p, p_1; q, q_1)$ of the type described in Chapter 4. Consider, in particular, the function $V(p, p_1; p_0, \hat{p}_0)$. This is a harmonic function on $\hat{\mathfrak{R}}$ with the same singularities as $H(p)$. Since it vanishes at the same point p_1 as $H(p)$ it must be identical with it and we have shown that

$$H(p) = \operatorname{Im}\left\{\int_{p_1}^{p} \sqrt{\mathscr{W}(p)}\, dp\right\} = V(p, p_1; p_0, \hat{p}_0). \tag{8.3.14}$$

In order to eliminate the unknown point p_1, we observe that

$$H(p) = \frac{1}{2}[H(p) - H(\hat{p})] \tag{8.3.15}$$

whence by (4.2.23)

$$H(p) = \frac{1}{2} V(p, \hat{p}; p_0, \hat{p}_0); \tag{8.3.16}$$

this formula is very similar to the representation (4.2.1) of the Green's function by means of the Abelian integrals on the double.

Thus, the uniformization of the closed surface $\mathfrak{R}$ in the Klein way may be performed as follows. One doubles the surface $\mathfrak{R}$ into $\hat{\mathfrak{R}}$ and determines on the new closed surface the Green's function $V(p, \hat{p}; p_0, \hat{p}_0)$. The line $V = 0$ will cut $\hat{\mathfrak{R}}$ into two discs $\mathfrak{M}$ and $\hat{\mathfrak{M}}$. $\mathfrak{M}$ can be mapped onto the unit circle in order to give the desired uniformization of $\mathfrak{R}$.

If $\mathfrak{R}$ has a boundary then $\hat{\mathfrak{R}}$ will have a boundary, too. It will therefore have a Green's function $G(p, q)$. We can then show as before that

$$H(p) = G(p, p_0) - G(p, \hat{p}_0) \tag{8.3.17}$$

and obtain immediately the possibility of cutting $\mathfrak{R}$ into the extremum domain $\mathfrak{M}$ by means of the zero lines of this harmonic function on $\hat{\mathfrak{R}}$.

Let us return now to our original problem of determining the maximum value of $\operatorname{Re} b_1$. We have characterized sufficiently the domain $\mathfrak{M}$ on which $\mathfrak{N}$ is mapped by the extremum function. We now give the numerical value for the maximum of Re b_1. We invert the power series development for the extremum function and find

$$z = \frac{1}{b_1} w + \cdots. \tag{8.3.18}$$

Then we express the Green's function $H(p)$ of $\mathfrak{M}$ in the form:

$$H(p) = \log \frac{1}{|z|} = \log \frac{1}{|w|} + \log |b_1| + O(|w|). \tag{8.3.19}$$

On the other hand, we have the representations (8.3.16) or (8.3.17) for $H(p)$ and by comparing coefficients we may find the value of b_1. Take, for example, the case of a surface $\mathfrak{R}$ with boundary. Let

$$G(p, p_0) = \log \frac{1}{|w|} + g(p_0) + O(|w|) \tag{8.3.20}$$

be the development of the Green's function of $\hat{\mathfrak{R}}$ near p_0 in terms of the given uniformizer w. The constant $g(p_0)$ depends on the choice of the uniformizer; it plays an important role in potential theory where it is sometimes called the capacity constant of $\hat{\mathfrak{R}}$ with respect to the point p_0 and the uniformizer w. From (8.3.17)—(8.3.20) we derive

$$\log |b_1| = g(p_0) - G(p_0, \hat{p}_0). \tag{8.3.21}$$

Thus, the problem of the first coefficient has been completely solved in terms of certain functionals of the covering surface $\hat{\mathfrak{R}}$.

In order to connect our result with classical questions of the theory of conformal mapping, we define a functional for all simply connected plane domains $\mathfrak{D}$ which contain the point at infinity. Let $z = \varphi(w)$ map the domain $\mathfrak{D}$ upon the circular domain $|z| > \varrho$ under the following normalization at infinity

$$z = \varphi(w) = w + \beta_0 + \frac{\beta_1}{w} + \frac{\beta_2}{w^2} + \cdots. \tag{8.3.22}$$

The radius ϱ of the image circle is a functional of $\mathfrak{D}$ called the mapping radius of $\mathfrak{D}$. Grötzsch and Pólya have considered the following problem:

Given an arbitrary bounded set of points in the w-plane, to find a continuum C which contains the given set and whose exterior $\mathfrak{D}$ has the least possible mapping radius ϱ.

Let $\mathfrak{R}$ be the w-plane from which the given point set has been removed. Let $\mathfrak{N}$ be the exterior of the unit circle in the z-plane. Consider all functions $w = f(z)$ which are schlicht in $\mathfrak{N}$, have at infinity the series development

$$w = f(z) = b_1 z + b_2 + \frac{b_3}{z} + \frac{b_4}{z^2} + \cdots, \tag{8.3.23}$$

and map $\mathfrak{N}$ into $\mathfrak{R}$. It is easily seen that the maximum of $\operatorname{Re} b_1$ for all these functions equals the reciprocal $1/\varrho$ of the minimal mapping radius ϱ in the preceding problem.

It is unessential that the normalization (8.3.23) is prescribed instead of (8.2.2); we can pass from one problem to the other by linear transformations. Thus, the extremum problem of Grötzsch—Pólya is answered by our result. It has already been treated by similar variational methods and an analogous characterization for the extremum domain has been obtained in special cases. It appears in the present treatment as a very special case of the coefficient problem for schlicht mappings of a Riemann surface $\mathfrak{N}$ into a Riemann surface $\mathfrak{R}$.

The function $z(p)$ defined by (8.3.10) is a polymorphic function on the surface $\mathfrak{R}$; that is, it undergoes a linear transformation if we continue it from some given point $p \in \mathfrak{R}$ along a closed path back to the same point. Moreover, it maps the boundary points of $\mathfrak{R}$ into points of the unit circumference. These properties are not characteristic for $z(p)$; consider, for example, the function $\zeta(p)$ which maps the universal covering surface $\mathfrak{A}$ of $\mathfrak{R}$ into the unit circle. This function will also have the same properties as $z(p)$. If $\mathfrak{R}$ is a planar surface with m boundary continua, there exists according to Riemann a function $\eta(p)$ which maps $\mathfrak{R}$ on a domain covering the unit circle m times, such that the image of each boundary continuum becomes the unit circumference. We now give a property common to all analytic functions $z(p)$ which are single-valued or polymorphic on a surface $\mathfrak{R}$ with boundary and which map each boundary continuum into a circular arc.

Consider the bilinear differential $\mathscr{L}(p, q) = \mathscr{L}_{\mathbf{M}}(p, q)$ of $\mathfrak{R}$; if both argument points p and q lie on the boundary of $\mathfrak{R}$ and if we use boundary uniformizers, $\mathscr{L}(p, q)$ will be real. Let $z(p)$ be a function on $\mathfrak{R}$ with the properties described above. We may use $z(p)$ as local uniformizer and write according to (4.1.5)':

$$\mathscr{L}(p, q)dp\,dq = \frac{dz(p)\,dz(q)}{\pi[z(p) - z(q)]^2} - l(z(p), z(q))dz(p)\,dz(q) \tag{8.3.24}$$

where $l(z, \zeta)$ is a regular analytic function of both arguments in the uniformizer neighborhood considered. Now let p and q lie near a boundary continuum of $\mathfrak{R}$; since this corresponds to a circular arc in the z-plane and since on a circular arc

$$\frac{1}{\pi}\frac{dz(p)dz(q)}{[z(p) - z(q)]^2} = \text{real}, \tag{8.3.25}$$

we derive from the reality of $\mathscr{L}(p, q)dpdq$ on the boundary of $\mathfrak{R}$:

$$l(z, \zeta)dzd\zeta = \text{real} \tag{8.3.26}$$

if z and ζ correspond to two points p and q on the same boundary continuum of $\mathfrak{R}$. In particular

$$l(z, z)dz^2$$

will be real on the boundary of the canonical domain. It is, therefore, a quadratic differential of $\mathfrak{R}$.

Let $z^*(p)$ be a local uniformizer of $\mathfrak{R}$ and let $z = \varphi(z^*)$ lead from it to the canonical variable z. We may express $\mathscr{L}$ in terms of the local uniformizer z^* in the form

$$\mathscr{L}(p, q)dp\,dq = \frac{dz^*(p)dz^*(q)}{\pi[z^*(p) - z^*(q)]^2} - l^*(z^*(p),\ z^*(q))\,dz^*(p)\,dz^*(q). \tag{8.3.24'}$$

Comparing (8.3.24) and (8.3.24)', we derive after easy calculation

$$l(z, z) = \frac{1}{6\pi}\{\varphi, z^*\} + l^*(z^*, z^*) \tag{8.3.27}$$

where

$$\{\varphi, z^*\} = \frac{\varphi'''}{(\varphi')^2} - \frac{3}{2}\left(\frac{\varphi''}{\varphi'}\right)^2, \quad \varphi' = \frac{d\varphi}{dz^*}, \tag{8.3.28}$$

is the Schwarz differential parameter. Since l^* can be calculated and since l is a quadratic differential, this result leads to a third order differential equation for each function $\varphi(z^*)$ which is polymorphic in $\mathfrak{R}$ and maps each boundary component into a circular arc.

In the case of a domain of genus 0 we may also consider the mapping of $\mathfrak{R}$ onto a plane domain bounded by m circles. If $\varphi(z^*)$ is the mapping function, we can again assert that it must satisfy the differential equation (8.3.27). This result shows the close relation between the mappings on circular domains and the bilinear differential $\mathscr{L}(p, q)$.

8.4. Canonical Cross-cuts on a Surface $\mathfrak{R}$

In the last section, we were led by an extremum problem for the coefficients of mapping functions to a cut system on a given surface $\mathfrak{R}$ which was composed of analytic arcs, made the cut surface to a disc and such that the disc could be mapped upon a circle with particularly simple behavior of the boundary under this mapping. In the topology of Riemann surfaces $\mathfrak{R}$, a crosscut or cycle on $\mathfrak{R}$ is determined only by its topological character; in particular a homotopic deformation of a cut is considered unessential.

It is, however, interesting to show that we may associate with each homotopy class of cuts one particular cut which can be characterized invariantly by an extremum problem. In order to formulate the extremum problem, we choose a fixed point $p_0 \in \mathfrak{R}$ and a fixed uniformizer $w(p)$ at this point. If $\mathfrak{M}$ is a subdomain of $\mathfrak{R}$ with boundary and if $G(p, q)$ is its Green's function, we have near p_0:

$$G(p, p_0) = \log \frac{1}{|w|} + g(p_0) + O(|w|). \tag{8.4.1}$$

We call $g(p_0)$ the capacity constant of $\mathfrak{M}$ at p_0 with respect to the uniformizer w.

We give now a curve Γ on $\mathfrak{R}$ which does not bound and ask for a curve C which is homotopic to Γ and such that the capacity constant $g(p_0)$ of the domain $\mathfrak{M} = \mathfrak{R} - C$ be a maximum. We have to show first that there exists a curve C for which the maximum value $g(p_0)$ is actually attained. For this purpose consider all competing domains $\mathfrak{R} - \Gamma$ and map their universal covering surfaces $\mathfrak{A}_\Gamma$ onto the unit circle so that the point p_0 goes into its center.

Let $p = p_\Gamma(\zeta)$ be the inverse mapping from the unit circle onto $\mathfrak{R} - \Gamma$. Consider finally the function $z(p)$ which maps the universal covering of $\mathfrak{R}$ itself upon the unit circle with the center corresponding again to p_0. Then the functions

$$z = \Phi_\Gamma(\zeta) = z(p_\Gamma(\zeta)), \quad \Phi_\Gamma(0) = 0, \tag{8.4.2}$$

are schlicht and bounded in the unit circle and form a normal family.

Let $a(\Gamma)$ be any continuous functional of Γ. We can then assert that there exists at least one admissible curve Γ for which $a(\Gamma)$ attains its least upper bound A. In fact, let Γ_ν $(\nu = 1, 2, \cdots)$ be a sequence for which $a(\Gamma_\nu)$ converges towards A. The functions $\Phi_{\Gamma_\nu}(\zeta)$ possess a subsequence $\Phi_\nu(\zeta)$ which converges uniformly in each closed subdomain of the unit circle towards a schlicht bounded function. Hence, there exists a uniformly convergent sequence of functions $p_{\Gamma_\nu}(\zeta)$ for which the corresponding sequence $a(\Gamma_\nu)$ converges towards the least upper bound A. Let $p_C(\zeta)$ be the limit of this sequence of functions; it maps the unit circle into the surface $\mathfrak{R}$ cut along an admissible curve C. For this particular curve C, $a(C)$ attain its maximum value A.

Having established the existence of an extremum cut C on $\mathfrak{R}$, we now characterize it by a variational method. We choose a Jordan curve γ on $\mathfrak{R}$ outside of a neighborhood of the extremum continuum C. We define on it an analytic reciprocal differential $R(p)$ which is orthogonal to all finite quadratic differentials on $\mathfrak{R}$. We perform a transformation of $\mathfrak{R}$ into itself by a cell attachment which preserves conformal type. According to Theorem 7.13.1 this deformation can be realized in terms of a local uniformizer $w(p)$ as follows:

$$w^* = w + \frac{\varepsilon}{2\pi i}\int_\gamma R(p_1)N(w, p_1)dw_1 - \frac{\varepsilon}{2\pi i}\int_{\tilde{\gamma}} \tilde{R}(\tilde{p}_1)N(w, \tilde{p}_1)d\tilde{w}_1 + o(\varepsilon), \tag{8.4.3}$$

where N is the variation kernel of $\mathfrak{R}$ and where we suppose (for the sake of definiteness) that $\mathfrak{R}$ has a boundary. We wish to hold the distinguished point p_0 fixed under this deformation and to have at p_0

$$\frac{dw^*}{dw}\Big|_{w=0} = 1; \tag{8.4.4}$$

this will permit us to use w^* as well as w in computing capacity constants at p_0. We impose, therefore, on $R(p)$ the additional restrictions

$$(8.4.5) \quad \frac{1}{2\pi i}\int_{\gamma} R(p_1)N(0, p_1)dw_1 = \frac{1}{2\pi i}\int_{\tilde{\gamma}} \tilde{R}(\tilde{p}_1)N(0, \tilde{p}_1)d\tilde{w}_1$$

and

$$(8.4.5)' \quad \frac{1}{2\pi i}\int_{\gamma} R(p_1)N'(0, p_1)dw_1 = \frac{1}{2\pi i}\int_{\tilde{\gamma}} \tilde{R}(\tilde{p}_1)N'(0, \tilde{p}_1)d\tilde{w}_1,$$

where the prime denotes the derivative of N with respect to the first argument.

By the deformation (8.4.3) the domain $\mathfrak{M} = \mathfrak{R} - C$ has undergone a transformation, too. Its conformal type has, in general, changed but we can determine the new capacity constant $g^*(p_0)$ by means of the variational formula (7.8.16) for the Green's function under cell attachment. If we use w as uniformizer at p_0, we find

$$(8.4.6) \quad g^*(p_0) = g(p_0) - \text{Re}\left\{\frac{2\varepsilon}{\pi i}\int_{\gamma} R(p_1)\left(\frac{\partial G(p_1, p_0)}{\partial p_1}\right)^2 dw_1\right\} + O(\varepsilon^2).$$

If we map the deformed domain $\mathfrak{R}^*$ back onto $\mathfrak{R}$ by the correspondence (8.4.3) and change over from the uniformizer w to w^* we do not affect the equation (8.4.6) since w and w^* as well as their derivatives agree at p_0. Thus, the extremum property of the curve C leads to the condition

$$(8.4.7) \quad g^*(p_0) \leqq g(p_0).$$

Because of the arbitrariness in the choice of the small real parameter ε, we conclude

$$(8.4.8) \quad \text{Re}\left\{\frac{1}{2\pi i}\int_{\gamma} R(p_1)\left(\frac{\partial G(p_1, p_0)}{\partial p_1}\right)^2 dw_1\right\} = 0$$

for every permissible choice of $R(p)$. The usual application of linear algebra leads, therefore, to the following condition. There exist two complex numbers λ_0 and λ_1 such that

$$\left(\frac{\partial G(p, p_0)}{\partial p}\right)^2 = \lambda_0 N'(p_0, p) + \bar{\lambda}_0 (N'(p_0, \tilde{p}))^- + \lambda_1 N(p_0, p) + \bar{\lambda}_1 (N(p_0, \tilde{p}))^- + \mathscr{Q}(p), \tag{8.4.9}$$

where $\mathscr{Q}(p)$ is a finite quadratic differential on $\mathfrak{R}$. For $p \in \mathfrak{R}$, the right side of (8.4.9) is a quadratic differential which has a double pole at the distinguished point p_0. Thus, we have

$$\left(\frac{\partial G(p, p_0)}{\partial p}\right)^2 = \mathscr{W}(p) \tag{8.4.10}$$

where $\mathscr{W}(p)$ is a quadratic differential on $\mathfrak{R}$ with a double pole at p_0. On the other hand, $(\partial G/\partial p)dp$ is a linear differential on $\mathfrak{M} = \mathfrak{R} - C$ which has a simple pole at p_0. We find, therefore, that the boundary C of $\mathfrak{M}$ relative to $\mathfrak{R}$ satisfies the differential equation

$$\mathscr{W}(p)\left(\frac{dp}{dw}\right)^2 = -1 \tag{8.4.11}$$

where w is an appropriate boundary uniformizer on C. Thus, C is an analytic curve in terms of uniformizers for $\mathfrak{R}$.

We can again give a geometric interpretation for the curve C analogous to our result in the last section. We observe that the function $\partial G(p, p_0)/\partial p$ is determined at each point of $\mathfrak{R}$ up to a $\pm$-sign under arbitrary analytic continuation. If we continue $\partial G/\partial p$ from the same initial value along two different paths to two points p_1 and p_2 lying opposite each other on the curve C, we will arrive at different determinations of the sign of $\partial G/\partial p$. In fact, by its definition, Green's function has always a positive derivative in the direction of the interior normal and since at p_1 and p_2 the sense of the interior normals is opposite, $\partial G/\partial p$ must have opposite sign. Thus, we see that $\partial G/\partial p$ is two-valued on $\mathfrak{R}$ and has the curve C as its branch line.

We now define a two-sheeted covering $\hat{\mathfrak{R}}$ of $\mathfrak{R}$; we consider the universal covering surface $\mathfrak{A}$ of $\mathfrak{R}$ and identify all points of $\mathfrak{A}$ lying over the same point $p \in \mathfrak{R}$ if the path connecting them cuts C an even number of times. All points over $p \in \mathfrak{R}$ which can be connected with p by a curve which cuts C an odd number of times are also identified and form the point $\hat{p}$ associated with p.

We assumed $\mathfrak{R}$ to have a boundary; we know, therefore, that $\hat{\mathfrak{R}}$

has a boundary, too, and possesses a Green's function $\hat{G}(p, q)$. Consider

$$G_1(p, p_0) = \hat{G}(p, p_0) - \hat{G}(p, \hat{p}_0). \tag{8.4.12}$$

This function is harmonic on $\hat{\mathfrak{R}}$ and has a positive logarithmic pole at the point p_0 and a negative logarithmic point at the associated point $\hat{p}_0$. It vanishes on the boundary of $\hat{\mathfrak{R}}$. Clearly, the Green's function $G(p, p_0)$ of the extremum domain $\mathfrak{M} = \mathfrak{R} - C$ has all the same properties on $\hat{\mathfrak{R}}$ and must, consequently, coincide with $G_1(p, p_0)$. Thus, we have proved:

THEOREM 8.4.1. *Let $\mathfrak{R}$ be a surface with boundary and Γ a closed curve on $\mathfrak{R}$ which does not bound on $\mathfrak{R}$. Consider the two-sheeted covering $\hat{\mathfrak{R}}$ of $\mathfrak{R}$ which is obtained from the universal covering surface $\mathfrak{A}$ of $\mathfrak{R}$ by identifying all points of $\mathfrak{A}$ which lie over the same point $p \in \mathfrak{R}$ and can be connected by a curve which cuts Γ an even number of times. Let $\hat{p}$ be the other point of $\hat{\mathfrak{R}}$ lying over the point p. Let $\hat{G}(p, q)$ be the Green's function of $\hat{\mathfrak{R}}$. Then the curve*

$$\hat{G}(p, p_0) = \hat{G}(p, \hat{p}_0) \tag{8.4.13}$$

is homotopic to Γ and is the curve C on $\mathfrak{R}$ which gives to $\mathfrak{R} - C$ the largest capacity constant at p_0.

It is easy to formulate the corresponding result for the case that $\mathfrak{R}$ does not have a boundary. Instead of $\hat{G}(p, p_0)$ we shall have to use the Green's function $\hat{V}(p, \hat{p}; p_0, \hat{p}_0)$ of the closed surface $\hat{\mathfrak{R}}$ and the extremum cutting C is just the zero line of this Abelian integral.

We may generalize our problem by prescribing curve systems Γ_ν on $\mathfrak{R}$ and ask for homotopic systems which maximize the capacity constant at some point $p_0 \in \mathfrak{R}$. The same reasoning can be applied and analogous results are found.

We deal next with the following extremum problem. Consider a surface $\mathfrak{R}$ and fix in it two points p_0 and p_1. Let $w(p)$ be a fixed local uniformizer at p_0 and $v(p)$ a given local uniformizer at p_1. The problem is to decompose $\mathfrak{R}$ by a system of crosscuts into two domains $\mathfrak{M}_0$ and $\mathfrak{M}_1$ such that $p_0 \in \mathfrak{M}_0$, $p_1 \in \mathfrak{M}_1$ and such that the sum of the capacity constant $g_0(p_0)$ of $\mathfrak{M}_0$ and of the capacity constant $g_1(p_1)$ of $\mathfrak{M}_1$ be as large as possible. One easily verifies that the question is significant, that is, that there exists at least one decomposition $\mathfrak{R} = \mathfrak{M}_0 + \mathfrak{M}_1$ such that the value of $g_0(p_0) + g_1(p_1)$ is the largest possible.

As before, we will characterize the extremum domains $\mathfrak{M}_0$ and $\mathfrak{M}_1$ by varying $\mathfrak{R}$ by means of a cell attachment and comparing the sums of the capacity constants before and after the variation. We select a curve γ_0 in $\mathfrak{M}_0$ and a curve γ_1 in $\mathfrak{M}_1$ and prescribe on γ_i a reciprocal differential $R_i(p)$ such that

$$\int_{\gamma_0} R_0(p)\,\mathcal{Q}(p)\,dw + \int_{\gamma_1} R_1(p)\,\mathcal{Q}(p)dv = 0 \tag{8.4.14}$$

for every finite quadratic differential on $\mathfrak{R}$. We vary $\mathfrak{R}$ by a simultaneous cell attachment at γ_0 and at γ_1 by means of the reciprocal differentials $R_0(p)$ and $R_1(p)$, respectively. The orthogonality condition (8.4.14) guarantees that the deformed surface $\mathfrak{R}^*$ will be of the same conformal type as $\mathfrak{R}$ and by Theorem 7.13.1 we can map back from $\mathfrak{R}^*$ onto $\mathfrak{R}$. We consider first the case that $\mathfrak{R}$ has a boundary. We have

$$\begin{aligned} w^* &= w + \frac{\varepsilon}{2\pi i}\sum_{\nu=0}^{1}\int_{\gamma_\nu} R_\nu(p_2)N(w, p_2)\,dw_2 \\ &\quad - \frac{\varepsilon}{2\pi i}\sum_{\nu=0}^{1}\int_{\tilde{\gamma}_\nu} \tilde{R}_\nu(\tilde{p}_2)N(w, \tilde{p}_2)\,d\tilde{w}_2 + o(\varepsilon). \end{aligned} \tag{8.4.15}$$

This mapping affects $\mathfrak{M}_0$ and $\mathfrak{M}_1$ and may be interpreted as a deformation by cell attachment of $\mathfrak{M}_i$ along the curve γ_i using the reciprocal differential $R_i(p)$. We want to keep the points p_0 and p_1 fixed under the shift (8.4.15) and also keep $dw^*/dw = 1$, $dv^*/dv = 1$, so that we may use w^* or w as local parameter in computing $g_0(p_0)$ and interchange v^* with v when determining $g_1(p_1)$. We require, therefore, the following conditions:

$$\sum_{\nu=0}^{1}\int_{\gamma_\nu} R_\nu(p_2)N(p_i, p_2)\,dw_2 = \sum_{\nu=0}^{1}\int_{\tilde{\gamma}_\nu} \tilde{R}_\nu(\tilde{p}_2)N(p_i, \tilde{p}_2)d\tilde{w}_2, \quad i=0,1, \tag{8.4.16}$$

and

$$(8.4.16)' \quad \sum_{\nu=0}^{1} \int_{\gamma_\nu} R_\nu(p_2) N'(p_i, p_2) dw_2 = \sum_{\nu=0}^{1} \int_{\tilde{\gamma}_\nu} \tilde{R}_\nu(\tilde{p}_2) N'(p_i, \tilde{p}_2) d\tilde{w}_2, \quad i=0,1.$$

Let $G_i(p, q)$ be the Green's function of the domain $\mathfrak{M}_i$; using the variational formula (7.8.16) for the Green's functions we find the following new values for the capacity constants:

$$(8.4.17) \quad g_\nu^*(p_\nu) = g_\nu(p_\nu) - \mathrm{Re}\left\{\frac{2\varepsilon}{\pi i} \int_{\gamma_\nu} R_\nu(p_2) \left(\frac{\partial G_\nu(p_2, p_\nu)}{\partial p_2}\right)^2 dw_2\right\} + o(\varepsilon),$$

for $\nu = 0, 1$. Because of the assumed extremum property of the domains $\mathfrak{M}_0$ and $\mathfrak{M}_1$ we deduce as before

$$(8.4.18) \qquad \mathrm{Re}\left\{\frac{1}{2\pi i} \sum_{\nu=0}^{1} \int_{\gamma_\nu} R_\nu(p_2) \left(\frac{\partial G_\nu(p_2, p_\nu)}{\partial p_2}\right)^2 dw_2\right\} = 0$$

for every pair of reciprocal differentials $R_\nu(p)$ which satisfy the conditions (8.4.14), (8.4.16) and (8.4.16)'.

The equation (8.4.18) leads in the usual way to equations for the quadratic differentials $[\partial G_\nu(p, p_\nu)/\partial p]^2$ on the surfaces $\mathfrak{M}_\nu$. Let $\varkappa_0$, $\varkappa_1$ and λ_0, λ_1 be four complex constants and define the quadratic differential on $\mathfrak{R}$:

$$\sum_{\nu=0}^{1} \left\{ \varkappa_\nu N(p_\nu, p) + \bar{\varkappa}_\nu (N(p_\nu, \bar{p}))^- + \lambda_\nu N'(p_\nu, p) + \bar{\lambda}_\nu (N'(p_\nu, \bar{p}))^- \right\}$$
$$(8.4.19) \qquad + \mathscr{Q}(p) = \mathscr{W}(p),$$

where $\mathscr{Q}(p)$ is a finite quadratic differential on $\mathfrak{R}$. $\mathscr{W}(p)$ has a double pole at p_0 and at p_1. From (8.4.18) and the side conditions (8.4.14), (8.4.16) and (8.4.16)' it follows that we can determine four constants $\varkappa_\nu$, λ_ν and a finite quadratic differential $\mathscr{Q}(p)$ on $\mathfrak{R}$ such that

$$(8.4.20) \qquad \left[\frac{\partial G_\nu(p, p_\nu)}{\partial p}\right]^2 = \mathscr{W}(p) \quad \text{for } p \in \mathfrak{M}_\nu.$$

We deduce from (8.4.20) that the boundary curves of the domains $\mathfrak{M}_\nu$ are analytic curves on $\mathfrak{R}$ satisfying the differential equation

$$\mathscr{W}(p)\left(\frac{dp}{dw}\right)^2 = -1. \tag{8.4.21}$$

It can easily be shown that there are no points on $\mathfrak{R}$ which are exterior to $\mathfrak{M}_0 + \mathfrak{M}_1$. Thus, the differential equation (8.4.21) determines a set of analytic arcs which cut the domain $\mathfrak{R}$ into two pieces $\mathfrak{M}_0$ and $\mathfrak{M}_1$ such that $\mathfrak{R} = \mathfrak{M}_0 + \mathfrak{M}_1$. Let us consider now a boundary arc γ of $\mathfrak{M}_0$; there are two possibilities regarding γ. It may separate $\mathfrak{M}_0$ from $\mathfrak{M}_1$ or it may be a division line between two subdomains of $\mathfrak{M}_0$. In other words, the edges of γ may be boundaries of $\mathfrak{M}_0$ and $\mathfrak{M}_1$ in the one case, or may both be boundary arcs of $\mathfrak{M}_0$ in the second. We want to show that the extremum property of $\mathfrak{M}_0$ excludes the second possibility. Let us suppose, in fact, that both edges of γ are boundaries of $\mathfrak{M}_0$; let us remove a subarc $\gamma_1 \subset \gamma$ and identify points on both edges of γ_1. In this way, $\mathfrak{M}_0$ will become a larger domain $\mathfrak{M}_0^*$ which still contains the point p_0 while $\mathfrak{M}_1$ will not be affected by the removal of an interior slit of $\mathfrak{M}_0$. Let $G_0^*(p, p_0)$ be the Green's function belonging to $\mathfrak{M}_0^*$. The difference function $G_0^*(p, p_0) - G_0(p, p_0)$ is regular harmonic in $\mathfrak{M}_0$ and non-negative on its boundary. But then it is positive inside $\mathfrak{M}_0$ by the minimum principle and, in particular, its value at p_0 will be positive. Thus, we have:

$$g_0^*(p_0) > g_0(p_0) \tag{8.4.22}$$

which shows that $\mathfrak{M}_0^*$ has a bigger capacity constant at p_0 than $\mathfrak{M}_0$. It leads, therefore, to a value

$$g_0^*(p_0) + g_1(p_1) > g_0(p_0) + g_1(p_1)$$

in contradiction to the assumed extremum property of $\mathfrak{M}_0$, $\mathfrak{M}_1$. Thus, we have shown that all boundary arcs of $\mathfrak{M}_0$ are also boundary arcs of $\mathfrak{M}_1$.

Since $\mathscr{W}(p)$ is single-valued on the whole surface $\mathfrak{R}$ we conclude from (8.4.20) that the differentials $\partial G_\nu/\partial p$ coincide on the boundary between $\mathfrak{M}_0$ and $\mathfrak{M}_1$, except possibly up to a $\pm$-sign. Since each Green's function has a positive derivative in the direction of the interior normal of its corresponding domain, we easily conclude

$$\frac{\partial G_0(p, p_0)}{\partial p} = -\frac{\partial G_1(p, p_1)}{\partial p} \quad \text{on boundary of } \mathfrak{M}_0, \mathfrak{M}_1. \tag{8.4.23}$$

Since, moreover, both Green's functions vanish on the common boundary, we recognize that $-G_1(p, p_1)$ is the analytic continuation of $G_0(p, p_0)$ across the boundary. Thus, $G_0(p, p_0)$ and $-G_1(p, p_1)$ form together a harmonic function on $\mathfrak{R}$ which is regular everywhere except at the points p_0 and p_1 where it has logarithmic poles of opposite sign.

Let $\mathcal{G}(p, q)$ be the Green's function of $\mathfrak{R}$; then we have obviously:

$$\mathcal{G}(p, p_0) - \mathcal{G}(p, p_1) = \begin{cases} G_0(p, p_0) \text{ in } \mathfrak{M}_0 \\ -G_1(p, p_1) \text{ in } \mathfrak{M}_1 \end{cases}. \tag{8.4.24}$$

Thus, we have expressed the harmonic function composed of G_0 and $-G_1$ in terms of the Green's function of $\mathfrak{R}$. The division line between $\mathfrak{M}_0$ and $\mathfrak{M}_1$ has the simple equation

$$\mathcal{G}(p, p_0) = \mathcal{G}(p, p_1) \tag{8.4.25}$$

which is the integral of the differential equation (8.4.21).

We are also able to calculate the maximum value of $g_0(p_0)+g_1(p_1)$ in terms of the Green's function $\mathcal{G}(p, q)$ of $\mathfrak{R}$. Let $\mathfrak{g}(p_0)$ and $\mathfrak{g}(p_1)$ be the capacity constants of $\mathfrak{R}$ with respect to p_0 and p_1 for the local uniformizers $w(p)$ at p_0 and $v(p)$ at p_1. We then derive from (8.4.24)

$$\begin{aligned} g_0(p_0) &= \mathfrak{g}(p_0) - \mathcal{G}(p_0, p_1), \\ g_1(p_1) &= \mathfrak{g}(p_1) - \mathcal{G}(p_0, p_1). \end{aligned} \tag{8.4.26}$$

Hence, we have proved:

THEOREM 8.4.2. *Let $\mathfrak{R}$ be a surface with boundary and $\mathcal{G}(p, q)$ its Green's function. Given two points p_0 and p_1 on $\mathfrak{R}$ with fixed local uniformizers w and v, let $\mathfrak{g}(p_0)$ and $\mathfrak{g}(p_1)$ be the capacity constants of $\mathfrak{R}$ at p_0 and at p_1. If $\mathfrak{R}$ is subdivided into any two domains $\mathfrak{M}_0$ and $\mathfrak{M}_1$ such that $p_0 \in \mathfrak{M}_0$ and $p_1 \in \mathfrak{M}_1$, we have the inequality:*

$$g_0(p_0) + g_1(p_1) \leqq \mathfrak{g}(p_0) + \mathfrak{g}(p_1) - 2\mathcal{G}(p_0, p_1) \tag{8.4.27}$$

for the capacity constants $g_0(p_0)$ of $\mathfrak{M}_0$ at p_0 and $g_1(p_1)$ of $\mathfrak{M}_1$ at p_1. Equality in (8.4.27) holds for the extremum decomposition of $\mathfrak{R}$ by means of the cut system

$$\mathcal{G}(p, p_1) = \mathcal{G}(p, p_0)$$

which divides $\mathfrak{R}$ into the two extremum domains $\mathfrak{M}_0$ and $\mathfrak{M}_1$.

Until now we have worked under the assumption that $\mathfrak{R}$ has a boundary. We can also carry through the same reasoning in the case of a closed surface $\mathfrak{R}$. We derive in exactly the same way the fact that $G_0(p, p_0)$ is the analytic continuation of $-G_1(p, p_1)$ across the common boundary of the domains $\mathfrak{M}_0$ and $\mathfrak{M}_1$. But since a closed Riemann surface does not possess a Green's function of the type $\mathcal{G}(p, q)$, we have to change our reasoning from this point on. We remark that $G_0(p, p_0)$ and $-G_1(p, p_1)$ may be considered as the single-valued real part of an Abelian integral of the third kind on $\mathfrak{R}$. Thus:

$$\text{Re}\,\{\Omega_{p_0p_1}(p)\} = \begin{cases} G_0(p, p_0) & \text{in } \mathfrak{M}_0 \\ -G_1(p, p_1) & \text{in } \mathfrak{M}_1. \end{cases} \tag{8.4.28}$$

The integral of the third kind $\Omega_{p_0p_1}(p)$ is determined only up to an additive constant. We may take any determination of an integral $\Omega_{p_0p_1}(p)$ and consider the lines on $\mathfrak{R}$ where

$$\text{Re}\,\{\Omega_{p_0p_1}(p)\} = \alpha, \qquad \alpha = \text{const.} \tag{8.4.29}$$

These lines will decompose $\mathfrak{R}$ into two domains $\mathfrak{M}_0$ and $\mathfrak{M}_1$ and their corresponding Green's functions will be

$$\begin{aligned} G_0(p, p_0) &= \text{Re}\,\{\Omega_{p_0p_1}(p)\} - \alpha \qquad &\text{in } \mathfrak{M}_0, \\ G_1(p, p_1) &= -\text{Re}\,\{\Omega_{p_0p_1}(p)\} + \alpha \qquad &\text{in } \mathfrak{M}_1. \end{aligned} \tag{8.4.30}$$

We recognize that the value $g_0(p_0) + g_1(p_1)$ is independent of the choice of α, so every decomposition (8.4.29) leads to an extremum domain. While the extremum decomposition is uniquely determined in the case of a surface with boundary, we have an infinity of extremum domains in the case of a closed surface.

In order to illustrate our result let us specialize to the case that $\mathfrak{R}$ is the sphere. We introduce w as universal uniformizer on $\mathfrak{R}$ and find the extremum domains by considering the lines:

$$\text{Re}\,\{\Omega_{p_0p_1}(p)\} = \log\left|\frac{w - w_1}{w - w_0}\right| = \alpha. \tag{8.4.31}$$

For each fixed α the plane is decomposed into two circular domains; the one which contains w_0 has the Green's function

$$G_0(w, w_0) = \log\left|\frac{w - w_1}{w - w_0}\right| - \alpha \tag{8.4.30$'$}$$

and the other, which contains w_1, has the Green's function

$$G_1(w, w_1) = \log \left| \frac{w - w_0}{w - w_1} \right| + \alpha. \tag{8.4.30)''}$$

Thus, we have proved:

THEOREM 8.4.3. *Let $\mathfrak{M}_0$ and $\mathfrak{M}_1$ be two plane domains which have no common points. If $w_0 \in \mathfrak{M}_0$ and $w_1 \in \mathfrak{M}_1$, we have the following inequality between the capacity constants $g_0(w_0)$ of $\mathfrak{M}_0$ at w_0 and $g_1(w_1)$ of $\mathfrak{M}_1$ at w_1:*

$$g_0(w_0) + g_1(w_1) \leqq 2 \log | w_0 - w_1 |. \tag{8.4.32}$$

The inequality (8.4.32) contains as a special case a theorem of the theory of schlicht functions in the unit circle which is due to Lavrentieff. Consider two schlicht meromorphic functions in the unit circle with the series developments near the origin:

$$w = f(z) = \sum_{\nu=0}^{\infty} a_\nu z^\nu, \qquad w = g(z) = \sum_{\nu=0}^{\infty} b_\nu z^\nu. \tag{8.4.33}$$

Suppose, moreover, that $f(z') \neq g(z'')$ for any points z' and z'' in the unit circle, that is, the images of the unit circle by means of $f(z)$ and $g(z)$ considered together are still schlicht over the w-plane. There arises the question of estimates for the coefficients of these schlicht function sets. Lavrentieff showed that the product $| a_1 b_1 |$ can be estimated if the values a_0 and b_0 are given. We will derive his result from inequality (8.4.32).

In fact, let $\mathfrak{M}_0$ be the image domain of the unit circle by means of $f(z)$ and let $\mathfrak{M}_1$ be the image by $g(z)$. It is easily seen that

$$G_0(w, a_0) = \log \frac{1}{|z|} = \log \frac{1}{|w - a_0|} + \log | a_1 | + O(| w - a_0 |) \tag{8.4.34}$$

and

$$G_1(w, b_1) = \log \frac{1}{|z|} = \log \frac{1}{|w - b_0|} + \log | b_1 | + O(| w - b_0 |). \tag{8.4.34)'}$$

Applying (8.4.32) with $w_0 = a_0$, $w_1 = b_0$, we find

$$\log | a_1 b_1 | \leqq 2 \log | a_0 - b_0 |, \tag{8.4.35}$$

that is,

$$|a_1 b_1| \leqq |a_0 - b_0|^2. \tag{8.4.35$'$}$$

One might continue the generalization of the concept of a schlicht function in a given domain $\mathfrak{D}$ by considering sets of schlicht functions $f_\nu(z)$ in $\mathfrak{D}$ such that no two image domains of $\mathfrak{D}$ overlap. In the theory of these "schlicht function vectors" the variational method can be easily applied.

8.5. Extremum Problems in the Conformal Mapping of Plane Domains

A particularly simple and interesting case of the general imbedding theory of a domain $\mathfrak{N}$ into a domain $\mathfrak{R}$ arises if $\mathfrak{N}$ is a domain in the complex z-plane with finite connectivity and $\mathfrak{R}$ is the complex w-plane. The theory of imbedding $\mathfrak{N}$ into $\mathfrak{R}$ becomes the classical theory of the schlicht conformal mappings of $\mathfrak{N}$. Our general variational method enables us to solve extremum problems connected with such schlicht mappings of the domain $\mathfrak{N}$.

It is convenient to introduce a few general concepts which will allow us to formulate a rather general result in extremum problems of the above nature. Let $\Phi[f]$ be a real valued functional defined for all functions $f(z)$ analytic in $\mathfrak{N}$. We assume that

$$\lim_{\varepsilon \to 0} \frac{1}{\varepsilon}\{\Phi[f + \varepsilon g] - \Phi[f]\} = \operatorname{Re}\{L_f[g]\}, \qquad \varepsilon \text{ real}, \tag{8.5.1}$$

exists for all analytic functions $g(z)$ in $\mathfrak{N}$ and that $L_f[g]$ is a complex linear functional of $g(z)$. $L_f[g]$ depends in general on $f(z)$; $\operatorname{Re}\{L_f[g]\}$ is called the functional derivative of $\Phi[f]$ for the argument function $f(z)$.

Suppose that we know that some functional $\Phi[f]$ is bounded for the class of all schlicht regular functions in $\mathfrak{N}$ and attains its maximum for at least one function of this class. We can then characterize the extremum functions by variational considerations. Suppose that the extremum function $f(z)$ maps the domain $\mathfrak{N}$ into a subdomain $\mathfrak{M}$ of $\mathfrak{R}$. We choose a Jordan curve γ in $\mathfrak{N}$ and determine a reciprocal differential $r(z)$ which satisfies the orthogonality condition

$$\int_{\gamma} r(z) Q(z) dz = 0 \tag{8.5.2}$$

for all finite quadratic differentials of $\mathfrak{N}$. Let Γ be the image in $\mathfrak{R}$ of the Jordan curve γ; we define on it the reciprocal differential

$$R(w) = r(z) \frac{dw}{dz} = r(z) f'(z) \tag{8.5.3}$$

and perform a variation of $\mathfrak{N}$ and $\mathfrak{R}$ by a cell attachment along γ and Γ as described in Section 7.14. Since $\mathfrak{R}$ does not possess any finite quadratic differentials, the conditions (8.5.2) guarantee that the conformal types of $\mathfrak{N}$ and $\mathfrak{R}$ are preserved under this deformation. If we map the deformed domains conformally back into the original domains, we obtain a new one-to-one relation between z and w which leads to a new schlicht function in $\mathfrak{N}$. By (7.14.10), we have

$$f^{\Delta}(z) = f(z) + \varepsilon h(z) f'(z) - \varepsilon H[f(z)] + o(\varepsilon) \tag{8.5.4}$$

with

$$h(z) = \frac{1}{2\pi i} \int_{\gamma} r(t) n(z, t) dt - \frac{1}{2\pi i} \int_{\gamma} (r(t))^{-} n(z, \bar{t}) d\bar{t} \tag{8.5.5}$$

and

$$H(w) = \frac{1}{2\pi i} \int_{\gamma} r(t) f'(t)^2 N(w, f(t)) dt. \tag{8.5.6}$$

The variation kernel $N(q, p)$ for the sphere was given in (7.10.2) and an arbitrary fixed point q_0 occurred in the formula. Since in most problems of conformal mapping the point at infinity plays a distinguished role in any case, we shall choose $w(q_0) = \infty$; then

$$N(w, \omega) = \frac{1}{w - \omega}. \tag{8.5.7}$$

If we calculate the value of $\Phi[f^{\Delta}]$ by means of (8.5.1) and (8.5.4), we obtain

$$\Phi[f^{\Delta}] = \Phi[f] + \varepsilon \operatorname{Re} \{L_f[hf' - H(f)]\} + o(\varepsilon). \tag{8.5.8}$$

Because of the extremum property of $f(z)$ and the arbitrariness in the choice of the small real quantity ε, we are led to

$$\text{Re}\,\{L_f[hf' - H(f)]\} = 0, \tag{8.5.9}$$

for every choice of $r(t)$, which agrees with the condition (8.5.2).

Because of the linear character of L_f, we may put (8.5.9) into the form

$$\text{Re}\left\{\frac{1}{2\pi i}\int_\gamma r(t)\,[A(t) + B(t) - f'(t)^2 C(f(t))]\,dt\right\} = 0 \tag{8.5.10}$$

with

$$A(t) = L_f[f'(z)n(z, t)], \quad B(t) = (L_f[f'(z)n(z, \bar{t})])^- \tag{8.5.11}$$

and

$$C(w) = L_f\left[\frac{1}{f(z) - w}\right]. \tag{8.5.12}$$

We observe that $A(t)$, $B(t)$ and $C(w)$ depend analytically on their arguments and that we have, on the boundary of $\mathfrak{N}$,

$$A(t)dt^2 = (B(t)dt^2)^- \tag{8.5.13}$$

Thus, $A(t) + B(t)$ is a quadratic differential of $\mathfrak{N}$.

The usual considerations allow us to deduce from (8.5.10) and (8.5.2) that

$$A(t) + B(t) - f'(t)^2\, C(f(t)) = Q(t) \tag{8.5.14}$$

where $Q(t)$ is a finite quadratic differential of $\mathfrak{N}$. Thus, (8.5.14) may be put into the simple form

$$C(w)dw^2 = Y(t)dt^2 \tag{8.5.15}$$

where $Y(t)$ is a quadratic differential of $\mathfrak{N}$.

If we use boundary uniformizers on $\mathfrak{N}$, we can deduce that the boundary of the image domain $\mathfrak{M}$ is composed of analytic arcs each of which satisfies the differential equation

$$C(w)\left(\frac{dw}{dt}\right)^2 \lesseqgtr 0. \tag{8.5.16}$$

We can show by the same reasoning which was applied in Section 8.2 that the extremum domain $\mathfrak{M}$ has no exterior points on $\mathfrak{R}$. After having shown by (8.5.16) that $\mathfrak{M}$ has an analytic boundary, we may apply to it a boundary variation of the Julia type described in Section 7.15. Using the notation of this section and formula (8.5.1), we obtain a function $f^{\Delta}(z)$ such that, using a uniformizer t on the boundary γ, we have:

$$\Phi[f^{\Delta}] = \Phi[f] - \operatorname{Re}\left\{\frac{\varepsilon}{\pi}\int_{\gamma} L_f[f'(z)n(z,t)]\nu(t)dt\right\} + o(\varepsilon). \tag{8.5.17}$$

By (8.5.11) this can be written as

$$\Phi[f^{\Delta}] - \Phi[f] = -\operatorname{Re}\left\{\frac{\varepsilon}{2\pi}\int_{\gamma}[A(t) + B(t)]\nu(t)dt\right\} + o(\varepsilon), \tag{8.5.17)'$$

and since $\nu(t)$ is orthogonal to all finite quadratic differentials of $\mathfrak{R}$:

$$\Phi[f^{\Delta}] - \Phi[f] = -\operatorname{Re}\left\{\frac{\varepsilon}{2\pi}\int_{\gamma} C(w)\left(\frac{dw}{dt}\right)^2 \nu(t)dt\right\} + o(\varepsilon). \tag{8.5.17)''$$

Since we are quite free to vary the boundary of $\mathfrak{M}$ with respect to the interior normal, that is to choose $\nu(t)$ positive, so long as the orthogonality to all finite quadratic differentials is satisfied, we see as in Section 8.2 that

$$C(w)\left(\frac{dw}{dt}\right)^2 \geqq 0 \text{ on each boundary curve of } \mathfrak{M}. \tag{8.5.18}$$

We thus arrive at the theorem:

THEOREM 8.5.1. *Let $\mathfrak{R}$ be a plane domain of finite connectivity; let $\Phi[f]$ be a real valued functional with the functional derivative* Re $\{L_f[g]\}$. *The schlicht regular function $f(z)$ which maximizes $\Phi[f]$ maps the domain $\mathfrak{R}$ onto a slit domain in the w-plane whose boundary curves are composed of analytic arcs satisfying the differential equation*

$$L_f\left[\frac{1}{f(z) - w}\right] w'(t)^2 = 1, \qquad w = w(t). \tag{8.5.19}$$

This theorem shows that in the case that $\mathfrak{R}$ is the sphere the nature of the original domain $\mathfrak{R}$ is rather unessential for the nature of the boundary slits of extremum domains. The differential equation (8.5.19) depends only on the nature of the functional encountered. We may also interpret the result of Theorem 8.5.1 in the following form. Let $\Phi[f]$ be the functional to be maximized. Introduce into it the function

$$f_1(z) = f(z) + \frac{\varepsilon}{f(z) - w} \tag{8.5.20}$$

and calculate

$$\Phi[f_1] - \Phi[f] = \mathrm{Re}\,\{\varepsilon\, C(w)\} + o(\varepsilon).$$

The extremum curves will then satisfy the differential equation

$$C(w)w'^2 = 1.$$

This result has been obtained previously by a quite different type of boundary variation in which it was shown that there actually exist schlicht functions in $\mathfrak{R}$ of the form

$$f_1(z) = f(z) + \frac{\varepsilon}{f(z) - w} + o(\varepsilon), \quad w = \text{boundary point of } \mathfrak{M} \tag{8.5.21}$$

which may be used as comparison functions [8a]. This method requires, however, a very penetrating study into the possible singularities of the boundary curves under conformal mapping.

Let us now illustrate the possibilities in applying Theorem 8.5.1. In order to be sure that a functional has a maximum within the family of all schlicht functions in $\mathfrak{R}$, we can frequently use the fact that important subclasses of schlicht functions form normal families. Let, for example, $z = 0$ lie in $\mathfrak{R}$; then it is well known that all functions $f(z)$ which are regular and schlicht in $\mathfrak{R}$ and have at $z = 0$ the normalization $f(0) = 0$, $f'(0) = 1$ form a normal family **S**. Thus, given any bounded functional on **S** satisfying (8.5.1), we can assert that there exists at least one function in **S** for which the functional attains its maximum value. But from this fact we can show the existence of a wide class of functionals which necessarily possess maxima in the family **F** of all regular schlicht functions in $\mathfrak{R}$.

In fact, let $\Phi[f]$ be a functional which is bounded on **S** and attains, therefore, its maximum in this family. If $f(z)$ is an arbitrary function of the class **F**, the transformation

$$f_0(z) = \frac{f(z) - f(0)}{f'(0)} \tag{8.5.22}$$

will transform it into an element of **S**. Thus, every functional $\Phi[f]$

bounded on **S** gives rise to a functional

$$\Psi[f] = \Phi[f_0] \tag{8.5.23}$$

which is bounded on **F** and attains its maximum there. We can readily calculate the functional derivative of $\Psi[f]$ if the functional derivative of $\Phi[f]$ is known. An easy calculation shows that

$$\Psi[f+\varepsilon g]=\Psi[f]+\varepsilon \operatorname{Re}\left\{ L_{f_0}\left[\frac{g'(0)}{f'(0)}(g_0(z)-f_0(z))\right]\right\}+o(\varepsilon) \tag{8.5.23)'$$

where $\operatorname{Re}\{L_{f_0}\}$ is the functional derivative of $\Phi[f]$ at the value $f_0(z)$ of the argument function.

Because of the close relation between the classes **F** and **S** most extremum problems for regular schlicht functions in $\mathfrak{R}$ are formulated for the class **S**. However, it must be observed that our variational treatment does not preserve the class **S**. Suppose that a function $f(z)$ of the class **S** maximizes a functional $\Phi[f]$, $f \in \mathbf{S}$. Since $f_0(z) = f(z)$ for all elements of **S**, we may equally well say that $f(z)$ maximizes $\Psi[f]$, $f \in \mathbf{F}$. Hence, by Theorem 8.5.1, the boundary curves of the extremum image $\mathfrak{M}$ satisfy the differential equation

$$L_f\left[\frac{f(z)^2}{f(z)-w}\right]\frac{w'^2}{w^2} = 1, \quad w = w(t), \tag{8.5.24}$$

as is easily seen by inserting the functional derivative of $\Psi[f]$ into (8.5.19).

We summarize:

THEOREM 8.5.2. *Let $\mathfrak{R}$ be a plane domain of finite connectivity; let* **S** *be the class of all regular schlicht functions $f(z)$ in $\mathfrak{R}$ which have at the point $z = 0 \in \mathfrak{R}$ the normalization $f(0) = 0$, $f'(0) = 1$. Let $\Phi[f]$ be a real functional satisfying* (8.5.1) *and bounded on* **S**. *Then $\Phi[f]$ attains its maximum in* **S** *for a function $f(z)$ which maps $\mathfrak{R}$ on a slit domain in the w-plane whose boundary slits satisfy the differential equation*

$$L_f\left[\frac{f(z)^2}{f(z)-w}\right]\frac{w'^2}{w^2} = 1, \quad w = w(t). \tag{8.5.24}$$

Let us consider, for example, the question of characterizing among all functions $f(z) \in \mathbf{S}$ that which attains at a given fixed point $z_0 \in \mathfrak{R}$

the largest possible value $\log | f(z_0) |$. Since $f(0)$ is prescribed to be zero, our problem is to determine for which function $f(z)$ the distance $| f(z_0) - f(0) |$ is maximal; the problem is therefore called the distortion problem of conformal mapping Since here

$$\Phi[f] = \log | f(z_0) |,$$

we obtain from (8.5.1)

$$L_f[g] = \frac{g(z_0)}{f(z_0)}.$$

Hence, by (8.5.24) the boundary curves of the extremum domain satisfy the differential equation

$$\frac{w'^2}{w^2} \cdot \frac{1}{1 - \dfrac{w}{f(z_0)}} = 1, \tag{8.5.25}$$

which can easily be integrated:

$$w(t) = f(z_0) \frac{4ce^t}{(1 + ce^t)^2}. \tag{8.5.25$'$}$$

The value of the constant of integration c will depend on the boundary component of $\mathfrak{N}$ considered and will, in general, be complex. There must be one boundary arc of $\mathfrak{M}$, however, which runs up to infinity since $\mathfrak{M}$ cannot contain this point. On this arc, c must necessarily be real and hence this particular boundary slit is a straight line lying on the ray from the origin $w = 0$ to the point $-f(z_0)$.

The further study of the extremum function $f(z)$ and the determination of the numerical value of $\log | f(z_0) |$ belongs to the theory of schlicht functions and will not be carried out here. We wished only to show the type of information which is available in that theory by means of the variational method.

As another example, we consider the series developments of all functions $f(z) \in \mathbf{S}$ at the origin:

$$f(z) = z + \sum_{\nu=2}^{\infty} a_\nu z^\nu. \tag{8.5.26}$$

We ask for the maximum value of the functional $\Phi[f] = \operatorname{Re} a_n$. Let

$$\frac{f(z)}{1-\frac{1}{t}f(z)} = z + \sum_{\nu=2}^{\infty} b_\nu\left(\frac{1}{t}\right) z^\nu; \tag{8.5.27}$$

the coefficients $b_\nu\left(\frac{1}{t}\right)$ can easily be computed and shown to be polynomials of degree $\nu - 1$ in the argument $\frac{1}{t}$ with coefficients which depend on $f(z)$. We clearly have

$$\frac{f(z)^2}{f(z)-w} = f(z) - \frac{f(z)}{1-\frac{1}{w}f(z)} = \sum_{\nu=2}^{\infty}\left[a_\nu - b_\nu\left(\frac{1}{w}\right)\right]z^\nu. \tag{8.5.28}$$

From Theorem 8.5.2 we easily derive that the function (8.5.26) which has the maximal value of Re a_n maps $\mathfrak{N}$ upon a slit domain in the w-plane whose boundaries satisfy the differential equation:

$$\frac{w'^2}{w^2}\left[a_n - b_n\left(\frac{1}{w}\right)\right] = 1. \tag{8.5.29}$$

This differential equation may be used as a starting point for a more penetrating study of the coefficient problem for schlicht functions in multiply-connected domains.

It should be observed in this connection that if t is a boundary point of $\mathfrak{M}$, the auxiliary function (8.5.27) also belongs to the class **S**. Hence, the extremum property of $f(z)$ leads to the inequality:

$$\operatorname{Re}\left\{b_n\left(\frac{1}{w}\right)\right\} \leqq \operatorname{Re}\{a_n\} \tag{8.5.30}$$

for all boundary points of $\mathfrak{M}$. This remark is very helpful in the study of the extremum domain $\mathfrak{M}$; in fact, the points on the w-plane for which

$$a_n = b_n\left(\frac{1}{w}\right) \tag{8.5.31}$$

are the critical points of arcs satisfying the differental equation (8.5.29). Let us ask, therefore, under what conditions a critical point w_0 of this type could lie on the boundary of the extremum domain $\mathfrak{M}$. By our previous remark, the function

$$f^*(z) = \frac{f(z)}{1 - \frac{1}{w_0} f(z)} \tag{8.5.32}$$

would also belong to the class **S** and possess by (8.5.27) the same n-th coefficient as the extremum function $f(z)$. But then $f^*(z)$ would itself be an extremum function and map $\mathfrak{N}$ on an extremum domain $\mathfrak{M}^*$ whose boundary slits satisfy

$$\left(\frac{w^{*\prime}}{w^*}\right)^2 \left[a_n - b_n^*\left(\frac{1}{w^*}\right)\right] = 1. \tag{8.5.33}$$

In order to evaluate the meaning of the simultaneous equations (8.5.29) and (8.5.33), we must investigate the relations between w and w^*, $b_n\left(\frac{1}{w}\right)$ and $b_n^*\left(\frac{1}{w^*}\right)$. The connection between $f^*(z)$ and $f(z)$ is most easily expressed in the form

$$\frac{1}{f^*(z)} = \frac{1}{f(z)} - \frac{1}{w_0} \tag{8.5.32$'$}$$

and correspondingly

$$\frac{1}{w^*} = \frac{1}{w} - \frac{1}{w_0}. \tag{8.5.32$''$}$$

Let w_1 be a boundary point of $\mathfrak{M}$; by (8.5.32)$''$ it corresponds to a boundary point of $\mathfrak{M}^*$:

$$\frac{1}{w_1^*} = \frac{1}{w_1} - \frac{1}{w_0} \tag{8.5.34}$$

To calculate $b_n^*\left(\frac{1}{w_1^*}\right)$, we consider

$$\frac{f^*(z)}{1 - \frac{1}{w_1^*} f^*(z)} = \left[\frac{1}{f^*(z)} - \frac{1}{w_1^*}\right]^{-1} = \left[\frac{1}{f(z)} - \frac{1}{w_1}\right]^{-1} = \frac{f(z)}{1 - \frac{1}{w_1} f(z)} \tag{8.5.35}$$

and compute its n-th coefficient. Hence, we have:

$$b_n\left(\frac{1}{w_1}\right) = b_n^*\left(\frac{1}{w_1^*}\right). \tag{8.5.36}$$

Thus, the differential equation (8.5.33) can be put into the form:

$$(8.5.37)\qquad \frac{w'^2}{w^2}\frac{w_0^2}{(w_0-w)^2}\left[a_n - b_n\left(\frac{1}{w}\right)\right] > 0, \quad w = w(t),$$

where $w(t)$ is the same representation of the boundary curves as that used in (8.5.29). Dividing (8.5.37) by (8.5.29), we obtain on all boundary curves of $\mathfrak{M}$ the relation

$$(8.5.32)\qquad \frac{w_0^2}{(w_0-w)^2} > 0, \quad w = w(t).$$

Thus, if there were a critical point w_0 on the boundary of $\mathfrak{M}$, all boundary slits would be rectilinear and lying on the line through $w = 0$ and the point w_0. We have thus proved that the boundary curves of the extremum domains in the coefficient problem are regular analytic arcs without any critical points.

The further discussion of the equation (8.5.29) was possible because of the existence of a finite transformation of the class **S** into itself, given by (8.5.27) with t on the boundary of $\mathfrak{M}$. While the variational method permits a comparison of the extremum function only with its immediate neighbors in the class, a finite transformation formula leads us to distant elements of the class and contains additional information. It is often possible to amplify the information gained by variation by using a simple transformation of the class of functions considered, analogously to the above treatment.

Another class of schlicht functions which is normal and of great interest in the theory of conformal mapping is defined as follows. Assume that the point $z = \infty$ lies in $\mathfrak{N}$; consider the class of all schlicht functions in $\mathfrak{N}$ which have at infinity the series development

$$(8.5.33)\qquad f(z) = z + a_0 + \frac{a_1}{z} + \cdots$$

and are regular analytic elsewhere in $\mathfrak{N}$. These functions form the class **F** of schlicht functions in $\mathfrak{N}$.

Since every variation (8.5.4) transforms a function of the class **F** into another function of the same class, we may apply the reasoning of this section immediately to extremum problems for the class **F**. We obtain:

THEOREM 8.5.3. *Let $\mathfrak{N}$ be a plane domain of finite connectivity containing the point at infinity; let* **F** *be the class of all schlicht functions in $\mathfrak{N}$ which are regular in $\mathfrak{N}$ except for a pole of type* (8.5.33) *at infinity. If $\Phi[f]$ is a real functional satisfying* (8.5.1) *and bounded on* **F**, *then it attains its maximum in* **F** *for a function $f(z)$ which maps $\mathfrak{N}$ on a slit domain in the w-plane whose boundary slits satisfy the differential equation*

$$L_f\left[\frac{1}{f(z)-w}\right]w'^2 = 1, \quad w = w(t). \tag{8.5.34}$$

Let us illustrate this theorem by some applications. Let us ask for the maximum value of the functional $\operatorname{Re}\{e^{-2i\alpha}a_1\}$ where α is a fixed real constant and a_1 is the coefficient of z^{-1} in the development (8.5.33). Applying (8.5.34), we see that the extremum function $f(z)$ maps $\mathfrak{N}$ onto a slit domain over the w-plane with the differential equation for the boundary slits:

$$w'^2 e^{-2i\alpha} = 1. \tag{8.5.35}$$

This differential equation can be integrated and leads to

$$w(t) = c + e^{i\alpha}t; \tag{8.5.35$'$}$$

that is, all boundary slits of the extremum domain $\mathfrak{M}$ are rectilinear segments with the direction $e^{i\alpha}$.

We see here a new aspect of the variational method, namely the possibility of giving existence proofs for certain canonical mappings. We may put an extremum problem for which the existence of an extremum function is ensured; we may then characterize the extremum function by variation and obtain in this way the existence of a function of the class with particular properties. For example, the preceding reasoning leads to the existence theorem:

There exists a function $f(z)$ of the class **F** *which maps the domain $\mathfrak{N}$ upon the w-plane slit along rectilinear segments in any prescribed direction.*

Let z_0 be a given fixed point in $\mathfrak{N}$; we ask for the maximum value of the functional $\operatorname{Re}\{e^{-2i\alpha}\log f'(z_0)\}$ for all functions $f(z)$ of the class **F**. Using (8.5.34), we see that the extremum function $f(z)$ maps $\mathfrak{N}$ upon a slit domain in the w-plane and each boundary slit $w(t)$ satisfies the differential equation:

$$\frac{w'^2}{[f(z_0) - w]^2} e^{-2i\alpha} = -1. \tag{8.5.36}$$

Integrating this equation we obtain the following parametric representation of the boundary slits:

$$\log [w - f(z_0)] = c + ie^{i\alpha}t, \quad w = w(t), \tag{8.5.37}$$

where the complex constant of integration c is, in general, different for different boundary slits. These curves are spirals with prescribed inclination around the point $f(z_0)$. For $\alpha = 0$, these spirals degenerate into circular arcs with center $f(z_0)$ and for $\alpha = \pi/2$ they become rectilinear segments pointing to the common center $f(z_0)$. We have thus proved the existence of a large class of canonical domains by raising an appropriate extremum question.

REFERENCES

1. R. Courant, *Dirichlet's Principle, conformal mapping, and minimal surfaces*, Interscience, New York, 1950.
2. G. Golusin, (a) "Interior problems of the theory of schlicht functions," *Uspekhi Matem. Nauk*, 6 (1936) 26—89. Translated by T. C. Doyle, A. C. Schaeffer and D. C. Spencer for Office of Naval Research, Navy Department, Washington, D. C., 1947. (b) *Some problems in the theory of schlicht functions*, Trudy Mat. Inst. Steklov, Moscow, Leningrad, 1949 (Russian). (c) "Method of variations in the theory of conform representation. I, II, III," *Mat. Sbornik*, 19 (1946), 203—236; 21 (1947), 83—117; 21 (1947), 119—132 (Russian, English summary).
3. G. Julia, *Leçons sur la représentation conforme des aires multiplement connexes*, Gauthiers-Villars, Paris, 1934.
4. P. Montel, *Leçons sur les fonctions univalentes ou multivalentes*, Gauthier-Villars, Paris, 1933.
5. A. C. Schaeffer and D. C. Spencer, *Coefficient regions for schlicht functions*, Colloquium Publications, Vol. 35, Amer. Math. Soc., New York, 1950.
6. A. C. Schaeffer, M. Schiffer, and D. C. Spencer, "The coefficient regions of schlicht functions," *Duke Math. Jour.* 16 (1949), 493—527.
7. M. Schiffer and D. C. Spencer, "The coefficient problem for multiply-connected domains," *Annals of Math.*, 52 (1950), 362—402.
8. M. Schiffer, (a) "A method of variation within the family of simple functions," *Proc. London Math. Soc.* (2), 44 (1938), 432—449. (b) "On the coefficients of simple functions," *Proc. London Math. Soc.* (2), 44 (1938), 450—452.
9. D. C. Spencer, "Some problems in conformal mapping," *Bull. Amer. Math. Soc.*, 53 (1947), 417—439.

9. Remarks on Generalization to Higher Dimensional Kähler Manifolds

9.1. Kähler Manifolds

To place the subjects discussed in the preceding chapters in a wider setting, we describe in this chapter certain special properties of Kähler manifolds of arbitrary complex dimension k. Since, as we shall presently show, a Riemann surface can always be made into a 1-dimensional Kähler manifold by the introduction of an appropriate metric, the theory of Riemann surfaces may be regarded as the special case $k = 1$ of the general theory of k-dimensional Kähler manifolds, and it is illuminating to see how some aspects of Riemann surfaces may be generalized and others not.

In this chapter proofs of some statements are omitted, references to literature being given instead. By omitting a few details we are able to give a general description of some aspects of Kähler manifolds without making the chapter too lengthy.

For the sake of completeness, we bring together in this section various known properties of complex manifolds.

A complex (analytic) manifold M^k of complex dimension k is a space to each point p of which there is associated a neighborhood $N(p)$ which is mapped topologically onto a subdomain of the Euclidean space of the complex variables $z^1, \cdots, z^k$. If $q \in N(p)$, the coordinates of q will be denoted by $z^i(q)$, $i = 1, 2, \cdots, k$. Wherever two neighborhoods intersect, the coordinates are connected by a pseudo-conformal mapping. We consider only those manifolds which are paracompact and Hausdorff.

Following [3] we introduce a conjugate manifold $\overline{M}^k$ which is a homeomorphic image of M^k in which the point p of M^k corresponds to the point $\bar{p}$ of $\overline{M}^k$ and the neighborhood $N(p)$ to $\overline{N}(\bar{p})$. Let Latin indices run from 1 to $2k$, and let

$$\bar{i} = i + k \pmod{2k}. \tag{9.1.1}$$

If $\bar{q} \in \overline{N}(\bar{p})$, we define

$$z^{\bar{i}}(\bar{q}) = (z^i(q))^-, \tag{9.1.2}$$

where $(z)^-$ denotes the complex conjugate of the quantity z. By means of (9.1.2) the neighborhood $\overline{N}(\bar{p})$ is mapped onto a domain in the space of the variables $z^{\bar{i}} = \bar{z}^i$, $i = 1, 2, \cdots, k$.

Now consider the product manifold $M^k \times \overline{M}^k$ whose points are the ordered pairs $(p, \bar{q})$, and let

$$z^i(p, \bar{q}) = \begin{cases} z^i(p), & i = 1, 2, \cdots, k, \\ z^i(\bar{q}) = (z^{\bar{i}}(q))^-, & i = k+1, \cdots, 2k. \end{cases} \tag{9.1.3}$$

Then

$$z^i(p, \bar{q}) = (z^{\bar{i}}(q, \bar{p}))^-, \qquad i = 1, 2, \cdots, 2k. \tag{9.1.4}$$

The product manifold $M^k \times \overline{M}^k$ is covered by the coordinates $z^i(p, \bar{q})$, $i = 1, 2, \cdots, 2k$. Introduce coordinates $x^i(p, \bar{q})$ by the formulas

$$\begin{aligned} z^i &= \frac{1+\sqrt{-1}}{2} x^i + \frac{1-\sqrt{-1}}{2} x^{\bar{i}}, \\ x^i &= \frac{1-\sqrt{-1}}{2} z^i + \frac{1+\sqrt{-1}}{2} z^{\bar{i}}, \end{aligned} \tag{9.1.5}$$

$$i = 1, 2, \cdots, 2k.$$

Then

$$x^i(p, \bar{q}) = (x^i(q, \bar{p}))^-, \quad i = 1, 2, \cdots, 2k. \tag{9.1.6}$$

On the diagonal manifold D^k of $M^k \times \overline{M}^k$ where $p = q$, we have

$$z^i = z^i(p, \bar{p}) = (z^{\bar{i}})^-, \quad x^i = x^i(p, \bar{p}) = (x^i)^-. \tag{9.1.7}$$

Thus D^k is covered either by the self-conjugate coordinates z^i, $z^{\bar{i}} = \bar{z}^i$, $i = 1, 2, \cdots, 2k$, or by the real coordinates x^i.

We shall be concerned mainly with the diagonal space D^k. A tensor A whose components are real when they are expressed in the real coordinates x^i will be called a real tensor. A real tensor A when expressed in self-conjugate coordinates z^i satisfies

$$A_{i \dots j}{}^{m \dots n} = (A_{\bar{i} \dots \bar{j}}{}^{\bar{m} \dots \bar{n}})^-. \tag{9.1.8}$$

Let unbarred Greek indices run from 1 to k, and write

$$\bar{\alpha} = \alpha + k, \quad \bar{\bar{\alpha}} = \alpha. \tag{9.1.8$'$}$$

Then (9.1.8) can also be written

$$A_{\alpha\bar{\beta}\ldots\gamma}{}^{\bar{\mu}\nu\ldots\lambda} = (A_{\bar{\alpha}\beta\ldots\bar{\gamma}}{}^{\mu\bar{\nu}\ldots\bar{\lambda}})^{-}. \tag{9.1.9}$$

The tensors properly associated with the original manifold M^k are the complex analytic ones whose indices range over values from 1 to k.

On D^k there is a "quadrantal versor" which is a real tensor $h_i{}^j$ satisfying

$$h_i{}^j h_j{}^l = \begin{cases} -1, & i = l, \\ 0, & i \neq l. \end{cases} \tag{9.1.10}$$

In self-conjugate coordinates z^i this tensor has the components

$$h_i{}^j(z) = \begin{cases} \sqrt{-1}, & 1 \leqq i = j \leqq k, \\ -\sqrt{-1}, & k+1 \leqq i = j \leqq 2k, \\ 0, & i \neq j, \end{cases} \tag{9.1.11}$$

or, in the real coordinates x^i,

$$h_i{}^j(x) = \begin{cases} 1, & i = \bar{j}, \quad 1 \leqq i \leqq k, \\ -1, & i = \bar{j}, \quad k+1 \leqq i \leqq 2k, \\ 0, & i \neq \bar{j}. \end{cases} \tag{9.1.11$'$}$$

The values (9.1.11) and (9.1.11)′ are pseudo-conformal invariants.

Given a vector φ_i, let

$$(I\varphi)_i = \varphi_i \tag{9.1.12}$$

be the identity transformation, and let

$$(h\varphi)_i = h_i{}^j \varphi_j \tag{9.1.13}$$

be rotation through a "quadrant". Given real numbers a and b, the operation $aI + bh$ applied to vectors corresponds to complex multiplication in which the reality of the vector is preserved. We have

$$(aI + bh)(cI + dh) = (ac - bd)I + (ad + bc)h. \tag{9.1.14}$$

In other words, the field obtained from the real vectors by adjoining the ope ator h is isomorphic to the complex number field.

Now suppose that D^k carries a Kähler metric g_{ij} of class C^∞. A Kähler metric is a Riemannian metric which satisfies the following two conditions:

$$\text{(a)}\quad g_{ij} = g_{pq} h_i^{\ p} h_j^{\ q},$$
$$\text{(b)}\quad D_p(h_i^{\ j} \varphi_j) = h_i^{\ j} D_p \varphi_j.$$

Here

$$D_p \varphi_j = \frac{\partial \varphi_j}{\partial z^p} - \left\{ \begin{matrix} q \\ jp \end{matrix} \right\} \varphi_q \tag{9.1.15}$$

denotes covariant differentiation, $\left\{ \begin{matrix} q \\ jp \end{matrix} \right\}$ being the coefficients of affine connection. Condition (a) states that the vectors φ_i and $(h\varphi)_i$ have the same length, while (b) states that the operators h and D commute: $Dh = hD$.

Let

$$h_{ij} = g_{jl} h_i^{\ l}. \tag{9.1.16}$$

Multiplying both sides of (a) by $h_r^{\ j}$ and summing on j from 1 to $2k$, we obtain

$$h_{ri} = g_{ij} h_r^{\ j} = g_{pq} h_i^{\ p} h_j^{\ q} h_r^{\ j} = -g_{pr} h_i^{\ p} = -h_{ir}. \tag{9.1.17}$$

Thus h_{ij} is skew-symmetric, and hence by (9.1.16)

$$h_{ij} h_p^{\ i} h_q^{\ j} = g_{jl} h_i^{\ l} h_p^{\ i} h_q^{\ j} = -g_{jp} h_q^{\ j} = -h_{qp} = h_{pq};$$

that is,

$$h_{pq} = h_{ij} h_p^{\ i} h_q^{\ j}. \tag{9.1.18}$$

In terms of self-conjugate coordinates z^i, the formula (9.1.18) shows by (9.1.11) that any non-zero component of h_{pq} is necessarily of the form $h_{\alpha\bar{\beta}}$ or $h_{\bar{\alpha}\beta}$. In other words, $h_{pq} = 0$ unless p and q are indices of opposite parity with respect to conjugation. A metric satisfying (a) is said to be Hermitian.

Condition (b) gives

$$h_p^{\ q} \left\{ \begin{matrix} p \\ i\ j \end{matrix} \right\} = h_i^{\ p} \left\{ \begin{matrix} q \\ p\ j \end{matrix} \right\}. \tag{9.1.19}$$

Taking $q = \alpha$, $i = \bar{\beta}$ and using self-conjugate coordinates, we obtain

$$\sqrt{-1} \left\{ \begin{matrix} \alpha \\ \bar{\beta}\ j \end{matrix} \right\} = -\sqrt{-1} \left\{ \begin{matrix} \alpha \\ \bar{\beta}\ j \end{matrix} \right\}, \quad j = 1, 2, \cdots, 2k.$$

Hence

$$(9.1.19)' \qquad \left\{\begin{matrix} \alpha \\ \bar{\beta}\ j \end{matrix}\right\} = \left(\left\{\begin{matrix} \bar{\alpha} \\ \beta\ \bar{j} \end{matrix}\right\}\right)^{-} = 0, \quad j = 1, 2, \cdots, 2k,$$

and therefore the only possible non-zero components of the coefficients of affine connection are those with all three indices of the same parity. Since

$$\left\{\begin{matrix} p \\ i\ j \end{matrix}\right\} = \frac{1}{2} g^{pq} \left[\frac{\partial g_{iq}}{\partial z^j} + \frac{\partial g_{jq}}{\partial z^i} - \frac{\partial g_{ij}}{\partial z^q}\right],$$

we conclude that

$$(9.1.20) \qquad \frac{\partial g_{\alpha\bar{\beta}}}{\partial z^\gamma} = \frac{\partial g_{\gamma\bar{\beta}}}{\partial z^\alpha} \quad \text{or} \quad \frac{\partial h_{\alpha\bar{\beta}}}{\partial z^\gamma} = \frac{\partial h_{\gamma\bar{\beta}}}{\partial z^\alpha}.$$

A 1-form φ on D^k is a differential form of the first degree

$$\varphi = \varphi_i dz^i,$$

where φ_i are the components of a covariant vector, the summation convention being used. A p-form, or exterior differential form of degree p, $p > 1$, is a sum of exterior products of 1-forms. Exterior multiplication, represented by the symbol $\wedge$, is associative, distributive, and satisfies (see [11])

$$\begin{aligned} &dz^i \wedge dz^j = -dz^j \wedge dz^i, \ dz^i \wedge dz^i = 0, \\ &a \wedge dz^i = dz^i \wedge a = a dz^i, \\ &dz^i \wedge a dz^j = a dz^i \wedge dz^j, \end{aligned}$$

where a denotes a scalar. Exterior multiplication has already been defined in Chapter 1 where, however, the symbol $\wedge$ was omitted for simplicity. In this chapter we adopt the notation current in tensor calculus. A p-form φ may be written in the form

$$(9.1.21) \qquad \begin{aligned} \varphi &= \varphi_{(i_1 \ldots i_p)} dz^{i_1} \wedge dz^{i_2} \wedge \cdots \wedge dz^{i_p} \\ &= \sum_{i_1 < \ldots < i_p} \varphi_{i_1 \ldots i_p} dz^{i_1} \wedge dz^{i_2} \wedge \cdots \wedge dz^{i_p}, \end{aligned}$$

where $\varphi_{i_1 \ldots i_p}$ is a skew-symmetric covariant tensor of rank p, or p-vector in the language of E. Cartan, and where the parentheses indicate that the indices are ordered according to magnitude.

Let

$$(9.1.22) \qquad \Gamma_{i_1\ldots i_p,\, j_1\ldots j_p} = \begin{vmatrix} g_{i_1 j_1} & \cdots & g_{i_p j_1} \\ g_{i_1 j_p} & \cdots & g_{i_p j_p} \end{vmatrix}.$$

Then

$$(9.1.23) \qquad \Gamma_{i_1\ldots i_p}{}^{j_1\ldots j_p} = \begin{vmatrix} g_{i_1}{}^{j_1} & \cdots & g_{i_p}{}^{j_1} \\ g_{i_1}{}^{j_p} & \cdots & g_{i_p}{}^{j_p} \end{vmatrix}$$

is just the Kronecker symbol which is usually denoted by $\delta^{j_1\ldots j_p}_{i_1\ldots i_p}$. We depart from the conventional notation in this instance for reasons of notational symmetry.

The differential $d\varphi$ of a p-form is the $(p+1)$-form

$$(9.1.24) \qquad d\varphi = (d\varphi)_{(i_1\ldots i_{p+1})} dz^{i_1} \wedge \cdots \wedge dz^{i_{p+1}},$$

where

$$(9.1.25) \qquad \begin{aligned} (d\varphi)_{i_1\ldots i_{p+1}} &= \Gamma_{i_1\ldots i_{p+1}}{}^{j(j_1\ldots j_p)} D_j \varphi_{(j_1\ldots j_p)} = \\ &= \frac{1}{p!} \Gamma_{i_1\ldots i_{p+1}}{}^{j j_1\ldots j_p} D_j \varphi_{j_1\ldots j_p}. \end{aligned}$$

Here

$$D_j \varphi_{j_1\ldots j_p} = \frac{\partial \varphi_{j_1\ldots j_p}}{\partial z^j} - \sum_{\mu=1}^{p} \begin{Bmatrix} i \\ j\ j_\mu \end{Bmatrix} \varphi_{j_1\ldots j_{\mu-1}\, i j_{\mu+1}\ldots j_p},$$

and we observe that

$$\Gamma_{i_1\ldots i_{p+1}}{}^{j j_1\ldots j_p} \begin{Bmatrix} q \\ j\ j_\mu \end{Bmatrix} = 0$$

since $\begin{Bmatrix} q \\ j\ j_\mu \end{Bmatrix} = \begin{Bmatrix} q \\ j_\mu\ j \end{Bmatrix}$. Hence in (9.1.25) we may replace covariant differentiation D_j by ordinary differentiation $\partial/\partial z^j$. We have

$$(9.1.26) \qquad d^2\varphi = d(d\varphi) = 0.$$

A form φ satisfying $d\varphi = 0$ is said to be closed, and a form $\varphi = d\psi$ is said to be exact. Formula (9.1.26) therefore states that an exact form is closed.

Let

$$(9.1.27) \qquad e_{i_1\ldots i_{2k}} = \Gamma_{i_1\ldots i_{2k}}{}^{1\,2\ldots 2k} \sqrt{\Gamma_{1\,2\ldots 2k,\, 1\,2\ldots 2k}},$$

and after de Rham [11] set

(9.1.28) $$*\varphi = (*\varphi)_{(j_1 \ldots j_{2k-p})}\, dz^{j_1} \wedge \cdots \wedge dz^{j_{2k-p}}$$

where

(9.1.29) $$(*\varphi)_{j_1 \ldots j_{2k-p}} = e_{(i_1 \ldots i_p) j_1 \ldots j_{2k-p}}\, \varphi^{(i_1 \ldots i_p)}.$$

We verify that

(9.1.30) $$**\varphi = (-1)^p \varphi,$$

and, for the scalar 1,

(9.1.31) $$*1 = e_{1\,2 \ldots 2k}\, dz^1 \wedge \cdots \wedge dz^{2k}.$$

Thus $*1$ is just the volume element.

The co-differential $\delta\varphi$ of a p-form φ is the $(p-1)$-form

(9.1.32) $$\delta\varphi = (\delta\varphi)_{(i_1 \ldots i_{p-1})} dz^{i_1} \wedge \cdots \wedge dz^{i_{p-1}}$$

where

(9.1.33) $$(\delta\varphi)_{i_1 \ldots i_{p-1}} = -(*d*\varphi)_{i_1 \ldots i_{p-1}} = -g^{ij} D_j \varphi_{i i_1 \ldots i_{p-1}}.$$

In contrast with the differential $d\varphi$, the co-differential involves the metric structure of the manifold in an essential way. We have

(9.1.34) $$\delta^2\varphi = \delta(\delta\varphi) = 0.$$

A form φ satisfying $\delta\varphi = 0$ is called co-closed; a form $\varphi = \delta\psi$ is said to be co-exact.

Let

(9.1.35) $$\omega = h_{(ij)} dz^i \wedge dz^j.$$

The condition (9.1.20) expresses the fact that ω is closed:

(9.1.36) $$d\omega = 0.$$

The condition (9.1.19), on the other hand, asserts that $D_l h_{ij} = 0$ and hence

(9.1.36)′ $$\delta\omega = 0.$$

Thus the form ω is both closed and co-closed.

The classical Laplace-Beltrami operator for p-forms is

(9.1.37) $$\Delta = \delta d + d\delta.$$

A p-form φ satisfying $\Delta\varphi = 0$ will be said to be harmonic, and one satisfying $d\varphi = \delta\varphi = 0$ will be said to be a harmonic field. From (9.1.36) and (9.1.36)′ we see that the 2-form ω is a harmonic field.

We recall that the Riemann curvature tensor

$$(9.1.38)\quad R^m{}_{ijl} = \frac{\partial}{\partial z^j}\begin{Bmatrix} m \\ i\ l \end{Bmatrix} - \frac{\partial}{\partial z^l}\begin{Bmatrix} m \\ i\ j \end{Bmatrix} + \begin{Bmatrix} p \\ i\ l \end{Bmatrix}\begin{Bmatrix} m \\ p\ j \end{Bmatrix} - \begin{Bmatrix} p \\ i\ j \end{Bmatrix}\begin{Bmatrix} m \\ p\ l \end{Bmatrix}$$

has the symmetries

$$(9.1.39)\quad \begin{cases} R_{hijl} = -R_{ihjl} = -R_{hilj}, \\ R_{hijl} = R_{jlhi}. \end{cases}$$

It also satisfies the Bianchi identity

$$(9.1.40)\quad R_{hijl} + R_{hjli} + R_{hlij} = 0.$$

The non-commutativity of covariant differentiation is expressed by the Ricci identity

$$(9.1.41)\quad (D_iD_j - D_jD_i)\varphi_{i_1 \ldots i_p} = -\sum_{\mu=1}^{p} \varphi_{i_1 \ldots i_{\mu-1}\, h\, i_{\mu+1} \ldots i_p} R^h{}_{i_\mu ij}.$$

In terms of geodesic coordinates y^i,

$$(9.1.38)'\quad R^m{}_{ijl} = \frac{\partial}{\partial y^j}\begin{Bmatrix} m \\ i\ l \end{Bmatrix} - \frac{\partial}{\partial y^l}\begin{Bmatrix} m \\ i\ j \end{Bmatrix}.$$

If the metric is Kählerian, then by (9.1.19)

$$(9.1.42)\quad h_m{}^n R^m{}_{ijl} = h_i{}^m R^n{}_{mjl}.$$

Thus, in self-conjugate coordinates $R^m{}_{ijl}$ is zero unless m and i have the same parity. In other words, $R_{hijl} = 0$ unless h, i are of different parity and also j, l. From (9.1.40) it follows that

$$(9.1.43)\quad R_{\alpha\bar\beta\gamma\bar\delta} = R_{\alpha\bar\delta\gamma\bar\beta} = R_{\gamma\bar\delta\alpha\bar\beta}.$$

In other words, indices of the same parity commute. Finally, any non-zero component of the Ricci tensor

$$(9.1.44)\quad R_{ij} = R^l{}_{ijl}$$

has indices of opposite parity.

9.2. Complex Operators

The tensors and operators considered in Section 9.1 are all real; in other words, the operators send a real tensor into a real tensor. Now we define the complex tensors and operators introduced in [7b].

As in [4] let

$$\underset{1,0}{\Pi}_i{}^j = \frac{1}{2}\,(g_i{}^j - \sqrt{-1}\,h_i{}^j). \tag{9.2.1}$$

The conjugate tensor is

$$\underset{0,1}{\Pi}_i{}^j = \underset{1,0}{\overline{\Pi}}_i{}^j = \frac{1}{2}\,(g_i{}^j + \sqrt{-1}\,h_i{}^j), \tag{9.2.2}$$

conjugates always being defined in terms of a real coordinate system. Let $\varrho + \sigma = p$, $\varrho \geqq 0$, $\sigma \geqq 0$, and set (compare [4])

$$\begin{aligned}\underset{\varrho,\sigma}{\Pi}_{i_1\ldots i_p}{}^{j_1\ldots j_p} = \Gamma_{i_1\ldots i_p}{}^{m_1\ldots m_\varrho\, n_1\ldots n_\sigma}\, \underset{1,0}{\Pi}_{m_1}{}^{r_1} \cdots \\ \underset{1,0}{\Pi}_{m_\varrho}{}^{r_\varrho}\, \underset{0,1}{\Pi}_{n_1}{}^{s_1} \cdots \underset{0,1}{\Pi}_{n_\sigma}{}^{s_\sigma}\, \Gamma_{(r_1\ldots r_\varrho)(s_1\ldots s_\sigma)}{}^{j_1\ldots j_p}.\end{aligned} \tag{9.2.3}$$

In self-conjugate coordinates

$$\underset{1,0}{\Pi}_i{}^j = \begin{cases} 1, & 1 \leqq i = j \leqq k, \\ 0, & \text{otherwise.} \end{cases} \tag{9.2.4}$$

Therefore, any non-zero component of the tensor

$$(\underset{\varrho,\sigma}{\Pi}\varphi)_{i_1\ldots i_p} = \underset{\varrho,\sigma}{\Pi}_{i_1\ldots i_p}{}^{(j_1\ldots j_p)}\varphi_{(j_1\ldots j_p)}$$

has precisely ϱ indices between 1 and k and σ indices between $k+1$ and $2k$. In other words

$$\underset{\varrho,\sigma}{\Pi}\varphi = \varphi_{(\alpha_1\ldots\alpha_\varrho)(\bar\beta_1\ldots\bar\beta_\sigma)}\,dz^{\alpha_1} \wedge \cdots \wedge dz^{\alpha_\varrho} \wedge dz^{\bar\beta_1} \wedge \cdots \wedge dz^{\bar\beta_\sigma}. \tag{9.2.5}$$

If $\varrho + \sigma = p > 2k$ or if either $\varrho < 0$ or $\sigma < 0$, we define $\underset{\varrho,\sigma}{\Pi}$ to be zero. We plainly have

$$\sum_{\varrho+\sigma=p} \underset{\varrho,\sigma}{\Pi} = \Gamma \tag{9.2.6}$$

and

$$\underset{\varrho,\sigma}{\Pi}\;\underset{\varrho',\sigma'}{\Pi} = \begin{cases} \underset{\varrho,\sigma}{\Pi}, & \varrho = \varrho',\ \sigma = \sigma'. \\ 0, & \text{otherwise.} \end{cases} \tag{9.2.7}$$

Thus (9.2.6) is an orthogonal decomposition of the identity operator Γ. Since

$$h_i{}^j = g^{jp}h_{ip} = -\,g^{jp}h_{pi} = -\,h^j{}_i, \tag{9.2.8}$$

we have

$$\underset{\varrho,\sigma}{\Pi}_{i_1\ldots i_p,\, j_1\ldots j_p} = (\underset{\varrho,\sigma}{\Pi}_{j_1\ldots j_p,\, i_1\ldots i_p})^- = \underset{\sigma,\varrho}{\Pi}_{i_1\ldots j_p,\, i_1\ldots i_p}. \tag{9.2.9}$$

If $\varphi = \underset{\varrho,\sigma}{\Pi}\,\varphi$ we say sometimes (after Hodge) that φ is of type (ϱ, σ).

We define next a complex covariant differentiator, namely

$$\mathscr{D}_i = \underset{1,0}{\Pi}_i{}^j D_j. \tag{9.2.10}$$

The corresponding contravariant differentiator is

$$\mathscr{D}^i = g^{il}\mathscr{D}_l = \underset{0,1}{\Pi}_j{}^i D^j = \overline{\underset{1,0}{\Pi}}_j{}^i D^j. \tag{9.2.10)'$$

The conjugate operators are

$$\overline{\mathscr{D}}_i = \underset{0,1}{\Pi}_i{}^j D_j, \tag{9.2.11}$$

$$\overline{\mathscr{D}}^i = \underset{1,0}{\Pi}_j{}^i D^j, \tag{9.2.11)'$$

It follows from (9.1.41) and the symmetry properties of the Kähler curvature tensor that

$$\begin{aligned}(\mathscr{D}^i\overline{\mathscr{D}}_i - \overline{\mathscr{D}}^i\mathscr{D}_i)\varphi_{j_1\ldots j_p} &= \sum_{\mu=1}^{p} \underset{1,0}{\Pi}_{j_\mu}{}^l R_l{}^h \varphi_{j_1\ldots j_{\mu-1}hj_{\mu+1}\ldots j_p} \\ &\quad - \sum_{\mu=1}^{p} \underset{0,1}{\Pi}_{j_\mu}{}^l R_l{}^h \varphi_{j_1\ldots j_{\mu-1}hj_{\mu+1}\ldots j_p}\end{aligned} \tag{9.2.12}$$

(complex form of the Ricci identity).

In the complex tensor calculus which we propose to use, the Hermitian operator Π replaces the symmetric identity operator Γ, and the complex differentiator $\mathscr{D}$ replaces D.

Formulas (9.1.25) and (9.1.33) may be written

$$\begin{aligned}(d\varphi)_{i_1\ldots i_{p+1}} &= \Gamma_{i_1\ldots i_{p+1}}{}^{j(j_1\ldots j_p)} D_j \varphi_{(j_1\ldots j_p)} \\ &= \Gamma_{i_1\ldots i_{p+1},\, j(j_1\ldots j_p)}\, D^j \varphi^{(j_1\ldots j_p)}, \\ (\delta\varphi)_{i_1\ldots i_{p-1}} &= -\Gamma_{ii_1\ldots i_{p-1}}{}^{(j_1\ldots j_p)} D^i \varphi_{(j_1\ldots j_p)} \\ &= -\Gamma_{ii_1\ldots i_{p-1},\, (j_1\ldots j_p)} D^i \varphi^{(j_1\ldots j_p)}.\end{aligned}$$

The complex analogues of these operators are

$$(9.2.13)\quad \begin{aligned}(\partial\varphi)_{i_1\ldots i_{p+1}} &= \sum_{\varrho+\sigma=p} \Pi_{\varrho+1,\sigma}\, d\,\Pi_{\varrho,\sigma} = \sum_{\varrho+\sigma=p} \Pi_{\varrho+1,\sigma}{}_{i_1\ldots i_{p+1}}{}^{j(j_1\ldots j_p)}\mathscr{D}_j\varphi_{(j_1\ldots j_p)} \\ &= \sum_{\varrho+\sigma=p} \Pi_{\varrho+1,\sigma}{}_{i_1\ldots i_{p+1},\, j(j_1\ldots j_p)}\mathscr{D}^j\varphi^{(j_1\ldots j_p)},\end{aligned}$$

$$(9.2.13)'\quad \begin{aligned}(\vartheta\varphi)_{i_1\ldots i_{p-1}} &= \sum_{\varrho+\sigma=p} \Pi_{\varrho,\sigma-1}\,\delta\,\Pi_{\varrho,\sigma} = -\sum_{\varrho+\sigma=p} \Pi_{\varrho,\sigma}{}_{ii_1\ldots i_{p-1}}{}^{(j_1\ldots j_p)}\mathscr{D}^i\varphi_{(j_1\ldots j_p)} \\ &= -\sum_{\varrho+\sigma=p} \Pi_{\varrho,\sigma}{}_{ii_1\ldots i_{p-1},\,(j_1\ldots j_p)}\mathscr{D}^i\varphi^{(j_1\ldots j_p)}.\end{aligned}$$

The conjugate operators have the forms

$$(9.2.14)\quad (\bar\partial\varphi)_{i_1\ldots i_{p+1}} = \sum_{\varrho+\sigma=p} \Pi_{\varrho,\sigma+1}\, d\,\Pi_{\varrho,\sigma} = \sum_{\varrho+\sigma=p} \Pi_{\varrho,\sigma+1}{}_{i_1\ldots i_{p+1},\, j(j_1\ldots j_p)}\overline{\mathscr{D}}^j\varphi^{(j_1\ldots j_p)},$$

$$(9.2.14)'\quad (\bar\vartheta\varphi)_{i_1\ldots i_{p-1}} = \sum_{\varrho+\sigma=p} \Pi_{\varrho-1,\sigma}\,\delta\,\Pi_{\varrho,\sigma} = -\sum_{\varrho+\sigma=p} \Pi_{\varrho,\sigma}{}_{ii_1\ldots i_{p-1},\,(j_1\ldots j_p)}\overline{\mathscr{D}}^i\varphi^{(j_1\ldots j_p)}.$$

The following identities are readily verified:

$$(9.2.15)\qquad *\Pi_{\varrho,\sigma} = \Pi_{k-\sigma,\,k-\varrho}\,*,$$

$$(9.2.16)\qquad *\partial = (-1)^{p+1}\vartheta *,\quad *\vartheta = (-1)^p\partial *,$$

$$(9.2.17)\qquad \partial^2 = 0,\quad \vartheta^2 = 0,$$

$$(9.2.18)\qquad \partial\bar\partial + \bar\partial\partial = 0,\quad \vartheta\bar\vartheta + \bar\vartheta\vartheta = 0.$$

We also verify that

$$(9.2.19)\quad \begin{aligned}(\Delta\varphi)_{i_1\ldots i_p} &= -D^iD_i\varphi_{i_1\ldots i_p} \\ &\quad - \sum_{\mu=1}^{p} R_{i_\mu}{}^h\varphi_{i_1\ldots i_{\mu-1}hi_{\mu+1}\ldots i_p} \\ &\quad - \frac{1}{2}\sum_{\mu,\nu=1}^{p} R_{i_\mu i_\nu}{}^{hi}\varphi_{i_1\ldots i_{\mu-1}hi_{\mu+1}\ldots i_{\nu-1}ii_{\nu+1}\ldots i_p}.\end{aligned}$$

In view of the properties of the curvature tensor for a Kähler metric, we have

$$(9.2.20)\qquad \Delta\,\Pi_{\varrho,\sigma} = \Pi_{\varrho,\sigma}\,\Delta.$$

Now we introduce a complex Laplace-Beltrami operator

$$(9.2.21)\qquad \Box = \bar\vartheta\partial + \partial\bar\vartheta.$$

Then

$$(9.2.22)\qquad \Delta = \Box + \overline{\Box}$$

where

$$\overline{\square} = \vartheta\bar{\partial} + \bar{\partial}\vartheta. \tag{9.2.23}$$

The following identities are readily seen to be valid:

$$\begin{gathered} *\square = \overline{\square}*, \quad \square = (-1)^p *\overline{\square}*, \\ \partial\square = \square\partial, \quad \bar{\vartheta}\square = \square\bar{\vartheta}. \end{gathered} \tag{9.2.24}$$

We note also the identity

$$\underset{\varrho,\sigma}{\Pi}\,\square = \square\,\underset{\varrho,\sigma}{\Pi}$$

which, in contrast with (9.2.20), is trivial. A calculation gives

$$\begin{aligned} (\square\varphi)_{i_1\cdots i_p} &= -\bar{\mathscr{D}}^i\mathscr{D}_i\varphi_{i_1\cdots i_p} \\ &\quad -\sum_{\mu=1}^{p}\underset{1,0}{\Pi}{}_{i_\mu}{}^{l}R_l{}^h\varphi_{i_1\cdots i_{\mu-1}hi_{\mu+1}\cdots i_p} \\ &\quad -\frac{1}{4}\sum_{\mu,\nu=1}^{p}R_{i_\mu i_\nu}{}^{hi}\varphi_{i_1\cdots i_{\mu-1}hi_{\mu+1}\cdots i_{\nu-1}ii_{\nu+1}\cdots i_p}. \end{aligned} \tag{9.2.25}$$

Taking conjugates in (9.2.25) and interchanging ϱ and σ, we find that

$$\begin{aligned} (\overline{\square}\varphi)_{i_1\cdots i_p} &= -\mathscr{D}^i\bar{\mathscr{D}}_i\varphi_{i_1\cdots i_p} \\ &\quad -\sum_{\mu=1}^{p}\underset{0,1}{\Pi}{}_{i_\mu}{}^{l}R_l{}^h\varphi_{i_1\cdots i_{\mu-1}hi_{\mu+1}\cdots i_p} \\ &\quad -\frac{1}{4}\sum_{\mu,\nu=1}^{p}R_{i_\mu i_\nu}{}^{hi}\varphi_{i_1\cdots i_{\mu-1}hi_{\mu+1}\cdots i_{\nu-1}ii_{\nu+1}\cdots i_p}. \end{aligned} \tag{9.2.25$'$}$$

Hence by (9.2.12) we have

$$\square = \overline{\square} = \frac{1}{2}\Delta. \tag{9.2.26}$$

If $p \geqq 2$, define

$$(\Lambda\varphi)_{i_1\cdots i_{p-2}} = -h^{(ij)}\varphi_{(ij)i_1\cdots i_{p-2}}, \tag{9.2.27}$$

while if $p = 0$ or 1, set $\Lambda\varphi = 0$. It may be verified that

$$(9.2.28) \qquad \Lambda\partial - \partial\Lambda = -\sqrt{-1}\,\vartheta; \; \Lambda d - d\Lambda = h^{-1}\delta h$$

where h is the operator

$$(h\varphi)_{i_1\cdots i_p} = h_{i_1\cdots i_p}{}^{(j_1\cdots j_p)}\varphi_{(j_1\cdots j_p)}$$

defined by

$$h_{i_1\cdots i_p,\, j_1\cdots j_p} = \begin{vmatrix} h_{i_1 j_1} \cdots h_{i_p j_1} \\ h_{i_1 j_p} \cdots h_{i_p j_p} \end{vmatrix};$$

Also

$$(9.2.29) \qquad \Lambda\delta = \delta\Lambda, \quad \Lambda\,\triangle = \triangle\,\Lambda,$$

$$(9.2.30) \qquad \Lambda\varphi = (-1)^p * (\omega \wedge *\varphi)$$

where φ is a p-form.

9.3. FINITE MANIFOLDS

A relatively compact subdomain B of the Kähler manifold M will be called a finite submanifold if each boundary point p of B has a neighborhood $N(p)$ in M in which real coordinates $u^1, u^2, \cdots, u^{2k}$ exist satisfying the following conditions: (i) each u^i is a function of the z^j of class C^∞, and the Jacobian $\partial(u^1, \cdots, u^{2k})/\partial(z^1, \cdots, z^{2k})$ does not vanish in $N(p)$; (ii) $N(p)$ is mapped topologically onto a subdomain in the u-space in such a way that the hyperplane $u^{2k}=0$ corresponds to the points of the boundary of B, the coordinates $u^1, \cdots, u^{2k-1}$ being local boundary parameters; (iii) the u^{2k}-curve is orthogonal to the hyperplane $u^{2k} = 0$ and hence g_{ij}, expressed in terms of the u^i, satisfies on the boundary the condition that $g_{i\,2k} = g^{i\,2k} = 0$ for $i = 1, 2, \cdots, 2k-1$. The coordinates u^i will be called boundary coordinates.

A finite manifold is either a finite submanifold with boundary or is a compact (closed) manifold.

The topological boundary operator will be denoted by b. In $N(p)$, $p \in bB$, set

$$t\varphi = \sum_{i_1 < \cdots < i_p < 2k} \varphi_{i_1\cdots i_p}\, du^{i_1} \wedge \cdots \wedge du^{i_p},$$

$$n\varphi = \sum_{i_1 < \cdots < i_p = 2k} \varphi_{i_1\cdots i_p}\, du^{i_1} \wedge \cdots \wedge du^{i_p}.$$

Then in $N(p)$, $\varphi = t\varphi + n\varphi$, but this decomposition will in general have geometrical meaning only on the boundary bB itself. However, if we choose u^{2k} to be geodesic distance from the boundary, $u^{2k} > 0$ in $N(p) \cap B$, then $t\varphi$ and $n\varphi$ are well-determined p-forms in a sufficiently small boundary strip $0 < u^{2k} < \varepsilon$ covered by the coordinate systems $u^1, \cdots, u^{2k}$. We verify that $*t = n*$, $t* = *n$.

The scalar product of two p-forms φ and ψ on a subdomain D of M is defined to be

$$(\varphi, \psi) = (\varphi, \psi)_D = \int_D \varphi \wedge *\bar{\psi},$$

and the corresponding norm is $\| \varphi \| = \sqrt{(\varphi, \varphi)}$. We plainly have

$$(\underset{\varrho,\sigma}{\Pi} \varphi, \psi) = (\varphi, \underset{\varrho,\sigma}{\Pi} \psi);$$

that is, Π is a symmetric operator.

If C^q is a differentiable q-chain on M^k with real coefficients and if φ is a $(q-1)$-form, we have the well-known Stokes formula

$$\int_{C^q} d\varphi = \int_{bC^q} \varphi \tag{9.3.1}$$

where bC^q denotes the boundary of C^q. In particular, if we take $q = 2k$ and $C^q = C^{2k} = B$, where B is a finite manifold, then for a p-form φ and a $(p+1)$-form ψ we have

$$\int_B d(\varphi \wedge *\bar{\psi}) = \int_B (d\varphi \wedge *\bar{\psi} + (-1)^p \varphi \wedge d*\bar{\psi}) = \int_B (d\varphi \wedge *\bar{\psi} - \varphi \wedge *\delta\bar{\psi})$$
$$= \int_{bB} \varphi \wedge *\bar{\psi}.$$

Thus

$$(d\varphi, \psi) - (\varphi, \delta\psi) = \int_{bB} \varphi \wedge *\bar{\psi}. \tag{9.3.2}$$

By specializing φ and ψ we derive at once from (9.3.2) the following well-known „real'' Green's formulas:

$$(d\varphi, d\psi) - (\varphi, \delta d\psi) = \int_{bB} \varphi \wedge *d\bar{\psi},$$

$$(d\delta\varphi, \psi) - (\delta\varphi, \delta\psi) = \int_{bB} \delta\varphi \wedge *\bar{\psi},$$

$$(\delta d\varphi, \psi) - (\varphi, \delta d\psi) = \int_{bB} (\varphi \wedge *d\bar{\psi} - \bar{\psi} \wedge *d\varphi), \tag{9.3.3}$$

$$(d\delta\varphi, \psi) - (\varphi, d\delta\psi) = \int_{bB} (\delta\varphi \wedge *\bar{\psi} - \delta\bar{\psi} \wedge *\varphi),$$

$$(\Delta\varphi, \psi) - (\varphi, \Delta\psi) = \int_{bB} (\varphi \wedge *d\bar{\psi} - \bar{\psi} \wedge *d\varphi + \delta\varphi \wedge *\bar{\psi} - \delta\bar{\psi} \wedge *\varphi).$$

Taking $\varphi = \underset{\varrho,\sigma}{\Pi}\,\varphi$, $\psi = \underset{\varrho,\sigma+1}{\Pi}\,\psi$ in (9.3.2), we obtain its complex analogue:

$$(\bar{\partial}\varphi, \psi) - (\varphi, \vartheta\psi) = \int_{bB} \varphi \wedge *\bar{\psi}. \tag{9.3.4}$$

From (9.3.4) we derive immediately the "complex" Green's formulas:

$$(\bar{\partial}\varphi, \bar{\partial}\psi) - (\varphi, \vartheta\bar{\partial}\psi) = \int_{bB} \varphi \wedge *(\bar{\partial}\psi)^-,$$

$$(\bar{\partial}\vartheta\varphi, \psi) - (\vartheta\varphi, \vartheta\psi) = \int_{bB} \vartheta\varphi \wedge *\bar{\psi},$$

$$(\vartheta\bar{\partial}\varphi, \psi) - (\varphi, \vartheta\bar{\partial}\psi) = \int_{bB} (\varphi \wedge *(\bar{\partial}\psi)^- - \bar{\psi} \wedge *\bar{\partial}\varphi), \tag{9.3.5}$$

$$(\bar{\partial}\vartheta\varphi, \psi) - (\varphi, \bar{\partial}\vartheta\psi) = \int_{bB} (\vartheta\varphi \wedge *\bar{\psi} - (\vartheta\psi)^- \wedge *\varphi),$$

$$(\Delta\varphi, \psi) - (\varphi, \Delta\psi) = 2\int_{bB} (\varphi \wedge *(\bar{\partial}\psi)^- - \bar{\psi} \wedge *\bar{\partial}\varphi + \vartheta\varphi \wedge *\bar{\psi} - (\vartheta\psi)^- \wedge *\varphi).$$

9.4. Currents

Let D be an arbitrary subdomain of M (which may coincide with M), and let $\mathbf{C}^p$ denote the space of p-forms φ which are of class C^∞ and have compact carriers relative to D (that is, vanish outside a compact subset of D). A p-current $T\,[\varphi]$ on D in the sense of de Rham [11] is a linear functional over the space $\mathbf{C}^{2k-p}$ which satisfies the following continuity restriction: for an arbitrary sequence of forms φ_μ, $\varphi_\mu \in \mathbf{C}^{2k-p}$, whose carriers are contained in a fixed compact subset K of D, where K is covered by a single coordinate system, $T\,[\varphi_\mu] \to 0$ $(\mu \to \infty)$ if φ_μ and each partial derivative tend uniformly to zero.

LEMMA 9.4.1. (*Partition of unity*). *Given a locally finite open covering* $\{U_i\}$ *of* D, *there exists a corresponding set of scalars* φ_j *such that:* (i) $\Sigma\,\varphi_j = 1$; (ii) $\varphi_j \in C^\infty$, $0 \leqq \varphi_j$ *everywhere, and the carrier of* φ_j *is contained in one of the open sets* U_i.

This is a standard lemma, easily proved (see [11]).

Given a $(2k - p)$-form φ of class C^∞ with a non-compact carrier, we say that $T[\varphi]$ is convergent and that

$$T[\varphi] = \sum_i T[\varphi_i \varphi]$$

if the series is convergent for each partition of unity.

The exterior product $T \wedge \psi$ of a p-current T with a q-form $\psi \in C^\infty$ is defined to be

$$T \wedge \psi[\varphi] = T[\psi \wedge \varphi],$$

and (after de Rham [11])

$$dT[\varphi] = (-1)^{p+1}T[d\varphi], \quad \delta T[\varphi] = (-1)^p T[\delta\varphi],$$

$$*T[\varphi] = (-1)^p T[*\varphi], \quad (T, \varphi) = T[*\bar{\varphi}].$$

We say that T vanishes at a point of D if there is a neighborhood N of the point such that $T[\varphi] = 0$ for all φ whose carriers lie in N. The carrier of T is the set of all points of D where T does not vanish. A p-current is said to be regular at a point of D if T coincides with a p-form of class C^∞ in some neighborhood of the point; otherwise T is called singular at the point. The set of singular points of T is called the singular set of T.

Given two p-currents S and T whose singular sets do not meet, there exist decompositions $S = S' + S''$, $T = T' + T''$, where S'', T'' are everywhere regular while the carriers of S' and T' do not meet. In this case we define

$$(S, T) = (S', T'') + (S'', T') + (S'', T''), \tag{9.4.1}$$

provided that the scalar products on the right converge. We observe that the scalar products on the right side of (9.4.1) are defined since the scalar product of a current and a form has already been given and since at least one of the factors in each scalar product is regular.

THEOREM 9.4.1. *If T is a p-current in $D \subseteqq M$ and if ΔT is regular at a point of D, then T is also regular at that point. If dT and δT are both regular at a point, then T is regular at that point.*

This theorem, often called Weyl's lemma, is proved in [11].

A current of degree 0 is a distribution in the sense of L. Schwartz [12], and an arbitrary p-current T may be represented as a formal differential form

$$T = T_{(i_1 \cdots i_p)} dz^{i_1} \wedge \cdots \wedge dz^{i_p}$$

where the coefficients $T_{i_1 \cdots i_p}$ are distributions in the space of local self-conjugate coordinates z^i. Hence the operator $\underset{\varrho, \sigma}{\Pi}$ may be applied to T:

$$\underset{\varrho, \sigma}{\Pi} T[\varphi] = T[\underset{k-\varrho, k-\sigma}{\Pi} \varphi].$$

We define

$$\bar{\partial} T[\varphi] = (-1)^{p+1} T[\bar{\partial}\varphi], \quad \vartheta T[\varphi] = (-1)^p T[\vartheta\varphi].$$

A p-current T is harmonic if $\Delta T[\varphi] = T[\Delta \varphi] = 0$. By Theorem 9.4.1 a harmonic current is equal to a harmonic form. Similarly, if $dT[\varphi] = \delta T[\varphi] = 0$, then T is equal to a harmonic field, that is, T is equal to a form ψ satisfying $d\psi = \delta\psi = 0$. If the p-current T is of type (ϱ, σ) and satisfies $\bar{\partial} T = \vartheta T = 0$ (or $\partial T = \bar{\vartheta} T = 0$), we say that T is a complex field.

If $T = \underset{\varrho, 0}{\Pi} T$, then $\vartheta T = 0$ automatically and the relation $\bar{\partial} T = 0$ implies that the coefficients of T are equivalent to holomorphic functions of the complex variables $z^1, z^2, \cdots, z^k$.

LEMMA 9.4.2. *If* $T = \underset{\varrho, 0}{\Pi} T$ *is holomorphic in a domain* $D \subseteqq M$, *then* $\delta T = 0$ *in* D.

By (9.2.28) we have $0 = (\Lambda \bar{\partial} - \bar{\partial} \Lambda) T = \sqrt{-1}\, \bar{\vartheta} T$, so $\bar{\vartheta} T = 0$. Since $\vartheta T = 0$, it follows that $\delta T = (\vartheta + \bar{\vartheta}) T = 0$.

9.5. HERMITIAN METRICS

It is easy to construct a positive-definite Hermitian metric on an arbitrary complex manifold M (which is paracompact and Hausdorff — hence normal).

In fact, let $\{U_i\}$ be a locally finite open covering of the complex manifold such that each U_i is covered by local coordinates $z_i^1, z_i^2, \cdots, z_i^{2k}$. Let $\{\varphi_i\}$ be a partition of unity corresponding to the covering $\{U_i\}$, and define

$$ds^2 = \sum_i \Big(\sum_{\alpha=1}^{k} \varphi_i \, | dz_i^\alpha |^2 \Big). \tag{9.5.1}$$

The metric (9.5.1) is plainly Hermitian.

Let h_{ij} be defined by (9.1.16) using the metric (9.5.1), and consider the 2-form ω defined by (9.1.35). In the special case $k = 1$, this 2-form is automatically closed ($d\omega = 0$) and therefore defines a Kähler metric. Hence:

LEMMA 9.5.1. *A Kähler metric may be constructed on an arbitrary Riemann surface.*

9.6. DIRICHLET'S PRINCIPLE FOR THE REAL OPERATORS

In this section we make no use of the complex structure, and the results are therefore valid for an arbitrary Riemannian manifold M of real dimension m. However, since m is not necessarily even, formulas (9.1.30) and (9.1.33) are to be replaced by

$$**\varphi = (-1)^{mp+p} \varphi, \tag{9.1.30$'$}$$

$$\delta\varphi = (-1)^{mp+m+1} * d * \varphi, \tag{9.1.33$'$}$$

where φ is assumed to be of degree p.

The expression

$$D(\varphi) = (d\varphi, d\varphi) + (\delta\varphi, \delta\varphi) \tag{9.6.1}$$

is the obvious generalization of the classical Dirichlet integral. Given a positive number s, we define

(9.6.2) $$D_s(\varphi) = D(\varphi) + s(\varphi, \varphi) = D(\varphi) + s \,\|\, \varphi \,\|^2.$$

Moreover, we write

(9.6.3) $$\Delta_s = \Delta + s.$$

Let $\mathbf{A}^p$ be the Hilbert space of norm-finite differential forms of degree p and let

$$\mathbf{A} = \sum_p \mathbf{A}^p.$$

If $\varphi \in \mathbf{A}$, then $\varphi = \psi^0 + \psi^1 + \ldots + \psi^m$ where $\psi^p \in \mathbf{A}^p$, and we define

$$\|\, \varphi \,\| = \sqrt{\sum_p \|\, \psi^p \,\|^2}.$$

We denote by $\mathbf{C}$ the subspace of $\mathbf{A}$ composed of forms of class C^∞ with compact carriers.

We say that a form φ is in the (closure of) the domain of the operator d if there exists a sequence $\{\varphi_\mu\}$, $\varphi_\mu \in C^\infty$, φ_μ, $d\varphi_\mu \in \mathbf{A}$, such that $\|\, \varphi - \varphi_\mu \,\|$ and $\|\, d\varphi_\mu - d\varphi_\nu \,\|$ tend to zero $(\mu, \nu \to \infty)$. If such a sequence exists with $\varphi_\mu \in \mathbf{C}$, we say that φ is in the domain of the operator d_c. We define the domains of δ, δ_c, in a similar fashion. If φ is in the domain of d, we write $\varphi \in d$; similarly for the other operators. Finally, we say φ is in the domain of the operator Δ if there exists a sequence $\{\varphi_\mu\}$, $\varphi_\mu \in C^\infty$, φ_μ, $\Delta\varphi_\mu \in \mathbf{A}$, $D(\varphi_\mu) < \infty$, such that $\|\varphi - \varphi_\mu\|$, $D(\varphi_\mu - \varphi_\nu)$ and $\|\, \Delta\varphi_\mu - \Delta\varphi_\nu \,\|$ all tend to zero $(\mu, \nu \to \infty)$.

We introduce the following spaces:

$$\begin{aligned}
\mathbf{G} &= \{\varphi \mid \varphi \in \Delta,\ D(\varphi, \psi) = (\varphi, \Delta\psi) \text{ for every } \psi \in \Delta\},\\
\mathbf{N} &= \{\varphi \mid \varphi \in \Delta,\ D(\varphi, \psi) = (\Delta\varphi, \psi) \text{ for every } \psi \in \Delta\},\\
\mathbf{H} &= \{\varphi \mid \varphi \in \mathbf{A},\ \Delta\varphi = 0\},\\
\mathbf{F}_c &= \mathbf{G} \cap \mathbf{H},\ \mathbf{F} = \mathbf{N} \cap \mathbf{H}.
\end{aligned}$$

Denote by $\mathbf{D}$ the closure, in the D_s-norm, of the space of forms φ of class C^∞ with $D_s(\varphi) < \infty$, and let $\mathbf{D}_c$ be the closure of $\mathbf{C}$ in the D_s-norm. Since any two norms $D_{s'}$, $D_{s''}$, $s' > 0$, $s'' > 0$, are equivalent, the spaces $\mathbf{D}$, $\mathbf{D}_c$ are independent of the choice of the positive number s.

If $\varphi \in \mathbf{G}$, we have $s\|\, \varphi \,\|^2 \leqq D_s(\varphi) = (\varphi, \Delta_s \varphi) \leqq \|\, \Delta_s\varphi \,\| \cdot \|\, \varphi \,\|$; that is, $\|\, \varphi \,\| \leqq \|\, \Delta_s\varphi \,\|/s$, and the same inequality is true for $\varphi \in \mathbf{N}$.

Let $\varphi \in \mathbf{G}$ or $\mathbf{N}$; then $\Delta\varphi = 0$ implies $D(\varphi) = 0$, and $\Delta_s\varphi = 0$ $(s > 0)$ implies $\varphi = 0$. Therefore

$$\mathbf{F} = \{\varphi \mid \varphi \in \mathbf{D},\ d\varphi = \delta\varphi = 0\},$$

and it will follow from (9.6.19) that

$$\mathbf{F}_c = \{\varphi \mid \varphi \in \mathbf{D}_c,\ d\varphi = \delta\varphi = 0\}.$$

Now let $[d\mathbf{C}]$ denote the closure, in the sense of the norm $\|\cdots\|$, of the space of forms $d\varphi$, $\varphi \in \mathbf{C}$, and let $[\delta\,\mathbf{C}]$ be defined in a similar way. Also, let $[\Delta\ \mathbf{C}]$ be the closure of the space of forms $\Delta\varphi$, $\varphi \in \mathbf{C}$. We have the following two well known formulas of orthogonal decomposition ([9a], [11]).

$$\mathbf{A} = [\Delta\ \mathbf{C}] + \mathbf{H}, \tag{9.6.4}$$

$$\mathbf{A} = [d\mathbf{C}] + [\delta\ \mathbf{C}] + \mathbf{F}. \tag{9.6.5}$$

These two formulas are easily proved using Theorem 9.4.1.

Given a form $\varphi \in \mathbf{A}$, we have by (9.6.4)

$$\varphi = \varphi_1 + \varphi_2$$

where $\varphi_2 \in \mathbf{H}$ and φ_1 is orthogonal to φ_2, that is, $(\varphi_1, \varphi_2) = 0$. We call φ_2 the harmonic component of φ and write $\varphi_2 = H\varphi$. We say that a current T of degree p is of class $\mathbf{A}$ if it is convergent (see Section 9.4) for every $\varphi \in \mathbf{A}^{m-p}$, $\varphi \in C^\infty$. For currents of class $\mathbf{A}$, we define $HT[\varphi] = T[H\varphi]$. Now let y be a point of M and consider the operator $1 = 1_y$ satisfying

$$(1, \varphi) = \varphi(y)$$

for every φ. Strictly speaking, the operator 1 applied to forms of degree $p > 0$ is not a current (since $1[\varphi] = (-1)^{mp+p}(1, *\varphi) = (-1)^{mp+p} * \varphi(y)$ is not a mapping into the reals), but it is a trivial extension of the notion of current and may be treated as though it were a current of class $\mathbf{A}$.

We define $H1[\varphi] = 1[H\varphi] = H\varphi(y)$, and then we have

$$(H1, \varphi) = (1, H\varphi) = H\varphi(y).$$

By Theorem 9.4.1, $H1$ is a symmetric harmonic double form and, applying the inequality of Schwarz, we obtain

$$|H\,\varphi\,(y)| \leqq \sqrt{(H1, H1)}\ \|H\varphi\| = K_y\ \|H\varphi\|.$$

This inequality shows at once that the space $\mathbf{H}$ is closed; and hence is a Hilbert space (the Hilbert space of norm-finite harmonic forms). The operator H denotes orthogonal projection onto **H**.

It follows that **F** is also closed, and hence **F** is the Hilbert space of norm-finite harmonic p-fields. We denote orthogonal projection onto **F** by F.

Let **B** denote the space of forms which belong to both the domain of d and the domain of δ; that is, $\mathbf{B} = \{\varphi \mid \varphi \in d, \delta\}$. We note the following lemma:

LEMMA 9.6.1. *The space* **B** *coincides with* **D**.

To prove this lemma, we have to show that, given any form φ in the domains of d and δ, there exists a sequence $\{\varphi_\mu\}$, $\varphi_\mu \in C^\infty$, $D_s(\varphi_\mu) < \infty$, such that $D_s(\varphi - \varphi_\mu) \to 0 \quad (\mu \to \infty)$.

We base the proof on formula (9.6.5). Let $\psi \in \mathbf{A}$; then

$$\psi = \psi_1 + \psi_2 + \psi_3, \quad \psi_1 \in [d\mathbf{C}], \quad \psi_2 \in [\delta\, \mathbf{C}], \qquad (9.6.6)$$
$$\psi_3 = F\psi \in \mathbf{F}.$$

If $\psi \in C^\infty$, then ψ_1, $\psi_2 \in C^\infty$. In fact, we regard (9.6.6) as a current formula and apply the operator $d\delta$ to it; we obtain

$$d\delta\psi_1 = \triangle\psi_1 = d\delta\psi \in C^\infty,$$

and hence, by Theorem 9.4.1, $\psi_1 \in C^\infty$. Similarly $\psi_2 \in C^\infty$.

Assume φ is in the domains of d and δ. Then there exist two sequences $\{\alpha_\mu\}$, $\{\beta_\nu\}$, such that $\|\varphi - \alpha_\mu\| \to 0$, $\|d\varphi - d\alpha_\mu\| \to 0$, and $\|\varphi - \beta_\nu\| \to 0$, $\|\delta\varphi - \delta\beta_\nu\| \to 0$. Let φ, α_μ and β_ν be decomposed according to formula (9.6.6):

$$\varphi = \varphi_1 + \varphi_2 + \varphi_3, \quad \alpha_\mu = \alpha_{\mu 1} + \alpha_{\mu 2} + \alpha_{\mu 3}, \quad \beta_\nu = \beta_{\nu 1} + \beta_{\nu 2} + \beta_{\nu 3}.$$

We define

$$\varphi_\mu = \beta_{\mu 1} + \alpha_{\mu 2} + F\varphi.$$

Then

$$\|\varphi - \varphi_\mu\|^2 = \|\varphi_1 - \beta_{\mu 1}\|^2 + \|\varphi_2 - \alpha_{\mu 2}\|^2 \to 0,$$

and

$$D(\varphi - \varphi_\mu) = \|d\varphi - d\alpha_{\mu 2}\|^2 + \|\delta\varphi - \delta\beta_{\mu 1}\|^2 \to 0$$

as $\mu \to \infty$. Thus $D_s(\varphi - \varphi_\mu) \to 0 \ (\mu \to \infty)$, and Lemma 9.6.1 follows.

If they exist, we define the Green's operator G_s of the manifold M to be the one-one linear mapping of **A** onto **G** whose inverse mapping is $\triangle_s$, and we define the Neumann's operator N_s to be the one-one linear mapping of **A** onto **N** whose inverse is $\triangle_s$.

If these operators exist, they are unique. For if $\varphi \in \mathbf{G}$, and $\triangle_s \varphi = 0$, then $\varphi = 0$. Similarly, if $\varphi \in \mathbf{N}$ and $\triangle_s \varphi = 0$.

THEOREM 9.6.1 [13c]. *For each positive s the Green's and Neumann's operators exist. They are symmetric operators and they satisfy*

$$D_s(G_s\varphi) \leqq ||\varphi||^2/s, \quad D_s(N_s\varphi) \leqq ||\varphi||^2/s. \tag{9.6.7}$$

Theorem 9.6.1 follows in a straight-forward fashion from Dirichlet's Principle, and we now give a proof which, in its main lines, does not differ from the classical Dirichlet's Principle.

Given $\beta \in \mathbf{A}$, write $\gamma = \beta/s$, and define

$$E_s(\varphi, \psi) = D(\varphi, \psi) + s(\varphi - \gamma, \psi - \gamma), \quad E_s(\varphi) = E_s(\varphi, \varphi).$$

The Dirichlet's Principle which we establish asserts the existence of a unique form α which minimizes $E_s(\varphi)$ and satisfies, perhaps in a generalized sense, the equation $\triangle_s \alpha = \beta$. If we restrict the forms in the minimum problem to those with compact carriers, then $\alpha = G_s \beta$ where G_s is the Green's operator while, if we place no restriction on the forms φ in the minimum problem, then $\alpha = N_s \beta$ where N_s is the Neumann's operator.

Let

$$e = e(s) = \inf E_s(\varphi), \ e_c = e_c(s) = \inf E_s(\varphi) \text{ for } \varphi \in \mathbf{C}. \tag{9.6.8}$$

It is obvious that

$$0 \leqq e(s) \leqq e_c(s) \leqq ||\beta||^2/s. \tag{9.6.9}$$

We have to show that the above minimum problems have solutions. Since the treatment is essentially the same in both problems, we consider the problem of minimizing $E_s(\varphi)$, $\varphi \in \mathbf{D}$. The minimizing properties will then show that the minimizing element is in the space **N**.

We base the proof on the following variant of B. Levi's inequality:

$$\sqrt{D_s(\varphi - \psi)} \leqq \sqrt{E_s(\varphi) - e} + \sqrt{E_s(\psi) - e}. \tag{9.6.10}$$

Let σ, τ be real numbers, $\sigma + \tau = 1$. Since $\sigma\varphi + \tau\psi - \gamma = \sigma(\varphi - \gamma) + \tau(\psi - \gamma)$, we have

$$E_s(\sigma\varphi + \tau\psi) = \sigma^2 E_s(\varphi) + 2\,\sigma\tau\, E_s(\varphi, \psi) + \tau^2 E_s(\psi) \geqq e;$$

that is,

$$\sigma^2[E_s(\varphi) - e] + 2\,\sigma\tau\,[E_s(\varphi, \psi) - e] + \tau^2[E_s(\psi) - e] \geqq 0.$$

Since the left side is homogeneous in σ and τ, this inequality is valid for arbitrary real numbers σ, τ and it follows that

$$|E_s(\varphi, \psi) - e| \leqq \sqrt{E_s(\varphi) - e} \cdot \sqrt{E_s(\psi) - e}.$$

Hence

$$\begin{aligned} D_s(\varphi - \psi) &= E_s(\varphi) - 2\,E_s(\varphi, \psi) + E_s(\psi) \\ &= [E_s(\varphi) - e] - 2[E_s(\varphi,\psi) - e] + [E_s(\psi) - e] \\ &\leqq [E_s(\varphi) - e] + 2\sqrt{E_s(\varphi) - e} \cdot \sqrt{E_s(\psi) - e} \\ &\qquad + [E_s(\psi) - e], \end{aligned}$$

and this is (9.6.10).

Let $\{\varphi_\mu\}$, $\varphi_\mu \in C^\infty$, be a sequence such that $E_s(\varphi_\mu) \to e$ $(\mu \to \infty)$. Then $D_s(\varphi_\mu - \varphi_\nu) \to 0$ $(\mu, \nu \to \infty)$ by (9.6.10), and hence by Lemma 9.6.1 there exists an $\alpha \in \mathbf{D}$ such that $e = E_s(\alpha)$. Since $E_s(\alpha + \varepsilon\varphi) \geqq e$ for every real ε, we conclude by the usual reasoning that the coefficient of ε in $E_s(\alpha + \varepsilon\varphi)$ must vanish, that is

(9.6.11) $$D_s(\alpha, \varphi) = (\beta, \varphi), \quad \varphi \in \mathbf{D}.$$

If $\varphi \in \mathbf{C}$, then

$$D_s(\alpha, \varphi) = (\alpha, \Delta_s\varphi).$$

Therefore α, regarded as a current, satisfies the equation

(9.6.12) $$\Delta_s\alpha[\varphi] = \beta[\varphi], \quad \varphi \in \mathbf{C}.$$

Theorem 9.4.1 remains valid if Δ is replaced by the operator Δ_s. In fact, the proof, which is based on the existence of a local elementary solution for the operator Δ_s, does not differ essentially from the proof given in [11] for the case $s = 0$. Hence, if $\beta \in C^\infty$, then α is equal to a form of class C^∞ and, in this case, $\Delta_s\alpha = \beta$ in the ordinary sense. Thus (9.6.11) becomes

(9.6.13) $$D_s(\alpha, \varphi) = (\Delta_s\alpha, \varphi), \quad \varphi \in \mathbf{D}.$$

The above reasoning applies equally well to the restricted minimum problem in which the competing forms belong to **C** and, in particular, the formulas (9.6.11), (9.6.13) are valid with **D** replaced by $\mathbf{D}_c$. We denote the solution of the restricted minimum problem by $G_s\beta$, that of the unrestricted problem by $N_s\beta$, and we show now that these operators are linear, bounded and symmetric, therefore self-adjoint.

It is sufficient to consider N_s, since the same reasoning applies to G_s. Let β_1, β_2 be two elements in the space **A**, and write $\alpha_1 = N_s\beta_1$, $\alpha_2 = N_s\beta_2$, $\alpha = N_s(\beta_1 + \beta_2)$. By (9.6.11) we have

$$D_s(\alpha - \alpha_1 - \alpha_2, \varphi) = 0, \quad \varphi \in \mathbf{D}.$$

If we choose $\varphi = \alpha - \alpha_1 - \alpha_2$, we conclude that $\alpha = \alpha_1 + \alpha_2$; that is,

$$N_s(\beta_1 + \beta_2) = N_s(\beta_1) + N_s(\beta_2).$$

Since it is obvious that $N_s(c\beta) = c\,N_s\beta$ for any real number c, the operator N_s is linear. Next, taking $\alpha = N_s\beta$, we have by (9.6.11)

$$s\,||\alpha||^2 \leqq D_s(\alpha) = (\beta, \alpha) \leqq ||\alpha|| \cdot ||\beta||,$$

and therefore $s\,||\alpha|| \leqq ||\beta||$. Thus

$$D_s(N_s\beta) = D_s(\alpha) \leqq ||\beta||^2/s.$$

In particular, N_s is continuous, and we may verify that the equation $\Delta_s\alpha = \beta$ is valid. In fact, let $\{\beta_\mu\}$, $\beta_\mu \in C^\infty$, $\beta_\mu \in \mathbf{A}$, be a sequence such that $||\,\beta - \beta_\mu\,|| \to 0$ $(\mu \to \infty)$, and let $\alpha_\mu = N_s\beta_\mu$. Then $\beta_\mu = \Delta_s\alpha_\mu$ and

$$||\,\alpha - \alpha_\mu\,|| \leqq ||\,\beta - \beta_\mu\,||/s, \; D_s(\alpha_\mu - \alpha_\nu) \leqq ||\beta_\mu - \beta_\nu\,||^2/s.$$

Hence $||\,\alpha - \alpha_\mu\,||$, $D(\alpha_\mu - \alpha_\nu)$, $||\,\Delta\alpha_\mu - \Delta\alpha_\nu\,||$ tend to zero $(\mu, \nu \to \infty)$, and it follows that $\alpha = N_s\beta$ is in the domain of Δ and that $\Delta_s N_s\beta = \Delta_s\alpha = \beta$. Finally, if φ, $\psi \in \mathbf{A}$, we have by (9.6.11)

$$D_s(N_s\varphi, N_s\psi) = (\varphi, N_s\psi).$$

Hence

$$(N_s\varphi, \psi) = (\varphi, N_s\psi). \tag{9.6.14}$$

This completes the proof of Theorem 9.6.1 so far as N_s is concerned.

In the restricted minimum problem we have an additional identity which is obtained as follows. Let $\varphi \in \mathbf{D}_c$, and let $\{\varphi_\mu\}$, $\varphi_\mu \in \mathbf{C}$, be a sequence converging to φ in the sense of the D_s-norm. If $\psi \in \Delta$, we have

$$D_s(\varphi,\psi) = \lim_{\mu\to\infty} D_s(\varphi_\mu, \psi) = \lim_{\mu\to\infty} (\varphi_\mu, \Delta_s\psi) = (\varphi, \Delta_s\psi)$$

and therefore

$$D_s(\varphi, \psi) = (\varphi, \Delta_s\psi), \quad \varphi \in \mathbf{D}_c, \quad \psi \in \Delta. \tag{9.6.15}$$

Since $G_s\beta \in \mathbf{D}_c$, we obtain

$$D_s(G_s\beta, \psi) = (G_s\beta, \Delta_s\psi), \quad \beta \in \mathbf{A}, \quad \psi \in \Delta. \tag{9.6.16}$$

This formula shows that $G_s\beta \in \mathbf{G}$, and the proof of Theorem 9.6.1 is completed.

If M is a finite manifold with boundary, Green's formula shows that the difference

$$D_s(G_s\beta, \psi) - (G_s\beta, \Delta_s\psi)$$

is the limit of integrals extended over the boundaries of a sequence of subdomains which converge to M, and these integrals involve the values of $G_s\beta$, $*G_s\beta$ on these boundaries. Formula (9.6.16) states that the limit of these boundary integrals vanishes and hence, in this sense,

$$t\,G_s\beta = n\,G_s\beta = 0. \tag{9.6.17}$$

Similarly, (9.6.13) shows that

$$t\,\delta N_s\beta = n\,dN_s\beta = 0. \tag{9.6.18}$$

In the case of scalars these formulas state that $G_s\beta$ vanishes at the boundary and that the normal derivative of $N_s\beta$ vanishes at the boundary, and therefore G_s, N_s have the boundary behavior of the classical Green's and Neumann's functions.

We have

$$\mathbf{F}_c = \{\varphi \mid \varphi \in \mathbf{A},\ \Delta G_s\varphi = \varphi - s\,G_s\varphi = 0\}, \tag{9.6.19}$$

$$\mathbf{F} = \{\varphi \mid \varphi \in \mathbf{A},\ \Delta N_s\varphi = \varphi - s\,N_s\varphi = 0\}. \tag{9.6.20}$$

In fact, if $\varphi \in \mathbf{A}$ and $\Delta G_s\varphi = \varphi - s\,G_s\varphi = 0$, then $\varphi \in \mathbf{G}$, $\Delta\varphi = 0$. Conversely, if $\varphi \in \mathbf{G}$, $\Delta\varphi = 0$, then $\varphi = G_s(\Delta_s\varphi) = s\,G_s\varphi$. Similarly for (9.6.20). From the minimum principle by which G_s, N_s were defined it follows that

$$F_c G_s = G_s F_c = F_c/s, \tag{9.6.21}$$

$$F N_s = N_s F = F/s. \tag{9.6.22}$$

Finally, we have the orthogonal decompositions:

$$\mathbf{A} = [\triangle \mathbf{G}] + \mathbf{F}_c, \tag{9.6.23}$$

$$\mathbf{A} = [\triangle \mathbf{N}] + \mathbf{F}, \tag{9.6.24}$$

where $[\triangle \mathbf{G}]$ denotes the closure under the norm $\|\cdots\|$ of the space of forms $\varphi = \triangle\psi$, $\psi \in \mathbf{G}$, and where $[\triangle \mathbf{N}]$ has a similar meaning with respect to the space $\mathbf{N}$. In fact, let $\varphi \in \mathbf{A}$, $(\varphi, \gamma) = 0$ for $\gamma \in [\triangle\mathbf{G}]$. Then $0 = (\varphi, \triangle G_s\psi) = (\varphi, \psi - sG_s\psi) = (\varphi - sG_s\varphi, \psi)$ for every $\psi \in \mathbf{A}$, and this implies that $\varphi \in \mathbf{F}_c$. Similarly, for (9.6.24).

9.7. Bounded Manifolds

A relatively compact subdomain M of a Riemannian manifold R will be called a *b*-manifold (bounded manifold). We have:

Theorem 9.7.1. [13b]. *Let M be a b-manifold imbedded in a Riemannian manifold R. Then $\mathbf{F}_c$ is the subspace of harmonic forms on R which vanish identically outside M.*

We define the Green's operator G_o of a manifold M to be the one-one linear mapping of $\mathbf{A} - \mathbf{F}_c$ onto $\mathbf{G} - \mathbf{F}_c$ whose inverse mapping is $\triangle$. If this operator exists, it is unique. For let $\varphi \in \mathbf{G} - \mathbf{F}_c$, $\triangle\varphi = 0$. Then $\varphi \in \mathbf{H} \cap \mathbf{G} = \mathbf{F}_c$, so $\varphi = 0$. The domain of the operator G_o may be extended to the whole space $\mathbf{A}$ by defining it to be identically zero on the space $\mathbf{F}_c$. Then G_o satisfies

$$(1 - F_c)\varphi = \triangle G_o\varphi, \quad F_c G_o \varphi = G_o F_c \varphi = 0, \quad \varphi \in \mathbf{A}. \tag{9.7.1}$$

We recall that an operator is completely continuous if it carries bounded sets into sets whose closure is compact.

Theorem 9.7.2. *A b-manifold possesses a bounded, symmetric completely continuous Green's operator G_o.*

The biharmonic Green's operator is the one-one linear mapping of $\mathbf{A} - \mathbf{H}$ onto $(\mathbf{G} \cap \mathbf{N}) - \mathbf{F}_c$ whose inverse mapping is $\triangle$. We extend this operator to the whole space $\mathbf{A}$ by defining it to be identically zero on $\mathbf{H}$, and denote it by B_o.

Theorem 9.7.3. *A b-manifold possesses a bounded biharmonic Green's operator* B_0.

We begin by deriving Theorem 9.7.3 from Theorem 9.7.2, and we observe first that Theorem 9.7.2 contains a solution (in the generalized sense) of the first boundary-value problem of potential theory. In fact, let φ be an arbitrary form in the domain of the operator $\triangle$ and set

$$\psi = \varphi - G_0 \triangle \varphi. \tag{9.7.2}$$

Then

$$\triangle \psi = \triangle \varphi - (\triangle \varphi - F_c \triangle \varphi) = F_c \triangle \varphi = 0.$$

On the other hand, $\psi - \varphi = - G_0 \triangle \varphi \in \mathbf{G}$, so ψ is a (generalized) solution of the first boundary-value problem. Since we may add to ψ any form in $\mathbf{F}_c$ without changing its boundary behavior, the solution is unique if and only if $\mathbf{F}_c$ contains only the form 0.

Now define the adjoint operators

$$B_0 = G_0 - G_0 H, \quad B_0' = G_0 - H G_0. \tag{9.7.3}$$

Given any form φ in the domain of the operator $\triangle$, set

$$\psi = (1 - H)\varphi - B_0' \triangle \varphi.$$

Then

$$\triangle \psi = \triangle \varphi - (\triangle \varphi - F_c \triangle \varphi) = 0, \quad H \psi = 0,$$

so $\psi = 0$. That is

$$(1 - H)\varphi = B_0' \triangle \varphi. \tag{9.7.4}$$

On the other hand, we obviously have

$$(1 - H)\psi = \triangle B_0 \psi, \quad \psi \in \mathbf{A}. \tag{9.7.5}$$

Let φ be an arbitrary form in the domain of $\triangle$, ψ an arbitrary form in $\mathbf{A}$. Since $B_0 \psi \in \mathbf{G}$, we have

$$(\psi, (1 - H) \varphi) = (\psi, B_0' \triangle \varphi) = (B_0 \psi, \triangle \varphi) = D(B_0 \psi, \varphi).$$

By (9.7.5)

$$(\psi, (1 - H) \varphi) = ((1 - H) \psi, \varphi) = (\triangle B_0 \psi, \varphi).$$

Therefore

$$D(B_0 \psi, \varphi) = (\triangle B_0 \psi, \varphi) \tag{9.7.6}$$

for every $\varphi \in \Delta$, $\psi \in \mathbf{A}$, and it follows that $B_0\psi \in \mathbf{N}$. Since $F_c B_0 \psi = 0$, B_0 is the biharmonic Green's operator whose existence is asserted in Theorem 9.7.3.

From Theorem 9.7.3 we obtain a solution of the first boundary-value problem for biharmonic forms. In fact, given any φ in the domain of Δ, let

$$\psi = \varphi - B_0 \Delta \varphi.$$

Then

$$\Delta\psi = \Delta\varphi - (\Delta\varphi - H\Delta\varphi) = H\Delta\varphi, \; \Delta^2\psi = \Delta\,\Delta\psi = 0,$$

so ψ is biharmonic. Since $\psi - \varphi \in \mathbf{G} \cap \mathbf{N}$, ψ is a (generalized) solution of the first boundary-value problem for biharmonic forms.

It remains to prove Theorems 9.7.1 and 9.7.2, and we base the proof of both on the following results:

Let R be an arbitrary Riemannian manifold of class C^∞. Given any point $q \in R$, there is a neighborhood $U = U(q)$ depending only on q such that, given an arbitrary integer ν, a number $s \geqq 0$, and a positive number η, there exists a double form $\gamma(x, y)$ with the following properties:

(i) the double form $\gamma(x, y)$ is defined for $x \in R$, $y \in U$, and $\gamma(x, y)$ is of class C^∞ in x, y if $x \neq y$. If $r(x, y)$ is the distance of the points x and y, $[r(x, y)]^{m-2}\, \gamma(x, y)$ is of class C^∞ for $x = y$.

(ii) for fixed $y \in U$, the carrier of $\gamma(x, y)$ has diameter less than η.

(iii) for $x \in R$, $y \in U$, we have $\Delta_s(x)\, \gamma(x, y) = g(x, y)$ where $g(x, y)$ is a double form of class C^ν in x and y.

(iv) the operator $(\varphi(x), d(x)\gamma((x, y))$ is a completely continuous transformation which maps forms $\varphi \in \mathbf{A}(R)$ into $\mathbf{A}(U)$. The same statement applies to $(\varphi(x), \delta(x)\gamma(x, y))$.

The form $\gamma(x, y)$ with the properties (i) — (iv) can be constructed by the method of Kodaira [9a]. Kodaira assumes that the metric is real analytic, but the infinite series used by him can be replaced by finite partial sums.

To prove Theorem 9.7.1, let $\varphi \in \mathbf{F}_c$ and extend φ over R by defining it to be identically zero outside M. There exists a sequence $\{\varphi_\mu\}$, $\varphi_\mu \in \mathbf{C}(M)$, converging to φ in the D_s-norm. Given a point $q \in R$, let $\gamma(x, y)$ be the double form associated with the neighborhood

$U = U(q)$. For $y \in U$ we have by Green's formula

$$\varphi_\mu(y) = D_s(\varphi_\mu(x), \gamma(x, y)) - (\varphi_\mu(x), g(x, y)),$$

As $\mu \to \infty$, we obtain as limit (in the sense of the norm $\| \cdots \|$)

(9.7.7) $\quad \varphi(y) = D_s(\varphi(x), \gamma(x, y)) - (\varphi(x), g(x, y)), \quad y \in U.$

Formula (9.7.7.) (with γ depending on s) is valid for each $s \geqq 0$. Since $d\varphi = \delta\varphi = 0$ in M and in $R - M$, the case $s = 0$ gives

$$\varphi(y) = -(\varphi(x), g(x, y)), \quad y \in U.$$

Thus $\varphi(y) \in C^\nu$ in U, and hence everywhere in R. This proves Theorem 9.7.1.

We base the proof of Theorem 9.7.2 on the following lemma:

LEMMA 9.7.1. *For every b-manifold the operator* G_s $(s > 0)$ *is completely continuous.*

In fact, since $G_s\varphi \in \mathbf{G}$, there exists a sequence $\{\psi_\mu\}$, $\psi_\mu \in \mathbf{C}(M)$, converging to $G_s\varphi$ in the D_s-norm, and we obtain formula (9.7.7) with φ replaced by $G_s\varphi$:

(9.7.8) $\quad G_s\varphi(y) = D_s(G_s\varphi(x), \gamma(x, y)) - (G_s\varphi(x), g(x, y)), \quad y \in U \cap M.$

Since M can be covered by finitely many neighborhoods U, it follows that G_s is a completely continuous transformation from $\mathbf{A}$ onto $\mathbf{G}$.

We remark that the operator N_s will not in general be completely continuous on b-manifolds. For if N_s were completely continuous, the space $\mathbf{F}$ would be finite dimensional.

To prove Theorem 9.7.2, we begin with the following observations concerning a completely continuous operator P, $\| P\varphi \| \leqq c \| \varphi \|$, $c > 0$. Let $0 < \lambda \leqq c$. Then

$$\| P\varphi - P\psi \| = \| (P\varphi - \lambda\varphi) - (P\psi - \lambda\psi) + \lambda(\varphi - \psi) \|$$

(9.7.9) $\quad \geqq \lambda \| \varphi - \psi \| - \| P\varphi - \lambda\varphi \| - \| P\psi - \lambda\psi \|.$

If $\{\varphi_\mu\}$ is a sequence of forms such that $(\varphi_\mu, \varphi_\nu) = \delta_{\mu\nu}$, then for $\mu \neq \nu$,

(9.7.10) $\quad \| P\varphi_\mu - P\varphi_\nu \| \geqq \sqrt{2}\,\lambda - \| P\varphi_\mu - \lambda\varphi_\mu \| - \| P\varphi_\nu - \lambda\varphi_\nu \|,$

and it follows that there is a positive number k such that $\| P\varphi_\mu - \lambda\varphi_\mu \| \geqq k$ for all but a finite number of values of μ. In fact, if $\| P\varphi_{\mu_i} - \lambda\varphi_{\mu_i} \| \to 0$ as $i \to \infty$ for some subsequence $\{\mu_i\}$, then the sequence $\| P\varphi_{\mu_i} \|$ could not contain a convergent subsequence, contrary to the hypothesis that P is completely continuous and the set $\{\varphi_\mu\}$ bounded. In particular, the space $\mathbf{E} = \mathbf{E}(P, \lambda) = \{\varphi \mid \varphi \in \mathbf{A}, P\varphi - \lambda\varphi = 0\}$ is a finite dimensional vector space (over the reals).

Consider next the space $\mathbf{E}' = \{\psi \mid \psi \in \mathbf{A}, (\psi, \varphi) = 0 \text{ for all } \varphi \in \mathbf{E}\}$. There is a positive number $a = a(P, \lambda)$ such that $\| P\psi - \lambda\psi \| \geqq a$ for all $\psi \in \mathbf{E}'$ with $\| \psi \| = 1$; that is,

$$\| P\psi - \lambda\psi \| \geqq a \| \psi \|, \quad \psi \in \mathbf{E}'. \tag{9.7.11}$$

For suppose $\| P\psi_\mu - \lambda\psi_\mu \| \to 0$ as $\mu \to \infty$, $\psi_\mu \in \mathbf{E}'$, $\| \psi_\mu \| = 1$. If $\{P\psi_{\mu_i}\}$ is a convergent subsequence of $\{P\psi_\mu\}$, then $\{\psi_{\mu_i}\}$ is also convergent, by (9.7.9). Let ψ be the limit of the ψ_{μ_i}. Then $\psi \in \mathbf{E}'$, $\| \psi \| = 1$ and $P\psi - \lambda\psi = 0$, that is, $\psi \in \mathbf{E}$, a contradiction.

To construct the operator G_o, we choose an arbitrary $s > 0$, and apply the above result with $P = G_s$, $\lambda = c = 1/s$; then $\mathbf{E} = \mathbf{F}_c$ by (9.7.19) and $\mathbf{E}' = \mathbf{A} - \mathbf{F}_c$ by (9.6.23).

Given $\beta \in \mathbf{A}$, let

$$m = \inf \| \beta - (\psi - sG_s\psi) \|, \quad \psi \in \mathbf{A} - \mathbf{F}_c, \tag{9.7.12}$$

and let $\{\psi_\mu\}$, $\psi_\mu \in \mathbf{A} - \mathbf{F}_c$, be a sequence such that

$$\| \beta - (\psi_\mu - sG_s\psi_\mu) \| \to m, \quad \mu \to \infty. \tag{9.7.12$'$}$$

Then $\|(\psi_\mu - \psi_\nu) - sG_s(\psi_\mu - \psi_\nu)\| \to 0$ $(\mu, \nu \to \infty)$ and, by (9.7.11), $\| \psi_\mu - \psi_\nu \| \to 0$ $(\mu, \nu \to \infty)$. Let φ be the limit of the ψ_μ. Then $\varphi \in \mathbf{A} - \mathbf{F}_c$, or $F_c\varphi = 0$, and $F_cG_s\varphi = 0$ by (9.6.21). The minimizing condition implies that $\alpha = \beta - (\varphi - sG_s\varphi)$ is orthogonal to $\psi - sG_s\psi$ for all $\psi \in \mathbf{A} - \mathbf{F}_c$, or $\alpha \in \mathbf{F}_c$. Then $\alpha = F_c\alpha$, or $\beta - (\varphi - sG_s\varphi) = F_c\beta$. Thus,

$$\varphi - sG_s\varphi = \triangle G_s\varphi = \beta - F_c\beta, \tag{9.7.13}$$

and we define $G_o\beta = G_s\varphi$. Further,

$$G_o\beta(y) = D_s(G_o\beta(x), \gamma(x,y)) - (G_o\beta(x), g(x, y)), \; y \in U \cap M, \tag{9.7.8$'$}$$

and it follows that the operator G_o is completely continuous. This completes the proof of Theorem 9.7.2.

The opposite of a b-manifold is a u-manifold (unbounded manifold) which is characterized by the property that, every form in the closure of the operator $\triangle$ belongs to $\mathbf{N}$. Let

$$\mathbf{H}_s = \{\varphi \mid \varphi \in \mathbf{A}, \; \triangle_s \varphi = 0\}. \tag{9.7.14}$$

On a u-manifold the spaces **G** and **N** coincide, $G_s = N_s$, and $\mathbf{H}_s$ contains only the form 0 for $s > 0$. Moreover,

$$\mathbf{H}(= \mathbf{H}_0) = \mathbf{F} = \mathbf{F}_c. \tag{9.7.15}$$

A compact (closed) manifold M may be regarded either as a b-manifold (without boundary) or as a u-manifold. In the case of a compact manifold, the operator G_0 coincides with the operator G of de Rham (see [11]).

9.8. Dirichlet's Principle for the Complex Operators

We assume now that M is a Kähler manifold with Kähler metric of class C^∞. The Dirichlet integral for the complex operators $\bar\partial$, ϑ is

$$D(\varphi) = (\bar\partial\varphi, \bar\partial\varphi) + (\vartheta\varphi, \vartheta\varphi), \tag{9.8.1}$$

and we define

$$D_s(\varphi) = D(\varphi) + \frac{1}{2} s(\varphi, \varphi) = D(\varphi) + \frac{1}{2} s\|\varphi\|^2. \tag{9.8.2}$$

The factor 1/2 is used in (9.8.2) since $\vartheta\bar\partial + \bar\partial\vartheta = \triangle/2$.

We define the domains of the operators $\bar\partial$, ϑ in the same way as for the operators d, δ. We denote by $\mathbf{A}^{\varrho,\sigma}$ the subspace of **A** composed of forms of type (ϱ, σ):

$$\mathbf{A} = \sum_{\varrho,\sigma} \mathbf{A}^{\varrho,\sigma}.$$

If $\varphi \in \mathbf{A}$, then

$$\varphi = \sum_{\varrho,\sigma} \Pi\,\varphi, \quad \|\varphi\| = \sqrt{\sum_{\varrho,\sigma} \|\Pi\,\varphi\|^2}.$$

Apart from a trivial factor 1/2 which occurs in the definitions of the spaces **G**, **N**, namely $D(\varphi, \psi) = (\varphi, \triangle\psi)/2$, the definitions of **G**, **N**, **H**, $\mathbf{F}_c$, **F** are the same as those given in Section 9.6.

We remark that

$$\mathbf{H} = \sum_{\varrho,\sigma} \Pi\,\mathbf{H}. \tag{9.8.3}$$

In fact, by (9.2.20), the operators $\triangle$ and $\underset{\varrho,\sigma}{\Pi}$ commute.

The complex analogue of the orthogonal decomposition formula (9.6.5) is valid:

$$\mathbf{A} = [\bar\partial \mathbf{C}] + [\vartheta \mathbf{C}] + \mathbf{F}. \tag{9.8.4}$$

The proof of this formula is the same as in the real case. By the same reasoning as that used to prove Lemma 9.6.1, but based on (9.8.4), we see that $\mathbf{B} = \{\varphi \mid \varphi \in \bar{\partial}, \vartheta\}$ may also be regarded as the space obtained by taking the closure, in the D_s-norm, of forms of class C^∞ which lie in the domains of $\bar{\partial}$ and ϑ.

We define the complex Green's and Neumann's operators as in the real case, and the method of Section 9.6 applies with trivial modifications to prove the complex analogue of Theorem 9.6.1, namely,

THEOREM 9.8.1. *For each positive s there exist complex Green's and Neumann's operators G_s, N_s These operators are Hermitian and they satisfy*

$$D_s(G_s\varphi) \leqq ||\varphi||^2/2s, \ D_s(N_s\varphi) \leqq ||\varphi||^2/2s. \tag{9.8.5}$$

As in the real case, it turns out that

$$\mathbf{F} = \{\varphi \mid \varphi \in \mathbf{A}, \varphi - sN_s\varphi = \triangle N_s\varphi = 0\}, \tag{9.8.6}$$

$$\mathbf{F}_c = \{\varphi \mid \varphi \in \mathbf{A}, \varphi - sG_s\varphi = \triangle G_s\varphi = 0\}. \tag{9.8.7}$$

In the case $\sigma = 0$, we have

$$\mathbf{F}^{\varrho, 0} = \{\varphi \mid \varphi \in \mathbf{A}^{\varrho, 0}, \varphi \text{ a holomorphic form of degree } \varrho\}. \tag{9.8.8}$$

Thus, on a Kähler manifold, the holomorphic differential forms are characterized as the eigen-forms of type $(\varrho, 0)$ of N_s which belong to the eigen-value s.

9.9. Bounded Kähler Manifolds

Assume that M is a b-manifold imbedded in a Kähler manifold R. It is obvious that $\mathbf{F}_c^{\varrho, \sigma}$ is composed of the harmonic forms of type (ϱ, σ) on R which vanish identically outside M.

LEMMA 9.9.1 [13a]. *The space $\mathbf{F}_c^{\varrho, \sigma}$ contains only the form 0 if either ϱ or σ has one of the extremal values 0 or k, where k is the complex dimension of M.*

Suppose, for example, that $\sigma = 0$, and let $\varphi \in \mathbf{F}_c^{\varrho, 0}$. Then φ is a holomorphic form of degree ϱ on R which vanishes identically outside M and therefore, by analytic continuation, it must vanish everywhere. Similarly, if $\sigma = k$, $\varphi \in \mathbf{F}_c^{\varrho, k}$, then $*\bar{\varphi} \in \mathbf{F}_c^{k-\varrho, 0}$, so $*\bar{\varphi}$ is holomorphic on R and vanishing outside M, therefore identically zero. The remaining two cases $\varrho = 0$, k are reduced to the preceding by taking conjugates.

Now let R be an arbitrary Kähler manifold, and let B be any finite submanifold of R. We can always choose a b-manifold M which contains the closure of B in its interior, and we denote the Green's operator of M by G_o. It is readily seen, by Theorem 9.4.1, that G_o is an integral operator with symmetric kernel $g = g(z, \zeta)$:

$$G_o\varphi(\zeta) = (\varphi(z), g(z, \zeta)). \tag{9.9.1}$$

Now let $\varphi \in \mathbf{F}$. Then we obtain from Green's formula by the usual reasoning,

$$(1 - F_c)\,\varphi(\zeta) = -2\int_{bB} [\varphi \wedge *(\bar{\partial} g)^- - (\vartheta g)^- \wedge *\varphi]. \tag{9.9.2}$$

Formula (9.9.2) is a Cauchy's formula for the space $\mathbf{F}$. If either ϱ or σ is equal to 0 or k, then $\mathbf{F}_c^{\varrho,\sigma} = 0$ by Lemma 9.9.1. In particular, if $\sigma = 0$, then $\vartheta g = 0$ automatically and we have simply

$$\varphi(\zeta) = -2\int_{bB} \varphi \wedge *(\bar{\partial} g)^-, \quad \varphi \in \mathbf{F}^{\varrho,0}. \tag{9.9.3}$$

In the ordinary case of Euclidean space of complex dimension $k = 1$, this formula reduces to the classical formula

$$\varphi(\zeta) = \frac{1}{2\pi i}\int_{bB} \varphi(z)\frac{dz}{z - \zeta}.$$

9.10. EXISTENCE THEOREMS ON COMPACT KÄHLER MANIFOLDS

In this section we assume that M is a compact Kähler manifold; then the Green's operator G_o coincides with the operator G of de Rham. We have

$$\underset{\varrho,\sigma}{\Pi} G = G \underset{\varrho,\sigma}{\Pi}, \ \Lambda G = G\Lambda, \ dG = Gd, \ \delta G = G\delta. \tag{9.10.1}$$

These formulas are all proved in the same way. For example, to show that G commutes with d, consider the difference $\psi = (Gd - dG)\varphi$. Since ψ is harmonic and, at the same time, orthogonal to harmonic forms, it is zero.

The formulas (9.10.1) show that G commutes with the complex operators ∂, $\bar{\partial}$, ϑ and $\bar{\vartheta}$.

We now consider the following "Cousin problem" for M. Let $\{U_i\}$ be a locally finite open covering of M, and let there be given in each U_i a current w_i of type (ϱ, σ) such that

$$\bar{\partial}(w_i - w_j) = 0, \ \vartheta(w_i - w_j) = 0, \text{ in } U_i \cap U_j. \tag{9.10.2}$$

The problem is to find a current w of type (ϱ, σ) defined over the whole of M, such that

$$\bar{\partial}(w - w_i) = 0, \ \vartheta(w - w_i) = 0 \text{ in } U_i. \tag{9.10.3}$$

Let

$$P = \bar{\partial} w_i, \ Q = \vartheta w_i \text{ in } U_i. \tag{9.10.4}$$

In view of (9.10.2) we see that P and Q are well defined, and the conditions of solvability of the problem are that

$$HP = HQ = 0. \tag{9.10.5}$$

If these conditions are satisfield, a solution is given by

$$w = 2G(\bar{\partial} Q + \vartheta P). \tag{9.10.6}$$

In fact,

$$\bar{\partial} w = 2G(\bar{\partial}\vartheta P) = \triangle GP = P - HP = P,$$
$$\vartheta w = 2G(\vartheta\bar{\partial} Q) = \triangle GQ = Q - HQ = Q,$$

and it follows that w satisfies (9.10.3).

Conversely, suppose that a current w of type (ϱ, σ) exists such that $\bar{\partial} w = \vartheta w = 0$ outside its singular set.

If we define

$$P = \bar{\partial} w, \ Q = \vartheta w, \tag{9.10.7}$$

the conditions (9.10.5) are satisfied and w is given by (9.10.6). The currents P and Q define the residues of w.

If we suppose that $\Lambda w_i = 0$, the solution w (if it exists) will also satisfy $\Lambda w = 0$. In fact,

$$\Lambda w = 2G(\Lambda \bar{\partial}\, Q + \Lambda \vartheta P)$$

where we have in U_i, by (9.2.28),

$$\Lambda\bar\partial Q = \bar\partial\Lambda Q + \sqrt{-1}\,\overline{\vartheta}Q = \bar\partial\vartheta\Lambda w_i + \sqrt{-1}\,\overline{\vartheta}\vartheta w_i = \sqrt{-1}\,\overline{\vartheta}\vartheta w_i,$$

$$\Lambda\vartheta P = \vartheta\Lambda P = \vartheta\Lambda\bar\partial w_i = \vartheta\bar\partial\Lambda w_i + \sqrt{-1}\,\vartheta\overline{\vartheta}w_i = \sqrt{-1}\,\vartheta\overline{\vartheta}w_i.$$

Since $\vartheta\overline{\vartheta} + \overline{\vartheta}\vartheta = 0$ by (9.2.18), we have $\Lambda w = 0$.

As a simple illustration, let M be a compact Riemann surface which, by Section 9.5, can be assumed to be a Kähler manifold of dimension 1. Let $q_1, q_2 (q_1 \neq q_2)$ be two points of M, z_1 a uniformizer with center at q_1, z_2 a uniformizer with center at q_2, and take a covering $\{U_j\}$ such that $q_1 \in U_1$, $q_2 \in U_2$, but neither q_1 nor q_2 lies in any other U_j. Let

$$w_j = \begin{cases} d \log z_1/2\pi i, & j = 1 \\ d \log z_2/2\pi i, & j = 2, \\ 0 & , j > 2, \end{cases}$$

Then the conditions (9.10.2) are satisfied with $\varrho = 1$, $\sigma = 0$, and we have

(9.10.8) $$P = q_1 - q_2, \quad Q = 0.$$

In fact, since w_j is of type (1,0), we have $\vartheta w_i = 0$ and therefore $Q = 0$. As for P, let φ be a 0-form of class C^∞ with carrier contained in U_1. Then

$$P[\varphi] = \bar\partial w_1[\varphi] = w_1[\bar\partial\varphi] = \int_{U_1} w_1 \wedge \bar\partial\varphi = \int_{U_1 - q_1} w_1 \wedge \bar\partial\varphi$$

$$= -\int_{U_1 - q_1} \bar\partial(w_1 \wedge \varphi) = -\int_{U_1} d(w_1 \wedge \varphi)$$

since $\partial(w_1 \wedge \varphi)$ is of type (2,0), therefore 0. Thus

$$P[\varphi] = -\lim_{\varepsilon\to 0} \int_{|z_1| \geqq \varepsilon} d(w_1 \wedge \varphi) = \lim_{\varepsilon\to 0} \int_{|z_1| = \varepsilon} w_1 \wedge \varphi$$

$$= \lim_{\varepsilon\to 0} \frac{1}{2\pi i} \int_{|z_1| = \varepsilon} \varphi \cdot \frac{dz_1}{z_1} = \varphi(q_1) = q_1[\varphi].$$

Similarly for the neighborhood U_2 of q_2.

Finally, let φ be any 0-form of class C^∞ in M. Then

$$HP[\varphi] = P[H\varphi] = H\varphi(q_1) - H\varphi(q_2) = 0$$

since $H\varphi$, a harmonic form of degree 0, must be a constant. It follows that there exists a current w of type (1,0) on M which is a holomorphic differential except at q_1, q_2 where it has simple poles with residues $+1$, -1 respectively.

These considerations may be generalized to a compact Kähler manifold M of arbitrary (complex) dimension k. In fact, suppose that S_1 and S_2 are two analytic subvarieties of M each of complex dimension $k-1$, and suppose that S_1 is homologous to S_2 (a condition which generalizes the statement that the residue sum vanish). In a neighborhood U_j, the subvarieties S_1 and S_2 are defined by minimal local equations $R_{1j}(z_j) = 0$, $R_{2j}(z_j) = 0$ respectively, where $z_j = (z_j^1, \cdots, z_j^k)$ refers to a local coordinate system. We define

$$w_j = \frac{1}{2\pi i}\{d \log R_{1j}(z_j) - d \log R_{2j}(z_j)\} \text{ in } U_j.$$

A computation similar to the above gives

$$P = S_1 - S_2, \; Q = 0. \tag{9.10.9}$$

The hypothesis that $S_1 - S_2$ is homologous to zero implies the existence of a current T such that $P = dT$, and then it follows that $HP = 0$. Hence there exists a current w of type (1,0) on M which is holomorphic in $M - S_1 - S_2$ and has a pole of residue $+1$ on S_1, a pole of residue -1 on S_2.

The Green's operator G combined with the notion of currents provides a systematic and powerful method for establishing the existence of meromorphic differential forms on compact Kähler manifolds.

9.11. The L-kernels on Finite Kähler Manifolds

In Chapter 4 we introduced the kernels $L(p, q)$ and $L(p, \bar{q})$ or, in our present notation, $L(z, \zeta)$ and $L(z, \bar{\zeta})$. These kernels formed the basis of the results obtained in Chapters 5 and 6, and the question arises whether these kernels can be defined on a finite Kähler manifold B. The answer is that they can be defined, but only in the special case $k = 1$ do they lead to a significant theory.

In fact, let

$$g = g_p(z, \bar\zeta) = \underset{\varrho,\sigma}{\Pi_z} g(z, \bar\zeta) \tag{9.11.1}$$

be the kernel of the operator G for forms of type (ϱ, σ), $\varrho + \sigma = p$, on the finite Kähler manifold B. It is readily seen that the difference

$$\partial_z g_{p-1} - \overline{\vartheta}_\zeta g_p \tag{9.11.2}$$

is harmonic and regular even at the point ζ. Define

$$L_p(z, \bar\zeta) = 2(\partial_z \bar\partial_\zeta g_{p-1} + \vartheta_z \vartheta_\zeta g_{p+1}), \tag{9.11.3}$$

$$L_p(z, \zeta) = 2(\partial_z \partial_\zeta g_{p-1} + \overline{\vartheta}_z \overline{\vartheta}_\zeta g_{p+1}), \tag{9.11.4}$$

where

$$\partial_z = \underset{\varrho,\sigma}{\Pi}\, d \underset{\varrho-1,\sigma}{\Pi}, \quad \partial_\zeta = \underset{\varrho,\sigma}{\Pi}\, d \underset{\varrho-1,\sigma}{\Pi}, \quad \bar\partial_\zeta = \underset{\sigma,\varrho}{\Pi}\, d \underset{\sigma,\varrho-1}{\Pi},$$

$$\overline{\vartheta}_z = \underset{\varrho,\sigma}{\Pi}\, \delta \underset{\varrho+1,\sigma}{\Pi}, \quad \overline{\vartheta}_\zeta = \underset{\varrho+1,\sigma}{\Pi}\, \delta \underset{\varrho,\sigma}{\Pi}, \quad \vartheta_\zeta = \underset{\sigma,\varrho}{\Pi}\, \delta \underset{\sigma,\varrho+1}{\Pi}.$$

By (9.10.1) we see that $L_p(z, \zeta)$ vanishes except in the two cases $\varrho = \sigma + 1$, $\varrho = \sigma - 1$, where

$$\begin{cases} L_p(z, \zeta) = 2\partial_z \partial_\zeta g_{p-1}, & \varrho = \sigma + 1, \\ L_p(z, \zeta) = 2\overline{\vartheta}_z \overline{\vartheta}_\zeta g_{p+1}, & \varrho = \sigma - 1. \end{cases} \tag{9.11.5}$$

In the special case $\varrho = k$, $\sigma = k - 1$, we obtain

$$L_{2k-1}(z, \bar\zeta) = 2\partial_z \bar\partial_\zeta g_{2k-2}; \quad L_{2k-1}(z, \zeta) = 2\partial_z \partial_\zeta g_{2k-2}. \tag{9.11.6}$$

If $k = 1$, these formulas agree with those given in Chapter 4.

We plainly have

$$L_p(z, \bar\zeta) = (L_p(\zeta, \bar z))^-, \quad L_p(z, \zeta) = L_p(\zeta, z); \tag{9.11.7}$$

$$\partial_z \bar\partial_\zeta L_{p-2}(z, \bar\zeta) = \overline{\vartheta}_z \vartheta_\zeta L_p(z, \bar\zeta); \quad \partial_z \partial_\zeta L_{p-2}(z, \zeta) = \overline{\vartheta}_z \overline{\vartheta}_\zeta L_p(z, \zeta); \tag{9.11.8}$$

$$*_z *_\zeta L_p = (L_{2k-p})^- \tag{9.11.9}$$

for both kernels. For $k > 1$ we cannot expect that $L_p(z, \zeta)$ will in general satisfy the equations $\partial_z L_p = \overline{\vartheta}_z L_p = 0$ since, for example, in the case $\varrho = 0$, $\sigma = 1$, the equation $\partial_z L_p = 0$ implies that $(L_p)^-$ is meromorphic and this is impossible because $L_p(z, \zeta)$ has a point singularity. In fact, these equations are not satisfied by either kernel if $k > 1$, and this is the reason why the kernels cannot be used in higher dimensions. However, the other properties of these kernels remain valid and we have:

LEMMA 9.11.1. *If* $\varphi \in \mathbf{F}^{\varrho, \sigma}$, *then*

$$(9.11.10) \quad (\varphi(z), L_p(z, \bar{\zeta})) = -\varphi(\zeta), \quad (\varphi(z), L_p(z, \zeta)) = 0.$$

The proof of this lemma is a generalization of that given in Chapter 4 for the case $k = 1$, and will be omitted.

We remark finally that, in the case $\varrho = k = 1$, $\sigma = 0$, two circumstances combine to make the kernels (9.11.6) meromorphic; (a) for scalars, $\triangle = 2\vartheta\bar{\partial}$; (b) if φ is of type (1, 0), then $\vartheta\varphi = 0$ implies that φ is holomorphic. If $k > 1$, the relation (a) remains true for scalars while (b) is true for forms of type $(k, 0)$. We have here an illustration of the extremely special character of Riemann surfaces.

9.12. Intrinsic Definition of the Operators

It is possible to give intrinsic definitions of all the operators introduced in Sections 9.1 and 9.2. For example, the operator $\bar{\partial}$ is characterized by the following four axioms: (i) linearity; (ii) anti-derivation, namely $\bar{\partial}(\varphi \wedge \psi) = \bar{\partial}\varphi \wedge \psi + (-1)^p \varphi \wedge \bar{\partial}\psi$ if φ is a form of degree p; (iii) for a scalar f, $\bar{\partial} f(w) = \underset{0,1}{\Pi}\, w(f)$, where w is a tangent vector of the manifold at the point being considered; (iv) for a scalar f, $\bar{\partial}\, df = -d\, \bar{\partial} f$. In (iii), the operator $\underset{0,1}{\Pi}$ is the projection operator associated with the direct-sum decomposition of the tangent bundle.

In terms of an arbitrary Hermitian metric, we can define an adjoint operator ϑ of $\bar{\partial}$, namely $\vartheta = -*\bar{\partial}*$, and we can introduce as Laplacian $\triangle = 2(\vartheta\bar{\partial} + \bar{\partial}\vartheta)$. Dirichlet's Principle remains valid for the Dirichlet integral defined by (9.8.1) in terms of the operators $\bar{\partial}$, ϑ, and we obtain Green's and Neumann's operators G_s, N_s. For a b-manifold M imbedded in an arbitrary complex analytic manifold R there exists a completely continuous Green's operator G_0, $\triangle G_0\varphi = \varphi - F_c\varphi$, where $\mathbf{F}_c = \{\varphi \mid \varphi \in \mathbf{D}_c, \bar{\partial}\varphi = \vartheta\varphi = 0\}$. In terms of the real operators d, $\delta = -*d*$, the Hermitian metric is Kählerian if and only if $\underset{\varrho,\sigma}{\Pi}$ commutes with $\delta d + d\delta$.

However, unless the metric is Kählerian, $\triangle$ will not in general be a real operator, and the operation of taking complex conjugates will not map the space of harmonic forms into itself. In particular, if M is a compact manifold, the relationship between the real harmonic forms satisfying $d\varphi = \delta\varphi = 0$ and the complex harmonic forms satisfying $\bar{\partial}\varphi = \vartheta\varphi = 0$ will not necessarily be simple.

REFERENCES

1. S. BERGMAN, (a) *Sur les fonctions orthogonales de plusieurs variables complexes avec les applications à la théorie des fonctions analytiques*, Mém. des Sci. Math. Vol. 106, Gauthier-Villars, Paris, 1947. (b) *The kernel function and conformal mapping*, Math. Surveys, No. 5, Amer. Math. Soc., New York, 1950.

2. S. BOCHNER, (a) "Remark on the theorem of Green," *Duke Math. Jour.*, 3 (1937), 334—338. (b) "Analytic and meromorphic continuation by means of Green's formula," *Annals of Math.*, 44 (1943), 652—673. (c) "On compact complex manifolds," *Jour. Indian Math. Soc.*, 11 (1947), pp. 1—21. (d) "Vector fields on complex and real manifolds," *Annals of Math.*, 52 (1950), 642—649. (e) "Tensor fields with finite bases," *Annals of Math.*, 53 (1951), 400—411.

3. E. CALABI, "Geometric imbedding of complex manifolds," *Annals of Math.*, 58 (1953), 1—24.

4. E. CALABI and D. C. SPENCER, "Completely integrable almost complex manifolds," *Annals of Math.* (to appear).

5. G. F. D. DUFF and D. C. SPENCER, (a) "Harmonic tensors on manifolds with boundary," *Proc. Nat. Acad. Sci., U.S.A.*, 37 (1951), 614—619. (b). "Harmonic tensors on Riemannian manifolds with boundary," *Annals of Math.*, 56 (1952), 128—156.

6. P. R. GARABEDIAN, "A new formalism for functions of several complex variables," *Jour. d'Anal. Math.*, 1 (1951), 59—80.

7. P. R. GARABEDIAN and D. C. SPENCER, (a) "Complex boundary value problems," Technical Report No. 16, Stanford Univ., California, April 27, 1951. (b) "A complex tensor calculus for Kähler manifolds," Technical Report No. 17, Stanford Univ., California, May 21, 1951.

8. W. V. D. HODGE, (a) *Harmonic integrals*, Cambridge Univ. Press, 1941. (b) "Differential forms on a Kähler manifold," *Proc. Camb. Phil. Soc.*, 47 (1951), 504—517.

9. K. KODAIRA, (a) "Harmonic fields in Riemannian manifolds," *Annals of Math.*, 50 (1949), 587—665. (b) "The theorem of Riemann-Roch on compact analytic surfaces," *Amer. Jour. of Math.*, 73 (1951), 813—875.

10. B. v. Sz. NAGY, *Spektraldarstellung linearer Transformationen des Hilbertschen Raumes*, Springer, Berlin, 1942.

11. G. DE RHAM and K. KODAIRA, "Harmonic integrals," Institute for Advanced Study, Princeton, 1950 (mimeographed).

12. L. SCHWARTZ, *Théorie des distributions*, Vols. I, II, Hermann, Paris, 1950—51.

13. D. C. SPENCER, (a) "Cauchy's formula on Kähler manifolds," *Proc. Nat. Acad. Sci., U.S.A.* 38 (1952), 76—80. (b) "Real and complex operators on manifolds," *Contributions to the theory of Riemann surfaces*, Annals of Mathematics Studies No. 30, Princeton Univ. Press, 1953. (c) "Dirichlet's principle on manifolds," Volume in Commemoration of the Seventieth Birthday of R. von Mises.

14. A. WEIL, „Sur la théorie des formes différentielles attachées á une variété analytique complexe," *Commentarii Math. Helvetici*, 20 (1947), 110—116.

15. H. WEYL, *Die Idee der Riemannschen Fläche*, Teubner, Berlin, 1923. (Reprint, Chelsea, New York, 1947).

Index

See also Table of Contents. An asterisk following a page number indicates a reference by number only to the bibliography at the end of the chaptre.